ICE Handbook of Ground Investigation

Geotechnical Field Activities and Laboratory Testing

emerald PUBLISHING ice Publishing

ICE Handbook of Ground Investigation

Geotechnical Field Activities and Laboratory Testing

Peter Reading and Miles Martin

Published by Emerald Publishing Limited, Floor 5,
Northspring, 21–23 Wellington Street, Leeds LS1 4DL.

ICE Publishing is an imprint of Emerald Publishing Limited

Other ICE Publishing titles:
ICE Handbook of Ground Investigation: Planning and reporting, geoenvironmental and nonintrusive techniques
Peter Reading and Miles Martin. ISBN 978-0-7277-6681-6
UK Specification for Ground Investigation
Association of Geotechnical and Geoenvironmental Specialists (AGS).
ISBN 978-1-7277-6523-9
Effective Procurement of Ground Investigation
Association of Geotechnical and Geoenvironmental Specialists (AGS).
ISBN 978-1-8366-2029-7

A catalogue record for this book is available from the British Library

ISBN 978-1-8354-9891-0

© Emerald Publishing Limited 2025

Cover photo: courtesy of Tim Newman and Thames Tideway. The image shows the Lambeth hiatus within the tunnel section, the detail of which is rarely seen in ground investigation sampling.

Commissioning Editor: Michael Fenton
Content Development Editor: Cathy Sellars

Production Editor: Emma Sudderick
Typeset by: KnowledgeWorks Global Limited
Index created by: David Gaskell

Dedication

To our wives, Véronique and Christine, who have endured many hours of writing and discussions on a regular basis over the duration of creating this book. They truly have the patience of saints, and without their support and encouragement we would not have completed the task.

Contents

Acknowledgments

We would like to thank the following organisations who have provided support, information and images to assist with the compilation of this handbook: Cambridge Insitu Ltd, Dando International Ltd, European Geophysics Ltd, GDS Instruments, Equipe Group, Geolabs Ltd, Geosense Ltd, Geotechnics, Geotechnical Engineering Ltd, Geo Observations Ltd, In Situ Site Investigation, K4 Soils Laboratory, Marton Geotechnical Services (MGS) Ltd, SatSense, Soil Engineering Ltd, Soil Instruments Ltd, SOCOTEC UK Limited, TerraDat (UK) Ltd and VJ Tech Ltd. Also, the individuals who have helped with specific images are mentioned in the figure captions.

In particular, we are grateful to SOCOTEC UK Limited for allowing us to draw on their immense archive as well as permitting us to use their computer storage and network systems to compile this book.

In addition, our special thanks go to Kim Beasley, who has provided the majority of the text and images for Chapter 3, and to Peter Hepton, whose detailed review of many of the chapters have made this handbook all the richer.

We would also like to express our gratitude to a number of individuals with whom we have had discussions about the contents of the book, and who have reviewed various parts of the text: Steve Tomlinson, whose expertise of drilling techniques is unparalleled and whose contributions and discussion over many years have been pivotal in the writing of Chapters 1 and 2; Daniel Griffith, for his review and advice on Chapter 1, and whose experience and knowledge we have drawn upon for the writing of this chapter; Joe Milner, who has provided much guidance regarding geophysical methods; Sheridan Elliot for the contributions and review of Chapter 4; and Ken Phaure, for many discussions over many decades regarding laboratory testing methods and protocols.

We are indebted to those who have helped us to shape the content and style of the handbook, including Peter Hepton, Andrew Ridley and Darren Ward; and to many others whose influence and inspiration has been significant to the authors while writing this book and more generally throughout their careers, including Tom Lunne, John Powell, Clive Dalton, David Norbury, Rory Mortimer, Jackie Skipper, Digby Harman, John Masters, Julian Lovell, Keith Spires, Robert Duck, Robert Craig, René David, Richard Carter, Marcus Matthews and Chris Clayton.

The concept of this handbook was born from a series of courses that the authors devised along with Equipe Group and Andrew Milne of Geotechnical Engineering Ltd, and was delivered at the Geotechnical Academy under the watchful eye and insatiable enthusiasm of Liz Withington.

About the authors

Peter Reading

Peter has over 50 years of experience in the geotechnical industry, mainly in the ground investigation sector. He started as a technician at the Soil Mechanics laboratory of Cementation and is now a director at PRGC Ltd and a visiting researcher at the University of Portsmouth. He is a UK registered ground engineering adviser. At Cementation, he transitioned from the laboratory to the office, supporting engineers as a correlator, calculating laboratory test results by hand. He gained early experience with computers for analysing results, including oedometer and effective stress testing, under the guidance of Professor Bromhead at Kingston Technical College (now Kingston University).

Peter then joined Taylor Woodrow Geotechnical section, Terresearch, where he was introduced to piling, ground anchors, grouting and sheet piling projects. In the early 1980s, he helped set up the first real-time visualisation and data capture logging system for effective stress triaxial testing under Harman's guidance. After the geotechnical division closed in 1992, Peter joined Exploration Associates as Principal Geotechnical Engineer. Following a merger with Soil Mechanics in 1997, he became Associate Director of the southern operation, overseeing three offices until 2008. He cofounded the Equipe Group, providing bespoke training for the geotechnical industry, and helped establish the trade show Geotechnica.

Peter began lecturing at Brunel University during this time and left Equipe in 2014 to start his consultancy and become a senior lecturer. In 2020, he moved to the University of Portsmouth to deliver the MSc Geotechnical Engineering degree, relinquishing the post in 2023. More recently, Peter has been delivering geotechnical courses for the Geological Society of London.

Peter's academic background includes a degree in Earth Science and Mathematics from the Open University and a PgDip from Surrey University. He is a Chartered Geologist, Chartered Scientist and UK Registered Ground Engineering Adviser. He has been involved in several industry initiatives to improve geotechnical knowledge and practice and is part of the AGS Ground Investigation Working Group. He has authored several technical papers.

Miles Martin

Miles is a Chartered Geologist and Registered Ground Engineering Specialist with over 35 years of experience in the ground investigation industry. He studied Geology for his degree at the University of Dundee (1988) and subsequently obtained an MSc in Geotechnical Engineering at the University of Surrey (2000). He is a Principal Engineer at SOCOTEC UK Limited.

Miles started his career as a technician in a geotechnical testing laboratory that, in the late 1980s, was principally undertaking projects associated with the London Docklands regeneration scheme. For Miles, a particularly interesting aspect of this period in his career was to be part of the progression from manual to digital data acquisition, as he contributed to the installation of a computer-controlled logging system to monitor the required long duration laboratory tests.

This initial spell in the laboratory was followed by 16 years spent primarily on site, allowing Miles to gain experience in a wide range of investigation, field testing and monitoring techniques, as well as in soil and rock description. This period also provided a wealth of opportunities for Miles to develop skills in the practicalities of ground investigation, such as site administration, work programming, logistics planning, client liaison, health and safety supervision and environmental risk management.

In parallel to these site-based pursuits, Miles has built up decades of experience in geotechnical and geoenvironmental investigation design, the formulation of sampling and testing strategies, the construction and refinement of ground models and risk assessments, data presentation and evaluation, and the interpretation of investigation findings.

In more recent years, Miles has focused increasingly on the creation and presentation of technical training courses, and the mentoring and professional development of junior colleagues, as well as on quality control processes such as the auditing of technical procedures.

Miles estimates that, in one capacity or another, he has been involved in at least 1500 projects to date, ranging from residential, commercial and industrial developments to highways, railways and bridges, pipelines and sewage treatment works. Having worked across all regions of the British Isles, Miles has gained an appreciation of how variable ground conditions can be and how critical this is to new developments. His fascination with this 'never-two-the-same' aspect of projects means he cannot imagine hanging up his boots and geology hammer any time soon ... there are just too many other sites out there waiting to be investigated!

emerald PUBLISHING   ice

Peter Reading and Miles Martin
ISBN 978-1-83549-891-0
https://doi.org/10.1108/978-1-83549-890-320251011

Introduction

'Forewarned is forearmed'

Throughout history, many buildings have been constructed with, by today's standards, a very limited understanding of the ground on which they stood. It is well documented that some historical buildings began to suffer structural distress even before construction was complete; the spires and towers of numerous churches, for instance, remaining distorted to this day.

One example is the Church of the Holy Saviour, in Puxton, north Somerset, whose fifteenth century tower apparently experienced subsidence during construction and ended up leaning significantly (Figure 0.1). It is thought that, for this reason, the tower was terminated at a lower height than originally intended. The superficial soils at Puxton consist of firm and stiff clay, and perhaps the stonemasons recognised this as a reasonable material in which to found the tower. What they were probably not aware of, however, and what a nearby borehole has subsequently confirmed, is that about two metres below the surface the clays become softer and contain layers of peat, now known to be a typical shallow-depth profile of the tidal flat deposits of the Somerset Levels.

The presence of these weak, highly compressible soils at depth, vulnerable to consolidation or even failure under substantial structural loads, is likely to be the cause of the difficulties encountered during construction of the church tower.

In other more severe cases, work may have been abandoned, or a collapse might have eventually occurred. On the other hand, the construction of countless buildings over the centuries has evidently been successful, many of these structures having stood the test of time. It is certain that a local knowledge of successes and failures would have provided a degree of understanding of what was likely to constitute promising ground for building and what should be avoided. In many cases, however, even with this rudimentary knowledge, an element of chance would still have been a factor in the success of a construction project.

The modern approach to design and construction is to mitigate that element of chance as much as possible. Developing a comprehensive understanding of the ground conditions beneath a site provides an opportunity to manage any properties of the ground that might affect the construction, either by some remedial action or by accommodating them in the design. For this reason, characterisation of the ground conditions is normally a principal goal for any ground investigation undertaken for a construction project.

Ground investigation is a relatively new science that has been developed over the past century from the earlier understanding of geology and the associated knowledge that soils and rocks behave in very different ways depending on their mineral assemblages and deposition. It has been described as 'the science of an art' (*Geodrilling*, 1979). The 'science' element is based on the knowledge that a successful engineering design will not only require suitable observations and data of the ground conditions to be collated, but also that there needs to be a method of processing and interpreting

this information in a way that will benefit the design process. It may be argued that the 'art' element refers to the need to be creative! The soils and rocks we must investigate are natural materials, with spatial variables that need to be understood, and often with hidden surprises that need to be discovered. The ground at depth is difficult to access and, as such, we typically only get to see a minute proportion of it in our investigations. The rest must be deduced, interpolated, imagined; and to do that with confidence, not to mention in three dimensions, may indeed be considered an art!

Progress in investigation techniques over the years has assisted substantially in our ability to provide a suitably comprehensive and accurate model of ground conditions. One significant step forward has been the development of Standards. Most investigation methods have gradually been incorporated into British, European or International Standards, resulting in greater confidence in the reliability of the data the various methods are producing. A personal demonstration of the benefit of a prescriptive approach to testing was observed using the humble and rather basic 'Standard Penetration Test': a series of in situ profiles of this index test, obtained at a site underlain by chalk, indicated one specific area that was giving somewhat poorer results than the others. This, it transpired, was an area close to a substantial dissolution feature in the chalk – a significant discovery that could then be accommodated into the construction design. With greater confidence that discrepancies are not linked to variations in method, we are able to deduce that they are more likely to indicate an anomaly in the ground.

Another exciting area of progress has been in technology. Data loggers can be used to measure and record movements in the ground, or track groundwater level fluctuations. Computers can monitor and even control long-running laboratory tests. Field data can be recorded digitally, using weatherproof mobile devices, and shared with minimal need for transcription, increasing both the accuracy of the information and the rapidity of its dissemination. Furthermore, thanks to the development of numerous hi-tech geophysical techniques, we can now peer into those unknown areas of the ground between our points of direct observation, giving us much greater understanding of the ground beneath the whole site area. Meanwhile, advances in computer software allow two- and three-dimensional models of the site to be generated from the substantial volumes of data these modern investigation techniques now produce; models that can be built as investigations progress, providing a powerful visual aid to assessment and planning.

One of the most notable aspects of technological progress, however, has been in communication. Gone are the days of field operatives needing a stack of envelopes and first-class stamps to post field records back to the office (or, in cases of greater urgency, driving off to find a telephone box)! Constant access to mobile communication, which can include video calling to illustrate issues being encountered on site, is now the norm. Internet access allows data captured on site, or in a laboratory, to be uploaded to the internet, potentially making it available to anyone in the world who may need to access it, effectively in real time. This digital communication also allows the automated data logging noted above to be accessed remotely, at any time, if needed.

Alongside these technological and communications revolutions, the relatively young discipline of geoenvironmental engineering has steadily grown and broadened investigation requirements, such that many investigation scopes now include significant geoenvironmental objectives alongside more traditional geotechnical aspects.

And, last but by no means least, advances in approaches to health and safety and environmental protection have allowed investigations to be carried out with ever greater mitigation of potential risks to ground investigation teams, and with a light environmental footprint too. This has been achieved through the development of effective hazard identification and control techniques, alongside much better awareness training and monitoring.

The overall result of the developments set out above is that ground investigation has progressively grown into a complex, hi-tech, multidisciplined industry, and one that has much to offer the vast majority of construction projects. These potentials can only be realised, however, by clients and their advisors being suitably aware of the project benefits that a comprehensive and high-quality ground investigation can deliver. Unfortunately, there is much inadequate investigation conducted in the UK, with the ground investigation market driven as much by price as by the pursuit of quality information. This is perpetuated because the complexity of the subject means it is often poorly understood, and the potential cost benefits to the project not fully appreciated. It is as well to note the following quote attributed to John Ruskin (1819–1900) that is as true today as it was more than 150 years ago:

It's unwise to pay too much, but it's worse to pay too little.

When you pay too much, you lose a little money – that is all.

When you pay too little, you sometimes lose everything, because the thing you bought was incapable of doing the thing it was bought to do.

The common law of business balance prohibits paying a little and getting a lot – it can't be done.

If you deal with the lowest bidder, it is well to add something for the risk you run.

By the same token, it is incumbent on ground investigation practitioners to demonstrate such benefits by producing clear and accurate data that make a tangible and positive difference to the project. This can only be achieved by a workforce of highly competent and enthusiastic investigation professionals. It is certainly easier to produce poor, unreliable information than to deliver good quality data. The latter require much skill and diligence, both in investigation planning and delivery. It also relies upon a sound knowledge of soils and rocks, how to assess their properties and, very importantly, how to test for these parameters without influencing the results. In addition, a full understanding of the reasons for doing the work is required, together with the relationship between the site, structure, ground and the parameters that will model the ground. Obtaining information of the highest standard is always the objective, as highlighted by Clayton and Smith (2013: p. 7):

> A competent ground investigation must be carried out for every building, construction or contaminated land project. Any construction professional who fails to carry out site investigation, or who plans to and carries it out to a low standard, ignoring accepted good practice as that contained in 'effective site investigation' will be considered negligent by fellow professionals.

The principal aim of this book and its companion volume (Reading and Martin, 2025) has been to create a practical guide to all aspects of routine ground investigation practice in the UK. The authors hope it will help to enhance knowledge in this essential subject for all parties that become involved in investigations, including clients, their consultants and contractors. As noted above, without a proper understanding of the benefits that a ground investigation can deliver to a project, it is unlikely that an effective approach to an investigation will be adopted, or the relevant techniques used.

These books describe most of the currently available methods and techniques that might be used during an investigation, providing both a technical manual and practical discussion of the methods, as well as addressing common misunderstandings and errors that might be faced.

The structure of these two books has been designed to take the reader for the initial concepts and design of the investigation through the various field processes, starting with nonintrusive reconnaissance and testing, intrusive investigation and sampling, to testing in the field and the laboratory, describing the features of the ground encountered, and finally reporting.

In a quest for accessibility and general readability, this book has been written in a more conversational style that is typically found in, say, a British Standard. For some elements, in particular field testing, the history around the development of a technique is included to explain why a test is conducted in its current form – for example, the standard penetration test that started life as a sampling tool and is currently used as an important measure of soil consistency. This approach is intended to provide context to the various methods, and also general interest for those who wish to know a little more than the nuts and bolts of current practice.

The authors have also used their many years of experience to record best practice, and is some cases forgotten practice – for example, the now rarely used method of waiting a few minutes once a tube sampler has been driven to allow the soil to swell against the wall of the sampling tube, thus increasing the likelihood of full sample recovery. In earlier times, the driller would often stop the drilling rig and have a cup of tea before removing the sample from the borehole. Today, such a methodical approach may be considered by some as time wasting, and the driller may well be driven by financial incentives to complete the borehole as quickly as possible, putting commercial gain first against quality sampling.

As ground investigation has become such an extensive subject, it is not possible to address all topics to the level they deserve. As such, for brevity, subject matter covered in suitable detail elsewhere has been summarised rather than recreated, and the reader directed elsewhere for additional detail. Reference is made to relevant standards, academic papers and existing authoritative publications.

There has been an attempt to rationalise terminology used to describe methods, and to unpick some of the general misconceptions that have worked their way into the vocabulary now in general use. In order to conform to current terminology and practice, the book title uses the term '*ground investigation*' instead of '*site investigation*', although it is recognised that, in common practice, the two descriptors are often used interchangeably. Similarly, the chapter discussing field testing was originally titled '*in situ testing*'. It is hoped that by adopting '*field testing*' some consistency is achieved between various publications; this also avoids the discussion as to the spelling of 'in situ', 'insitu' or 'in-situ', all of which may be found in earlier publications.

So, what about the future? The short answer is that our industry needs to ensure that highly educated and skilled practitioners will continue to be available. While advances in digital data collection and three-dimensional visualisation techniques, noted above, are providing new insights into ground behaviour, skills in the management and interpretation of that data are still needed, together with the appropriate sampling, field and laboratory testing that are needed to provide actual parameters for design. Ground models, whether they are presented in a low-tech or hi-tech fashion, are only as good as the information used to compile them. It is easy to make much of a little data; in some cases, giving a false impression of the ground conditions and an unjustified level of confidence in what is effectively a poor design. Common consequences of these inadequacies include construction delays, contractual claims and increased project costs.

In this respect, a note of caution is given to the potential overreliance on geotechnical parameters that may be found in published literature. It is true that there are many publications that provide typical parameters that might be adopted for preliminary design purposes. Such data should not be used for final design, however, without appropriate verification that they are applicable to the specific site under consideration. While published data can be useful to provide an indication of typical ranges of a certain parameter, it will often be found that the actual conditions are far from those anticipated. An example would be parameters used in the prediction of foundation settlement, where actual values should be used for critical analysis rather than those obtained from published data. Even with site specific input parameters, limitations in conventional analytical methods, and the difficulty in modelling aspects such as drainage paths, which influence the rate and amount of settlement, mean that actual settlements can be quite different to those predicted. There is a higher risk, however, that this discrepancy would be unacceptably large should calculations be based only on published data.

Similarly, it is recommended that the approach of using correlations between various properties is adopted only sparingly, particularly when it comes to parameters that are critical to engineering design. Unfortunately, this trend, which typically relies upon the combination of basic test methods and published correlations, is often used as a low-cost alternative to undertaking more appropriate techniques to establish the required parameter directly. An example might be the prediction of soil stiffness properties by correlation with basic shear strength or standard penetration test results, rather than incorporating suitable techniques into the investigation. While this is an understandable approach where no other is available, its routine use should be discouraged, as it often leads to a more conservative design approach and is therefore ultimately not cost-effective.

There is much talk about artificial intelligence (AI) potentially replacing physical investigation. Currently, however, it appears more likely that that it would provide a powerful tool that could be used to enhance, rather than replace, existing methods. As discussed above, in the context of current methods of ground modelling, the accuracy of the model depends on the quantity and quality of the input parameters, and any future AI application would be equally dependent on accurate data.

When it comes to modelling ground conditions, the problem very often lies in the detail, and while the power of AI, for example, may be able to guide us to potential risks, it will not be able to precisely locate features that might contribute to that risk. This is largely because the basic ground profile beneath most parts of the UK is quite unpredictable. As noted by Terzaghi (1936: p. 13): 'it is unfortunate that the ground was made by nature and not by man'.

In terms of detail, dissolution features in chalk and limestone are a case in point. We know they are a risk in these materials, but the only way to be confident that such features are not located beneath the intended structure is to conduct an investigation of the entire site, typically using a combination of methods such as geophysics to scan the whole site and a grid of boreholes to investigate where anomalies are indicated. It is difficult to imagine being able to move away from this well-established approach with any confidence in the foreseeable future.

Equally, returning to our Puxton church example given at the start of this introduction, compared with when the church tower was constructed, we may now have a better understanding of the general geology in the Puxton area, and knowledge of the risk that soft alluvial soils may pose to shallow foundation stability. Nevertheless, we are still no nearer to understanding the actual thickness of the alluvium at that specific point, nor whether there are lateral variations in the proportion of peat lenses the alluvium contains, nor whether a local erosion channel might have resulted in a variable depth to good founding strata beneath the church footprint. We are no nearer to understanding these details until, of course, we undertake a site specific, targeted investigation.

So, powerful planning, analytical and design tools should always be embraced by our industry and used to assist us in any way possible, and that includes AI. After all, having weighed up the demanding engineering requirements of a modern construction project, together with the potential variations in ground conditions that Mother Nature might have up her sleeve, any form of intelligence (artificial or not) may well consider a given site and earnestly conclude:

'You're gonna need a bigger ground investigation!'

emerald PUBLISHING   ice Publishing

Peter Reading and Miles Martin
ISBN 978-1-83549-891-0
https://doi.org/10.1108/978-1-83549-890-320251002

Chapter 1
Intrusive methods of investigation

1.1.   Introduction

While much can be gained from nonintrusive investigation techniques, such as geophysics, it is rare that these methods can be relied upon exclusively to provide all the information required by the designer. Therefore, the use of some form of intrusive investigation will be necessary, typically in the form of trial pits and/or boreholes, to confirm the expected geology and obtain samples for both geotechnical and geoenvironmental testing.

The decision regarding what investigation methods to conduct is complex (see Chapters 1 and 2 in *ICE Handbook of Ground Investigation – Planning and Reporting, Geoenvironmental and Non-Intrusive Techniques* (Reading and Martin, 2025)), and consideration should be given to what information is required. Once this has been defined then other decisions, such as how deep the intrusive holes need to be and what sampling and testing will be required, will need to be addressed. In addition, practical constraints such as the existing access to the site may govern what method or equipment would be suitable to deliver the necessary information.

There are many well-established intrusive investigation techniques in current use. However, the decision regarding which methods to use is often made by financial factors and availability of plant rather than the technical necessity and contribution to the project. This chapter will explain what can be expected from the various methods and how to decide what to do to provide the best results for a particular project and conditions. Table 1.1 details the intrusive methods in common use today; some are more effective than others, depending on the situation and the information required. For completeness, all are included here with some indication of their effectiveness and both advantages and disadvantages. There is no one method that will suit all and often the ground investigation will need to use a combination of techniques to provide the necessary information for design.

However, it is as well to remember that: 'If we do not know what we are looking for it is unlikely that we will find it' (Anon).

There are other intrusive methods not included in Table 1.1 that have, from time to time, been used but these are not considered to be suitable for ground investigation and are therefore not discussed. These include flight auguring, hollow stem auguring and flush boring. While these techniques are able to put a hole down, the material recovered is highly disturbed and it is not possible to know exactly what depth the material comes from, nor describe the soils fully in terms of structure and consistency. The methods also cause significant disturbance to the borehole wall, which may influence any instrumentation that might be considered (see also BDA *Manual of Rotary Drilling* (Tomlinson, 2021)).

Other methods, including penetrometers and other pushed cone tests that are sometimes referred to as exploratory holes, are discussed in Chapter 5.

Table 1.1 Commonly used intrusive investigation methods (continued on next page)

Method/ equipment	Depth range	Achievable sample quality	Installations	Notes
Trial pitting by hand digging	Shallow, <1.5 m, unsupported, but can be deep, >1.5 m, only with shoring.	Good. Description of soil can be detailed. Can obtain block samples.	Possible but not usually done.	Deep pits are expensive, require design of shoring. Ground water can limit depth achieved.
Trial pitting by machine excavation	Generally limited by reach of machine; ranges from 3.5 m with standard backhoe to 6 m with a 360-degree excavator (can be dug deeper if machine digs in).	Good bulk samples, Class 3, 4 and 5.	Not usually done.	Large area of disturbance.
Hand augering	Up to 4 m.	Class 2 to 3 with driven sample tubes.	Yes, but unlikely to extend below water table.	Limited by water, hole stability and physical effort.
Window sampling	6 to 8 m. Limited by ground water, hole stability and obstructions.	Class 4/5. Tube may block off in granular soils. Soil often compressed as it is sampled	Only if hole remains stable.	Only able to penetrate soft or firm soil. Requires a reduction in hole diameter with depth.
Dynamic sampling (drop weight)	Typically, 6 m in favourable ground conditions.	Class 3 to 4. Thick-walled tube gives high sample disturbance. UT100 may be used at shallow depths.	Limited depth can be cased. Generally, only if hole remains open and stable.	Pull out limited. Needs to reduce diameter with each tube driven to limit friction. Can be configured to conduct standard penetration tests (SPTs).
Dynamic sampling hydraulic percussive	Up to 20 m in favourable ground.	Class 2 to 4. Class 2 only obtained in firm cohesive soils. Can be used to take UT100 samples for Class 1. Usually set up with thick-walled barrels, giving high sample disturbance.	Can be cased. Most installations possible.	Usually used in conjunction with higher class sampling methods such as rotary drilling. Can be configured to conduct SPTs.

Table 1.1 Continued

Method/ equipment	Depth range	Achievable sample quality	Installations	Notes
Cable percussion boring	Usually to depths of up to 60 m but can drill deeper (maximum of about 100 m).	All sample classes can be obtained. Class 1 samples with UT100s.	Casing enables a wide range of instrumentation to be installed.	Many tests can be conducted. Sampling is rarely continuous.
Rotary drilling	Holes can be drilled to several hundreds of metres, limited by power of the rig.	Can be Class 1 with right method and combination of tooling, flush and bit type.	Casing enables a wide range of instrumentation to be installed.	Many tests can be conducted. Sampling can be continuous.
Sonic drilling	Only limited by size of rig.	Will achieve full recovery, but samples are highly disturbed (Class 3–5). Most rigs can be adapted to use conventional coring methods.	Casing can be inserted enabling hole to remain open for installations.	Will drill through most materials. Can be swapped to conventional coring when rock encountered or to obtain thin-walled samples in suitable soils.

1.2. Trial pitting

1.2.1 General

The simplest form of subsurface investigation is to dig a pit. Trial pits (occasionally referred to as 'trial holes' or 'test pits') are temporary excavations dug either by machine or hand depending on their objectives, and in some cases depending on the working area or access available.

Trial pits are a good means of investigating shallow deposits of cohesive and dry granular soils. They enable a visual inspection of the ground to be conducted, so that features such as its structure, the presence of foundations, buried services and evidence of contamination can be described, potentially in three dimensions. They also enable undisturbed samples, or large bulk samples, to be recovered and, where appropriate, field tests to be undertaken.

Although trial pits only really vary in terms of their dimensions, depth and means of excavation, they are typically subdivided into different types based on their various objectives. Some terms in general use for different types of pits are discussed in the following section.

It is important to determine the objectives of the trial pitting exercise during the investigation design and define the type of trial pits required.

1.2.2 Trial pit types

1.2.2.1 Inspection pits and utility pits

Inspection pits, sometimes referred to as 'service pits', are shallow excavations, typically hand dug from ground level (i.e. without accessing the pit), and therefore generally no more than about 1.2–1.5 m deep. The principal objective of an inspection pit is to verify the absence of buried services, prior to undertaking deeper intrusive investigation such as a borehole or penetrometer test. Most buried services will be found within this depth, and a simple inspection pit will therefore typically meet its objective. There are a few exceptions, however – for example, in cases where ground levels have been changed, where a gravity sewer has needed to be constructed at a greater depth, or where trenchless techniques have been used to take a service beneath a road or railway. Under such circumstances, deeper excavation may be required, or an alternative location selected for the proposed borehole.

Figure 1.1(a) shows an example of a borehole having been positioned in an inspection pit clear of a drain that has been found during its excavation.

Inspection pits may also be used to find shallow depth structures like buried concrete slabs or foundations, or some other feature such as tie bars to retaining walls. Figure 1.1(b) shows an example of a shallow foundation exposed in an inspection pit.

In BS 5930 (BSI, 2015) a distinction is made between inspection pits, typically intended to help avoid services, and 'utility pits and trenches' that are used to provide a positive identification of buried services. This positive identification may be part of the services clearance process in preparation for the ground investigation (see Section 1.2.3), or the mapping out of existing services may be one of the investigation objectives, to inform the proposed construction phase (see Figure 1.2).

Figure 1.1 (a) Drain encountered in inspection pit, allowing a borehole to be positioned so as to avoid damage; (b) shallow foundation exposed in hand-dug inspection pit (images courtesy of SOCOTEC)

(a)   (b)

1.2.2.2   Trial pits

The generic term 'trial pit' is frequently used for routine excavations, primarily intended to provide information on ground conditions, together with an opportunity for sampling and testing. All observations must be made from ground level. Under no circumstances should entry into a trial pit be made unless specific precautions have been taken, as discussed in Section 1.2.3.

While hand digging may be used if only shallow-depth information or sampling is required, such as for a topsoil survey, trial pits are generally machine dug and extended to the maximum depth feasible with the excavator used. This is typically up to 2 m for a mini excavator, about 3.5–4.5 m for a wheeled excavator and about 6 m for a larger tracked excavator. Achievable depths will also depend on the ground conditions and excavation size.

Trial pitting offers a rapid and relatively cost-effective way of obtaining large amounts of information on shallow depth ground conditions (see Figure 1.3). For example, it is not uncommon to

Figure 1.2 Utility trench excavated by hand to record locations and depths of various buried services (image courtesy of SOCOTEC)

Figure 1.3 Routine 3 m deep trial pit in clay-dominated soils (authors' image)

be able to complete six to eight routine, backhoe excavator dug trial pits per day, although this is likely to be much reduced if complex ground conditions need to be recorded, or where there are significant distances between pit locations. When considering such work, it is essential to engage an excavator operator who can demonstrate experience in investigation works or similar activities. Ideally, the excavator will be provided with a range of buckets of different sizes and types.

As observations are made exclusively from the surface, with deep trial pits it becomes increasingly difficult to see and measure features accurately, not least because of reduced levels of available light. Nevertheless, there may be particular circumstances where the use of deep trial pits could provide valuable information. A clear change in material type, for example, may be seen in the arisings being brought to the surface and, if excavation is halted at that point, the approximate boundary depth may be determined by measuring the depth of the pit. Similarly, if a bedrock surface is known to be present at depth, an excavator driver is likely to be able to 'feel' it, and probably hear the scraping of the bucket on the rock, even if direct observation might be difficult.

The width of a trial pit is usually limited to a single bucket width, 0.6 m typically being used for a wheeled excavator. A wider pit may be excavated to make observations from the surface easier, particularly at depth. Larger pits, however, remove a greater volume of soil, and are therefore slower to excavate and backfill. Trial pit lengths are usually no longer than is necessary to reach the required depth, the length and depth generally being similar in value, although pits can naturally be extended if required.

Safety issues regarding making trial pit observations and measurements from ground level are considered in Section 1.2.3.

1.2.2.3   Trial trenches

A trench is essentially a long trial pit (see Figure 1.4). The distinction between a 'pit' and a 'trench' is not specifically defined but, where excavated by machine, a trench is likely to require successive moves of the excavator in order to extend its length. Trenches can be excavated in sections (effectively a series of trial pits, positioned end-to-end) where having the full required length of trench open at once might be inconvenient or restricting.

Trenches are often used to find and record features such as strata changes, where lateral as well as vertical variations can be determined. They can also be useful in finding buried features, the locations of which are known only approximately – for example, the subsurface remains of a demolished building or, in the case of utility trenches, to enable services to be mapped across the site (see Figure 1.2).

As both vertical and lateral variations should be described, the features observed may be more easily recorded on sketch sections, rather than on just a conventional log. Measurements of lateral distances will need to be related to a datum point, such as one end of the trench, in the same way that depths need a datum, typically ground level.

Trenches are often shallower than typical trial pits, although not necessarily. Nevertheless, their overall large dimensions mean that consideration needs to be given to control of the volume of arisings they can create.

The inspection of a trench from the ground surface can also be hazardous, owing to potential instability of the long side (see Section 1.2.3).

Note that the turf has been removed and kept separate from the spoil, and a ground protection sheet has been laid to heap the spoil on.

1.2.2.4   Observation pits

The term 'observation pit' is generally used for a pit intended for entry by personnel. In such cases, health and safety issues are therefore paramount, and the use of observation pits should only be considered when other forms of investigation would be unable to meet the specific investigation objectives. The range of hazards are discussed in Section 1.2.3.

The reason that physical access may be required is typically so that a close inspection of the ground can be undertaken, to record details that may not be possible from the surface. Examples include the measurement of natural features in situ – for example, discontinuity orientations and apertures – or to assess the condition of manmade structures, such as foundations or retaining walls. Physical access might also be required to undertake certain field tests, such as plate loading tests, at a particular depth or within a specified stratum, or recover samples that need to be taken in situ, such as block samples.

Observation pits can be relatively small or very large, depending on their purpose and any site constraints. Figure 1.5 shows a large observation pit excavated to enable a detailed description of the ground in situ.

Where very deep observations are required, the use of a shaft might be more suitable. These are usually formed by hand excavation and take the form of a circular section, the sides being supported by successive concrete manhole rings that are sunk by self-weight, thus providing full support over the excavated depth. Sampling and observations are made of the material at the base of the shaft or that exposed in the lower part of the shaft wall.

1.2.3   Safety

Holes in the ground are inherently dangerous. The safety of those involved in the trial pitting activity, as well as anyone else who may come into contact with the process, is absolutely paramount,

Figure 1.5 Observation pit with benched faces, to facilitate a detailed inspection of the ground. Entry and egress facilitated by a shallow slope at the far end (image courtesy of SOCOTEC)

and if trial pitting cannot be carried out safely then an alternative form of investigation should be sought. While this handbook is not intended to be a comprehensive health and safety manual, the need for adequate precautions cannot be overstated.

The digging process presents a range of hazards, as does the excavation itself, and no trial pitting should be undertaken without a full risk assessment. Further information on risks associated with excavations, together with related inspections and risk assessments, is given on the Health and Safety Executive (HSE) website. The Construction (Design and Management) Regulations 2015 (HSE, 2015) should also be consulted in respect of the assessment of risk relating to excavations.

While risk assessments should always be specific to the particular circumstances relating to a given site, hazards that will need to be addressed for a trial pitting exercise should include, but not be limited to, the following.

Buried services: As with all intrusive investigation, before starting a trial pit it is essential that the work area is checked for services. A suitably robust, multifaceted approach should be undertaken, as set out in the HSE document HSG47 (HSE, 2014) and PAS128:2022 (BSI, 2022b). This should always include the sourcing and checking of up-to-date service drawings, the use of a cable avoidance tool (CAT) and signal generator (Genny), supplemented by ground probing radar (GPR) as necessary. A site inspection should also be carried out to identify visual evidence on the ground of the presence of services.

It is essential to locate all services in the relevant site area before any intrusive work is attempted. In this way, excavations can be planned to avoid damage to them.

If a trial pit is to be carried out in the vicinity of an identified service, it is prudent to excavate the pit parallel to the alignment of the service where possible. This avoids one end of the trial pit getting too close to the service, and also mitigates the risk of crosscutting any unmarked services that may be running parallel to the known one.

Precautions to take in the excavation process are discussed in Section 1.2.4.

Overhead services: In addition to buried utilities, the presence of overhead communications and power lines will need to be considered if machine excavation is planned, both in respect of the working area and access routes.

If work close to overhead power lines is unavoidable, the electricity company should be consulted for advice. Depending on the type of cable, there may be an option to shroud them to enable work to be undertaken safely. All work should be undertaken in accordance with Guidance note GS6 in INDG258 (HSE, 2013).

Working near plant: Working around mechanical excavators is always hazardous. The pit logger/ engineer who is directing the works should discuss the required work with the machine driver. A series of simple hand instructions should be agreed so that the driver can be made aware of what the logger requires. The pit logger should stand at the end opposite the excavator and be always visible to the driver (see Figure 1.6), and should avoid position movements that have not been communicated to the driver. It is essential that the logger should not enter the digging arc while digging is in progress. The logger should give a clear signal to stop digging and instruct that the bucket is rested before looking at the spoil or into the pit. This simple procedure will substantially reduce the risk of injury.

In respect of observation pits, no personnel should access the pit, or work in the pit, while the excavator is in operation.

On sites where there are other workers or members of the public nearby, barriers should be erected around the trial pit and excavator to prevent anyone approaching the working area too closely.

In addition, on sites where there are moving vehicles, or other plant, the risk of being struck should be considered in respect of the selected trial pitting areas. Again, this risk can be mitigated by using temporary barriers to make the work area more visible and to reduce the risk of the logger inadvertently stepping out of the work area, potentially into a zone of vehicle movements.

Figure 1.6 Mechanical excavation of trial pit (or trench, in sections), with observations and measurements being made from more stable, short end of pit, in view of the driver (authors' image)

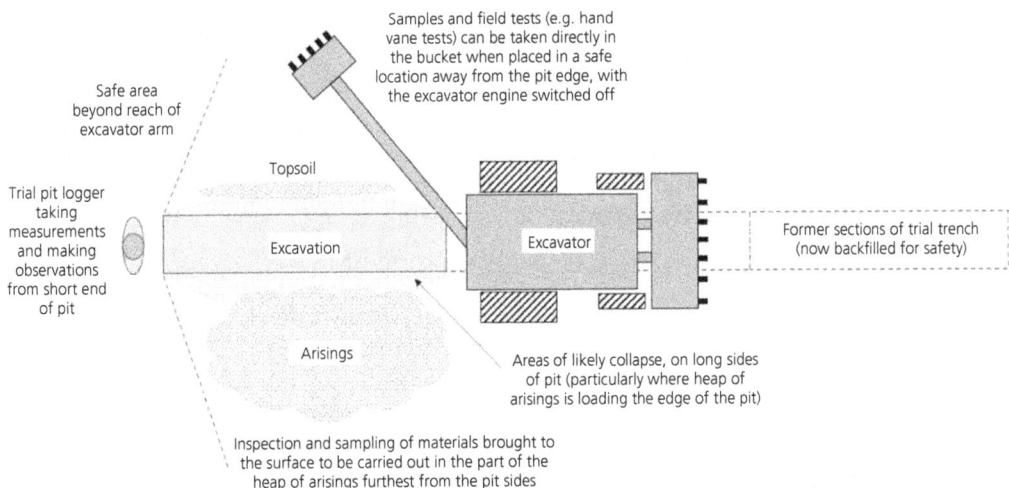

Slips, trips and falls: There is a risk of slips, trips and falls in the area of a trial pit, including tripping over items on the ground, such as samples, pieces of equipment and arisings from the pit, particularly those including large fragments that can roll to the foot of the spoil heap. Slips may occur from wet ground in the working area, particularly wet clay that may have been exposed or excavated. There is also a potential risk of falls, as it is possible to lose your footing at the edge of even the smallest of excavations. In larger excavations, consideration needs to be given to falling in the pit completely, as discussed below.

Work at height: Working in proximity of a pit of any depth should be considered as working at height, and appropriate risk assessment and methods must be in place for this work. In particular, great care should be taken by the logger whenever the pit needs to be approached for making observations or taking measurements or photographs. This should only be done from the short end of the pit opposite the excavator (see Figure 1.6), which reduces the risk of falling into the pit should the logger lose their footing. Further guidance on working at height is available on the HSE website.

In addition to a fall into the open pit, one hazard is the ground adjacent to the pit giving way, potentially taking with it anyone stood next to the pit. The potential collapse of an excavation side is possible at any time, and in any soil type. Clay soils are perhaps especially dangerous in this respect, as pits in clay can typically be cut vertically, and they give the impression of being stable. However, clays tend to fail suddenly, and it is often seen that the whole excavation side collapses in one event. The long sides of the pit are much more vulnerable to this instability, and this is another reason why the logger should only approach the short end of the excavation.

With long trenches, observing everything from one end may be impractical, but this can be managed by excavating and backfilling the trench in sections, so that each section is of a length that can be reasonably viewed from the more stable end, rather than needing to approach the side (see Figure 1.6).

Examination of the materials in the arisings should be made on the side of the heap furthest from the pit, as the area between the arisings and the pit is often particularly vulnerable to collapse. Sampling and testing (e.g. hand vane tests) may be carried out on soil in the excavator bucket, before it is tipped out, so long as the bucket is placed at a safe distance from the pit and the excavator engine has been switched off.

Where there is a possibility of other workers or members of the public approaching the excavation, the use of temporary fencing may be required. Should the pit need to be left open overnight – which should be avoided whenever possible – it should be suitably protected by fencing, together with signs warning of the excavation and lighting.

Personnel entry: The base of an excavation is a potentially dangerous environment, and the authors' view, in line with the general risk hierarchy, is that personnel entry into pits should be avoided wherever possible, and all other methods of obtaining the required information should be considered seriously before such an exercise is attempted.

The golden rule is ON NO ACCOUNT ENTER AN UNSUPPORTED PIT. The reason for this is that all trial pit walls will fail in time, often within less than 30 min of being excavated. Should someone be in the pit at the time of failure, even bending down or crouching in a relatively shallow pit, the soil could cover them, at least partially, and under this amount of soil it may not be possible to expand the chest cavity to breath. One cubic metre of soil weighs approximately two tonnes and, considering that when an excavation fails it is generally on the side where the excavated soil has been mounded, both the material forming the side of the pit, and the spoil above, may fall into the pit.

Essentially, for any excavation to be sufficiently safe for access, the walls should either be supported artificially, by shoring, or they should be battered back to an appropriate angle so they will be self-supporting.

Where shoring is used, this can take the form of trench boxes or steel sheet piles/trench sheets, placed in the pit and held in position by whalers and props. This method is complex and, as noted in the *UK Specification for Ground Investigation* (ICE, 2022), design calculations for such excavation support need to be undertaken by a suitably qualified and experienced person and approved by a qualified Temporary Works Designer. The exercise needs machines to lift the trench boxes or sheets into place, and so a lifting plan will be required, and operatives must be appropriately trained. When using this method, one or two sheets may be left out such that observation windows are available for inspection and sampling of the soil. If the design cannot accommodate this then it may mean access is only available at the ends of the pit. In many cases, the fact that inspection is not possible for the whole pit or face can be a disadvantage.

Any shored trial pits should be treated as a confined space, and the risk assessment will need to cover all relevant aspects of confined-space working. For example, all personnel accessing a shored pit should be appropriately trained for confined-space work, all activities in the pit should be supervised by personnel at ground level who should be trained in support and recovery, and an appropriate rescue plan should be in place.

In addition, it is essential that the air quality in the pit is monitored. The presence of noxious gases should never be ruled out and appropriate oxygen depletion and noxious gas warning monitors should be provided, together with breathing apparatus if the risk assessment concludes that a risk is present. All work should follow the recommendations presented in the HSE document INDG258 (HSE, 2013).

The alternative to shoring would be to step the excavation of the pit to ensure stability of the pit sides, as was done for the observation pit shown in Figure 1.5. Pits of this size require a lot of space, which is generally only afforded on open undeveloped sites. As with shoring, this approach also needs planning and design. It cannot be assumed, for example, that a slope benched at, say, 1:1 would be adequate. This would effectively form a 45-degree slope that, although potentially stable under ideal circumstances, would be excessively steep in many cases. Depending on the soil types and groundwater conditions, slopes as shallow as 10–20 degrees may be required in order to achieve suitable stability. Again, the approval of a qualified Temporary Works Designer will be needed, which may require the appointment of a third-party specialist if the other parties involved at the time of the investigation cannot undertake this role.

For many years there was an assumption that where a pit was only shallow, say a metre deep, it would be stable and safe. As soil can be very unpredictable, however, this is no longer considered a safe approach, and stability should always be controlled either with shoring or with slopes cut to a batter. The one exception is mentioned in BS 5930 (BSI, 2015), which indicates that unsupported large area pits of no more than 1 m deep may be entered, provided that the risk assessment has concluded that they are likely to remain stable. It is emphasised, however, that this only applies to excavations of 'large area' and should certainly not be adopted for pits of standard width. The principle is that work in large excavations can be carried out on the excavation floor at a safe distance from the cut slope. As such, should a failure of the slope occur, no personnel in the pit would be harmed. An example is shown in Figure 1.7, where discontinuity surfaces are being logged in the

floor of the pit, remote from the excavation edge, which can be seen at the top of the photograph. Temporary barriers may be used to ensure workers maintain a safe distance.

In most cases, the reason for the excavation is to discover the properties of the soil or rock in question. It is therefore most unlikely that a risk assessment, or the design of shoring or temporary slopes, can be conducted with sufficient confidence unless it is based on a knowledge of the ground and groundwater present at the site. Consequently, excavations intended for personnel entry may need to be undertaken as a supplementary phase of investigation, once a sufficient understanding of the site has been gained.

Where personnel are entering an excavation, it is essential that there is always provision for access and egress. In a shored pit this is likely to be by secured ladders, while in an excavation with battered sides it can either be by way of a suitably shallow slope, see Figure 1.5, or a stairway, as in Figure 1.8.

Figure 1.8 Entry and egress to observation pit facilitated by an aluminium stairway (image courtesy of SOCOTEC)

1.2.4 The trial pitting process

1.2.4.1 Selecting the pit location

Trial pit locations will initially be determined to suit the investigation objectives, and then possibly adjusted slightly, or sometimes substantially, in consideration of the following factors.

- Presence of services (buried or overhead).
- Difficulty of reinstatement – for example, selecting an area of soft landscaping rather than concrete or macadam surfacing.
- The pit, or excavator if used, potentially obstructing other site activities, or access.
- The effect the ground disturbance may have on the future development of the site (i.e. it is important to avoid digging pits close to the possible locations of future shallow foundations).
- Potentially undermining any existing structures or services, unless they are known to be redundant.

The selected location can be confirmed once the appropriate PAS128 survey has been undertaken satisfactorily. If a permit to dig system is being used, which is recommended, the permit must be issued prior to starting excavation.

1.2.4.2 Excavation by hand

Hand excavation methods should be used whenever a pit is being dug to check for services (inspection pits) or to provide a positive location of services (utility pits or trenches). It may also be the simplest, and safest, option should shallow trial pits be required for near-surface sampling, for instance.

Hand tools mainly consist of spades, shovels and scissor shovels (post-hole diggers), together with picks or bars for dense materials, or for levering out coarse fragments from the base of the pit. All hand tools should have insulated handles for additional protection should an electrical cable be damaged.

Excavation should always be carried out as gently as possible, so as not to inadvertently damage a pipe or cable just beneath the visible surface of the excavation. Where a service has been located and is to be exposed by excavation, digging alongside the service is recommended, rather than directly above it (HSE, 2014), and then it should carefully be uncovered by extending the pit laterally.

During excavation the CAT and Genny should continue to be used to provide ongoing guidance.

In some instances, evidence can be found below the surface that helps indicate when an excavation may be in the proximity of a service. This includes lateral variations in materials above the pipe or cable, the installation trench having been backfilled with a different type of soil, bedding materials around the service itself, such as sand or fine gravel, and markers placed specifically to warn of the presence, such as plastic tape or tiles.

Preventing accidental damage to services is more difficult where dense or coarse materials need to be penetrated. The temptation to use greater force, with more aggressive tools such as picks and bars, may negate the benefit of hand excavation altogether. In recent times, however, technology has been developed to remove the soil around services using vacuum extraction. The method is aided by high velocity jetting of air or water to loosen soil particles and using the vacuum to suck up the debris. This method greatly reduces the risk of damaging utilities, as well as being relatively quick. It is particularly useful where there are several cables and pipes placed in close proximity to each other, where hand excavation would be difficult. Such methods require access by larger items of plant, such as air compressors or vacuum trucks, and so may not be feasible in all locations. When using this method consideration should be given to backfilling. The material removed is often unsuitable, being a slurry. Thus, additional material may need to be provided to facilitate backfilling the excavated area.

1.2.4.3   Excavation by machine

Larger excavations, which generally include routine trial pits, trenches (apart from utility trenches) and observation pits, are generally excavated by machine.

Excavators come in many sizes and forms, from small rubber-tracked mini-diggers to wheeled excavators with extending back-hoe (Figure 1.9(a)) and steel track-mounted 360-degree rotating excavators (Figure 1.9(b)). The larger machines generally produce a much larger size trial pit and consequently much more spoil. This must be planned because the bigger the pit and the spoil heap, the more restricted access and movement around the work site will become. It should also be noted that the larger excavators will require a low loader to deliver them to site. The selection of the most appropriate excavator for a given project will depend on factors such as the required depth and width of excavations, site surface conditions, the amount of tracking needed between locations and possible access restrictions. For example, steel tracked machines cannot be tracked on public roads, because they may cause serious damage to the road surfacing, and this may make their use restrictive on certain projects.

Owing to the plan area of such pits, hand digging in the zone possibly occupied by services is unlikely to be practical. Therefore, following an appropriate PAS128 service clearance procedure, mechanical excavation typically begins from the ground surface. At shallow depths, this should be carried out carefully, removing thin layers of material at a time, checking for any evidence of services, such as clean sand or gravel used for pipe bedding, or warning tapes or tiles. At this stage, an excavator bucket without teeth can be used as an additional safeguard against damaging a pipe

Figure 1.9 (a) Excavation of a trial pit with a wheeled excavator; (b) using a tracked machine; (c) trial pit width typically equal to width of bucket (images courtesy of D. Beskeen and C. Payton, SOCOTEC)

(a)   (b)   (c)

or cable, should one be encountered. Beneath the zone in which services might be found, standard excavator buckets with teeth are likely to be used.

In terms of logging the ground conditions, the preferred method is to use the excavator bucket to remove the soil in thin layers, thus enabling the soils to be identified and described as they are encountered and allowing all changes to be captured. The excavator should be stopped regularly to allow material in the bucket, or on the heap of arisings, to be examined safely. Where a change in material type is noted the excavation process should be halted immediately, allowing the depth of the strata boundary to be measured. If this is not visible by looking down the pit, then the depth of the pit should be measured to give an approximation.

While it is expedient to log strata as the pit is dug, it may not be possible to see complete units of soil until the pit excavation has been completed. For example, in fluvial deposits there are often discrete layers of gravel and sand interbedded with thin clay or silt. It is important to capture these structures in the trial pit description.

Samples should be taken as digging proceeds, either directly from the bucket, or from the heap of arisings after the bucket is emptied. This is easier to do as each layer is retrieved and will avoid cross contamination between samples.

Standard trial pits may be used to conduct certain field tests for which there is no requirement to enter the pit. For example, hand vane tests can potentially be undertaken where cohesive materials are brought to the surface in a reasonably intact state. The material should first be assessed for any signs of disturbance, and hand vane testing (see Chapter 5) can then be conducted on any masses that appear intact, before the excavator bucket is emptied. In addition, infiltration testing can be conducted once the pit and sampling has been completed (see also Chapter 5).

In observation pits that are designed for personnel entry, other tests such as California Bearing Ratio or plate testing may be conducted during the excavation process. Once the pit has been dug to the required depth for the particular test, digging is halted and the selected precautions to allow personnel entry are put in place. When safe to do so, the relevant testing is conducted, and then excavation can be continued once the testing has been completed. If this method is adopted, it is essential to ensure the pit is stable before starting testing, which may require temporary shoring or battered excavation sides (see Section 1.2.3).

1.2.4.4 Removal of strong materials

Mechanical excavators are typically able to remove most soils and weak rocks to attain the depth of excavation required. Minor obstructions encountered in machine dug pits, such as thin layers of brickwork, concrete or beds of rock, can often be broken up and removed by a toothed bucket and a skilled machine operator.

In some cases, pits may be required to be dug into and beneath hard surfaces such as pavement, roads or concrete slabs. When located outside, the use of an excavator with a hydraulic breaker may prove suitable, although caution is required to ensure that the breaking out does not damage utilities in the vicinity.

The breaker can cause cracking and other disturbance in the materials of the surrounding area, which may be problematic if the surface is required to be reinstated to a pre-excavation condition. Hand-held breakers, used either with a compressor or generator, can provide a somewhat neater finish, particularly when used with a spade tip rather than a pointed tip. An alternative might be to saw cut the sides to provide a very neat edge (see Figure 1.1(a)).

Where a surface concrete slab is of substantial thickness, and in particular if it is reinforced, the use of stitch drilling may be considered. With this method, the perimeter of the required excavation area is cored with a series of small diameter core holes using a single barrel diamond corer, each drilled to intersect the previous one (see Figure 1.47). Once the entire perimeter has been cored the concrete may be broken up and removed by a hand-held breaker. Alternatively, the block can be divided into smaller blocks by further stitch drilling. On removal of the concrete, the pit can be continued by standard excavation methods.

For small, hand-dug pits, such as an inspection pit for a borehole, a hand-held breaker or possibly a concrete saw is likely to be used for such work. Alternatively, consideration might be given to rotary coring at a moderately large diameter – for example, 300 mm – forming a hole through which the pit can then be excavated (see Section 1.3.5.10).

1.2.4.5   Trial pitting and groundwater

Where groundwater seepage is noted in a trial pit, excavation should stop and details of the strike recorded, such as its depth, location in the pit, and its relationship, if any, to strata changes.

If the materials in which the seepage is encountered consist of silt, sand or clean gravel, excavation below the water table may be difficult, or impossible, owing to the likely collapse of these materials from the trial pit walls. If this is the case, it is recommended that no further excavation be attempted, and a different method of investigation should be sought. Continued excavation may result in disturbance to a large area with little technical gain.

Even where excavation below the groundwater level is possible, it may not be desirable, particularly if permeability is such that the lower part of the pit fills with water. The presence of water will probably obscure any further features encountered, meaning that deeper excavation may be of little value. In addition, a trial pit that has encountered a substantial volume of groundwater can be difficult to backfill without causing a dangerous, very soft, often slurry filled pit, this possibly presenting a hazard to people or livestock. If this is the case, it is essential that the pit area is adequately fenced, and warning signage put in place, something that may not be easy to arrange had the situation not been anticipated. It may be several days before the pit can be successfully and adequately backfilled, often requiring several revisits to the location to ensure the area is perfectly safe before the fencing is removed.

An alternative approach would be to pump water out of the pit into a bowser, for disposal. If the materials are of high permeability, however, this may not reduce the volume of water in the pit significantly. In situations where the management of groundwater may be difficult, or where its impact on backfill may result in a hazard, a decision may be made to terminate a trial pit where a significant water strike may be encountered.

1.2.4.6   Photographs

Recording the excavation findings by photographing the pit faces and arisings on completion of the pit is a routine requirement. All photographs should include a graduated scale, such as a survey staff (Figure 1.1(b)) or pole (Figure 1.3). As with any site photograph, it is also good practice to include a photograph board that identifies the site, the date of excavation and the details of the trial pit, such as reference number and depth (Figure 1.7). The photograph board should also include a photographic colour chart and grey scale.

A camera with a high-quality zoom lens may also be used to pick up finer detail that, together with inspection of the soils removed from the pit, enables good quality information to be obtained. It is good practice to keep a photographic record of the condition of the general location around the pit, both before digging begins (precondition) and after the excavation has been backfilled (postcondition).

Difficult lighting conditions, with high contrast between the light levels in the pit and at the ground surface, sometimes makes obtaining informative photographs a challenge. Digital cameras mounted on a boom, or 'selfie stick', provide a good way of capturing images of the faces without the need to enter the pit. Furthermore, in recent years cameras capable of taking a 360-degree photograph have been developed, and these can potentially be useful in providing a complete record of the excavation, albeit somewhat distorted, as shown in Figure 1.10.

Photographs from drones may also be considered for trial pits that may be difficult to access or view from a conventional standpoint.

Figure 1.10 Example of a 360-degree photograph of the walls of a deep trial pit (image courtesy of SOCOTEC)

1.2.4.7 Dealing with arisings

Digging should begin by removing the topsoil and surface vegetation, or otherwise the layers of hard surfacing if applicable. These surface materials should be stored separately from the rest of the excavated soil for replacing when the pit has been completed (as shown in Figure 1.4). In some circumstances, the topsoil might be removed over a large area so that the materials excavated from depth do not mix with the topsoil around the pit. This may also be required for aesthetic reasons, particularly where materials like chalk are excavated, to prevent a white scar being left on the surface when pitting has been completed.

Alternatively, the ground surface beneath the proposed heap of arisings could be protected, particularly if the ground is at risk of being contaminated, by placing heavy gauge plastic sheeting or a tarpaulin on the ground, ensuring the excavated materials do not mix with the surface soils. Plywood boards placed on top of the sheeting can also be useful as this prevents snagging of the plastic and makes the backfilling process easier.

Ideally, the heap of arisings should not be placed too close to the edge of the pit, as the weight of soil can increase the likelihood of collapse of the excavation side. From a practical point of view, however, anything more than about a metre or so is likely to make excavation and backfilling more time consuming.

1.2.4.8   Backfilling and reinstatement

Trial pits are generally backfilled with their arisings, and these should be replaced in the same stratigraphic sequence as they occur in the ground, finishing with the topsoil (where present). The materials should be placed into the pit in layers, and each layer should be compacted using the excavator bucket. Even with this treatment, owing to a bulking effect, it is often found that the soil cannot be compacted sufficiently to place all of the arisings back in the pit. Where this is the case, any excess material should be mounded over the top of the pit and left to settle naturally. If the excess spoil is removed, it may be found that the backfill material settles over time and leaves a depression, which might be a hazard. Should it be essential that the backfill materials settle as little as possible, perhaps to enable the surface to be reinstated without delay, it may be necessary to backfill the pit with imported granular fill rather than the excavation arisings. Under normal circumstances, however, this would be avoided owing to the trial pit arisings then becoming waste materials needing suitable disposal.

Where pits are dug in areas that require full reinstatement of the surfacing, such as paved areas, the use of a small compacting roller or vibrating plate may be used at ground level. Even so it is recommended that the area be left to settle for a few days before final surface replacement is undertaken. Any trial pits undertaken in the highway will need to be reinstated in accordance the New Roads and Street Works Act (NRSWA) 1991, by operatives who are appropriately NRSWA trained.

Reinstatement of the removed surface may be required to restore the structure to its original form with the top finish matching the original. In some cases, where sections of reinforced concrete slabs are removed, it will be necessary to provide reinstatement of the steel reinforcement. This may necessitate breaking out around the several of the remaining bars to allow new reinforcing bars to be tied to them, thus providing continuity and structural integrity. The concrete can then be made good using a mix that provides the same strength to the original. Where the concrete forms a road pavement, it is often necessary to insert steel dowels to tie the replaced concrete with the surrounding concrete slabs.

1.3.   Boreholes

1.3.1   General

Boreholes have been used for many centuries as a means of accessing water, and exploiting minerals, as well as for construction; the Romans having used tripod rigs to install wooden piles. The first true investigation boreholes, which were the forerunners of current day methods, were developed to find oil during the nineteenth century.

Modern ground investigation uses many different forms of borehole to investigate the geology and geotechnical properties beneath a site, enabling identification of soil and rock, samples to be obtained, field tests to be carried out and instruments to be installed. There are numerous influences on the success of any particular method, including ground conditions, available working space and the type of information required, such that no one method will suit all situations.

The most basic forms use handheld equipment (see Section 1.3.2), but more commonly mechanical means of creating boreholes are used. The various mechanical methods commonly adopted today fall into four main categories: dynamic, percussive, rotary and sonic. The following sections consider the various methods in order of increasing complexity and generally increasing cost. The terminology used for the various methods does need to be accurate otherwise there is a risk of misinterpretation.

Details of casing sizes and tooling for dynamic sampling, cable percussion and rotary drilling are fully documented in BS EN ISO 22475-1, Annex C (BSI, 2021).

1.3.2 Boreholes using handheld equipment

There are several methods of forming a borehole using handheld equipment, all operating in a similar fashion, requiring one or two people to either push or turn various tools into the ground. The most commonly adopted tool, which can still be found in use, is the hand auger. Although rarely used in modern investigations, they might still be useful in locations that are not possible to access with powered equipment. The augers are relatively light and portable, but in use they do have limitations, requiring significant physical effort to penetrate to more than a few metres in firm cohesive soils, and being almost impossible to progress through granular soils. The methods are suited to soft or firm cohesive soils, where the hole will remain stable for the time it is bored. The materials being removed from the hole can be used for soil description, and it is possible to obtain limited samples for laboratory tests.

As no support is provided for the hole that is formed, these methods are generally unsuitable for loose granular soils that are prone to collapse, particularly where they are below the water table.

The hand auger uses 19 mm dia. (3/4-inch BSP) steel pipe connected in 1 m lengths using a threaded sleeve connector. An auger is attached to the first section and the borehole is advanced by screwing it into the ground. The auger is withdrawn when the auger body is full of soil. The soil retained within the auger may be sampled to allow further inspection and laboratory testing. It is good practice to describe the soil as it is recovered and then retain the soil as small, disturbed samples or bulk samples. Such samples are Quality Class 3 to 4 (see Chapter 2 for sample classes).

Tube samples may be obtained by attaching a slide link hammer to a 38 or 50 mm dia. sample tube. The slide link hammer is connected to the rods and lowered to the base of the hole where it can be driven to its full length by blows from the hammer. Once driven to the sample length, the rods are withdrawn with the sample tube. The quality of the tube samples obtained can be Class 2 but are often Class 3 or 4.

It is generally found that the deeper the bore is taken the greater the physical effort required to progress the borehole. In the authors' experience, the maximum depth achieved using this method has been 6 m when used in firm to stiff clay, but this is unusual and normally depths of around 3–4 m might be achieved.

Other variations of the hand auger system are available, including the peat sampler and Russian auger; both are only suited to soft soils and would generally be used where the ground is unsuitable for plant access (see also Chapter 2 for methods of sampling).

1.3.3 Boreholes by continuous dynamic sampling methods

1.3.3.1 General

There are several variants of advancing a borehole where a dynamic force, either delivered by repeated blows from a percussive hammer or a dropped weight, is used to drive a hollow tube into the ground. These methods, which both form a borehole and allow samples to be taken from the recovered material, have become popular owing to their relative low cost and simple operation.

The method is derived from a technique known as 'window sampling', which uses a handheld pneumatic hammer, similar to a road breaker, to drive a thick-walled tube sampler into the ground. The tube has slots cut into the side (the 'windows') to enable access to the soil that has been forced into the hollow tube once the tube has been brought back to the surface. Sampling depth can be extended by the addition of rods between the hammer and the sampling barrel. In general, use of this system has declined, largely owing to health and safety reasons as the hammer presents a hazard in terms of hand arm vibration syndrome (HAVS). Nevertheless, as it uses portable, hand-held equipment, the method is still occasionally adopted where access for other plant is not possible.

Variations on this original method were subsequently developed, both in terms of barrel design and in the technique of driving the barrels into the ground. The introduction of plain cylindrical barrels, lined with a an inner rigid PVC tube, meant that the sample could be pulled out of the barrel horizontally, and therefore 'windows' were no longer needed. This became known as 'windowless' sampling, to distinguish it from the original method. To drive these barrels into the ground, a drop weight mechanism on a chain lift can be used, this mechanism being mounted on a short mast on a small, tracked or wheeled rig. In recent years, an alternative approach of driving such barrels using a percussive hammer mounted on a rotary rig has become a routine sampling method.

In BS 5930 (BSI, 2015), a distinction is made between the barrel types (window and windowless) as well as the equipment types (hand-held percussive, drop weight and percussive hammer). Nevertheless, there remains a degree of confusion in the industry, with incorrect terms often being employed.

The authors' preferred terminology, to readily distinguish between these methods, is as follows.

(1) **Window sampling** – when referring to the method using a handheld pneumatic hammer to drive sampling barrels on rods. The barrel has slots or 'windows' cut into its side for access to the sample. See Section 1.3.3.2.
(2) **Dynamic sampling drop weight** – carried out using a small tracked or wheeled machine with a short mast. The driving mechanism is a drop weight, carried on a chain lift, that generally drives steel sampling barrels with a PVC liner. This method is described in Section 1.3.3.3.
(3) **Dynamic sampling hydraulic percussive** – similar to the dynamic sampling drop weight method, using the same type of barrel, but driven by rapid blows from a hydraulic driven hammer powered by a rotary drilling rig. See Section 1.3.3.4.

In recent times, these definitions have become blurred and the handheld device has been used to drive the liner barrel assembly while the window sample tubes can be driven by both the drop weight and hydraulic percussive method. There are also some small tracked rigs that are capable of using a hydraulic percussive hammer.

1.3.3.2 Window sampling

Window sampling, which employs driving a tube with longitudinal cutaway sections, or windows, as shown in Figure 1.11, was developed as an efficient and economic method of obtaining an indication of the soils that are present near the surface.

Originally used in exploration geology, it was employed for tasks such as the determination of overburden thickness. Its use extended into environmental sampling during the 1980s where it was found to be an efficient way of obtaining a continuous record of the shallow depth ground conditions, as well as allowing small, disturbed samples for environmental testing to be obtained. Although its use has extended into geotechnical data collection, the quality of sample and the uncertainty over the actual depth from which the material is recovered makes this method unsuitable for anything other than an indicative assessment of the strata sequence.

The thick wall of the sampler causes considerable disturbance to the soil both ahead of the sampler and as the soils enter the sampler. Often the soil becomes compressed owing to the high wall friction that increases as the sampling barrel is driven further into the ground, the result of which shortens the recovered column of soil compared with the driven length. The driving force causes shears to develop in cohesive soils, increases density and alters the water content. Coarse particles such as gravel tend be broken by the process of driving, which radically alters the particle size distribution. Particles that are more competent can wedge and block-off the tube mouth, preventing further soil from entering the tube, therefore making depth determination unreliable. For the various reasons outlined above, the method is to be considered poor. However, the main reason for its decline in use is due to health and safety concerns linked to manual handling, moving parts and vibrations generated during operation.

Figure 1.11 A selection of window sampler tubes (image courtesy of D. Rackley, SOCOTEC)

1.3.3.3 Dynamic sampling drop weight

Dynamic sampling drop weight is carried out using a small, lightweight tracked unit that gives ease of access and a small working footprint; see Figure 1.12. These small mobile rigs are quick to set up and relatively simple to operate. The system uses a simple drop weight or trip hammer to drive an open sample barrel into the ground, as shown in Figure 1.13.

The weight is tripped at the top of the travel allowing it to free fall onto an anvil. The anvil is located onto the top of the drive rods that are fitted onto the sample barrel. The barrel, typically 1 m long, has an inner plastic liner, usually between 1–2 mm thick, which is held in place by a lip on the cutting shoe screwed onto the bottom of the barrel (the assembly is shown in Figure 1.14).

Figure 1.12 Dynamic sampling drop weight rig; relatively lightweight and compact (image courtesy of D. Rackley, SOCOTEC)

Figure 1.13 Dynamic sampling; the sampling tube is connected to the anvil – the drop weight and chain drive is behind the protective cage at the top of the photograph (image courtesy of D. Rackley, SOCOTEC)

Figure 1.14 Dynamic sample tube assembly – tube with head assembly (top); inner plastic liner with retainer, core catcher and cutting shoe (bottom) (image courtesy of D. Rackley, SOCOTEC)

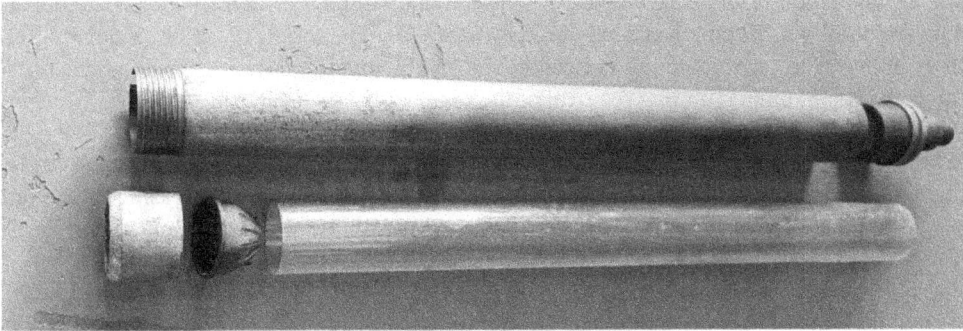

Figure 1.15 Stages in the construction of a dynamic sampling drop weight borehole (authors' image)

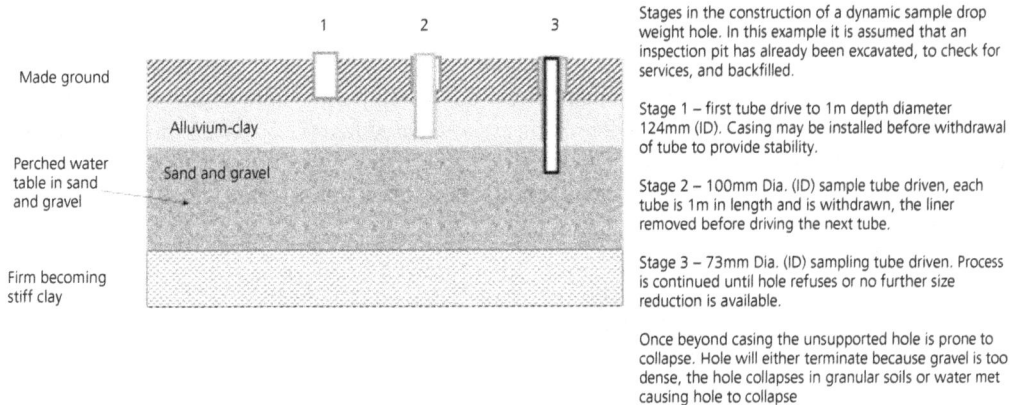

Stages in the construction of a dynamic sample drop weight hole. In this example it is assumed that an inspection pit has already been excavated, to check for services, and backfilled.

Stage 1 – first tube drive to 1m depth diameter 124mm (ID). Casing may be installed before withdrawal of tube to provide stability.

Stage 2 – 100mm Dia. (ID) sample tube driven, each tube is 1m in length and is withdrawn, the liner removed before driving the next tube.

Stage 3 – 73mm Dia. (ID) sampling tube driven. Process is continued until hole refuses or no further size reduction is available.

Once beyond casing the unsupported hole is prone to collapse. Hole will either terminate because gravel is too dense, the hole collapses in granular soils or water met causing hole to collapse

The area ratio, Ar, will generally range from about 37% to more than 100% (see Chapter 2: Geotechnical sampling).

The barrels are driven into the ground by blows from the dropped weight. The soil is retained in the barrel, within a plastic liner. Once the barrel has been driven to its full length of 1 m below the base of the borehole, the assembly is completely withdrawn from the borehole, the barrel disassembled and the liner enclosing the sample withdrawn. A fresh liner is inserted into the assembly, and another 1 m long drive rod is added. The whole assembly is reinserted into the borehole and the process repeated. Figure 1.15 details the stages in the construction of a borehole using this method.

The barrel diameter is usually reduced progressively as the hole depth increases. This is done to reduce the friction on the outside of successive tubes, which can otherwise make driving difficult. Diameters of the drive assembly do vary between manufacturers. Typically, the tube internal diameters range from 124–33 mm. At shallow depth, where the borehole is at a suitable diameter, it may be possible to obtain a UT100 sample (see Chapter 2). It should be noted that as the diameter of the sampler reduces the area ratio increases, which in turn causes greater disturbance to the recovered sample. Area ratios of 37%–103% are typical for the size ranges mentioned above. The process of reducing the diameter of successive sample tubes results in progressively smaller volumes of sample being retrieved.

The main limitation to the depth of drive is the ground stiffness or density. Often the borehole reaches 'refusal' (i.e. the practical depth limit) by about 5 or 6 m, and rarely achieves depths greater than 10 m, even in favourable ground conditions. The limited pull-back from the light rig also restricts the depth, often requiring removal of the sample tube by hand using jacks.

The method also relies on the bore remaining open and stable while the sampler is removed from the borehole. Some configurations operate with a duplex system that will allow casing to be inserted as the sampler is driven. While this system will provide stability to the borehole, the casing is a larger diameter and can be difficult to remove on completion of the borehole.

Although there are several configurations of drop weight and drop height, ranging from 10 kg falling 500 mm to 63.5 kg falling 750 mm, it is general practice to use only the heavier configuration of 63.5 kg falling 750 mm. This heavier setup gives the same energy as that required to conduct a standard penetration test (SPT) (see Chapter 5: Geotechnical field testing), and the rig can therefore be used to carry out SPTs as well as sampling. Where undertaken, SPTs are typically carried out at 1 m intervals – that is, between the successive tube samples.

As dynamic probing (see Chapter 5: Geotechnical field testing) can also be undertaken with these rigs, the two techniques of sampling and probing are often used together – for example, continuing with dynamic probing from the base of a dynamic sampler borehole, or carrying out a dynamic probe adjacent to the borehole. If the latter combination is used, they should be carried out at least 0.5 m apart (ICE, 2022).

While the dynamic sample drop weight method is routinely used for shallow depth ground investigations, its various limitations should be considered when assessing the information this technique provides. The main disadvantages with this system may be summarised as follows.

- Driving in granular soils can block the sample barrel and hence little or no soil is recovered.
- Coarse granular soils will usually present an impenetrable obstruction and will halt progress.
- The reducing diameter with depth limits the amount of soil recovered and hence the mass of material available for testing.
- The area ratio of the sample tube increases as the diameter reduces; thus soil disturbance increases. Soil disturbance results in generally low sample quality (Class 3 or 4).
- In some soils, the driving process will compact the soil into the sample tube, which may lead to inaccurate strata boundary depths.
- Casing is of limited help and is therefore not usually used.

1.3.3.4   Dynamic sampling hydraulic percussive

In recent times, rigs have been manufactured that offer more than one method of constructing the borehole using a dual head. On modern ground investigation rotary rigs, a rotary drilling head can be accompanied by a dynamic sampling head. This combination allows driving by percussive methods to penetrate made ground and superficial soils with a sampling barrel similar to that used in the method described in the previous section. The main difference between this hydraulic percussive method of driving and the drop weight method is that the barrels are usually driven at a single diameter because the heavier and more powerful rig is able to pull the barrels out of the borehole after each drive length has been completed. The use of this method will generally provide a reasonable quality of sample, particularly in cohesive soils (possibly Class 2). It also provides a borehole of sufficient diameter to enable rotary drilling to follow.

The hydraulic hammer system is driven by the rig hydraulics and can deliver blows of a constant energy. The method enables the blows to be delivered efficiently to provide a near continuous driving action.

The greater power available from a rotary rig, and the ability to use the rig winch to remove the sampler and rods, usually means the hole can be progressed to a much greater depth, 20 m or more, and at a larger diameter (approximately 100 mm) than can be achieved by the smaller, lighter rigs operating the drop weight system.

Other than that, the sample barrels are similar to those of the dynamic sampling drop weight system, again suffering from a very high area ratio. In addition, most of the limitations listed in the previous section also apply to this technique, although anecdotal evidence would suggest that the sample quality is slightly improved by the faster driving this method allows.

1.3.4 Cable percussion boring

1.3.4.1 Background to the cable percussion method

Cable percussion boring, sometimes referred to as cable tool, shell and auger or light cable percussion boring, has been used for many different purposes over many centuries. More recently (c. 1900), the method has been adapted from agricultural uses such as well boring to various civil engineering applications. Around the start of the twentieth century, tripod rigs – which progressed a borehole using gravitational means – were being used to install piles. This is not the first appearance of these techniques because it is commonly believed that the Romans used a similar method to install wooden piles.

It was during the late nineteenth century that engineers began to appreciate the need to understand the ground upon which they were building, rather than relying simply on engineering judgment. This gave rise to the desire to carry out research of a particular site to determine the nature of the soil and the best way to form foundations and other civil engineering structures on and in the ground. At the same time, great strides were being made in the field of geology, which is still the core subject at the heart of ground investigation and soil mechanics. The early boreholes were often logged by the driller and used local terminology for the materials found.

The photograph shown in Figure 1.16 was taken in 1958 and shows that even then some rigs were still being operated by hand, diesel engines to power the rig being a relatively new addition at that time. Many of the tools seen around the rig are similar to those commonly used today. A modern version is shown in Figure 1.17.

At the start of the twentieth century, engineers began to form a consensus and both the drilling methods, and soil and rock logging started to become more consistent. However, it was not until the 1930s that our understanding of soil mechanics and the implications of ground conditions were truly appreciated.

Today, apart from the cable percussive rigs now being powered by an engine, and the introduction of crucial health and safety modifications, the method is almost unchanged from these early practices. While cable tool boring has many critics, the method has the advantage of being able to drill relatively large diameter holes that enables progress through granular soils such as alluvial sand

Figure 1.16 This photograph was taken in 1958 and shows a cable percussion tripod rig drilling a borehole on Brompton Road, London (note the hand operated winch) (authors' image)

Figure 1.17 A modern cable percussive rig (image courtesy of SOCOTEC)

and gravel, which are not suited to other methods of forming a borehole such as rotary drilling. This has ensured that the method has persisted in the UK where in many locations the near surface ground conditions comprise soils rather than rocks. It is notable that until the 1990s this method of drilling was the main method used to form boreholes in all soils, rotary rigs generally only being deployed to drill in competent rock.

1.3.4.2   The method of cable percussive boring

For modern ground investigation, this method is used to form boreholes and take samples in soils such as clay, sand and gravel. Typically, the material retrieved from the borehole is disturbed Class 3 (see Chapter 3). However, this method remains in use because of its relatively low cost, and ability to readily change tooling such that field tests may be conducted in the borehole along with high quality, Class 1 sampling. This is despite concerns over manual handling and uncertainty regarding towing the rig.

Generally, the technique is not suited to forming boreholes in rock, although some harder materials can be penetrated where they are found to be relatively thin or weak. In most cases, this type of boring will terminate on rock and other methods of extending the borehole will be required to continue the borehole, such as rotary drilling. It is common for a cable percussion borehole to be used to start a borehole, drilling through granular soils, in preparation for a rotary borehole to follow on. Casing of an appropriate size will be left in the ground, allowing the rotary casing to be inserted through the cable tool casing.

Cable percussion boring is a highly specialist activity that requires a well-trained and skilled operative to perform. When engaging a cable tool operator, it is important to consider their qualifications and experience.

Typically, the rig is some 8.8 m long in its towing position and is generally towed to site by a 4x4 vehicle. The rig length to weight ratio is very close to that permitted to be towed by conventional 4x4 vehicles. Compliance with highways legislation should be checked for specific vehicle combinations. The rig is erected by using the rope tethered to the samson bar to raise the legs (see Figure 1.19(b)), enabling them to be deployed forward of the rig, further raising pulls the rig to its operational position when the stays and stabilisation bar are fitted. Typically, the rig is some 6.8 m high with a free drop below the crown wheel of 5.6 m. These dimensions are for the Dando 2000 (see Figure 1.29).

Once erected, the method uses tools connected to the wire rope that are dropped under gravity into the ground thus enabling the removal of the soil, which is forced into the tube. By repeating this action, a borehole can be formed. With the right equipment and a skilled operative, depths up to 100 m are achievable in suitable ground conditions. The deeper the borehole the slower the rate of boring, and in modern practice reverting to other methods is usually made for depths more than about 60 m.

The basic boring process is carried out using a wire rope wound onto a drum (Figure 1.18(b)) that is rotated by a winch motor. The rope passes over a pulley or crown wheel that is fixed on the top of a tripod frame formed by the rig legs (Figure 1.18(a)). The drum rotational speed is controlled by a clutch and a brake, and boring is carried out by winding the rope to the top of the rig tripod with tools attached. The clutch is then released, which allows the rope and tools to fall freely under gravity. The clutch is used to engage the motor and wind the rope around the drum and hence bring the tools to surface. The brake is used to control the position of the end of the rope, or any tools attached.

The rope has a shackle on the free end (Figure 1.19(a)), which is formed to a loop, and is used to attach various tools. These are used to advance the hole, to provide stability to the hole, to carry out in situ testing and take samples. The diameter of the rig rope should be adequate and is usually 16 mm, with a safe working load of 3 tonnes. By adding pulley blocks, the pull-back force can be increased. It is important to check the rope for wear each day.

Figure 1.18 Elements of a cable percussion drilling rig: (a) crown wheel assembly; (b) winch drum and clutch showing guarding of moving parts (authors' images)

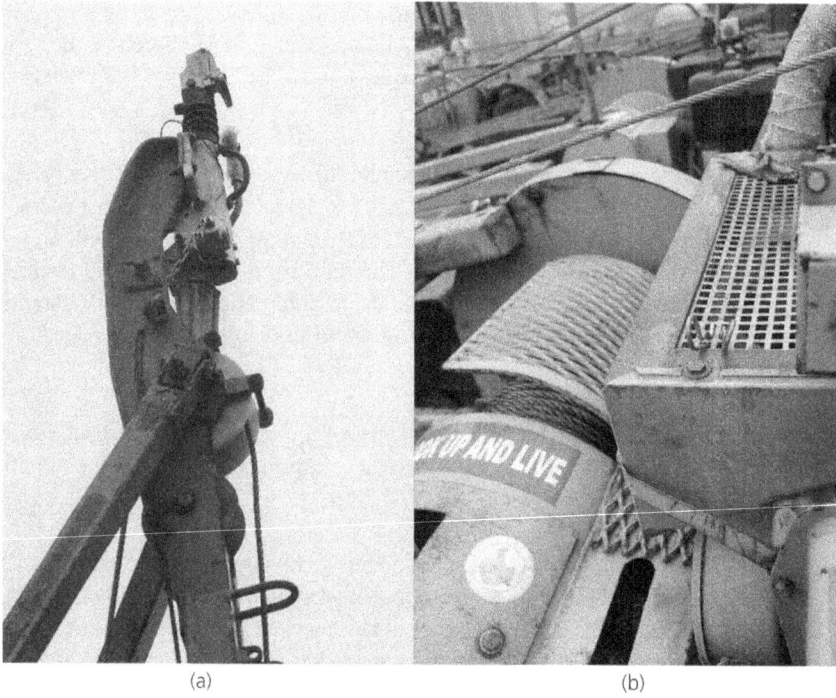

(a)　　　　　　　　　　　　(b)

Figure 1.19 Cable percussive drilling rig and operation: (a) the rope eye with a shackle connection, (authors' image); (b) rig being erected using the rope and pulling on the samson bar located at the rear of the rig (image courtesy of L Dando, SOCOTEC)

(a)　　　　　　　　　　　　(b)

Figure 1.20 shows some of the more commonly used tools. Often tools are used with additional weights to provide more energy to the drop and improve the stability of the free-falling tools, increasing the depth of penetration with each drop and also helping to keep the borehole straight.

1.3.4.3   Cable percussive tools

Some of the original tools used for cable percussion boring have fallen into disuse. The shell is still in use today, but the auger is no longer used. Commonly, the auger was formed of an open sided tube with a scooped cutting edge. The auger would be lowered to the base of the hole and used by turning it by hand using a rod spanner or 'bitch', making it possible to gather up soft and loose material at the base of the hole that had been broken up by chisels or the clay cutter. The tools commonly used today are described below.

THE SHELL

The shell (Figure 1.20(a)) is a tube approximately 1 m long with a renewable cutting shoe screwed onto the bottom end. The top of the tool has a removable swivel and eye threaded onto the shell, which can be connected the shackle on the end of the rope.

The shell is used to progress the borehole in any soil except clay, and is ideally suited to boring through granular soils. The shell is fitted with a clack, a hinged flap which is now usually made of steel but traditionally would have been made of leather. The clack operates as a simple valve, opening to allow material to enter the tube as the shell is dropped into the soil at the base of the borehole, and closing as the shell is lifted, thus preventing the soil from falling out. This action will often allow water to escape as the shell is lifted, and fine soils such as fine sand and silt may well be lost by this action.

The shell will often be repeatedly lifted and dropped over a small height when at the base of the hole to develop a pumping condition referred to as surging, which aids the movement of the soil into the tube by repeatedly opening and shutting the clack valve. This process can cause a similar condition to piping (see Section 1.3.4.6), the process resulting in a significant disturbance of the soil with a loss of the finer particles, which can seriously alter the particle size distribution of any samples taken. This condition has been known to provide misleading information in the design of dewatering schemes, to costly effect.

Figure 1.20 Cable percussion tooling: (a) lower end of shell, showing clack. (b) clay cutter (authors' images)

(a)   (b)

CLAY CUTTER OR SECTION TOOL

The clay cutter (see Figure 1.20(b)), as its name implies, is used in cohesive soils and has replaced the original auger. The clay cutter is similar to the shell in construction, although it is an open-ended tube (i.e. without a clack) and has slots on opposite sides of the tube that enables the operative to knock the clay out of the tube after it has been recovered from the borehole (see Figure 1.21(a)). Often the clay cutter will leave disturbed clay at the base of the borehole because it is not equipped with a clack. It takes some considerable skill to clean the base of the borehole using the clay cutter and remove all disturbed material, as is required if an undisturbed sample is to be taken, or field test carried out.

CROSS BLADE CLAY CUTTER OR STUBBER

The cross-blade clay cutter, or 'stubber' (see Figure 1.21(b)), is a relatively recent addition (*c.* 1985) to the cable percussion rig tooling. It is used in cohesive soils and is favoured by drillers because, being smaller and lighter, it is much quicker to pull in and out of the borehole and is simpler and quicker to empty when compared with other tools such as the standard clay cutter. The clay caught in the three sections of the stubber can be quickly knocked out of the tool. The main disadvantage with this tool is that because it is light and short it can wander in the borehole as it falls, which may result in boreholes that are not straight.

Figure 1.21 Cable percussion tooling (a) Removing clay from a clay cutter. (b) cross-blade clay cutter (authors' images)

(a)   (b)

CHISELS

Chisels are heavy, solid steel tools used to break up hard ground and obstructions met in the borehole (Figure 1.22). There are two types of chisels used in common practice: the cross head chisel and the Californian chisel. The choice of either of these chisels is one of preference rather than any demonstrable difference in the result. The adoption of 'hard boring techniques', which may include the use of a chisel to advance the borehole through 'hard strata', is often seen as a contentious activity. The difficulty arises in the definition of hard strata. Hard boring may be recorded because the rate of advance falls below 0.5 m/h using conventional tools and may not necessarily require the use of a specific chisel. Depending on the ground conditions, progress may be made more easily using the casing and shell to break up hard ground, rather than a chisel that will require the tool to be regularly swapped with the shell to advance the hole. This method is more successful at progressing in mudstones and other semiplastic materials.

The definition given in the *UK Specification for Ground Investigation* (ICE, 2022) gives three criteria, of which the first and at least one of the other two should be met to classify the material as a hard stratum. These are as follows.

- Boring with normal appropriate tools cannot proceed at a rate greater than 0.5m/h.
- 100 mm dia. samples cannot be driven more than 300 mm with 50 blows from the drive hammer.
- An SPT shows resistance in excess of 35 blows per 75 mm.

Figure 1.22 (a) Cross head chisel; (b) California chisel (authors' images)

(a)   (b)

1.3.4.4 Casing

Casings are steel tubes, of various diameters, which may be driven into the ground as the borehole progresses. Casing is used to keep the borehole open, either preventing collapse of the hole in granular soils or stopping the hole from closing in when in cohesive soils. Casing is required to seal groundwater entry into the borehole or to provide a protective seal when boring through contaminated soils (see 1.3.7.6). The tubes are threaded, with a male thread at one end and a female thread at the other, so they can be screwed together to enable the casing 'string' to extend to whatever depth is required. The first tube, or 'lead length', is fitted with a cutting shoe at its base, while the top length of casing is fitted with a drive head. The drive head has a hole on opposite sides which allows a steel rod (knocking bar) to be pushed through, which can be used to hang the casing or to drive the casing by repeatedly lifting dropping the tools onto the knocking bar. The casing has a larger internal diameter than the tools to enable the tools to pass through it without jamming. Although casing lengths can range from 0.5 to 3 m, it is general practice to limit the length to 1.5 m, or even 0.5 m for larger diameter casing. Longer sections of casing are heavy and pose risks from manual handling.

As casing becomes damaged, it is common to cut off the damaged section and re-thread the end. As such, casing should not be regarded as having fixed lengths.

The casing is driven into the ground as the borehole is advanced, but it should not be driven ahead of boring tools, otherwise it can cause significant soil disturbance below the borehole. The first length of casing is known as the lead length and will be fitted with a tapered cutting shoe with a tapered cutting edge. Casing clamps are used to hold the casing when it is being removed once the borehole has been constructed to its final depth (see Figure 1.23(b)).

In complex strata, and typically in deep boreholes, the casing diameter will be started as large as possible and the diameter progressively reduced in a telescopic, or nested, fashion. Nesting is the term given where several strings of casings of reducing diameters are used one inside the other. Casing sizes are reduced in this way to reduce friction and provide support or to ensure an effective seal when drilling through water bearing strata and, in particular, to prevent migration of contamination.

Commonly, casings can be found in 150 mm, 200 mm, 250 mm and 300 mm dia. sizes, although some suppliers still supply the imperial equivalents. Table 1.2 provides the dimensions of commonly used cable percussive casings and tools. A few investigation companies may also be able to drill in even larger casing sizes, although most of these have been used for installing piles rather than for investigation. Each casing size is machined to fit inside the larger casing without jamming. It should be borne in mind that several strings of casings and greater depth require a rig powerful enough to handle and remove significant weight and overcome the potential friction of the casing against the soil.

It is important to consider that each change of casing diameter will require a change in the size of the tooling. Each casing size or string extends from ground level passing through the successive strings to the base of the hole where it is taken further until the next string is introduced. In order to drill in several strings in one borehole requires significant logistical planning as, for example, a 30 m borehole with three strings of casing will require the three sets of tools and as much as 70 m of casing. The steel casings and tools are heavy, and consideration should be given to mechanical handling equipment, to manage them at the drilling site.

Figure 1.23 (a) A stillage of casing; (b) casing clamps (authors' images)

(a)   (b)

Table 1.2 Nominal casing sizes for cable percussion boring

Nominal size	Casing OD (mm)	Casing ID (mm)	Tooling size (mm)
150 mm (6 inch)	168	149	140
200 mm (8 inch)	219	200	194
250 mm (10 inch)	273	250	241
300 mm (12 inch)	324	301	291

OD = outside diameter; ID = inside diameter

1.3.4.5 Example of advancing a cable percussion borehole through a sequence of strata

The following section, illustrated by Figure 1.24, describes the stages in the construction of a cable percussion borehole.

Figure 1.24 Sequence for construction of a cable tool borehole in varied strata (authors' image)

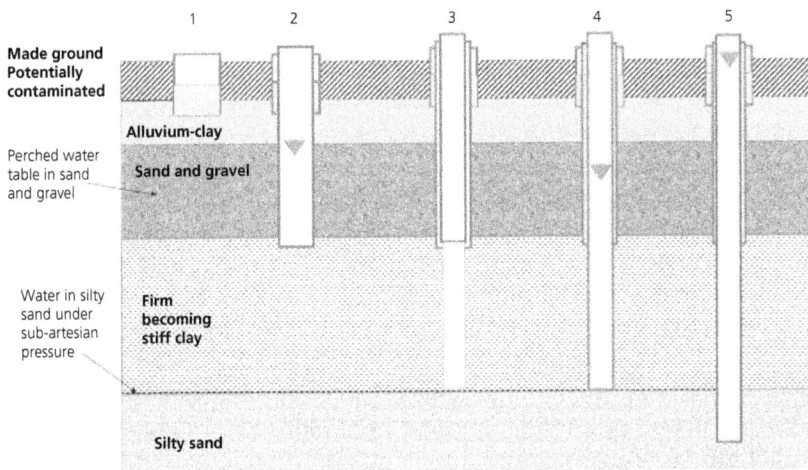

Stage 1 – 250mm diameter casing inserted through made ground to top of alluvial clay. Bentonite seal installed and left to cure before recommencing boring.

Stage 2 – Insert 200mm diameter casing through bentonite seal and bore through sand and gravel using a Section Tool with a Clack and advancing casing to provide support to borehole. Water level kept high to ensure "blowing" conditions do not occur.

Stage 3 – Seal 200mm diameter casing into clay. Drill uncased through the clay at 200mm diameter using a Clay Cutter or Stubber, hole maintained dry (a little water may be used to aid removal of clay).

Stage 4 – On encountering sand water will rise up the borehole. If depth of clay is known 150mm diameter casing should be inserted before meeting sand to enable the high water pressure within the sand to be controlled. If depth is not known it is important to have water avaliable before it is needed and casing ready to install. High water pressure within the sand will cause the sand to rise up inside the borehole. It is essential to maintain control of the water level to enable the borehole to be progressed into the sand.

Stage 5 – Once water is under control boring into the sand can commence ensuring the casing is at the base of the borehole at all times and ensuring water in the borehole is above the sub-artesian water level. (It is advisable to continue boring with only short pauses for breaks to ensure the equilibrium is maintained.)

During boring at each stage appropriate sampling and testing may be conducted.

In this example, the ground conditions consist of made ground underlain by an alluvial clay with relatively low permeability. A layer of sand and gravel follows, which is water bearing, underlain by stiff clay that is almost impermeable. Finally, the borehole is terminated in silty sand.

To plan the construction of the borehole, it is important for the driller to know the potential sequence of strata that might be encountered.

Because the gravel may be a minor aquifer, the driller may be required to form a seal, often termed 'aquifer protection' (see Section 1.3.7.6), to prevent any possible contaminated water from the made ground migrating down into the sand and gravel. To do this, the borehole will need to be started with a sufficiently large casing diameter to accommodate smaller diameters of casing as the borehole is progressed. For example, **250 mm dia. casing** is inserted to into the top of the alluvial clay. This cohesive soil forms an aquiclude and hence will seal the base of the hole, preventing water and leachates migrating below this level. The casing should be inserted a short distance into this stratum, approximately 0.5 m, after which an aquifer protection seal can be made inside the casing using bentonite clay.

Using the **200 mm dia. tools**, along with the 200 mm dia. casing, the hole is progressed through the bentonite seal and alluvial layer, and into the sand and gravel beneath. The section tool (clay cutter) will be used through the Alluvium while the shell will be adopted to progress through the sand and gravel. While drilling in this stratum, potable water will be added and maintained in the borehole at a level above the ground water level to ensure a positive head and prevent piping (discussed in detail in Section 1.3.7.5).

Once the firm to stiff clay has been met the 200 mm dia. casing will be driven into the top of this stratum to provide a seal against groundwater entry (while this is not strictly intended as aquifer protection, it controls the amount of groundwater in the borehole, and helps keep water in the different aquifers separate). The borehole will be progressed using the clay cutter at 200 mm diameter but without casing, as an open hole, for as long as the borehole remains stable. Once the silty sand is met it is unlikely that the borehole will stay open without support, and so the **150 mm dia. casing** will be inserted through the 200 mm and 250 mm casing strings. This casing will pass easily through the open hole in the clay because it is of a smaller diameter than the tools the driller has used to bore through the clay. The casing will advance following the hole through to the final depth within the silty sand. Water will again need to be added to the borehole to ensure a positive head is maintained and to prevent piping.

The above example is a general case; often a borehole might be drilled with a single string of casing, particularly if made ground is absent or uncontaminated and the ground conditions are less complex. The complexity of the expected ground conditions will dictate the number of strings of casings and the necessary tooling that will be required; thus planning is paramount.

1.3.4.6 General considerations

Cable percussion boring has many advantages, including its ability to handle several casing strings and to progress in both granular and cohesive materials (i.e., superficial soils of all types). The borehole can be progressed through water bearing strata and advanced through relatively hard strata. Samples can be taken, and tests undertaken. The use of casing allows the borehole to be taken through most soils other than rock. This flexibility has ensured this relatively simple and cheap method has persisted in use for more than a century with little change to the process.

The cable tool rig has several variants and can be used in a modular form, which allows the rig to be broken down into small handleable parts; those individual parts are more easily carried to locations that are difficult to access with the full-sized rig. The modular rig can then be rebuilt at the borehole location.

An additional advantage of a modular rig is that, where necessary, sections of the legs can be removed, thus reducing the overall height of the rig, which can then operate in areas of restricted space, or low headroom, sometimes as little as 2.5 m (see Figures 1.25 and 1.26).

Situations of low headroom are often in enclosed spaces, such as inside buildings, particularly basements, or enclosed courtyards. Under such circumstances the diesel engine can be replaced with either an air winch or an electric motor to avoid having to deal with the exhaust fumes (which may otherwise need to be removed from the working area using extraction fans and ducting). It should be noted that the lower height of the rig requires drilling tools and casings to be shorter, which in turn means more handling and slower progress.

The customised Dando 100, shown in Figure 1.27, is both a low headroom version and tracked, making access to restricted locations much easier.

The main disadvantages of the cable percussion rig are that the method does disturb the strata and in granular soils the disturbance is significant, leading to the particle size distribution being seriously affected. In particular, the process of dropping the shell and driving the casing can break up larger particles, which increases the relative proportion of smaller particles. The particle size that can be successfully sampled using the shell is limited by the size of the clack aperture. This in turn is dependent on the diameter of the tools being used; roughly the largest particle that could be sampled is about 30% of the borehole diameter. Larger particles will be broken up and sampled as fragments, as will particles that do not fully align into the shell. In addition, because water must be

Figure 1.25 A modular rig working in an area of restricted access (image courtesy of SOCOTEC)

Figure 1.26 A modular rig working in a location of low headroom (image courtesy of SOCOTEC)

Note that the ceiling has been removed to provide more headroom and ventilation tubing to remove exhaust fumes.

Figure 1.27 A custom modified cable tool rig for restricted headroom, carried on a tracked platform (image courtesy of SOCOTEC)

used to bore in granular soils there is a significant quantity of water trapped in the shell with the granular soil; as the shell is raised up the borehole some of this water escapes and with it significant quantities of finer material may be washed out. By losing both coarser particles owing to fragmentation and finer particles owing to washout, the particle size distribution of the recovered soil may be significantly different to that seen in situ.

The soil in the shell is bought to surface and tipped out of the tube by upending the shell using a tipping hook. The soil and water should be caught in a suitable receptacle or in a tray to enable a sample to be collected. When sampling from this material, any water should be included in the sample because this will invariably contain fine material. The sample should be left to stand for a while so that the suspended fines can settle, after which excess water may be drained away.

In granular soils it is necessary to add water to the borehole to enable the soil to be brought to the surface. In water bearing soils it is essential to keep the water in the borehole slightly above the water table, a condition known as water balance. This produces a positive head in the borehole and prevents adverse hydrostatic conditions known as 'blowing' or 'piping' (see Section 1.3.7.4). In fine sand and silt, it may be necessary to use an undersized shell where the diameter is smaller than the standard shell. This is because the standard-sized shell can develop suction pressures as it is withdrawn which, when drilling in fully saturated fine sand and silt can draw material into the borehole and casing inducing significant loss of ground around the borehole. This effect will loosen the soils below the borehole, influencing tests such as SPTs. The soil and water mix rising up the casing under this pressure difference is a condition referred to as 'blowing'. An undersized shell will negate this phenomenon, to some extent. Ideally, the shell should have a diameter approximately 90% of the casing internal diameter.

By using casings it is possible to prevent the borehole from collapsing, and also control water entry into the borehole, therefore allowing the borehole to be deepened; this is a significant advantage over the use of methods that are only able to install a couple of lengths of casing such as dynamic sampling drop weight.

It is possible to take Class 2 and sometimes Class 1 'undisturbed' samples of cohesive soils in a cable percussion borehole, and these are discussed in more detail in Chapter 2: Geotechnical sampling. Samples can include thin-walled pushed samples and piston samples that can provide high quality Class 1 samples, although disturbance ahead of the borehole may affect the samples taken using these methods. The commonly used open tube samples comprise U100s and UT100s, which are driven using a sliding link hammer (Figure 1.28). The driving process and geometry of the open tubes can produce variable quality of samples (Chapter 2). Other typical samples include bulk samples, taken from soils that are disturbed when removed from the borehole, and small-disturbed samples in tubs. Samples for chemical and geoenvironmental analysis are also readily taken from cable percussion boreholes (Reading and Martin, 2025).

The cable percussion borehole provides the ability to undertake in situ (field) testing. The various field tests, including those conducted in boreholes as well as those independent of boreholes, are discussed in Chapter 5. These can include strength related tests, such as borehole vane tests and SPTs, groundwater testing such as variable and constant head permeability as well as various down-hole geophysical tests.

It has become common practice that the cable percussion borehole is used to prebore to competent cohesive soils where rotary drilling may begin/start. In general, the cable tool borehole will be completed using 200 mm dia. casing, which is sufficient to allow rotary S size casing to be inserted.

Figure 1.28 Sliding link hammers used to drive open tube samples (authors' image)

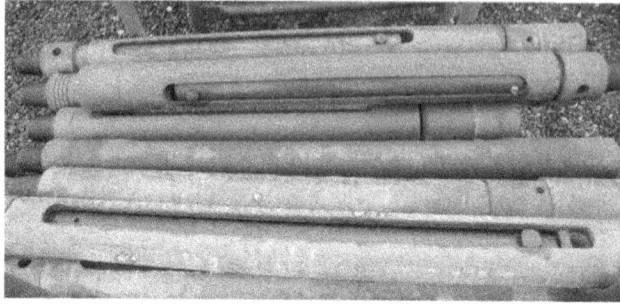

Although the cable percussion method is very flexible, some concern has been expressed regarding safety aspects of the rig. There are moving parts that cannot be guarded, such as where the tools are dropped into the borehole and pass through the casing. This is where the drilling assistant needs to work to help guide the tools into the casing. Pinching a finger is a too common occurrence. Guarding is now mandatory for the drum and clutch mechanism because these are moving parts (see Figure 1.18(b)). Modifications are also fitted to most rigs to prevent the rope jumping off the top wheel or crown wheel. Prior to this modification this was also a common occurrence, posing a risk to the operative who would need to climb the mast to free the rope.

Furthermore, there are general concerns regarding the manual handling of the boring equipment, much of which weighs more than should be lifted unaided. For example, the SPT assembly weighs 115 kg; most of this is fixed at one end of the tool making moving and loading the tool particularly challenging. Most rigs are towed behind a pick-up truck with a length of some 8.8 m (see Figure 1.29).

Figure 1.29 Rig (Dando 2000) dimensions: (a) length when in travelling position, 8.483 m; (b) height when erected, 5.669 m (images courtesy of Dando International Drilling Equipment)

(a)   (b)

The weight to length ratio is close to the permitted towed trailer limit and certainly exceeds this if tools are loaded onto the rig for transportation. These limitations may require a support vehicle, or additional site deliveries, to provide the drilling crew with the appropriate range of equipment for the job.

1.3.5   Rotary drilling

1.3.5.1   Overview

Rotary drilling has its roots in mining but has been adapted for civil engineering, and in particular ground investigation, as the designs of deeper foundations, tunnels and other deep structures have brought about a requirement for high quality information on the geotechnical characteristics of bedrock.

Rotary drilling typically employs a rig with a mast and a drilling head. The head produces a rotational torque that turns a series of rods, the lower end of which is attached to a drilling bit. Cuttings produced by the drilling bit are removed using a flushing medium, which is a fluid that may consist of water, drill muds, polymers, air, air-mist or foam. These are discussed further in Section 1.3.5.8. While some boreholes are drilled just to carry out in situ tests or install instrumentation, most are drilled to obtain core samples of the strata, enabling detailed inspection and testing at the surface.

For ground investigation purposes, rotary boreholes are commonly drilled vertically but, where needed, may also be drilled at an inclination. Inclined boreholes are used for many reasons, such as to investigate a void or enable core to be taken at a particular location, perhaps to intercept a proposed tunnel alignment. Inclined holes may facilitate the investigation of ground that cannot be accessed from vertically above, typically owing to a natural or man-made obstruction. They can also be used to intercept inclined strata to help determine the dip and dip direction of features such as bedding or discontinuities (see Figure 1.30).

Effective rotary drilling requires the skilled use of a combination of several factors, such as the selection of drill bit type, drill speed, torque, flush type and circulation fluid recovery. It is now

Figure 1.30 Rotary drilling rig being set up to drill an inclined borehole, with an excavator assisting with the handling of the drill rods (image courtesy of SOCOTEC)

possible to obtain a continuous record of these critical parameters (drilling parameter recording (DPR)), which can significantly contribute to the borehole record (see also Chapter 3).

1.3.5.2 Rotary open hole drilling

Open hole drilling is carried out with or without casing and uses a full face bit to form a hole. This relatively rapid, low-cost technique may be used to advance a borehole through overburden materials until bedrock is encountered, investigate the ground for readily identifiable features such as mine workings or natural cavities, or drill to a specific depth of interest – for example, the level of a proposed tunnel, prior changing to coring techniques. It may otherwise be used to create a suitable hole for certain field tests, or in which to install some kind of monitoring instrumentation, and under some circumstances it may simply be needed to clear debris from a section of borehole that has collapsed.

The type of bit used is dependent on the rig power, the depth of borehole, the drilling method and the rock type being drilled. A suitable drilling flush is needed to remove cuttings from the bit face, to prevent clogging and to keep the bit cool (drilling flush is discussed in Section 1.3.5.8).

Open hole drilling uses either a conventional rotary action or a 'down-the-hole' (DTH) hammer, which combines the rotary action with high frequency percussion, to remove rock material.

There are three main bit types used in conventional rotary open hole drilling, as shown in Figure 1.31. The rock-roller was first developed in the USA, the very first patent on a rock-roller bit that had two cones was taken out by Howard Hughes for drilling oil wells. The 'tricone' design, with three intersecting steel cones with hardened steel teeth, appeared in the 1930s and is the commonest form in use today. The design is varied, depending on the rock type being drilled, by altering the spacing and length of the teeth.

Drag bits are used in softer materials such as soft limestone and chalk, shale and weakly cemented sandstone. The polycrystalline diamond (PCD) bit has replaceable cylindrical cutters that have micron sized synthetic diamonds set in tungsten carbide. The more expensive PCD bits have greater durability and a longer life compared with the standard bits.

Figure 1.31 From left to right: rock-roller or tricone bits, drag bits and a polycrystalline diamond (PCD) bit (images courtesy of MGS)

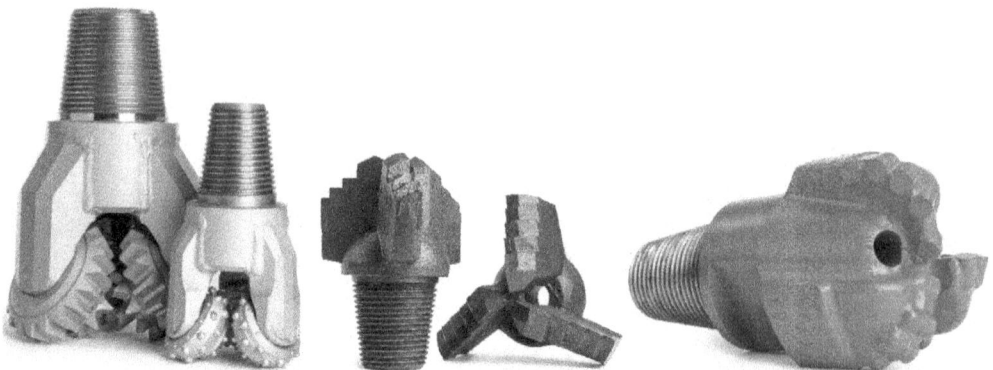

Adding a percussive element to the rotary drilling is particularly effective in stronger rocks, in a similar way that a small hand drill may use a hammer action to penetrate harder materials. The percussion can be provided by a pneumatic or hydraulic hammer, that can either be located at the top of the drill string, as part of the drilling rig, or at the base of the drill string, just behind the drilling bit. Although several variations can be used, 'top-hammer drilling' generally uses the rig hydraulics to operate the hammer, meaning that a separate flushing system is needed to keep the borehole clean. With DTH hammers, it is generally beneficial for them to be pneumatic, so that compressed air can be used both to operate the hammer and also to flush out the borehole.

In ground investigation boreholes, the DTH hammer is used more commonly as it has a few advantages over top-hammer drilling, including providing better control over the straightness of the borehole, and applying much less stress to the drilling rods. One potential disadvantage, however, is that if ever a borehole collapse were to result in drilling equipment not being recoverable, a DHT hammer would be a costly piece of equipment to lose.

While the use of a pneumatic DTH hammer can be a rapid method of progressing an open hole into strong rock, there are a few drawbacks, including the need for an air compressor powerful enough to operate the hammer efficiently and keep the borehole suitably clear of debris. In addition, the technique can produce a large amount of dust, meaning that it may not be suited to some urban sites or, at the very least, measures will need to be put in place to collect the dust at the surface or suppress it in the borehole (typically by injecting some water into the air flow). Dust generation could also be avoided using a hydraulically driven DTH hammer. These operate at lower pressures and can therefore progress the borehole with less disturbance of the ground surrounding the borehole, as well as avoiding the problem of dust creation. However, the technique requires a substantial amount of water, either from a clean supply, or by recirculating and cleaning water being returned from the borehole, and this may be less feasible in some locations.

The drilling bit used with a rotary percussive hammer generally consists of a nickel-alloy steel body, with a flat or domed face into which are fitted tungsten carbide studs, of various shapes. In operation, the bit is rotated such that with each hammer blow the studs are directed onto a slightly different area of rock, thereby breaking up the surface at the base of the borehole.

With the exception of the reverse circulation technique, discussed below, the flushing medium used in open hole drilling is pumped down the hollow drilling rods and returns to the surface by way of the annular space between the drilling rods and the borehole wall, or casing bringing the rock chippings to the surface (see Figures 1.32(a) and (b)).

When drilling an open hole, stabilising collars are provided to ensure the bore remains straight. The diameter of the stabilisers should be close to the hole diameter. The drill flush needs to be of sufficient velocity to keep the bit free and clear of debris.

For ground investigation, an open hole might be drilled to carry out in situ down-hole testing or to install instrumentation, and these activities would typically benefit from a general understanding of the ground conditions the open hole has passed through. Although core samples are not being recovered, in such boreholes it may be possible to distinguish between the main strata from the cuttings. Owing to the mixing of materials in the borehole and the time lag for chippings to return to the surface, it is generally difficult to determine the strata boundary positions with any real accuracy. The method of reverse circulation, discussed below, can improve this because the up-hole velocity is constant.

Figure 1.32 Open hole drilling methods: (a) rock roller; (b) DTH hammer; (c) reverse circulation (authors' image)

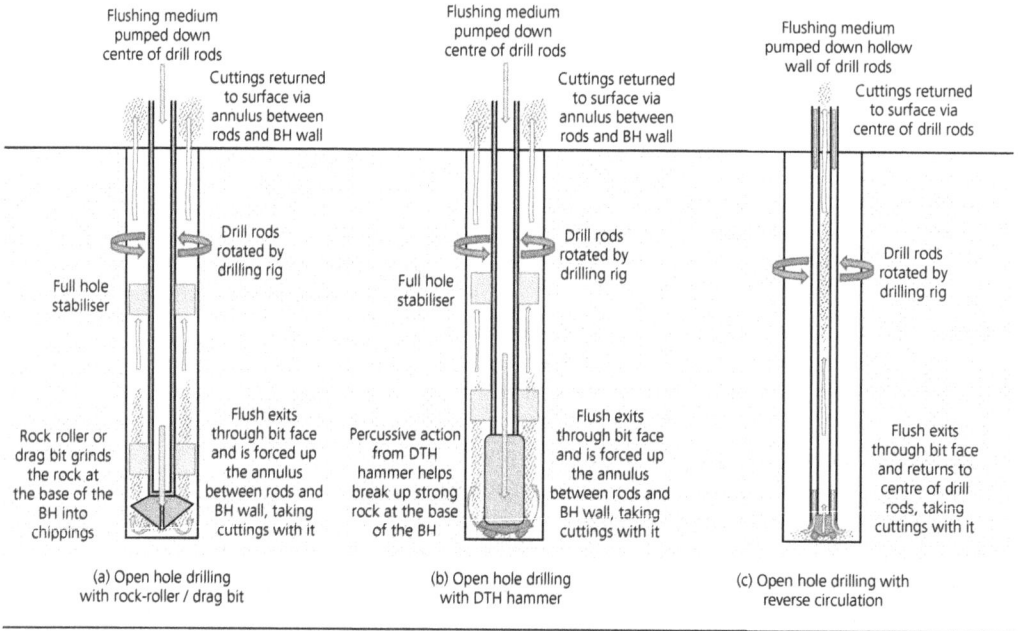

Flushing medium pumped down centre of drill rods

Cuttings returned to surface via annulus between rods and BH wall

Flushing medium pumped down centre of drill rods

Cuttings returned to surface via annulus between rods and BH wall

Flushing medium pumped down hollow wall of drill rods

Cuttings returned to surface via centre of drill rods

Drill rods rotated by drilling rig

Drill rods rotated by drilling rig

Drill rods rotated by drilling rig

Full hole stabiliser

Full hole stabiliser

Rock roller or drag bit grinds the rock at the base of the BH into chippings

Flush exits through bit face and is forced up the annulus between rods and BH wall, taking cuttings with it

Percussive action from DTH hammer helps break up strong rock at the base of the BH

Flush exits through bit face and is forced up the annulus between rods and BH wall, taking cuttings with it

Flush exits through bit face and returns to centre of drill rods, taking cuttings with it

(a) Open hole drilling with rock-roller / drag bit

(b) Open hole drilling with DTH hammer

(c) Open hole drilling with reverse circulation

If the borehole is unstable, casing may be required to ensure the hole does not collapse (see Section 1.3.5.4).

1.3.5.3 Reverse circulation

In some circumstances, open hole drilling may be conducted with the use of reverse circulation drilling. This method pumps the flush down the hole between the casing and drill rods, the flush passing into the bit and through the ducts in the bit and is fed back up to surface through the centre of the drill rods (see Figure 1.32(c)). This method removes the drill cuttings in a regular flow and because the flow is consistent, the cuttings appear at surface without mixing and can therefore be logged (sometimes termed 'mud logging'). Samples of the cuttings can be taken for identification and records. This technique is widely used for mineral exploration, although less so for ground investigation.

1.3.5.4 Casing in open hole drilling

Open hole drilling is often conducted in overburden materials and is then followed by conventional coring once the deeper bedrock has been reached. Such overburden materials may be variable in nature, possibly water bearing, and relatively unstable, and it is frequently necessary to line this section of the borehole with casing to maintain stability. It may be possible to drill an open borehole and then, once the required depth has been reached, lower a smaller diameter casing into the hole. This, however, relies on the borehole staying open long enough for it to be cased, and in many materials is unlikely to be successful.

An alternative approach is to drill and case the hole simultaneously. Often referred to as an 'overburden drilling system', or by various manufacturers' names, this method uses a drilling bit that pulls the casing down behind it as the borehole progresses. The difficulty with this method is that

Figure 1.33 Open hole casing systems: (a) and (b) using a eccentric bit and (c) and (d) using a concentric bit (authors' image)

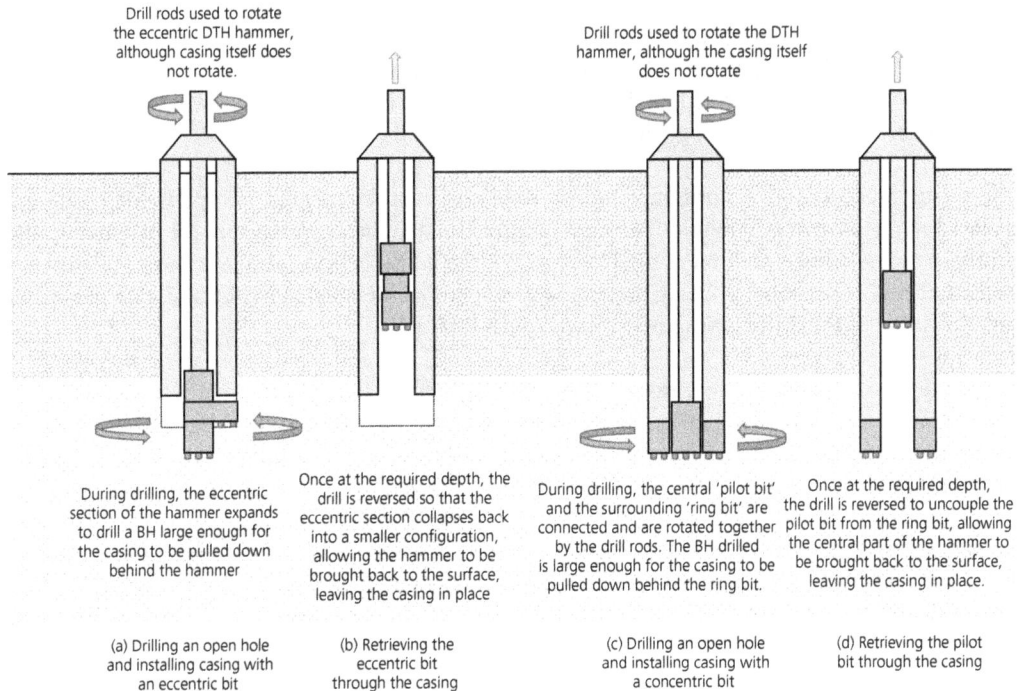

Drill rods used to rotate the eccentric DTH hammer, although casing itself does not rotate.

Drill rods used to rotate the DTH hammer, although the casing itself does not rotate

During drilling, the eccentric section of the hammer expands to drill a BH large enough for the casing to be pulled down behind the hammer

Once at the required depth, the drill is reversed so that the eccentric section collapses back into a smaller configuration, allowing the hammer to be brought back to the surface, leaving the casing in place

During drilling, the central 'pilot bit' and the surrounding 'ring bit' are connected and are rotated together by the drill rods. The BH drilled is large enough for the casing to be pulled down behind the ring bit.

Once at the required depth, the drill is reversed to uncouple the pilot bit from the ring bit, allowing the central part of the hammer to be brought back to the surface, leaving the casing in place.

(a) Drilling an open hole and installing casing with an eccentric bit

(b) Retrieving the eccentric bit through the casing

(c) Drilling an open hole and installing casing with a concentric bit

(d) Retrieving the pilot bit through the casing

the drill bit needs to create a hole of sufficient diameter to accommodate the casing, but then also be retrieved back to the surface through that casing.

Two different methods have been developed to achieve this. The first uses an eccentric drill bit, which is essentially a DTH hammer with a section of the hammer that increases in width when drilling, and then retracts, by a reversal of the rods rotation direction, when it needs to be brought to the surface (see Figures 1.33(a) and (b)). Although proving to be highly successful in many ground conditions, if the system is not kept adequately flushed there is some risk in sand-dominated soils, of the hammer becoming jammed in the casing, and even potentially a risk of lengths of casing become unscrewed while rotating the rods in an effort to free the hammer.

The second method uses a pair of concentric drill bits, a central 'pilot pit' surrounded by a circular 'ring bit'. When drilling, these are linked and work together, but can then be separated, again by a reversal of the rods, such that the pilot bit can be brought back to the surface, leaving the ring bit in place (see Figures 1.33(c) and (d)).

In both the eccentric and concentric systems, the casing does not rotate and is essentially pulled down into position by the hammer.

1.3.5.5 Rotary core drilling

Although rotary core methods were originally reserved for drilling rock and materials that other intrusive methods were unable to penetrate, in modern ground investigation rotary core drilling is employed more frequently and is often used to obtain high quality cores from complex fluvial deposits, over-consolidated cohesive strata and weak rocks, such as chalk, as well as much harder rock types.

There are several ways of drilling to obtain cores. The simplest method is to use separate casings and core barrel. This technique, often called conventional coring, is similar to cable percussive drilling in that the hole is formed, in this case using the core barrel, and support is provided as and when needed by installing casing when the core barrel is out of the borehole.

The coring process involves drilling a hollow steel core barrel, typically 1.5 or 3 m long, into the material at the base of a borehole. The core barrel is rotated through rods that are attached to the drilling head. A tubular drill bit at the lower end of the barrel cuts an annular slot as the barrel is rotated, the 'core' effectively being the remnant of material, delineated by the slot, that passes up into the barrel as the drilling progresses (see Figure 1.34(a)).

When the core barrel is full it is extracted by withdrawing all the rods to bring it to surface. This requires each rod, normally 3 m in length, to be unscrewed from the rod string, which is a time-consuming process. Each rod in turn is clamped at the rod joint enabling the upper rod to be removed while ensuring the core barrel is not dropped back down the borehole.

In common with open hole drilling, discussed above, consideration always needs to be given to the risk of materials in the borehole wall becoming dislodged (see Figure 1.34(b)), possibly making it difficult, or even impossible, to retrieve the barrel. This could result in the potential loss of

Figure 1.34 Conventional rotary coring and casing (authors' image)

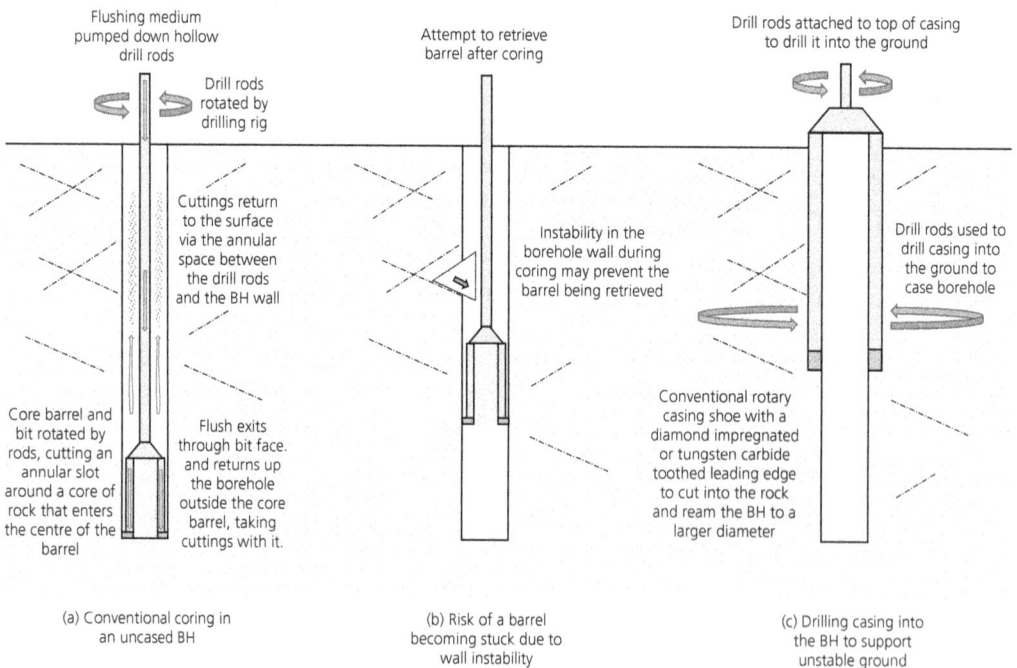

(a) Conventional coring in an uncased BH

(b) Risk of a barrel becoming stuck due to wall instability

(c) Drilling casing into the BH to support unstable ground

expensive equipment, and the premature termination of the borehole, which may then need to be started again in a nearby location. Such a situation is conventionally mitigated by casing the borehole. It should be noted, however, that the inside diameter of the casing must be at least that of the borehole, so as to accommodate the barrel that is forming the hole. As such, the casing needs to be drilled into the ground, using a casing head attached to the drill rods, thereby reaming the borehole out to a larger diameter as it is installed (see Figure 1.34(c)).

For this reason, the leading edge of a rotary casing shoe incorporates a drilling bit (see Figure 1.35), typically diamond impregnated for stronger rocks or otherwise with tungsten carbide teeth for weaker rocks.

Casing advance should always follow behind the coring process and should not extend below the cored depth, to ensure that the material being cored is not disturbed. The installation of casing can be time consuming, slowing down the overall borehole progress, and should therefore be factored into the drilling programme wherever possible.

An alternative to the conventional coring approach is the wireline technique, as shown in Figure 1.36. This method uses a system whereby both the core barrel and the casing are advanced together, therefore avoiding some of the potential issues of separate coring and casing processes described above. The casing is set just behind the core bit, and the coring process is achieved by the rotation of the whole casing string. There is an inner core barrel, mechanically latched onto the casing so that as the casing is drilled into the base of the borehole the core passes up into the barrel. When coring reaches the required depth, the inner barrel can be recovered to the surface, leaving

Figure 1.35 Rotary casing shoe with a lead length on top incorporating a casing drilling bit (authors' image)

the casing and drill bit in place. This is achieved by dropping an 'overshot' tool down the borehole on the winch rope (hence the term 'wireline'). The overshot tool locates onto the top of the barrel, releasing it from the casing, and enabling it to be lifted out of the borehole. This process removes the need to pull the entire drill string from the borehole and, consequently, this method is much quicker than conventional drilling and produces higher quality of core in most situations.

Whether using a conventional coring or wireline technique, the borehole can only be progressed in short sections, the length of each 'core run' being limited by the physical length of the core barrel. Barrel lengths are generally either 1.5 m or 3 m and, while it is possible to core using a 3 m barrel, the use of this long barrel has manual handling issues owing to the excessive weight, especially when full of core, and particularly when drilling to obtain larger diameter cores. Consequently, ground investigation boreholes are more frequently drilled using a 1.5 m long barrel. Furthermore, there are several reasons why the driller may terminate a core run prior to the barrel being full – for example, if the bit appears to have 'blocked off' and is no longer cutting properly, or a short run is intentionally being used to improve recovery. As such, while typically core runs will be 1–1.50 m in length, some may be much shorter.

When each run is complete, the barrel is brought to surface, either on the drilling rods in conventional coring, or on the winch with the wireline technique, and is finally suspended over the borehole on the winch rope. The support operative will then 'shoulder' the barrel and carry it to a set of trestles where it is laid in a horizontal position for dismantling and recovery of the core (see Figure 1.37).

Figure 1.36 Wireline drilling (authors' image)

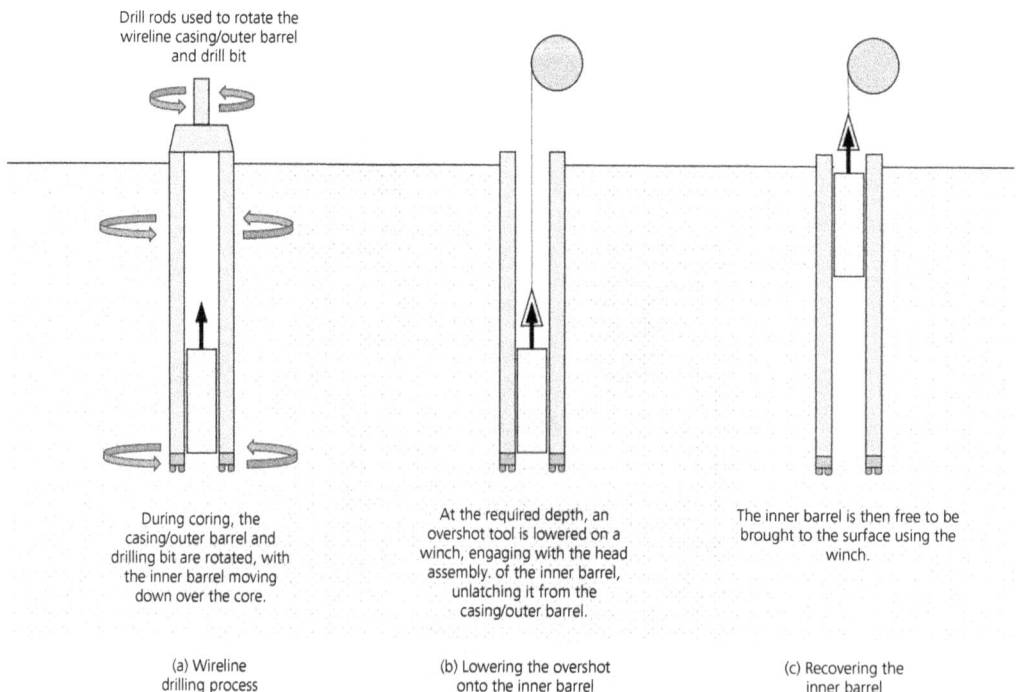

Drill rods used to rotate the wireline casing/outer barrel and drill bit

During coring, the casing/outer barrel and drilling bit are rotated, with the inner barrel moving down over the core.

At the required depth, an overshot tool is lowered on a winch, engaging with the head assembly. of the inner barrel, unlatching it from the casing/outer barrel.

The inner barrel is then free to be brought to the surface using the winch.

(a) Wireline drilling process

(b) Lowering the overshot onto the inner barrel

(c) Recovering the inner barrel

When transferring drilling rods and the core barrel to the trestles, some of the weight is taken by the winch rope, but this still requires a significant weight to be taken by the operative. Although rarely seen on rigs in the UK, a mechanical rod handling system can avoid unnecessary manual handling problems.

Once at the surface, the core barrel can be dismantled to expose the core, which is removed and placed in either wooden or plastic core boxes. The use of coreliner, a clear plastic liner within the core barrel (see Section 1.3.5.6), improves the core condition and makes the transfer from the barrel to the core box easier. Core is conventionally placed in the core box with the depth increasing from left to right, top to bottom, as writing is laid out on a page (see Figure 1.38).

The driller will mark the core with the top and bottom depths of each core run and will ideally use spacers to separate the runs. In order to use the box more efficiently, where a run length exceeds the length of the box, it will either be separated into two sections at a natural fracture (as with the first core run in Figure 1.38), or if the core is not fractured the driller may cut the core so that it fits the box. Any artificial cuts should be clearly identified, for clarity.

There are a number of reasons why the length of core recovered from a given run may be different to the run length. The length of core recovered may be shorter than the run – for example, if a

Figure 1.37 Drilling rods and barrel ready for use supported on trestles (image courtesy of SOCOTEC)

natural or artificial void has been encountered, if there has been some erosion of the material during the coring process, or if some core has fallen out of the barrel when the barrel was being lifted from the base of the borehole. Where 'core loss' from a particular run is owing to a section being left in the borehole, that length of core may be picked up as part of the following core run, possibly giving an excess of core compared with the run length (known as core gain). These issues will be assessed as part of the logging process (Reading and Martin, 2025). It should be noted, however, that owing to such core recovery difficulties, the top and bottom depths of the core runs may not represent the actual depths in the ground at which the core was situated; they are simply the depths the core barrel covered while recovering the section of core presented.

Once in the boxes the core can either be logged on site or transported and logging at the office or laboratory of the investigation contractor. It is always preferable to log core on site, as this ensures minimal damage from transportation. However, this does require some forward planning to provide a suitable dry space for the activity (see Reading and Martin (2025), where site set up and core logging is discussed in detail). Logging on site also enables the core to be cleaned of drilling fluid and photographed with minimal delay, which is beneficial, as these activities should be carried out as soon as possible once the core is out of the ground.

Core quality, that is to say the proportion and undisturbed nature of the material recovered, can be affected by drilling parameters such as rotation speed, flush medium, bit pressure, hold back force and bit type. The driller controls most of these parameters, and it is therefore essential to use a skilled operator/driller who understands how the many variables can influence the core quality. The driller is able to adjust the torque and rate of the rotating bit, the type of flush medium and the flush pressure, which are also essential elements of the drilling process.

If the proportion of core recovery is poor, the core run length is conventionally reduced so that adjustments to the coring method can be made in an attempt to improve the recovery. Typically, specifications will require minimum core recovery per run of 95%. While this is desirable, it is by no means achievable in all materials, a realistic approach should be adopted with respect to core recovery.

Figure 1.38 Core box showing core with depth markers (authors' image)

In some strata it may not be possible to obtain a high recovery purely because the material type does not respond to coring. For example, partially cemented silty sandstone will readily break up turning the material to silty sand, which may be washed away by the drill fluid with resultant very low core recovery (see also Section 1.7.5). In general, very weak rocks and coarse soils – that is, sand and gravel – are particularly challenging in terms of core recovery. High contrasts between materials, such as interbedded strong limestone and stiff clay, can also be difficult, as can relatively weak materials that contain stronger concretions, such as 'claystone' in clay or flints in chalk. Such masses can become lodged in the coring bit, rather than being cut, and then erode softer materials as the core run is advanced. Such issues will be assessed as much as possible by the driller and the core logger.

Managing core quality requires a high degree of skill from the driller and should be planned in advance. In addition to drilling technique, ensuring the right core barrel assembly, core bit and flush medium is used is essential, and there is no one combination which will suit all strata. Various options of barrel, bit and flush are considered in the following sections.

1.3.5.6 Core barrels

For ground investigation boreholes, the core barrel is generally a double tube barrel, in that there is an outer steel barrel that rotates and has a bit attached that cuts the rock away, and an inner steel barrel that does not rotate and receives, retains and protects the core, which is held from slipping from the barrel by a core catcher and springs (see Figure 1.39).

In conventional coring, the whole barrel is brought to the surface when full of core, whereas with wireline drilling the outer barrel is effectively an extension of the casing and remains in the borehole while the inner barrel is withdrawn.

Mazier originally designed a triple tube system, although the barrel was only used in conventional drilling and the assembly of the barrel was cumbersome and intricate. In recent decades, however, barrel design has developed, and many now accommodate a clear plastic liner, known as a 'coreliner', inside the inner of the two steel barrels. These are known as 'triple tube' core barrels, even although the plastic liner is single-use and removed with the core. The coreliner offers further protection to the materials in the core barrel and, as it is removed and placed in the core box with the

Figure 1.39 (a) Barrels with core retaining springs and (b) core catcher (authors' images)

(a)   (b)

core, helps protect the core during transportation and storage too. Where coreliner is used, end caps can be placed at each end of the recovered core in each run. In addition, coreliner allows core to be pulled out of the barrel horizontally, whereas with a basic double tube barrel the core needs to be tipped out, sometimes needing encouragement with a hammer. Without coreliner there is therefore a greater risk that core is broken and disturbed during the extraction process.

Figure 1.40 shows a triple tube barrel on trestles with the coreliner shown in position ready for the bit to be put onto the barrel. To drill using coreliner requires a different size bit to provide core of a slightly smaller diameter to that obtained without coreliner. Table 1.4 gives core and coreliner sizes.

A summary of some of the commonest core barrels in use in the UK is given in Table 1.3. Some of these have been designed as triple tube barrels, while other essentially double tube barrels can be modified to accept coreliner, therefore converting them to triple tube.

As with much of the drilling equipment in use today, there is no international agreement for the sizes of core barrels, drilling bits or other drilling equipment such as rods, and this has resulted in a confusingly large number of sizes, methods of notation and configurations. Some persist with imperial measurements and use letters to denote the type and size of equipment while others use a metric system. EN ISO 22475-1, Annex C (BSI, 2021) provides an overview of the most popular sizes found in use in the UK, which is an amalgamation of several different approaches. In addition, the standard also provides diagrams of tooling and bit designs for various materials. The Eurocode recognises two main types of core barrel: 'Metric' series and the 'W' series, both of which are detailed in Annex C of EN ISO 22475.

The selection of a suitable barrel size is important, as research has shown that the larger the diameter of the borehole the better the quality of the core. However, there is a trade off with this approach in that the larger the core the heavier it will be, and transportation can become an issue, requiring mechanical handling methods to move the core. In addition, larger diameter core will require a more powerful and larger rig to accommodate the larger equipment, which all increases the overall

Figure 1.40 A triple tube barrel, comprising the outer steel barrel threaded for attaching the drilling bit, the inner steel barrel and then a third tube: a plastic coreliner to aid core removal (authors' image)

Table 1.3 Rotary core barrels: configuration and approximate sizes

Barrel type	Barrel configuration	Hole diameter range (mm)	Core diameter range (mm)	Remarks
B series	Single tube	46–146	42–123	Fits into metric and W series casing
TT series	Ultra-thin-walled double tube	46–56	35.3–45.2	Fits into metric casing
T2 series	Thin-walled double tube	66–101	51.7–83.7	Fits into metric casing
T6 series	Thin-walled double tube	101–146	57–123	Fits into metric casing
T6S series	Double tube, split inner tube	86.3 to 146.3	57.7–115.7	Fits into metric casing
WF series	Thick-walled double tube	98.8–199.3	76.2–165.1	Fits DCDMA casing sizes B, N.
412F & HWAF	Triple tube	107.2–99.2	74.7–71.4	Air flush
Geobore P	Triple tube	127.7	83	
Geobore S	Triple tube	146	102	

Table 1.4 Relationship between the core barrel type and the diameter of core recovered (continued on next page)

Core barrel type	Standard core diameter (mm)	Core diameter using coreliner (mm)	Coreliner inside diameter (mm)	Coreliner outside diameter (mm)	Special core lifter case required?
T2-66	51.7	48.0	50.3	52.3	Yes
T2-76	61.7	58.0	60.3	62.3	Yes
T2-86	71.7	68.7	70.4	72.4	No
T2-101	83.7	80.7	82.4	84.4	No
T6-76	57.0	53.0	56.5	58.5	Yes
T6-86	67.0	63.0	65.0	67.0	Yes
T6-101 T6H	79.0	76.0	78.5	80.5	No
T6-116	93.0	90.0	92.5	94.5	No
T6-131	108.0	103.5	106.5	109.5	Yes
T6-146	123.0	118.0	121.0	124.0	Yes
NWL	47.6	45.1	47.0	49.0	Yes
HWL	63.5	61.1	63.0	65.0	Yes
PWL	85.0	83.1	86.0	88.0	Yes
HWF	76.2	73.0	74.4	76.4	No

Table 1.4 Continued

Core barrel type	Standard core diameter (mm)	Core diameter using coreliner (mm)	Coreliner inside diameter (mm)	Coreliner outside diameter (mm)	Special core lifter case required?
PWF	92.1	87.0	89.8	92.3	No
SWF	112.8	107.0	110.5	113.5	No
UWF	178.0	170.0	172.5	177.0	No
412F	74.7	73.0	76.8	78.8	No
Mazier 101	72.0	72.0	74.0	76.0	No
GBS	102.0	102.0	106.5	109.5	No

cost. Typically, for site investigation, Geobore S or similar wireline systems are used and give cores of approximately 102 mm in diameter and produce an overall borehole diameter of 146 mm. This is considered to be a reasonable size and rig power compromise.

In addition to core quality, the selected borehole diameter will affect the speed of drilling. Smaller diameter core is quicker to drill but there is a risk that the core quality will be poor, particularly when drilling in jointed rock.

Choosing the correct method to optimise core recovery may often require trials to arrive at the best combination for a particular strata sequence. In some cases, it may be necessary to change the configuration mid hole. If this is expected, the use of conventional drilling techniques is often found most suitable because the casing is run separately to the core barrel, and it will therefore only be necessary to alter the configuration of the barrel. However, if a wireline system is used, the whole assembly will probably require removal from the borehole.

1.3.5.7 Rotary drill coring bits

There are many varieties of bit design with variations in materials and cutting agent (i.e. the hard points that actually cut into the rock), as well as the position and shape of the cutting surface and position of the flushing ports.

Bits are divided between the following.

- TC – Tungsten carbide set. Plain studs of tungsten carbide are set within the cutting surface. These are suited to soft, low abrasive materials.
- GTS – Geotechnical saw-toothed carbide set. The saw-toothed arrangement cutting as a saw rather than grinding down the material being drilled.
- PCD – Polycrystalline diamond (PCD). Can be set as surface or impregnated.
- TSD – Thermally stable diamond (TSD). As PCD but using higher quality diamonds.

For the very hardest materials, carbonado, black diamond can be used, which is an impure form of diamond where the crystal includes graphite and amorphous carbon.

Table 1.5 provides a guide to the type of bit that would be required depending on the strata being drilled. A typical surface set bit is shown in Figure 1.41 and an impregnated bit in Figure 1.42.

Table 1.5 The relationship between the rock bit type series classification and approximate rock hardness

Rock hardness

Soft	Soft to medium	Medium hard Low abrasivity	Medium hard High abrasivity	Hard Low abrasivity	Hard Medium to high abrasivity	Ultra-hard
Clay	Sand	Greenstone	Medium sandstone	Hard Limestone	Taconite	Pyritic
Soft shale	Shale	Soft sandstone	Siltstone	Quartzite	Jasper	Hematite
Chalk	Marble	Dolomite	Limestone	Pegmatite	Quartz	Conglomerate
Gypsum	Salt	Andesite	Hard shale	Hematite	Rhyolite	Taconite
Weakly cemented materials	Medium limestone	Basalt		Peridotite	Granite	
				Schist	Chert Gneiss	
				Serpentine	Gabbro	
				Dolomite	Basalt	
				Marble		
				Andesite Nerite Diorite		

TC

GTS

PCD

TSD

Surface set stones per carat

10/15 20/25 30/40 40/60 60/80

Impregnated bit type number

2 4 6 8 10 9 10 10

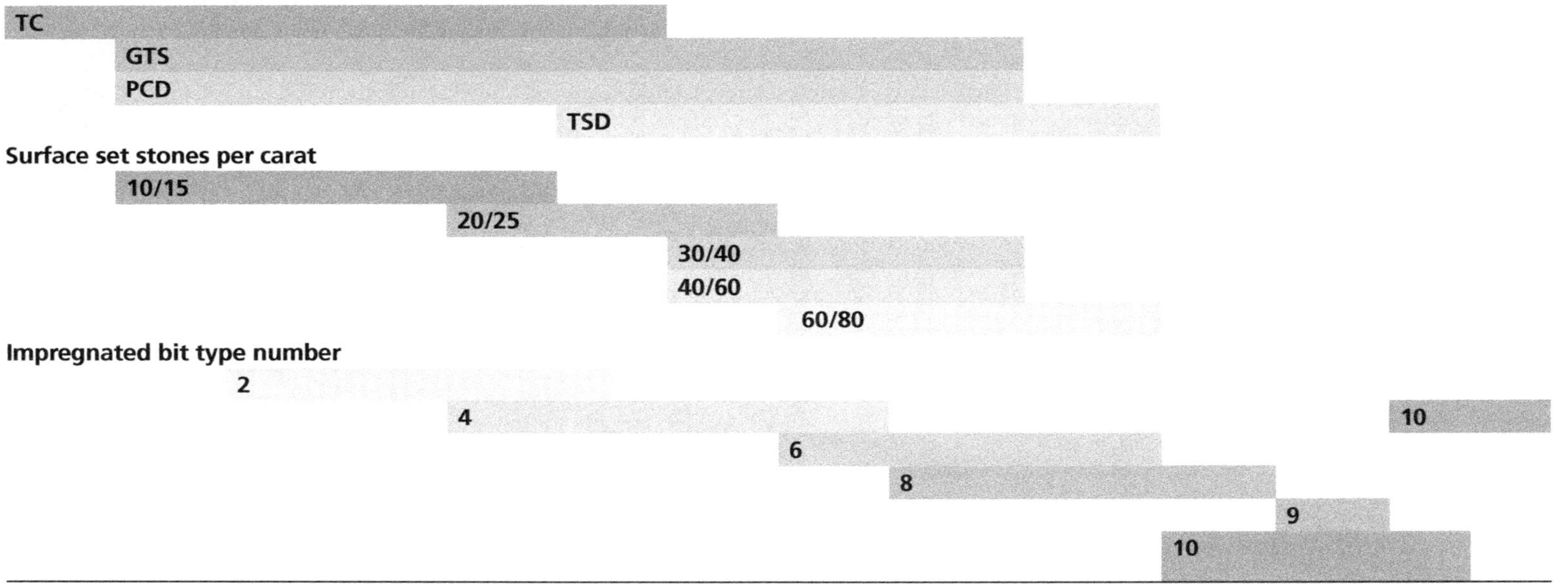

It is considered that 'A good driller can make a poor bit last days while a poor driller may well destroy a good bit in minutes.'

The bit design needs to be able to cut the rock cleanly while ensuring the cuttings are removed efficiently and the bit does not become too hot. The matrix must hold the cutting agent, usually diamonds or tungsten carbide inserts, while resisting the shock from the drilling process. Key to the effectiveness of a particular bit is the design and position of the waterways and the shape of the cutting face. A range of different bit sets and waterway locations are shown in Figure 1.29. In general, the waterways are normal; that is, discharging between the core and the bit. This means that the flush medium is in contact with the core that, if susceptible, will soften. The alternative is to use face or bottom discharge, which means the majority of the flush will flow away from the core and outside the bit. These bits are preferred in soft and cohesive soils, although they do need some care from the driller to avoid clogging. This can be avoided by holding back the drill string so that core is cut more slowly, or by using stepped or tapered bit profiles.

Surface set bits, shown in Figure 1.41, have diamonds set on the surface of a medium; the diamonds are medium- to fine-sized. When the surface diamonds are worn away, the bit is unable to cut.

Impregnated bits, shown in Figure 1.42, have industrial diamonds impregnated throughout the matrix material – for example, manufactured diamonds are set into a tungsten carbide matrix. The diamonds are fine and evenly distributed through the matrix so that, as wear takes place, a new set of diamonds are exposed. These bits will generally last longer but are more expensive.

Impregnated bits are defined by a classification from 1 to 10; Series 1 is used for softer materials such as softer limestone and chalk, for example, while Series 10 is used for the hardest rocks such as quartzite. In general, the density of the diamonds set within the bit increases with higher series number.

Some high-quality bits have the diamonds set by hand. However, these can be very expensive and so most diamond bits used for ground investigation are impregnated or surface set. Often the bits are faced with studs that are either impregnated with diamonds or are of tungsten steel (see Figure 1.43).

Figure 1.41 A surface set bit (image courtesy of MGS)

Figure 1.42 An impregnated bit (image courtesy of MGS)

Figure 1.43 Some core bit configurations: (a) Tungsten insert studs normal discharge; (b) Tungsten inserts with face discharge; (c) impregnated diamond with normal discharge (authors' images)

(a)   (b)   (c)

The size of the diamonds is generally required to be smaller the harder the rock. This is so the stress on the diamonds is spread more evenly as the rock hardness increases.

1.3.5.8 Rotary drilling flush

Flush is needed to keep the drilling bit lubricated, free from debris and clogging, as well as keeping the bit cool. If this is not controlled in conjunction with the drill speed, bit wear will be fast and excessive, causing slow drilling progress. In addition, the fluid can provide support to the borehole wall and help prevent collapse. The fluid is normally delivered through the hollow drill rods, into the drill bit, and out through flushing ports set into the face of the bit. The fluid with suspended cuttings then passes between the barrel and the casing up to the surface. Water-based flush can be collected in tanks for disposal or reuse.

In some respects, the simplest flushing medium to use is air, as water is not always readily available at all drilling locations. However, it requires a high velocity to bring cuttings up to the surface (denser fluids can achieve this at much slower rates), and this can be quite abrasive to weaker rocks. In addition, a powerful, and therefore large, air compressor needs to be transported to the borehole location. For these reasons, the flush medium often used is clean water. However, again, this may not be suitable in some strata, and therefore 'mist', a combination of air and water, may be used instead. Where borehole stability becomes an issue the use of a drilling mud, a mixture of water and bentonite clay, or polymer, a synthetic thickener added to water, may be employed as an alternative to casing the hole. These give weight and viscosity to the fluid and adhere to the borehole wall to reduce collapse or rapid loss of the drill fluid in highly porous strata. Alternatively, a foaming agent can be added to the water.

Air, mist and foam provide little or no lubrication or friction reduction, hence bit wear using these fluids can be high. It is also more difficult to control the return with these forms of flush which is often undesirable with respect to environmental protection. Figure 1.44 shows a rig using air flush on encountering water in the borehole. It should also be considered that some polymers are undesirable when drilling into the water table and can be a source of contamination, meaning that permission to use these needs to be sought from the Environment Agency. Similarly, the use of air flush is likely to be unsuitable for drilling in coal measures, as the high pressures required can encourage mine gases to migrate through fractures and potentially present a risk to occupied buildings at the surface. In such strata, drilling methods need to be agreed with the Coal Authority, who will issue a permit that details various requirements, including the flushing medium.

The key to a successful flush is the up-hole velocity. The optimum up-hole velocity when using water flush is of the order of 40cms/sec. This can be controlled by reducing the hole volume by

Figure 1.44 Rotary drilling using air flush on encountering ground water – volume of water in hole increased rapidly (image courtesy of Geotechnics)

increasing the size of the drill rods, which in turn increases the flow. Other options include using a larger pump or by adding a mud or introducing air with a foaming additive.

The use of air-based flushes (air, mist or foam) is a good way to reduce the amount of liquid in the borehole, although the returns are more difficult to control and will often spill out or blow out of the hole. This can be undesirable in restricted environments and urban areas where there is a requirement to keep the site clean. For water-based flushes the use of an inverted bucket, cut to fit over the rods, or a mat through which the drill rods pass at surface, can provide a degree of control, but muds, polymers and water are more easily controlled using a 'gooseneck' above the casing head. The gooseneck enables fluid to be channelled into settling tanks where the drill fluid can be cleaned and then reused, as shown in Figure 1.45.

Water as a flush medium requires the drill cuttings to be fine. If they are too coarse (and therefore heavy) they will not be held in suspension. However, adding a mud or polymer will increase the viscosity of the flush, which will then be able to hold larger particles in suspension. This also aids the borehole wall stability, whereas clean water may permeate into the strata causing it to soften which can result in slumping or collapse. A good polymer will also help reduce friction on the drill string and rate of wear, by mitigating abrasion on the rods and drilling parts. When using drilling additives such as bentonite, barites or polymers, it is essential to allow sufficient time for the mix to fully hydrate; these additives provide stability to the borehole and reduce friction on the drilling tools.

It is poor practice to omit the installation, wherever possible, of a recycling system that allows reuse of the drilling fluid. This can consist of the following.

- **Settling tanks** – a simple series of tanks where the drill fluid flows from one tank to the next and allows soil and rock particles to settle out, progressively cleaning the drill fluid ready for return to the borehole.

Figure 1.45 Rotary core drilling in a public area: (a) controlling the drill fluid using a gooseneck and (b) settling tanks keeps the drill site clean and tidy (images courtesy of SOCOTEC)

(a)   (b)

- **Shale shaker** – a mechanical separator consisting of vibrating grids that remove the solid particles and return the clean fluid.
- **Hydro cyclones** – a bank of cyclones used to separate the drilling cutting from the fluid.

In all cases, the solids can be disposed of to a registered waste site. It is generally required to classify this waste to ensure it is sent to an appropriate waste management site, thus reducing the environmental impact but also making the drilling process much more efficient and economic. Where additives are used the condition of the return drill fluid will need to be checked and potentially adjusted to provide the required drill fluid composition for reuse.

1.3.5.9   Core orientation

During coring, although the inner core barrel is designed not to rotate with the outer barrel, they are prone to a certain amount of movement, which can therefore rotate the core from its original position in the ground. As such, it is not usually possible to determine the orientation of discontinuities, such as joints, from normal coring. There are a number of systems that will mark magnetic north on the top of the core, thus enabling the dip direction of joints to be determined. These methods are not infallible, however, and it is considered that DTH cameras provide a much better method to orientate joint sets. This technique is discussed in detail in Chapter 3.

1.3.5.10   Rotary drilling into structures

The sections above have focussed on rotary drilling techniques primarily designed to investigate the ground. In some ground investigations, however, it is also necessary to undertake drilling specifically to investigate an existing structure, such as a retaining wall, floor slab or road pavement. Such investigations may involve coring into masonry, reinforced concrete or layers of macadam, and the core holes may be vertical, inclined or even horizontal, depending on the circumstances. They can be used to determine, for example, the thickness of a wall, the condition of the materials that form the wall and, to a certain extent, the ground behind it.

Where vertical holes are required, it may be possible to use a large rig as described in the above sections, although it will usually be more cost effective to use a small, lightweight drill for this type of work. Such boreholes typically do not need to extend to a great depth, generally being limited to the thickness of the structure or pavement being investigated, and so using just a core barrel, with no additional drilling rods, is usually sufficient.

As the materials to be cored are often reasonably strong, a very simple, single tube core barrel is generally used (see Figure 1.46(b)). To provide an axial force to the barrel while coring, the drills are often bolted to the floor or wall, depending on the direction of drilling (see Figure 1.46), and the barrel advanced by a manual winding mechanism.

Rotary coring may be used when a larger opening is required in concrete by stitch drilling (see Figure 1.47). This is achieved by drilling several holes that slightly overlap, allowing the concrete to be neatly removed. This method might be used when access through the concrete is required for a trial pit, or a field test, or when a borehole is to be drilled through a floor over a basement, something that might be considered particularly when there is no easy way to gain access into the basement with a drilling rig. It can also apply to where a borehole is to be drilled through the deck of a jetty. When conducting this type of borehole preparation, consideration needs to be given to recovery of the core. If the location is above a basement the core can be allowed to drop to the basement floor, provided the area has been made safe. A stack of cardboard boxes makes a good

Figure 1.46 (a) Horizontal drilling into a concrete quay wall (image courtesy of D. Partridge, SOCOTEC); (b) single tube coring barrels (image courtesy of MGS)

(a)

(b)

Figure 1.47 Stitch drilling to remove a section of a thick concrete slab (image courtesy of SOCOTEC)

cushion to catch the core. When the hole is on a jetty this method may not be suitable because there is a risk that the core will drop onto the bed at the point where the borehole is to be drilled. Ideally, the core should be retained, which may be achieved by supporting the underside of the core or catching the core as it drops on a suitable craft positioned appropriately, or in suitably strong netting located underneath. An alternative method to retain the core is to insert a Hilti bolt into the centre of the core, enabling a lifting eye to be fitted. To compete this, the hole should be drilled to within 10 mm of the base of the section. Inserting a lifting eye allows a winch rope to be connected with shackle to the core. The core can be freed with a sharp tap from a sledgehammer, which will crack the final piece of concrete, thus freeing the core that can then be recovered.

Breaking, sawing or coring out hard materials takes time. Where such obstructions are known to be present, such as concrete at the ground surface, or where the presence of former foundations are expected – for example, the use of breakers, sawing or coring to assist in the advance of the hole should be planned for, and the associated time and cost accommodated.

1.3.6   Sonic drilling

Sonic drilling is a relatively recent addition to ground investigation drilling techniques. The method has several advantages over more traditional techniques, but also has some significant disadvantages. The method uses a drilling head that produces a high frequency oscillation, in the region of 80 to 150 hertz. Drilling uses a barrel and rods similar to conventional rotary drilling. The barrel is a single steel tube, with or without a liner. Some systems are able to use a liner, but the sample quality is still generally poor.

The barrel is advanced by vibration and will penetrate a wide range of materials including wood, steel and concrete. In addition, in most materials the entire core is recovered. The system does not generally require a flush, enabling drilling to be carried out dry, making the system very clean. In difficult ground conditions, water flush is added to cool the drill bit. The system is very successful in granular soils, where full recovery can be achieved, something that is not normally the case using more traditional drilling techniques.

The sample extraction from the barrel causes significant disturbance, the soil being removed by tapping or vibrating the barrel using the drill head to make the soil fall out. With the barrel suspended, the soil is caught in a plastic sampling bag providing a continuous disturbed sample.

It is notable that the sample is warm when it is extracted, which is due to heat being transferred from the drilling barrel generated by the high energy vibrations used by the technique. These factors make the sample quality relatively poor, and unsuitable for many types of laboratory test. The heat, for example, changes the water content such that the strength and density are altered. Detailed logging is also difficult because much of the original soil structure is destroyed by the vibration as the drill advances and is further disrupted by the method of removal from the barrel. Nevertheless, because the entire profile is recovered, samples of granular soils are typically of very good quality because the fines are generally not lost, as would be the case with conventional drilling and coring. This means that accurate particle size distribution analysis is likely to be achievable.

The main advantages of the sonic system are that it will drill through most obstructions, as demonstrated by the core shown in Figure 1.48. Here, coring was conducted into a dock apron. The apron is beneath a 1.5 m thick layer of gravel and a mixture of man-made materials, including steel and concrete fragments, that had built up over recent times. The apron is constructed of concrete with brick and stone sets and founded on the London Clay Formation. The entire profile was sampled, complete, in a single core run. The core achieved full recovery of the apron construction and the underlying London Clay Formation. However, the latter was highly disturbed, being both softened and stretched by the drilling and removal process.

Sonic drilling is particularly useful when forming a borehole in glacial till, as these materials can comprise gravelly clay with cobbles and boulders. Such materials may also include large pockets

Figure 1.48 A single core run through a dock gate apron coring through brickwork, concrete and into the London Clay Formation, using sonic drilling (authors' image)

of sand. Drilling through sequences such as these is difficult with either cable tool methods, where the boulders may well prevent progression of the borehole, or with rotary drilling where the sand may similarly halt the borehole progress. Sonic drilling, on the other hand, is likely to recover all of these material types. The clay and sand recovery may well be disturbed, but the whole sequence can be sampled.

Most systems are able to turn off the sonic oscillation when required, and revert to conventional coring, or they can be switched to conventional rotary drilling by using a dual function head. This enables the borehole to be continued in stronger strata, or when a higher quality sample is required. Sonic methods are best used to drill to the depth of interest quickly and efficiently. Once at a required depth, the drilling system may be changed to a conventional coring or sampling technique using a flush system and double or triple tube core barrel, thus utilising the advantages of both sonic and conventional drilling.

In summary, sonic drilling provides a useful technique, particularly when combined with other methods. It can provide good samples of granular soils for particle size distribution testing, although not in an undisturbed state, meaning that density cannot be measured. Samples of cohesive materials are very poor, owing to the high energy imparted to the soil, which produces heat and means that the water content cannot be reliably measured. In addition, the high energy will destroy the soil structure.

The depths to strata boundaries and other material changes can be difficult to determine accurately, mainly owing to the method of withdrawing the sample from the core barrel – that is, by holding the barrel in a vertical position and knocking the barrel with a hammer to dislodge it into a plastic bag or a section of guttering. The quality of sample obtained using sonic drilling is generally Class 3, although when the system is changed to another drilling method it is possible to recover Class 1 samples.

1.3.7   Some considerations when drilling a borehole

1.3.7.1   General

This section discusses some of the challenges encountered while trying to put down a borehole. These points are mainly general and will apply to any method unless stated otherwise.

1.3.7.2 Drilling using more than one method

Where the geology changes, or requirements are such that one method of drilling cannot complete the required borehole to depth, it is common practice to change methods in the same borehole. This is frequently the case where superficial soils are present over bedrock, where the borehole might be started using cable percussion methods until rock is met, and then followed by rotary coring in the rock. The borehole needs to be drilled such that the casing size taken down to rockhead is sufficient to accommodate the rotary casing that will pass through it.

Depending on the nature of the superficial strata, it may be possible to start rotary coring at a higher level. In recent times, for example, the use of rotary coring methods in stiff clays has become common practice, whereas previously coring would have been reserved only for materials that could not be penetrated by other techniques.

The sequence of drilling using cable percussion boring with rotary follow-on is shown in Figure 1.49.

Figure 1.49 Sequence for drilling a cable tool borehole with rotary follow-on (authors' image)

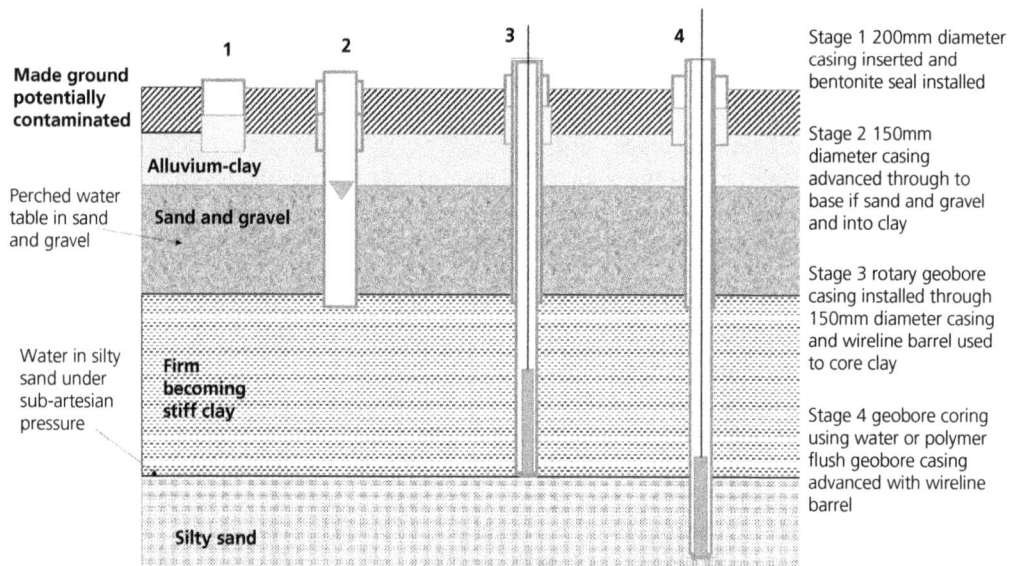

1.3.7.3 Drilling over water

Drilling boreholes over water requires adequate planning, particularly if it is not possible to access the craft carrying the rig directly from the bank. Such holes can only be drilled by rigs capable of handling casing and, therefore, typically these would either be cable percussive rigs or rotary rigs. Both require a stable platform of sufficient size to allow all necessary equipment, welfare and support to be set out on the craft. Figure 1.50 shows a typically congested platform, illustrating the need for good organisation of plant and equipment.

It is essential to plan the borehole construction to ensure that all equipment needed is on board and is safely secured. The craft can either be a floating platform or a jack legged platform. Floating craft might be used where there is limited change in the water level while jack legged platforms are typically

Figure 1.50 Over water drilling (cable percussion and rotary) on a jack legged platform (images courtesy of SOCOTEC)

Figure 1.50 Over water drilling (cable percussion and rotary) on a jack legged platform (images courtesy of SOCOTEC)

used in water where the tidal range is significant. In any event, the operation requires careful planning, considering emergency evacuation, access and egress with a suitable location for transfer of staff and equipment, and adequate support with a vessel and crew on station while work is being undertaken on the platform. A suitable location will also be required to lift the equipment and rigs onto the craft.

Invariably, it will be necessary to obtain permissions from marine organisations before any works can take place.

Drilling is carried out in the same way as drilling on land, with the exception that conductor casing is used from the deck to the bed level. This is usually of a larger diameter to the drilling casing. Actual drilling depths would normally be determined from the top of the conductor casing. When using a floating platform where the water level changes, the depths for drilling are usually measured from bed level, although it should be recognised that the bed level may not be a particularly stable datum in all cases.

Where a floating platform is used in tidal conditions, it will be necessary to add and remove lengths of casing as necessary. In any event, it is normal to work shifts ensuring 24-h working on the borehole. This is often adopted purely because the cost of the platform is such that there is a desire to minimise the hire time.

Some investigations will require the use of cone penetration testing (CPT, see Chapter 5) alongside a conventional cable percussion or rotary rig. This is used to provide information in softer sediments and allow a magnetometer to be run behind the cone. Where this is the case, it is useful to have two access locations through the vessel (i.e. 'moon pools') within 2 m of each other, such that the CPT does not coincide with the borehole location. In general, the more rigs required to drill the holes the larger the platform will need to be. It is prudent to plan the positioning of plant where rigs will need to be swapped onto and off the hole, to minimise plant movements on the deck.

1.3.7.4   Drilling fine-grained granular soil below the water table

When groundwater is present while drilling in granular soils, and in silty fine sand, it is essential to keep the level of water in the borehole above the level of the groundwater, as this provides a pressure balance that will prevent a negative head forming at the base of the hole. If this is not done, there will be a significant risk that the water will rise rapidly in the borehole and bring with it fine particles such as sand and silt. This is generally known as blowing or piping conditions. The result is that progress can be very slow or halted completely. In addition, if more than one string of casing is installed the sand can be forced into the annulus between the two casing strings, which may result in them becoming jammed. The result of this is that the two strings of casing will pull out of the hole locked together, which can make them difficult to separate and can be very time consuming.

Where piping does occur, it is of concern that a significant zone of looser soil can develop if this condition is allowed to persist; it will affect sample quality and in situ testing results. It is therefore imperative to have sufficient quantities of water available before such conditions develop.

In some instances the casing will be very tight in the ground. This is owing to changes in effective stress in the soil surrounding the borehole induced by the action of drilling. Freeing this can be achieved by leaving the casing for a while, introducing water into the hole and between casings or by vibrating or knocking the casing to reduce the pore pressure. In extreme situations, the casing can become stuck fast and may snap in the borehole. Depending how this parting occurs will depend on how, or even if, the casing can be retrieved (see also Section 1.3.7.8.). It is possible to control this by keeping the two casings free by systematically moving the casings to ensure they are loose. Ideally, drilling should not be stopped, and in acute cases consideration might be given to 24 h working.

1.3.7.5 Drilling in contaminated ground

Where the ground is found to be contaminated, it is essential that contaminated soil and water is contained and disposed of in a controlled manner. Depending on the type of contamination anticipated, containment is likely to require ground protection, in and around the rig, together with adequate measures to prevent splashing and dispersion of the contaminated material. The example in Figure 1.51 shows a site using heavy gauge plastic sheeting to protect the ground, and plastic curtains covering fence panels to contain splashes.

Operatives are required to be suitably equipped to protect them from contact with the contaminated spoil and water. In accordance with a project specific risk assessment (Reading and Martin, 2025) personal protective equipment (PPE) may include items such as boots, gloves, disposable overalls and face mask, or even respiratory apparatus where necessary.

Figure 1.51 Ground protection and splash protection, ready for drilling to begin (courtesy of SOCOTEC)

To avoid cross contamination between boreholes, the site may require an area where contaminated equipment can be washed down with a jet washer, see Figure 1.52. This area should ensure the fluid is retained to allow suitable disposal. Similarly, potentially contaminated soil and groundwater will need to be appropriately segregated and stored ready for appropriate registered disposal. Figure 1.52 shows an example of an intermediate bulk container (IBC) being used to collect hydrocarbon contaminated sludge and water to send for suitable disposal.

Not all contamination is as shown in these photographs, and a different approach is likely to be needed under different circumstances. It is essential that a desk study is conducted to assess if contamination might be present and, if so, what measures would be applicable. This should be determined by a rigorous risk assessment process.

When forming a borehole through contaminated soils, it is usually required to form an aquifer protection seal between the contaminated soil and the uncontaminated soils and ground water. This is achieved by extending the casing through the base of the contaminated layer and into the top of a low permeability layer and then forming a bentonite seal inside the casing.

The seal is about 1 m thick and should be allowed to set before proceeding. It should be noted that the bentonite seal is intended to replace an aquiclude, and this technique would not be effective if no aquiclude is present (e.g. if potentially contaminated made ground was underlain directly by a sand or gravel layer). Once the bentonite is in place, the casing may then be withdrawn to the top of the low permeability soil to allow the bentonite to fill the borehole under the casing. The casing should remain in place until the rest of the borehole has been completed. The borehole should be advanced through the bentonite seal using a smaller casing size, for example the original casing might be 250 mm dia. and then the borehole would be continued with 200 mm dia. casings.

Figure 1.52 (a) Area to wash down equipment and (b)–(c) measures to contain contaminated water for disposal (courtesy of SOCOTEC)

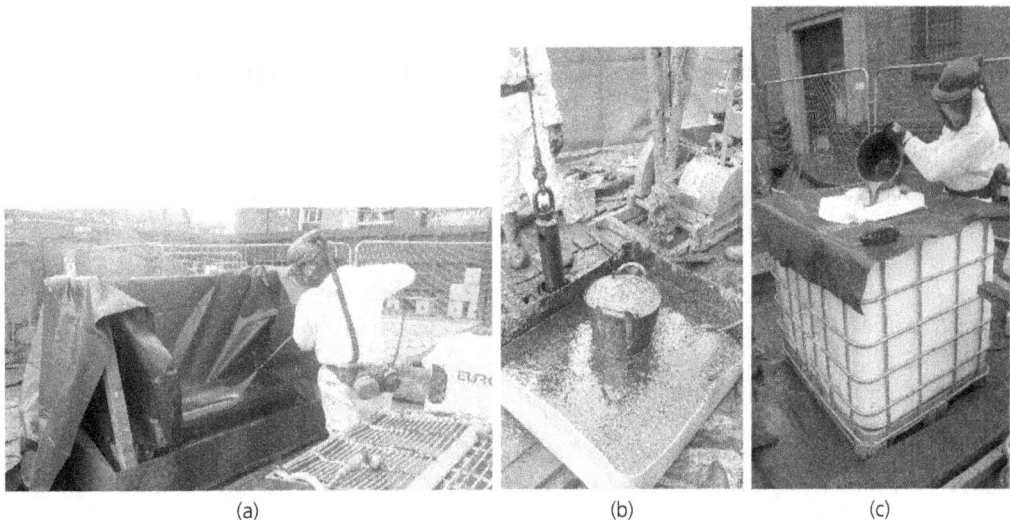

(a)　　　　　　　　(b)　　　　　　　　(c)

To protect the seal from breaking up where the borehole is to be continued by rotary drilling, it is recommended that the rotary casing is fitted inside the second string of casing such that the rotary casing does not come into contact with the bentonite seal. If the casing is in contact with the bentonite seal, the seal will be disrupted when the rotary casing is advanced in the borehole. For the above example, the rotary casing of 146 mm (Geobore S) would be run through the 200 mm dia. casing.

1.3.7.6   Artesian water •

The majority of boreholes encounter groundwater. Most water strikes are relatively stable, but in some boreholes, groundwater will rise up the hole and settle at a level either below ground level, known as a 'subartesian', or will rise above ground level and water will flow from the borehole, known as an artesian water strike. Artesian and subartesian water strikes are controlled by the hydrogeology and geology of the region. A good desk study should be able to predict if the ground water is likely to behave in this manner. Where artesian water is thought to be present, the borehole will need to be designed with this in mind. In some cases, however, artesian water comes as a surprise to both the site team and the designer. Sealing artesian flows should not be rushed, and time needs to be given for monitoring, planning and sealing the borehole.

Groundwater is a valuable resource and by law artesian water flows have to be controlled and the flow stopped. Groundwater cannot be discharged to a water course or drain, unless a discharge consent is in place, and so artesian flow generally needs to be controlled while drilling continues to the final depth, after which the hole can be sealed.

Clause 4.4 of the UK Specification for Ground Investigation (ICE, 2022) states: 'When artesian water is encountered, the Contractor shall cease progressing the hole immediately, inform the Investigation Supervisor and attempt to contain the artesian head by extending the casing above ground level by as much as is practical.'

When artesian groundwater conditions occur, it is important that details of the strike are recorded, as this is useful information for the Ground Investigation Designer and is needed to enable the flow to be effectively controlled.

The information required should include the following.

- Depth below ground level at which the artesian conditions were first encountered.
- Depth and diameter of the casing in the borehole.
- Details of the ground conditions and water strikes.
- Whether water flow is contained within the borehole casing, or whether it is flowing up the outside of the casing. In other words, is the temporary casing providing a seal to contain the flow?
- Estimation of excess hydrostatic head of water above ground level.

Estimating the artesian head can be done simply by adding casing above ground level. This is normally limited to 3 m with a cable percussion rig but could be up to about 4.5 m with a rotary rig (depending on the rig type). This should always be risk assessed, particularly in respect of working at height. If possible, the borehole should be left overnight for the water level to stabilise.

An assessment of the permeability of the ground, or flow rate of water is useful, and this can be achieved by monitoring the rise of water in the casing with a dip meter and recording it as for a rising head permeability test.

If the temporary casing has not formed a seal, the flow and excess head can be difficult to estimate. In a rotary borehole, a method that can be successful is to use a packer, which can be installed just above the strike and the excess head and flow recorded in the drill rods.

The Designer may require water samples for assessment of water quality.

If the borehole is left overnight, or for a period while decisions are made and additional equipment procured, the flow should be contained by adding casing above ground level. It is preferable not to seal the casing at ground level, as this can increase water pressure on this seal potentially causing it to fail.

If required, it is possible for a borehole to be drilled below an artesian water strike, although this is much easier to do if the strike is expected, enabling the appropriate provisions to be made beforehand. In such cases, the borehole has to be designed to control the excess groundwater during drilling, and the effective sealing of the flow on completion. It is important to obtain available information on the ground and groundwater conditions in the premobilisation phase. Often the Designer is aware of the presence of artesian water from a previous phase of investigation on the site, and the former borehole logs and water records should be made available.

Some control methods are outlined below.

If a confining layer, normally of clay (i.e. an aquiclude), is present above the aquifer, then a temporary steel casing can be driven into it, taking care that the underlying aquifer is not penetrated. This casing needs to be of large enough diameter to allow the planned diameter at the bottom of the hole to be achieved (allowing for casing reductions if required). It is normal practice in a rotary hole to dry drill (i.e. without flush) the last 0.25–0.50 m into this aquiclude, and in a cable percussion hole to drive the casing rather than surge it. These techniques should form an effective seal. As a guide, this method is normally suitable for excess heads up to 4.5 m, and generally with low flows up to 0.25 l/s.

If artesian flows in excess of the above are anticipated, it may be prudent to consider installing, within the casing, a permanent plastic or steel liner by grouting. This will require the borehole to be constructed at a larger diameter to provide sufficient radial annulus suitable for grouting. The minimum grout annulus specified is generally 50 mm and the liner will need to be centralised in the hole. The grout will need to be installed by tremie to the base of the hole. To prevent the grout filing the borehole inside of the liner the bottom end needs to be sealed. In a rotary borehole, this can be achieved using a screw-on plastic cap or putting a cement mortar mix in the base of the liner several days before installation. If the borehole is going to be continued by cable tool methods, then a 1 m plug of sand can be place in the base of the liner with 1 m of bentonite above it. The rest of the liner can then be filled with water to ground level. The grout must have a short setting time and develop strength quickly, so the seal remains intact when the hole is continued. Usually, a pure cement grout mixed at minimum solids to water ratio (25 kg of cement to 11 l of water) is used. Samples of each grout mix should be taken so the set can be observed. The grout is normally left overnight to set, but 24 h is preferable to prevent disturbance. The casing lengths, both the steel outer and plastic inner, should be installed so that a thread is left at ground level for the fixing of control caps/valves and so forth. If plastic casing is used, glue-on couplings can be fitted to UPVC pipe. Generally, HDPE pipe is not suitable for this work as it cannot be glued.

Boreholes in artesian groundwater conditions constructed by rotary drilling methods can use a weighted drill mud to counterbalance the excess hydrostatic head. The weight (specific gravity or particle density, SG) of the mud needs to be increased so that it exerts a positive pressure on the borehole walls in the aquifer so that it exceeds the hydrostatic head of the groundwater at that depth. Water has an SG of 1 and the weighting agent barites has an SG of about 4.3 to 5. Therefore, adding barites to water will increase the drill fluid weight, although an additional suspension agent is also needed. The suspension agent commonly used is bentonite, a clay mineral. It should be noted, however, that the type of bentonite required is different to that used for grouting and needs to be obtained from a drilling mud manufacturer. Because bentonite is a clay mineral, consideration should be given to any requirements for in situ permeability testing and water monitoring installations, as these may be compromised by the presence of bentonite. Some polymer-based muds will also keep barites in suspension, and if consideration is given to their use advice should be sought from the supplier.

If boreholes locate unexpected artesian water, work should stop, and an assessment of the borehole should be determined so that an informed decision can be made on whether the borehole can be continued to the planned depth or if it should be sealed and abandoned. The decision will be based on whether it is possible to control the flow and the disposal of the excess water. Where the borehole is to be continued, sealing will be more difficult than if it were preplanned and the overall event will probably result in additional cost and potential damage to adjacent land and properties. It is not unusual for boreholes to be sealed and abandoned when artesian water is encountered unexpectedly, and it is considered this is generally the correct decision. At the very least, this creates a secure situation enabling further borehole drilling to be planned, along with the necessary approach to deal with the artesian strike methodically.

1.3.7.7 Working on sloping ground

Where investigation is required on sloping ground, there are a number of ways this can be achieved – for example, by using a rig where the mast can be angled to provide the borehole at the required inclination. However, if the ground is sloping more than 10–15 degrees, the operation of the rig and handling of equipment becomes difficult and starts to pose a safety risk. It advised that alternative solutions are sought. This might include provision of a scaffold platform or the use of a slope climbing rig.

Rigs have been incorporated into tilting platforms that enable them to be used on slopes. These slope climbing platforms can accommodate dynamic sampling and rotary rigs (see Figure 1.53(a)). To provide adequate working space, a second platform can accompany the platform mounted with the rig. It may be necessary to tether the platform to ensure that it does not move while drilling is taking place. An alternative is to construct a scaffold platform to provide a safe working area (see Figure 1.53(b)).

As a further alternative, rig masts can be mounted into a frame that may be deployed using a long reach boom, as shown in Figure 1.54.

When working requires moving around on sloping ground, it is important that access is planned particularly for rotary and dynamic sampling rigs mounted on a tracked chassis. These rigs tend to have a high centre of gravity and will overturn if they are tracked across a slope. This should be avoided, and the access route should be planned to track up the slope rather than traverse across the slope.

Figure 1.53 (a) A rotary rig mounted on a slope climbing working platform chassis; (b) a scaffold platform with a cable tool rig (images courtesy of SOCOTEC)

(a)

(b)

Figure 1.54 Dynamic sampling rig mounted on long reach tracked vehicle (image courtesy of SOCOTEC)

1.3.7.8 Recovering lost equipment

Equipment may be lost down a borehole for several reasons, the most common being when the borehole wall collapses or fragments of rock or gravel wedge in the borehole trapping tools. On the other hand, a tool itself may fail, such as when threads strip, leaving casing or sample tubes in the ground, or when the wire rope snaps, for example.

Most drillers gather a selection of tools that can be used to recover equipment lost down the borehole. These are often fashioned by the driller themselves and might include grapples to snag the broken end of a rope, or tools to jam inside casing or hooks to catch onto lost equipment.

When a borehole collapses, which is something that commonly occurs when a borehole is left for some time, or is progressed too far without casing, then it may need to be cleaned out to its original depth if further progress is required. For the driller, this is clearly a situation to be avoided as it may be time consuming and, unless there are strong mitigating circumstances, potentially work that cannot be charged for. When cleaning out a borehole in this way, it is normal not to take samples or conduct tests until the previous depth has been achieved. Where casing has become stuck in the ground, hydraulic jacks can sometimes be used to free it, as shown in Figure 1.55. There is a risk, however, that the casing will come apart, leaving some sections deep in the ground. Recovery can be achieved by over-drilling in a larger diameter, although in most cases this would not be cost effective, and the lost lengths would be abandoned, provided they are not in a position that might affect the proposed works.

Figure 1.55 Jacking out casing that has become stuck (image courtesy of SOCOTEC)

1.3.7.9   Unexploded ordnance (UXO)

Where UXO is considered to be a risk, it is common practice to run a magnetometer sonde into the borehole to detect the presence of ferrous material. Where this is undertaken, it is necessary to use stainless steel rather than mild steel casing so as not to interfere with the magnetometer readings. The magnetometer will provide information approximately 2 m below the base of the borehole and in a cylindrical zone around the base of the borehole. If UXO is a potential risk for some depth, it will be necessary to insert the magnetometer sonde into the borehole several times to ensure progression of the hole is safe.

1.3.7.10   Finishing the borehole

Once the borehole has been drilled to its final depth either instrumentation will be installed, and the surface finished with an access chamber (see Chapter 4), or the borehole will be backfilled. Backfilling should take place as the casing is removed. The borehole should not be allowed to collapse. The detail for the backfill will be dependent on the strata that have been bored. Should any of the strata drilled be an aquifer, then the backfilling must be sufficient to ensure that the borehole does not provide a pathway for contaminants. Many boreholes are backfilled with spoil (i.e., arisings from the drilling process), although this is not recommended because the spoil will not be consistent and can arch, causing collapse and surface settlement at a later date. The *UK Specification for Ground Investigation* (ICE, 2022) recommends that boreholes are grouted using a grout mix of 1:1:5 by weight of cement, bentonite and water. If the borehole is to be grouted the mixed grout should be placed through a tremie pipe at the base of the hole, which is slowly raised as the grout is placed.

Other methods might be to backfill with material of similar properties to the natural soils or to insert low permeability seals where materials that are forming aquicludes are present.

Commonly bentonite pellets are used rather than grouting. With sufficient water these will hydrate in approximately 15 min. The pellets should be introduced slowly to avoid arching. When the borehole is dry, or there is insufficient water in the borehole, it will be necessary to add water while the pellets are poured to ensure they fully hydrated.

The final metre of the borehole should be restored to as close to the original surface as possible. Where the borehole is in a sensitive area where settlement might cause a hazard, it is prudent to incorporate measures to prevent a void migrating to surface. Placing a paving slab or inverted road cone at the base of the inspection pit and then completing the final stage of backfill should reduce the risk of any subsequent depression forming.

emerald
PUBLISHING

ice

Peter Reading and Miles Martin
ISBN 978-1-83549-891-0
https://doi.org/10.1108/978-1-83549-890-320251003
Emerald Publishing Limited: All rights reserved

Chapter 2
Geotechnical sampling

2.1. Introduction

Obtaining samples from the ground is a key element of any ground investigation. Samples are taken to enable close inspection and description of the materials encountered and for laboratory testing for both geoenvironmental and geotechnical purposes. This chapter deals with obtaining samples of soil and rock for inspection and geotechnical testing. Samples taken for geochemical and geoenvironmental analysis are discussed in the *ICE Handbook of Ground Investigation – Planning and Reporting, Geoenvironmental and Non-Intrusive Techniques* (Reading and Martin, 2025).

The type of sample obtained will be dependent on the required usage; samples might be required for several different reasons such as record purposes, description and testing with tests for very different characteristics being sought at the laboratory. In any event, it is essential to understand the type of sample required during the planning stages of the investigation, to ensure the appropriate quality and quantity of material is obtained (see also Reading and Martin, 2025). Quality class is discussed below, but it should be appreciated that even if the method, equipment and operator's skill are as good as they can possibly be, it cannot be guaranteed that the desired quality class of sample will be obtained.

Sampling options are dependent on the material type – for example, in clay soils it is possible to retrieve high quality near undisturbed core samples, while in most coarse soils, and particularly gravel, it is impossible to obtain an undisturbed sample and often it is difficult to even obtain a representative sample. The sampling strategy should take account of these practical considerations.

Undisturbed sample – one showing little or no visible effect of the sampling process, with structure and properties unchanged from those seen in the field.

Representative sample – one that contains all the constituent parts of the material as present in the field but may be physically disturbed with significant changes to structure and properties such as water content and density. Ideally, the method of extracting the sample from the ground and the process of sampling should not preferentially include or exclude particles that form the material.

Table 2.1 shows the suggested masses of soil required for testing that will provide a representative sample based on the maximum particle size. To provide a sufficient sample is not possible with some sampling methods – for example, dynamic sampling drop weight, where the volume of sample obtained reduces with depth. From such samples it may not be possible to obtain a subsample of a given stratum that is of representative size, and this is particularly difficult to achieve if the stratum is thin. It is important that subsampling from continuous tube samplers avoids mixing material from different strata, care being taken to ensuring that samples do not cross strata boundaries.

Table 2.1 Suggested minimum sample mass required for various laboratory tests.

Test	Coarsest fraction in the particle sizes present		
	SAND and finer	Fine & medium GRAVEL	Coarse GRAVEL
Water content	100 g	350 g	4 kg
Index properties (liquid & plastic limits)	550 g	(1.1 kg)	(2.2 kg)
Particle density (small pycnometer)	300 g	600 g	600 g
Particle density (gas jar)	1.5 kg	2 kg	4 kg
Particle size distribution (sieve test)	150 g	2.5 kg	17 kg
Grading (sedimentation – pipette)	250 g	250 g	250 g
Chemical (Sulphate, Organic, Chloride & pH)	150 g	600 g	3.5 kg
Compaction (20 mm down Proctor mould)	25 kg	25 kg	(25 kg)
Compaction (37.5 mm down CBR mould)	80 kg	80 kg	(80 kg)

Table compiled from a combination of BS 1377 (BSI, 1990) and BS EN 1997-2 (BSI, 2007a); where masses are shown in parenthesis, the values are derived from conversion of imperial masses.
Pycnometer is adopted from BS EN ISO 17892-3, earlier documents have used Pyknometer.

In recent times, drilling rigs have been developed to enable flexibility with more than one sampling method achievable during the drilling process, in some cases enabling more than one sampling category to be obtained. For example, this might be achieved by changing the tooling, such as piston sampling (Class 1) and thick wall sampling (Class 2) conducted using a cable percussion boring rig. Similarly, a dual head rig that might have a dynamic sampling hydraulic percussive capability (Class 2) and a rotary head (possible to achieve Class 1) on the same mast, enabling a switch from one method to the other when ground conditions dictate.

Where samples are intermittent, such as when using cable percussion methods of progressing the borehole, it is important that they are taken to an ordered regime. However, the frequency should be changed if the ground conditions are such that important elements of the strata would be missed by an intransigent frequency. It is always good to remember that the ground investigation comprises research of the ground. While it is possible to anticipate the material types and their general properties that might be encountered, the material being investigated might be found to be different, requiring a different sampling approach. It is therefore essential to be flexible and prepared to change methodology such as sample type and/or frequency in response to what is being found. All too often important information will be ignored or missed by following a rigid sampling specification.

For example, when using cable percussion methods it is common practice, and roughly in accordance with the UK Specification for Ground Investigation (ICE, 2022), to adopt a rather rigid sampling strategy with tube or bulk samples being taken at 1 m intervals to 5 m and then at 1.5 m intervals below 5 m, and small disturbed samples taken at intermediate depths. This is sometimes refined, actually following the specification more closely, to take a sample at the start of each new stratum with further samples being taken at the intervals given above depending on the depth the change is met. Often the tube samples will be taken alternatively with standard penetration tests (SPTs). If this strategy is adopted, it is useful to change the sequence between boreholes such that in one borehole the SPT is conducted first and in the next the tube sample is taken first. Ideally, the depth of the first sample should also be changed, so rather than starting at 1.5 m, for example, the first is carried out at 2 m. This strategy, although requiring careful management, will allow a more useful analysis of the results of testing when they are plotted against depth, providing a well distributed range of results.

All samples should be clearly labelled with details such as the site location, exploratory hole identification and depth, and for bulk, tube and core samples the depth range should be included (i.e. both the top and bottom of the sample). The label should also include the date the sample was taken. It is becoming common practice to use a portable label printer to produce sample labels. Their use will remove many of the transcription errors and illegible handwriting issues that often result in difficulty identifying samples later in the investigation process. In any event, it is essential that the labelling is clearly readable and is weatherproof. Where label printers are used, it is common to provide a barcode giving each sample a unique reference. This avoids any transcription errors, particularly where the label production is an integral part of the logging record. Details of labelling requirements may be found in BS EN ISO 22475-1 (BSI, 2021).

When obtaining any sample, care should be taken to avoid cross contamination caused by mixing the sample with material from other horizons, this being particularly relevant when sampling from trial pits or when there is contamination within the soil. Procedures should be adopted to ensure that the sample obtained is as representative as possible of the stratum being sampled (see Reading and Martin (2025) regarding obtaining geoenvironmental samples).

2.2. Sample quality
2.2.1 Sample quality classification
The choice of sampling method will depend on the quality of sample required. The achievable sample quality is defined by five sampling categories, or methods of sampling, which can deliver one of five quality classes. For tube and core samples, the actual quality achieved is defined using the tube or core quality index.

The five sampling categories are annotated as category A to E (BS EN ISO 22475-1 (BSI, 2021) and BS EN 1997-2 (BSI, 2007a)). Each sampling category is able to provide samples of a given quality, as shown in Table 2.2. Sample quality is defined by a quality class that ranges from near undisturbed, Class 1, to highly disturbed, Class 5. Each sampling category may achieve samples of lower quality, thus sampling using a Category A method may achieve samples of quality Classes 1 to 5. The use of any one category method does not guarantee that samples will be of the highest quality class for that category.

The importance of sample quality has been recognised for many years, with many researchers describing how poor sample quality and disturbance can seriously affect the results of tests to determine the soil properties. However, it is only in very recent times that a structured approach has been developed. The idea of a quality classification was first discussed by Rowe (1972), this work being based on an earlier sample quality grading developed in Germany by Idel *et al.* (1969). Although these researchers developed a five-class system, its general use has largely been

Table 2.2 Sampling categories

Sampling category	Achievable quality class	Typical sampling method
A	1 to 5	Driven tube sampling using thin-walled tubes Rotary cored with triple tube barrel Block samples Sherbrook sampler Laval sampler
B	2 to 5	Driven tube sampler using thick-walled tubes Rotary cored with double or single tube barrels Hydraulic percussive dynamic sampling
C	3 to 5	Dynamic sampling drop weight
D	4 and 5	Sonic drilling SPT (S) Sampling from hollow stem auger
E	5	Window sampling Augering Wash boring

ignored by practitioners until relatively recently with the publication of Eurocodes, in particular BS EN ISO 22475-1 (BSI, 2021), which defines the features seen in samples for each of the five quality classes. The document also provides a quality index for tube samples to be assessed at the time of logging. Where this takes place will be dependent on whether testing is required, thus logging and the index assessment may be undertaken in the field, logging store or laboratory depending on the requirements for the individual sample. This requires planning and individuals at all stages who are trained in the process of determining the quality index. The approach for rock cores follows a similar format.

The five sample quality classes in common use today are detailed in Table 2.3, which is based on BS EN ISO 22475-1 (BSI, 2021). Importantly, this document also provides criteria against which the sample quality may be determined. Table 2.3 provides an indication of what might be achieved, although actual results may be very different to those suggested, this particularly being the case where sampling sensitive soils or those with large particle size ranges. In addition, sampling below the water table ideally requires water balance to ensure movement of water does not contribute to the sample disturbance. Keeping a water balance ensures that the water pressure in the borehole is either equal to or slightly greater than the water pressure in the ground at the depth the sample is being taken. Disturbance can be minimised by keeping a positive pressure in the borehole to negate the risk of up-surging or piping, which can bring an influx of soil particles into the borehole (sometimes referred to as adverse hydrostatic conditions).

2.2.2 Sample preservation

Key to the onward transportation of samples and their subsequent suitability for testing is the means by which they are contained and protected from external influences.

The extent to which samples will require protection will ultimately depend on their intended use. At the time of sampling, this may not be known. It is therefore best practice to adopt the best quality of sample protection, sealing and transportation. The following is intended as a guide to the

Table 2.3 Sample quality classes with suitable sample types and methods and achievable laboratory tests

Quality class	Sample condition	Test for which sample may be suitable	Typical sampling methods and sample types to achieve quality class
1	Little or no disturbance, structure unchanged, water content and density unchanged No geometric distortion Strength and compressibility unaffected	Strength deformation Stiffness Density porosity permeability Particle size distribution Water content Plasticity index	Rotary core (triple tube) UT100 Block sample Thin-walled sampler Piston sampler Lavel sampler Sherbrook sampler
2	Minor disturbance, no signs of drilling disturbance, slight water softening on wall of sample. Fabric and discontinuities show little or no sign of disturbance. Sample is representative Some minor geometric distortion, density and water content unaffected	Density Porosity permeability Particle size distribution *Water content* Plasticity index	U100 Rotary coring (double tube)
3	Evidence of disturbance to structure, particles show realignment Density altered, water content and particle size unaffected	*Particle size* distribution *Water content* Plasticity index	*Dynamic sampling hydraulic percussive* Thick-walled tube samples
4	Highly disturbed structure altered density, water content changed	Plasticity index *Particle size distribution*	Trial pit Disturbed samples from dynamic sampling drop weight Single tube rotary coring Sonic coring
5	Completely disturbed sample may be unrepresentative Boundaries unclear Loss of fines Particle size distribution altered by loss of fines or grain crushing	Broadly soil boundaries and soil type	Disturbed samples from window sampling Wash boring Hollow stem auger

Classes 1–3 include identification of strata boundaries, sequence, weathering and discontinuities.
Classes 1–4 include identification of strata boundaries and sequence of layers.
Test and method shown italicised may be unreliable depending on material type and method.

various methods available. Most are suggested in BS EN 22475-1 (BSI, 2021) and BS 5930 (BSI, 2015); each method may not be suitable for every sample type.

Preservation may require several factors to be considered such as maintaining the particles together to ensure the sample is representative, so preserving its intact state such that particles cannot be repositioned during handling and transportation. This ensures that properties such as water content and density are not changed. The environment may need to be controlled to avoid temperature and humidity variation, as well as light and in some cases oxidation. Depending on the sample type and its intended use, any one or combination of these reasons may need to be considered.

Plastic bags are useful sample containers provided they are strong and do not tear or deform with the required weight of soil contained within them and, as such, they should be specifically manufactured for industrial use. Some practitioners use household bags, particularly zip seal food bags, instead of tubs or jars for small disturbed samples. These are considered to be unsuitable because the zip seal is not reliable, and the bags tend to be made from relatively thin plastic. Plastic bulk bags are mainly used to contain and transport disturbed samples of soil, such as mixed soils with sand and gravel. With the exception of heat sealing, which is generally not practical on site, it should be assumed that it is not possible to seal bags sufficiently to maintain the water content of the sample. The neck of the bag is usually tied either using a bag tie or by tying the neck by knotting the bag opening. Neither method can reliably provide a watertight seal. If it is required to obtain the water content of the sample at the time of sampling, it is recommended that a sample is taken either using a tub or pail with a snap-fit airtight lid (see below). Furthermore, such containers need to be filled so that there is no sizable air void that might allow evaporation or oxidation to occur. There has been some debate over labelling of bag samples. While it is useful to write the sample details onto the bag itself, using an indelible marker, these details can be difficult to read or may get smeared. When faced with several tens of bags, however, it is much easier to identify the required sample if the label is fixed to the neck with a bag tie. Such labels should be water and tear resistant. A duplicate of this label should be placed inside the bag, as a backup should the external label become detached.

Sealing wax is probably the most effective way of sealing a sample and is used for tube and liner samples where the sample is retained and transported within the tube or liner to ensure the natural water content is preserved. The best wax is a mix of 15% beeswax with paraffin or resin. It is applied as a liquid and allowed to set, and so will fill undulations and gaps between the sample and its container. The wax will not shrink and may only melt away from the sample if exposed to excessive heat. Storage recommendations dictate that the sample should be stored at a temperature of between 5–25°C. Within this temperature range, the wax should be a stable solid and as such will ensure the water content is unchanged for many months. Although the sealing and protective properties of wax are excellent, its use has declined owing to issues of hot works on sites. It is the authors' opinion that the advantages outweigh the disadvantages and, provided good risk assessment has been made and the heat source required to melt the wax is subject to good procedures, its use should be recommended. Should a burner with a naked flame be an issue, there are electric melting pots that will run from a vehicle battery.

Metal discs/waxed paper discs are sometimes used on the ends of tube samples before sealing with wax, as described above. The introduction of a metal or paper disc with a diameter only slightly smaller than the tube will ensure wax does not penetrate into the sample. This is particularly useful if the sample is granular in nature or is fractured or fissured. The same issues regarding hot works apply in respect of the wax that is used in conjunction with these discs.

Plastic tubs and pails should be made of robust material and should have snap-fit airtight lids or have threaded caps that are also capable of providing an airtight fit. Typically, tubs hold approximately 1–2 kg of soil, although large tubs are also available. When sampling the soil should occupy as much of the tub as possible. In this way, by excluding as much air as possible, the sample should retain the water content at the time of sampling for many months. When labelling the tub or pail, it is essential to label the container and useful to label the lid as well. These samples are often used to log the recovered soil, a process that will often require a number of lids to be removed at the same time with the potential risk of transposing lids to the wrong container.

Glass jars were in common use before the 1980s when they were replaced by plastic tubs. They do pose a risk of breakage and thus replacement with plastic was a natural safety progression, although in recent times the desire to employ more sustainable processes and avoid single use plastic has seen a revival of glass containers that can be completely recycled. In any event, the lids that are screwed on require a rubber sealing ring to ensure it is airtight.

End caps are commonly used to seal the end of the tube and liner samples. Metal end caps were originally used to protect the threads of the steel or aluminium U100 (U4) sample tube. Sealing would be achieved using wax on each end of the sample. The need for these was largely discontinued when plastic and rubber end caps were introduced, along with the plastic liner system. In many cases, the same plastic end caps would be used, irrespective of the actual tube diameter, being forced onto metal tubes should their use be required. With the reluctance to use wax on site, plastic liners were adopted solely to seal the tubes. This practice persists today. The end caps must be manufactured for the specific diameter of the tube or liner for which they are intended. In many cases, single size caps are to be found on site and these are used across a range of sample tubes and liners. Even sample tubes and liners that are manufactured and supplied as the same size have variations in their diameter between the different suppliers, thus end caps are often seen to be poorly fitting. It is recommended that there should be at least two methods of sample preservation if end caps are to be used. This would ideally be a combination of wax and the end cap, but the use of duct tape can provide a suitable watertight seal provided it is applied in the manner described below.

O-ring sealing caps or packers are primarily used for rigid or metal sampling tubes whereby the sealing ring is located around the periphery and housed between two steel plates that expand the sealing against the wall of the borehole. The packer just fits into the ends of the sample tube. When the plates are pulled together using a threaded rod, which is attached to the lower plate, the rubber sealing ring is pushed tightly against the wall of the sample tube. This provides a watertight seal. While this method of sealing the sample is very effective, it is rarely seen because the packer is relatively expensive and would mainly be used for high profile work where preservation of water content is considered to be paramount.

Duct tape is generally used to ensure the endcaps on liner samples do not come loose. Providing the tape is applied to extend around the cap and onto the tube, an effective seal may be achieved. The tape should be wound for several turns ensuring there are no creases. Even the smallest crease could potentially provide a pathway for water to escape or to enter the sample. While duct tape is robust, the tape may sometimes lose adhesion or be snagged, which could compromise the sealing properties. It is recommended that duct tape on its own may not be a suitable method of sealing and should, where possible, be used in conjunction with a second method.

Cling film on its own is not considered to be a suitable method of sealing a soil sample for any length of time, although when used as a rapid seal to prevent moisture loss during the sample preparation process, it does provide a very practical and effective way to protect the sample. Cling film may be used in the first stage of subsampling, to ensure the sample remains in a supported intact form. It may be used to wrap a soil or rock sample initially, which will be further protected with foil and then wax soaked muslin.

Aluminium foil is, again, not suitable on its own to protect soil or rock samples and would only be used in a series of measures to protect samples, particularly core, when subsampled.

Tube samples (e.g. UT100s) may be transported in the sampling tube, or the plastic liner used in the sampling process (e.g. dynamic samples), and these should ideally be retained in a rigid wooden or plastic box such that they cannot move. They should be cushioned to reduce the effects of vibration. If tubes contain soft clay, silt and sand, they should be transported and stored in a vertical position with the top of the sample uppermost, as shown in Figure 2.1.

Rock core samples should also be housed within liners and any space packed to protect the core from movement during transportation.

Figure 2.1 Collection of UT100s with plastic end caps being stored upright (image courtesy of SOCOTEC)

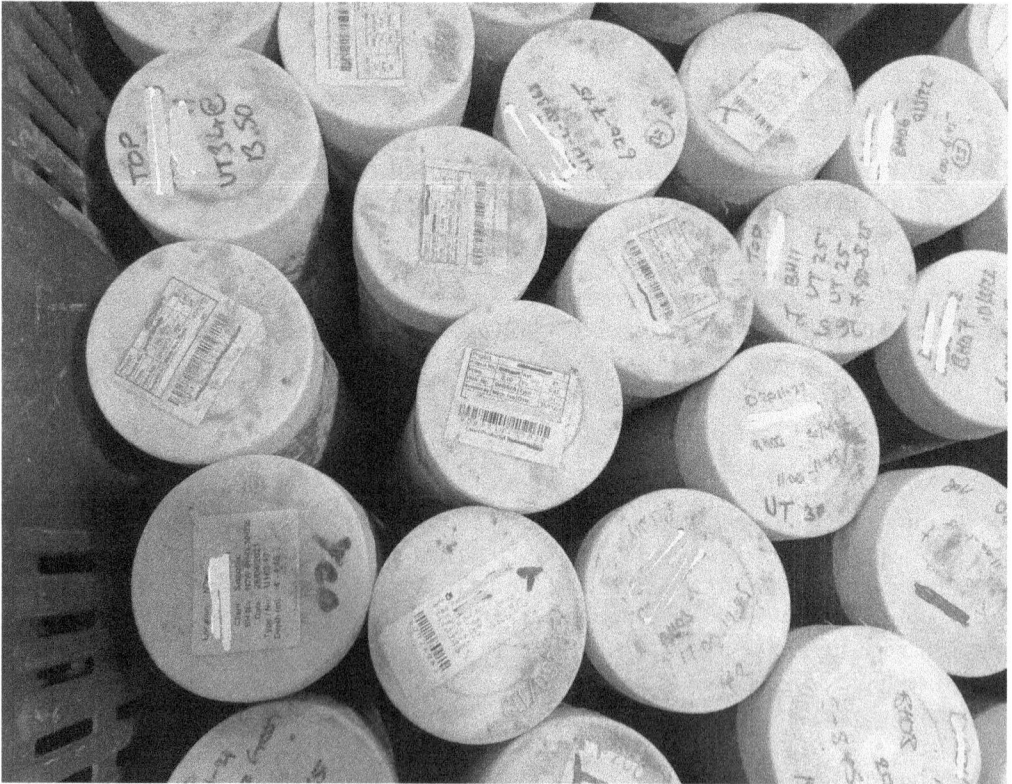

Packing should also be provided in tube samples to prevent the material from moving if the sample does not fully occupy the space in both ends of the tube. Typically, crushed paper, or soil of the same type as in the tube, may be packed into the end of the tube, ideally with wax or other inert sealing material to prevent contact between the sample and the packing material. The packing materials should be inert, so as not to affect the sample, and for this reason some materials, such as expanding foam, are not suitable.

2.2.3　Labelling

Sample labelling is an important part of the sampling process and requires a certain amount of planning to ensure that each sample's identification is clear and unique. The following points should be considered.

- It is essential that all samples are labelled so that they can be identified at any time in the future.
- Labels need to contain a suitable amount of information – project name and reference, exploratory hole reference, sample type and number, sample depth (top and bottom depths for samples taken over a depth range, such as tube samples, bulk disturbed samples and SPT split spoon samples) and sampling date.
- It is important for labels and the information they contain to be weatherproof. If handwritten, indelible marker pens should be used, otherwise ink can wash off or fade in sunlight, and the labels should be made from plastic or waxed paper that will not tear easily or disintegrate in water. The one exception to this is where per- and polyfluorinated alkyl substances (PFAS) are being investigated, when an alternative marker will be required.
- Most samples need to be labelled twice. For convenience, a label is usually placed where it is easily visible, such as tied to the top of a bag, on the lid of a tub or on the end cap of a tube sample. This is inadequate on its own as it is too easy for labels, lids or end caps to become muddled up once they have been removed from sample containers. It is therefore essential to ensure a second label is placed on or inside the sample container itself as a backup. For sample tubs, this would typically be placed on the side of the tub. For bag samples, a label can be put into the bag along with the soil, or alternatively the sample information can be written on the outside of the bag. For tube samples, a second label can either be attached to the outside of the tube or, for samples sealed with wax, a label can be embedded in the wax. This can also be used to show which end is the top of the sample (the label always being placed at the top).
- Where sample containers or end caps are reused from one project to another, care should be taken to ensure that any information pertaining to a former sample is erased before reuse. This will avoid potential confusion in the event of the newer sample information becoming detached.
- Some tablet-based, digital logging systems now available can output sample information to a mobile label printer (see Figure 2.2). This has advantages over handwritten labels, with fewer transcription errors and the ease of producing multiple labels. They also typically include a barcode that enables digital scanning of sample reference numbers for sample management purposes.
- Where orientation of a sample is relevant, this should also be included on the attached labels – for example, an arrow indicating the top of a tube sample or core sub-sample, or a north arrow on the top of a block sample.

Of the list given above, it is highly recommended that onsite logging, using a computer system and a portable site printer to produce the labels, is adopted. This will ensure that the sample labels

Figure 2.2 Typical label printed on site using a mobile label printer (image courtesy of Equipe Group)

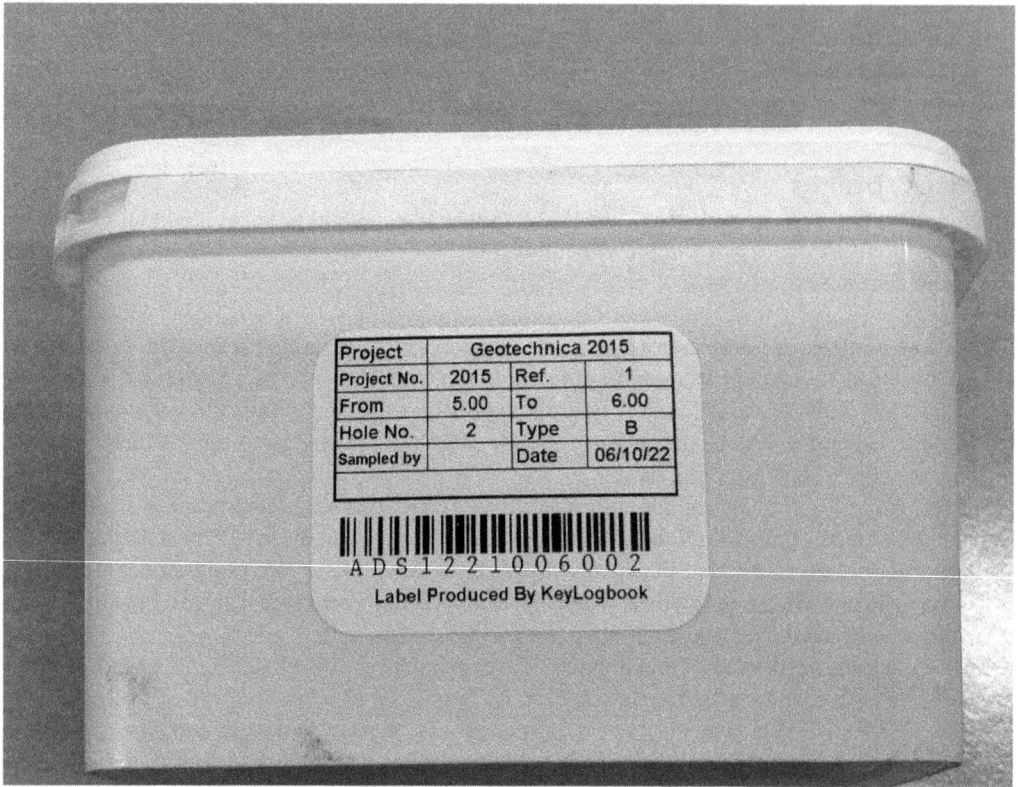

match the sample details provided on the logs and will avoid issues such as 'loss' of samples owing to poor labelling.

2.2.4 Subsampling

Subsampling is a particularly important process that involves identifying, removing and protecting sample sections from a continuously sampled borehole. Subsamples are generally selected to provide samples for onward testing, in particular where the original liner cannot be relied upon to allow transportation of the samples to the laboratory without significant alteration of density, water content or intactness.

Normally, for intact core this would involve selecting continuous sections of core with a length of at least three times the core diameter. The subsample needs to be sufficiently intact such that it can be cut from the whole sample and then wrapped and sealed. If the material is likely to break up with this handling, then the section of liner should be cut with the sample retained inside the cut section. The subsample can then be sealed within the cut section of liner. Alternatively, disturbed samples can be selected where the core is not sufficiently intact or where only disturbed samples are required.

When choosing sections for subsampling it is important that the subsample does not cross strata boundaries. Where there is a material change within a length of core, subsamples of each material should be selected. Subsampling is usually conducted on site and shortly after sampling has taken place. The subsampling requirement normally adopts a general approach, such as to select one subsample per core run. Such a sampling strategy, if rigidly adhered to, may miss sections of core of significance, and it is therefore beneficial for the person carrying out the sampling to have a good understanding of the objectives of the process. It is always a good policy to subsample more sections than may be needed. As sampling is conducted, the engineer or technician choosing each subsample should provide notes detailing the nature and quality of each subsample.

It should be borne in mind that the subsampling process generally selects the best section of the sample and would avoid poorer, less competent sections that are more difficult to preserve. It is useful if the person subsampling is also the person describing the samples recovered. They will therefore be in a position to make comments on the representativeness of each of the subsamples with respect to the overall core condition. The subsample details, such as the location and depth of the top and base of the subsample, should be recorded on the log and on a label, together with a unique reference number. The top of the subsample should be identified. In the event that subsampling needs to be undertaken prior to the full logging process, each subsample should be described in full, so that its description may be incorporated into the log by the logger.

Once each subsample has been selected it should be immediately sealed. This initial sealing and protection can be provided by wrapping in cling film, with a sample label being incorporated into the cling film wrapping. At this stage, the subsample may be temporarily set aside while the remaining samples from the same borehole are selected.

The following stages are often conducted on a batch of subsamples rather than individually, so it is more economic and energy efficient. It is essential that these selected samples are individually labelled and protected from changes in temperature and unnecessary movement. Each subsample is wrapped in foil and then completely sealed in muslin soaked in melted wax, as shown in Figure 2.3. While the wax is still soft a second label may be waxed onto the subsample. Once the

Figure 2.3 Core sample in plastic film, waiting to be wrapped in foil and wax-soaked muslin (images courtesy of SOCOTEC)

Label shows depth range, project and borehole details, and arrow pointing to top of sample.

wax has solidified the subsample is ready for onward transportation. Ideally, the subsamples for each borehole may be retained in a core box of suitable size for ease of handling. The core box should also be labelled to identify the samples contained therein.

Strong intact subsamples of rock, such as granite, sandstone and limestone, do not require the same level of protection, although samples of what might appear to be intact shale and mudstone do require protection because these may degrade with time if the core is affected by temperature and humidity changes.

2.3. Sample types

2.3.1 General

The following sections discuss the different sample types, and the methods commonly used to obtain them in the UK. It will be noted that in many cases the terminology has become confused, often with the sample being referred to as the method that is used to obtain the sample. For example, the term 'dynamic sample' refers to a tube sample with the sample retained in a plastic liner. However, many other methods deliver a sample that is retained in a plastic liner. The following sections use the terminology in common use today and, as far as possible, unpick the cross terminology used to describe the sample obtained and its quality. The following discussion of sample types is not an exhaustive list. It describes the sampling methods in common use, and other less common methods are also included, some of which can be found described in current standards and publications. There are many historical methods that have fallen into disuse and these have been excluded either because they are very complex, or the equipment is difficult to find. Most are precursors to modern less complex methods.

BS EN ISO 22475-1 (BSI, 2021), along with other sections of the codes, uses acronyms to define each sample type. Because of the large number of sample types, the authors prefer to use the descriptive name for the sample type and have not adopted the acronyms. For completeness, a list of these acronyms is provided at the end of this chapter.

2.3.2 Disturbed samples

2.3.2.1 General

Disturbed samples are generally obtained by collecting a sample from soil arising either from a trial pit spoil heap, the soil retrieved from a borehole, or at a shallow depth directly from a small hand-dug pit. They are generally distinguished by their size, being referred to as either small disturbed or bulk disturbed samples.

2.3.2.2 Small disturbed samples

Small disturbed samples are gathered from all types of exploratory hole, mainly as a means of obtaining a record of the material type and for some classification testing. These are sometimes referred to as tub samples or jar samples, which references the container type typically used. On an exploratory hole record, the notation 'D' is commonly adopted to represent the small disturbed sample. Small disturbed samples may be taken in most soil types, although in soils that contain gravel or coarser particles a larger sample (i.e. bulk sample) would be required to provide a more representative amount of material. The small disturbed sample is Class 3 provided it is stored in an airtight container, as shown in Figure 2.4.

Small disturbed samples are obtained with all forms of intrusive investigation. They may consist of the cutting shoe material from an undisturbed sample and as an intermediate sample taken between larger samples, thus giving better continuity when describing the soils encountered.

Figure 2.4 Small disturbed samples in plastic tubs with snap-fit airtight lids (image courtesy of SOCOTEC)

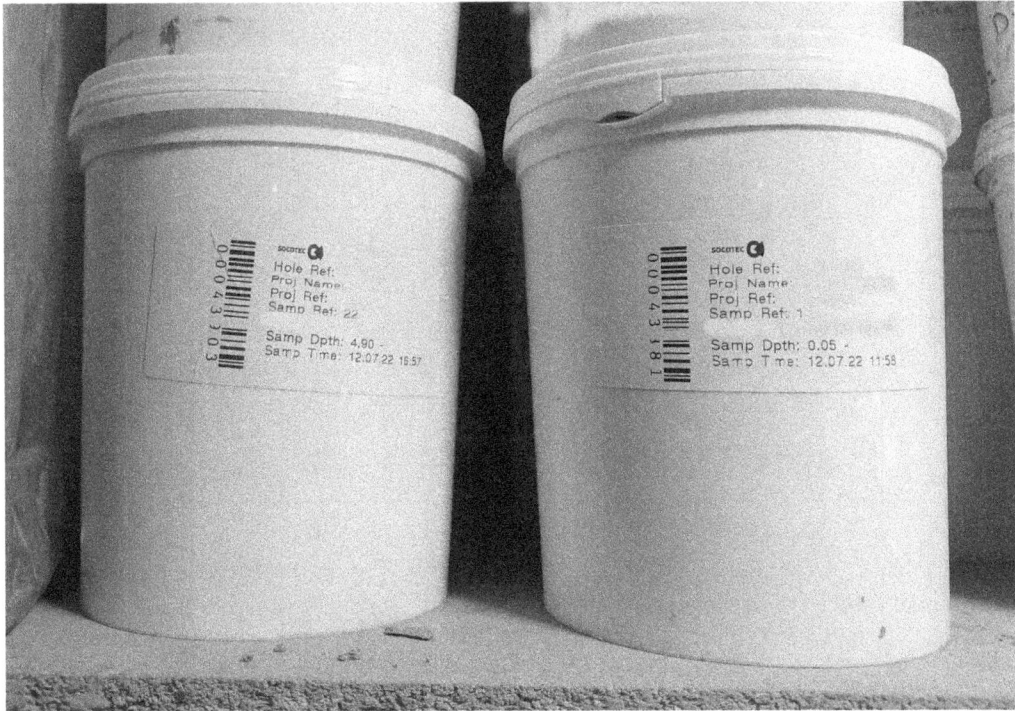

The sample container should be rigid and have an airtight lid. Ideally, containers should have a snap-fit or screw-on lid. The container size should be sufficient to hold a sample of approximately 1–2 kg (the *UK Specification for Ground Investigation* (ICE, 2022) proposes a minimum of 1 kg). Small disturbed samples are often taken for record and soil description but can also be used for soil classification tests where the mass of soil required is less than 1–2 kg, such as water content and plasticity index testing. The sample should completely fill the tub, excluding as much air as possible, so the water content can be preserved.

Some practitioners use small plastic bags such as food zip bags. This practice cannot be recommended because they are easily damaged and are not fully airtight. Thus, water content cannot be reliably measured and as such would render the samples as Class 5. The use of such bags is primarily to reduce cost and deviates from the main reason for taking these samples, which is to provide a testable sample of suitable quality and to provide an easy means of confirming the soil description.

2.3.2.3 Bulk disturbed samples

Bulk disturbed samples (see Figure 2.5) are used to retain a greater mass of disturbed soils encountered in trial pits and boreholes. Such samples are disturbed and can only be Class 3 or lower if they are of insufficient size and/or unrepresentative.

Bulk disturbed samples are typically held in strong plastic bags. These can generally hold up to about 20 kg of soil, and it is noted that the *UK Specification for Ground Investigation* (ICE, 2022)

Figure 2.5 A series of bulk samples taken from a trial pit (image courtesy of SOCOTEC)

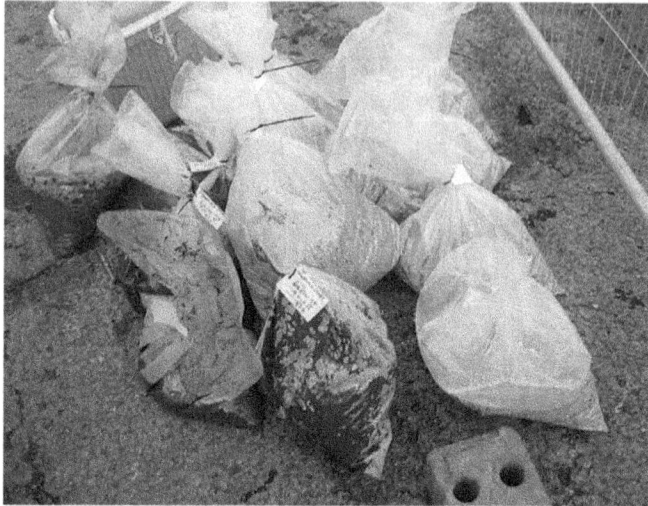

indicates that a typical bulk sample should be at least 15 kg. In some cases a larger sample may be required, particularly if the assessment of soil for earthworks is envisaged. If this is the case, more than one bulk bag may be taken from the same soil horizon, forming a 'large bulk disturbed sample'. The *UK Specification for Ground Investigation* (ICE, 2022) proposes a minimum of 40 kg for a large bulk sample. The bulk sample is annotated on the exploratory hole record as B and large bulk disturbed samples as LB.

Bulk samples can be used in all soil types although some forms of investigation, such as dynamic sampling, may not recover sufficient material for the sample to be appropriate for some tests.

The main use of the bulk sample is to obtain representative samples for soil description and the types of laboratory testing where a greater mass of material is required, such as sieve analysis and compaction testing. The aim of using bulk bags is to enable a sufficiently representative mass of soil to be taken for laboratory testing. For some tests, this may require very large samples to be taken, as shown in Table 2.1.

When obtaining bulk samples from trial pits (see Chapter 1), it is essential to avoid the mixing of materials between individual strata. Discrete bucket loads of the selected horizon should be tipped from the excavator bucket onto a tarpaulin or plastic sheet. Samples can then be taken from these selected heaps to avoid mixing.

Sampling of granular soils is generally conducted using a random selection of the excavated material often from the side of the trial pit spoil heap, although this may result in a biased sample because coarser particles will tumble down the heap and collect at its base. However, if a more representative sample is required, this may be achieved by forming the soil into a cone and then quartering the cone, as shown in Figure 2.6. All material from two opposing quarters should be retained. If large amounts of soil are required, several quarters can be taken. Although this method is rarely used in the field, it is a most efficient and accurate way of obtaining a representative sample.

Figure 2.6 Cone and quartering for representative sampling of granular material (authors' image)

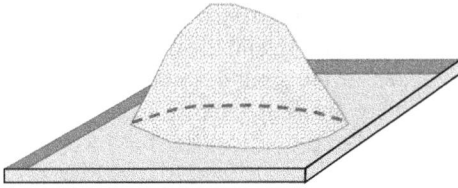

Figure 2.6 Cone and quartering for representative sampling of granular material (authors' image)

Step 1: Heap full sample in a suitable tray

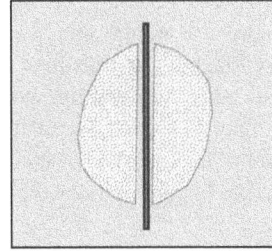

Step 2: From above, divide sample into two halves

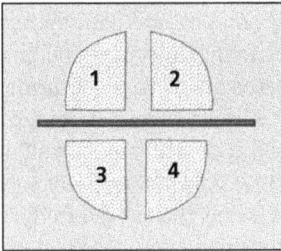

Step 3: Divide each half to produce four quarters

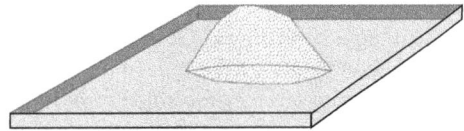

Step 4: Combine material from two opposing quarters (i.e., 1 and 4, or 2 and 3) to create test specimen

If more than one sample is taken at one level (e.g. a large bulk sample), this should be recorded on the exploratory hole log and the bags should be labelled accordingly – for example, B6 (1 of 2) and B6 (2 of 2).

It should be noted that water content might not be preserved in the bulk bag because it is not possible to achieve a complete seal with any form of bag tie. If water content is important to the investigation, a separate small disturbed sample with a sealed lid, as described in Section 2.3.2.2, should be taken to provide an accurate water content of the sample at the time of sampling.

Bulk samples may be taken from other types of exploratory hole, such as window sampling, dynamic sampling and sonic drilling. Where these subsamples are removed from tube samples, care should be taken not to sample across strata boundaries, thus mixing soil types. While the bulk sample may be representative of the material present in the tube, it must be considered that material recovered in this way may not be representative of the stratum from which the tube sample has been obtained. In addition, the mass of soil obtained in this sampling process may not meet the typical mass requirements of the sample type being specified (e.g. a bulk sample may be less than 15 kg owing to a lack of material recovered).

2.3.3 Block samples

Block samples may be obtained by predominantly hand methods in either shallow or shored pits using the methods discussed below. Other large samples, described by many standards as block samples, use mechanical methods that may be powered by a rig or possibly using a hand mechanism to rotate blades. These semi and fully mechanical methods have been included in Section 2.3.7.

Hand-cut block sampling potentially provides a Class 1 sample. The notation commonly used on exploratory hole records is 'BS'. Best results are achieved with soils that have some clay content that are capable of standing vertically for the duration of the sampling process. Cutting the sample is aided if gravel is not present. The samples are usually cut in shallow pits where the base of the pit is wide enough for an operative to work comfortably and safely. The pit needs to be stable, with a suitable distance between the work area and the pit edges, or the ground cut back with stepped or sloping sides to ensure the safety of the persons obtaining the sample (see Chapter 1, Section 1.2.3). The pit must be dry. In inclement weather, it is useful to provide shelter for the operative and the sample (a pop-up gazebo is ideal).

Block samples can provide the least disturbed of all sample types. They comprise cubes of the material being sampled that are cut by hand in situ. The sample is typically a 300 mm cube but may be larger. The sample is carefully cut, gradually removing the surrounding soil until the required depth, normally 300 mm, has been reached. The process requires meticulous care and patience. First the top surface is prepared flat and level, then the sides are pared away from the surrounding soil leaving a cubic block. Once the sample has been fully cut on all four sides it is covered with plastic sheeting and waxed muslin (see Section 2.2.2). Then a wooden or steel box is built around the sample, any voids being filled with Plaster of Paris or expanding polystyrene foam such that the sample is securely held. When the sides have been fixed, the base may then be cut from the soil either using a thin bladed knife or wire saw. The sample is then lifted or slid onto the box base. The top of the box is fixed in place and the sample is ready for transportation (see Figure 2.7).

Typically, block samples are obtained in shallow trial pits, although they can be obtained in pits at any depth. Deeper pits will require shoring or benching, and there should be adequate space for the person cutting the sample to access each side of the sample such that they can work comfortably and safely. Consideration should be given to moving the sample, which is typically heavy (often

Figure 2.7 (a) A block sample under preparation; (b) a block sample completed and wrapped ready for the top of the box to be placed (images courtesy of SOCOTEC)

(a)

(b)

greater than 50 kg), and could require at least two people, or mechanical assistance, to lift. A risk assessment should always be undertaken to address the safety issues related to sampling, including accessing and working inside the pit, and the manual handling of the heavy sample.

The main disadvantage is that the sample can be easily disturbed by the paring process. The process can take a considerable time and may result in a poor sample after an hour or more of careful work. However, where successfully cut, high-quality Class 1 samples can be achieved, particularly in clay soils. It is especially important to prevent water content changes in the sample as it is being prepared, and ideally this type of sample should be taken when the weather is dry but not too hot. In any event, the sample will undergo stress relief and changes to the pore water pressure, both of which can be restored in laboratory tests for properties such as strength and deformation.

2.3.4   Tube samples

2.3.4.1   Tube sample types

Tube samplers may be pushed or hammered into the ground, and more recently vibrated into the ground as is the case with sonic drilling. Some methods can obtain near continuous samples, such as rotary coring, sonic drilling and dynamic sampling, although sample quality can vary considerably. The final quality of the sample obtained is dependent on the composition of the material being sampled, the design of the tube and how the sample is treated once it has been removed from the ground, including how the sample is sealed, transported, stored and extruded from the tube.

There are numerous types of sample tubes, which can be divided into two main groups: thick-walled and thin-walled. Each of these may be deployed in a variety of ways giving a diverse range of sample sizes, diameters and lengths, as well as samples of quality classes ranging from Class 1 to Class 5. Choosing the right sampling method is key to delivering the necessary information and parameters required for the proposed works. It is a decision that needs to be taken at a very early stage in the planning of an investigation because it will dictate the methods of intrusive investigation to ensure the required outcome.

Early attempts to obtain cores of soil were developed using the tube sampler system devised by the Raymond Group in the USA. This thick-walled sampler evolved into the SPT in common use today. Initially, thick-walled tubes were driven into the ground, and to aid removal of the soil the tube split into two halves allowing easy access to the sample. The tubes were small diameter and often became obstructed internally in stiff clay or gravelly soils. The samples obtained in most cases were highly disturbed, mainly owing to the tube geometry and the fact that it can take many blows to drive the sampler to depth.

Tube sampling development has resulted in several tube designs. Commonly, the tube samples were referred to as 'undisturbed' samples with the notation 'U' used to define them on borehole logs. The use of the term 'undisturbed' is somewhat unfortunate because many of the tube samplers in common use do not provide a truly undisturbed sample. The early tube samples in use before 1974 were four inches in diameter, were made from standard BSP pipe, and were referred to as U4s. Subsequently, with the adoption of the metric system, the samples were renamed U100s, denoting a tube sample of 100 mm dia. Tubes of smaller diameter have also been manufactured, including 75 mm and 38 mm, for example. It was recognised at an early stage that, in order to obtain useful samples for testing in the UK, it is generally more successful to take as large a diameter sample as is practicable. This was mainly owing to the presence of gravel-sized particles that were likely to block off in smaller diameter tubes, and this led to the adoption of 100 mm dia. as a norm. This diameter was found to be convenient when drilling in the smallest hole diameter in common use of six inches (150 mm), a practice that persists today.

Research (Lunne *et al.*, 1997a, 1997b) has supported the practical preference for larger diameter samples, demonstrating that the larger the diameter the more representative the test result when samples are used for strength testing, provided the sample is also a Class 1 sample. Lunne *et al.*'s work compared the results for laboratory and high-quality field tests along with tests conducted on block samples.

A study of the effects of tube geometry on the quality of tube samples has shown that there are several factors that will affect the outcome. These include the area ratio (Ca), the taper angle of the cutting shoe (α) and the relationship between the inside diameter of the cutting shoe with the inside clearance of the sample tube (Ci).

Area ratio, Ca = cross sectional area of cutting shoe annulus/cross sectional area of sample %
Thus Ca = $[(D_2^2 - D_1^2)/D_1^2] \times 100$ (%)
And the inside clearance ratio is given by Ci = $[(D_3^2 - D_1^2)/D_1^2] \times 100$ (%)
Where D_1 = inside diameter of cutting shoe D_2 = greatest outside diameter of cutting shoe
D_3 = inside diameter of the sample tube or liner

It should be noted that 'Ca' is the notation for area ratio adopted by Eurocodes, although in many publications area ratio is given the notation 'Ar'. Table 2.4 shows typical tube geometry for common samplers. Figure 2.8 shows tube sample types.

Many practitioners have recognised the effect of the sampler geometry on sample quality and have used the area ratio and an efficient tube design (Clayton and Siddique, 2001). However, it has only been since the publication of Eurocode in 2005 (BS EN ISO 22475-1) that such details have been adopted in specifications and general methodology. It is now commonly accepted that in order to obtain good quality samples using tube samplers the area ratio should be less than 15%. The taper on the cutting shoe should be no more than 5 degrees and the tube should ideally be pushed or driven with a few blows, although how many blows a few represents is still not clearly defined. In order to reduce the risk of splitting and burring, the cutting shoe angle is increased to around 30 degrees for a small distance at the end of the taper, as can be seen in Figure 2.10.

Table 2.4 Typical area ratio and angle of cutting edge for some common samplers

Tube type	Area ratio (Ca)	Angle of cutting edge (α)
UT100	15%	5°
U100 (formerly U4)	27%	30°
U100 with plastic liner tube	47%	30°
Dynamic sampler	50% to >100% depending on diameter	30°
Thin wall (aluminium)	8%	5°
SPT split spoon	50%	45°

Figure 2.8 UT100 (note square thread); U100; U100 liner system tube; U75 thin-walled tube (image courtesy of D. Rackley, SOCOTEC)

The maximum internal clearance should be no more than 0.5%. Other requirements for open tube samplers include the provision of an over drive space at the top of the tube, this allowing any softened material to be accommodated rather than be compressed as the sample enters and rises up the tube. Similarly, a nonreturn valve is provided to allow any air or water that is present above the sample as it rises within the tube to escape, again without causing a pressure to be developed.

Quality determination of tube samples and core is defined in Tables H3 and H4 in BS EN ISO 22475-1 (BSI, 2021). Prior to this, there was no agreed defined classification of sample disturbance. Table 2.5 combines the key elements of Tables H3 (rotary core sample classification) and H4 (tube sample classification). These determinations should be made at the time of logging, assuming this will take place either before or simultaneously with laboratory testing. As noted earlier, the classification process must be carried out by trained loggers conversant with the procedure and able to recognise the features described within the tables.

It should be noted that the sample quality is not completely dependent on the area ratio; some samplers have a high area ratio but can still produce high-quality samples – for example, foil samplers.

2.3.4.2 Cores and tube samples – causes of disturbance

Many issues that are encountered regarding the sampling process and subsequent sample disturbance can be avoided by employing good drilling practices and using well maintained drilling equipment. This equipment should be compatible with the correct internal diameters of the sampling liner and barrel. It is also essential that core catchers and nonreturn valves are not worn and are compatible with the barrel or tube system being used. Daily checks of the rods and other drilling equipment should be carried out to ensure they are not bent or damaged, and that the slide-hammer used to drive open sample tubes is not bent and operates freely. In addition, the number of weights needs to be adjusted for the soil strength to ensure repetitive energy from the slide-hammer is delivered to the sampling mechanism.

Table 2.5 Combined quality index for tube and core samples (after Tables H3 and H4 in BS EN ISO 22475-1 (BSI, 2021) and Norbury and Powell (2022)

Quality index	Descriptor	Probable quality class	Signs of disturbance	Coarse particles driven through matrix / matrix loss	Bedding and fissures	Evidence of overuse of water / induced softening	Evidence of drying out
1	Very poor	3/4	Severe disturbance	Clearly evident	Bedding and fissures show severe signs of disruption	Signs of severe softening	Extreme drying out clearly evident
2	Poor	3	Signs of significant disturbance	Evidence of particle realignment	Some disturbance of structure through sample and concentrated at periphery	Some softening of sample penetrating into sample body	Signs of shrinking and cracking
3	Average	2/3	Some evidence of method of sampling	Localised evidence	Some disturbance generally confined to sample periphery	Localised softening of sample	Localised shrinking and/or cracking
4	Good	1/2	No obvious evidence	No obvious evidence	No particular disturbance, slight downward turn at sample periphery	No obvious signs Ends of sample and periphery slightly softened	No obvious evidence
5	Very good	1	No signs	No signs	Intact, no signs of disturbance	None	None

Even with compliance to the above requirements, sample disturbance can occur through the life cycle of the tube sample. An understanding of how and where such effects will occur can help reduce the severity of sample disturbance.

DURING DRILLING, BEFORE SAMPLING

Stress relief – caused by the drilling process while forming the borehole. At the base of the borehole the vertical stress will be reduced to zero. This can be mitigated when rotary drilling by the use of a heavy drill fluid, but it will not be eliminated. Stress relief can induce swelling in clay soils where water or air penetrates the soil. If sampling is carried out relatively rapidly the effect is generally small. However, if a borehole is left overnight, for instance, swelling might be significant, as will be the effect on total and effective stress. It is recommended that the first action on returning to a borehole at the start of a shift should be to progress the hole at least three borehole diameters below the overnight level before conducting any sampling or testing.

Piping – disturbance created in the ground around the base of the borehole owing to a change in the head of water within the borehole compared with the surrounding area. A lower head in the borehole will induce flow into the borehole which, within granular and particularly relatively fine-grained granular soil, will be sufficient to induce soil particles to move into and up the borehole, changing density and soil structure in the zone around and below the borehole. This can be avoided by maintaining a positive head within the borehole. However, an action such as the use of the shell, by its nature, will induce some inward flow of particles, generally known as 'boiling' or 'piping'. Further details are given in Chapter 1.

Compaction – caused by driving tools and casing into the ground. The force applied forms a pressure bulb ahead of the base of the borehole, and much of this effect can simply be eliminated by a slower, more methodical drilling process. In particular, casing should not progress ahead of drilling below the base of the borehole.

DURING SAMPLING

Stress relief – many samplers have a slightly larger internal dimeter to the internal diameter of the cutting shoe. This is designed to assist sample retention, but it also allows stress relief that causes expansion of the sample inside the sample tube.

Remoulding – the friction of the sample against the wall of the tube causes remoulding and distortion of the sample. This changes across the tube diameter, such that the centre of the sample advances into the tube at a faster rate to the soil at the circumference. The fact that remoulding occurs close to the sample edges is one reason smaller diameter samples tend to contain a greater proportion of disturbed material than do larger diameter samples.

Alteration – The outer edge of the sample is generally in contact with water in the borehole or, where used, drilling fluid as the sample enters the tube, and any laminae may slightly separate owing to friction. This creates a slight suction which causes water to migrate towards the outer edge of the sample, which both distorts the structure and alters the water content across the sample.

Sample loss – this mainly occurs as the sample is extracted from the borehole when the friction of the sample with the sides of the tube is less than the tensile force required to sever the sample from the soil at the base of the borehole. This can be avoided to some extent by using a core catcher and by waiting for a few minutes before removing the sample from the borehole to allow the sample to expand against the wall of the sampling tube (see 'Stress relief' above). In earlier times, the driller would often stop the rig and have a cup of tea before removing the sample from the borehole.

Gouging – occurs where a soil contains gravel or lithorelicts of rock fragments within a weaker matrix, such as glacial till or weathered mudstone. The larger hard fragments can become caught on the cutting shoe as the rest of the sample enters the tube, causing scoring and gouging which distorts the sample and destroys the structure. This can result in a Class 1 sampling method achieving no more than a Class 3 sample – as seen in Figure 2.9. Similar effects are seen if the cutting shoe is burred.

AFTER SAMPLING

Stress relief – this can occur during extrusion of the sample from the tube and will materialise as swelling of the sample as it emerges from the tube.

Water content changes within sample – this commonly occurs particularly if the sample is coarser grained and is stored on its side, when water may migrate within the sample.

Water softening – this occurs when water from the drilling process is allowed to come into contact with the sample, either around its periphery or at the ends. Invariably, the sample will undergo stress changes and potentially pore suctions will develop, particularly in cohesive soils, which may draw in free water causing softening of the sample.

Loss of water – will occur over time but is particularly rapid if the sample is not properly sealed. The most effective method of sealing the sample is to use wax and not to rely on end caps that may not fit tightly or may trap dirt between the tube and cap causing a poor seal. Loss of water content will significantly affect stiffness, compressibility, shear strength and bulk density.

Disturbance during transportation – handling of samples as they are moved from the drilling site to the laboratory, or other temporary storage location, should be conducted carefully; samples can be seriously damaged if they are dropped or thrown. Often temperature changes will occur during transportation and samples may also be subjected to vibration induced by road surfaces and the vehicle used to move the samples. All should be minimised and avoided wherever possible. Rotary core and samples in liners should be packed into rigid boxes, using packing materials to limit movement, such as foam cut-outs, sections of cardboard or other materials, which will protect the samples from physical shock while in transit. All sensitive soils should be transported in a vertical position and also packed in such a way as to prevent physical shock.

Disturbance during extrusion and preparation – samples in metal tubes should be extruded from the tube in the same direction as they entered the tube (i.e. Pushing from the base). Ideally, the

Figure 2.9 Dynamic sample, showing a typically disturbed nature of the material where gravel in the clay has resulted in gouging during the sampling process (authors' image)

ram used to push the sample from the tube should be a close fit into the tube and should operate in one smooth action. When the friction between the sample and the tube exceeds the applied thrust, samples should be carefully parred from the top of the tube until the friction has been overcome. On no account should the tube be hit with a hammer in an attempt to encourage the sample to extrude. Ideally, the sample should be supported at all times by extruding directly into the preparation mould.

Temperature changes – significant changes in temperature should be avoided at all stages, in particular temperature extremes. Freezing may result in ice lenses forming in the samples, which will alter the distribution of water within the sample and can seriously disrupt the sample's structure and density. Excessive heat can cause water content changes or expansion of the sample, which again will destroy sensitive structures. Ideally, samples should be stored at the same temperature as that of the ground, which would typically be between 5–15 degrees. In any event, sensitive soils such as silt and normally consolidated clay will require careful handling and avoidance of temperature changes during storage and transportation. Such soils can easily be altered and once this has occurred their original state will not be retrievable.

Chemical changes – this is particularly significant where samples are taken from within the anoxic zone, where oxygen from the atmosphere has not penetrated. On retrieval, air may be drawn into the sample causing significant chemical changes. This is particularly significant when sampling organic soils which will readily accept oxygen, causing both a chemical and colour change. It is essential that samples that may be affected are sealed as quickly as possible.

Some sample tubes, especially steel tubes, will react over time with the chemical composition of the sample and corrosion may be accelerated at the boundary between the sample and tube. This is particularly noted where soils are acidic, and the reaction can make sample extrusion difficult. To avoid this, samples should be tested as quickly as possible after they have been taken.

Duration between sampling and testing – Delays can enable chemical reactions to take place between the tube material and sample. In addition, high-quality tube samples retain their in situ effective stress for a short period once removed from the ground, but with time the pore water pressure will dissipate to zero, reducing the effective stress. For effective stress type laboratory tests, the period of time between sampling and test set-up (where the test specimen is placed into a test cell and the in situ stresses restored) should be kept to an absolute minimum. While in special circumstances it might be possible to reduce the time to a few hours, this will rarely be practical. Provided the sample is securely encased in the sample tube, and sealed, the normal duration between sampling and testing will be approximately 2–3 weeks.

2.3.5 Thin-walled open tube samplers
2.3.5.1 General
BS22475-1 (BSI, 2021) gives details of the tube geometry to ensure that disturbance from insertion of the tube is minimal. Key to this is an area ratio of less than 15%. Typically, the tube is fitted with a cutting shoe, screwed onto the base of the tube, or otherwise the lower end of the tube includes an integral cutting edge. The inclusion of a cutting shoe has the effect of locally increasing the cross-sectional area of the tube (where the shoe and tube are threaded together) and hence increases the area ratio; thin-walled tubes with an integral cutting edge therefore provide a more desirable tube geometry. In practice, the angle of the taper is increased to between 20 to 30 degrees close to the cutting edge to avoid an easily damaged feather edge. Figure 2.10(a) shows the design of

Figure 2.10 (a) A Standard U100 cutting shoe and a UT100 cutting shoe with 5-degree taper and 30 degrees at edge; (b) a UT100 with cutting shoe damaged by presence of gravel (images courtesy of SOCOTEC)

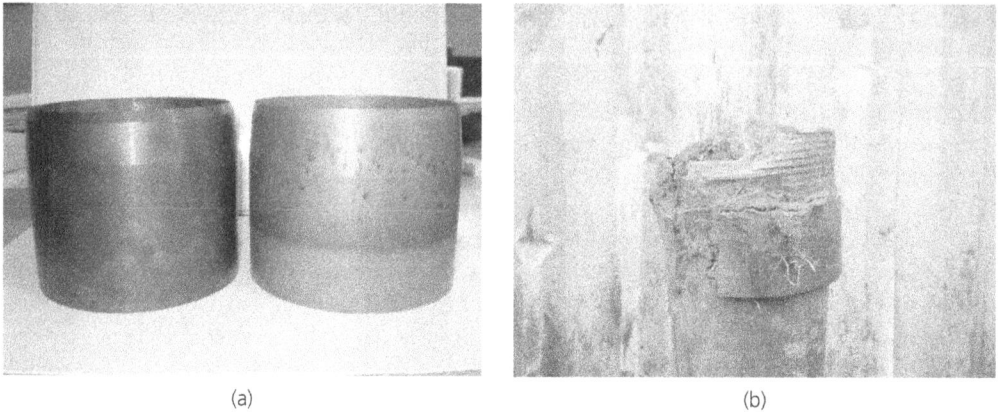

(a)

(b)

the UT100 cutting shoe, alongside a standard U100 cutting shoe. The thinner profile is not always beneficial, particularly where gravel is present, as shown in Figure 2.10(b).

To reduce wall friction between the sample and the tube as driving takes place, some tubes are designed to have a slightly larger internal diameter than the cutting shoe. This feature reduces the risk of blocking the tube, which can occur when wall friction becomes excessive. It also helps retain the sample as it is being withdrawn from the borehole by allowing a slight elastic expansion as the sample enters into the tube. Many designs also include a sample retaining mechanism or core catcher comprising a plastic retaining collar between the base of the sample tube and the cutting shoe. Other coring methods use a split metal ring housed in a taper within the cutting shoe or drill bit that fully retracts during sampling and slides down as the sample is removed from the borehole.

All sample tubes should be removed from the sampling assembly as soon as possible after they have been brought to the surface, the ends of the soil sample being trimmed so as not to protrude from the tube. Ideally, the ends of the tube should be waxed using a paraffin wax, or sealed using material with similar properties, to provide a secure airtight seal (see Section 2.2.2).

Thin-walled samplers have an area ratio of less than 15% and would generally produce a Class 1 sample.

Ideally, thin-walled tubes should be pushed into the ground in one smooth action, rather than being driven into the ground with a hammer. When used in conjunction with cable percussion drilling methods, this can be achieved using a pulley system, or the pull down on a rotary rig, or using the stroke from a cone penetrometer testing (CPT) rig. All these methods will enable the sample tubes to be smoothly pushed into the ground. Whichever method of applying the downward force is used, the wall of the tube used should be thin when compared with the area of the sample. Typical thin-walled tubes are aluminium and have an area ratio as low as 8% and a cutting edge of no greater than 5 degrees (see Table 2.4). The sample length can vary from 0.45 m to 1 m; however, the length to diameter ratio should be less than 20. Typically, diameters of 50 mm, 75 mm and 100 mm are used. A 100 mm dia. tube of 1 m length is commonly used and satisfies the length to diameter ratio requirement.

Thin-walled tubes of this design are suitable for use in soft and firm soils and recover good samples where the soil is cohesive. However, the thin cutting edge can be easily damaged if gravel is present. If used in sand or silt, recovery can be poor owing to high pore pressures developed at the cutting edge, which tend to block off the tube. Because the samples taken are usually soft it is essential the tubes are transported and stored in a vertical position. Failure to do this will often result in an air pocket forming along the length of the sample, as the material settles, and the resultant disturbance of the entire sample.

2.3.5.2 Piston samplers

Piston samplers provide a very high-quality sample, generally Class 1, most being thin walled with an area ratio of less than 10%. They are generally pushed into the ground rather than hammered. Sample tubes are either 0.5 or 1 m in length, the hydraulic piston that drives the tubes having a maximum 1 m stroke. Typically, they are used in soft or firm cohesive alluvial soils. They are represented on borehole records by the letter P.

Piston samplers come in two configurations: the fixed piston and free piston. The piston sampler is a development of the thin-walled tube and was developed to aid recovery, particularly in soft cohesive, silty or sand soils. In both types, the tube fits around the piston and is fixed in place by three or more grub screws that locate into holes at the top of the tube. The whole assembly is lowered to the base of the borehole, and then the tube is pushed into the ground while the piston remains on the top surface of the sample – that is, the base of the borehole. The piston develops a vacuum at the top of the sample that aids retention of the material in the tube. The base of the piston has a convex face and rubber ring that provide an airtight seal with the sample tube. A simple ball valve in the headworks allows any air or water above the piston head to be expelled as the tube is pushed into the ground. The piston face remains static during this process.

A potential problem is that most samplers do not have a core retention system other than the vacuum developed at the top of the sample by the piston. In soft clay soils, the suctions that develop as the sample tube is extracted from the borehole can be equal to the internal suction at the top of the sample, resulting in the sample being stretched in the tube. This can be a particular issue with the use of longer tubes.

The main advantage of the piston sampler is that only a small amount of drilling fluid can come into contact with the top of the sample, the space being occupied by the piston at all times.

The fixed piston is rigidly held on the drilling rods and locates into the sample tube head once it has been driven to the full depth.

The free piston sampler has an independently held piston on a set of rods that run inside those that hold the sample tube. The two are locked together with a reverse thread until the sampler is in place at the base of the borehole, the face of the piston located at the lower end of the sample tube. The two parts are unlocked by rotating the inner rods clockwise, thus releasing the piston. The sample tube is then pushed into the ground. Once the tube has been pushed to its full depth, and the piston is now positioned at the top of the sample tube, the system is relocked by twisting the rods counterclockwise and the whole assembly can be removed from the borehole.

Once retrieved from the borehole, the sample tube can be removed from the assembly by undoing a relief valve in the head, this action releasing the vacuum that develops at the top of the sample.

Storage and transportation should follow that described above, with the sample remaining vertical at all times.

2.3.5.3 UT100

The UT100 is able to provide a Class 1 sample in most cohesive soils. It is typically shown on borehole records as UT, denoting an undisturbed thin-walled tube. The '100' refers to the approximate diameter (in mm) of the sample obtained.

Samples should not be stored for long periods prior to testing as the high strength steel used to manufacture them is prone to rusting. The chemical reaction changes the water content and possibly the chemical composition of the sample, and for this reason some tubes are made from aluminium.

The UT100 sample tube was designed in the 1990s from a collaboration between Soil Mechanics (now SOCOTEC) and Norwest Holst (now Soil Engineering). The design was carried out in response to the desire to comply with research and Eurocodes that, at the time, were showing the traditional driven sample tubes were not compliant. They aimed to provide a robust tube that would be more appropriate to stiff soils and could be driven using the traditional methods of drilling such as cable percussion and dynamic sampling. Until that time, with the occasional exception of piston samples in soft soils, both of these drilling methods were using thick-walled samplers that were not capable of achieving a Class 1 sample.

The main objective was to provide a design for a steel body that could incorporate a removable and hence reusable cutting shoe while using the traditional sliding link hammer method of driving. This, by its very nature, requires a robust tube and cutting shoe. The design adopted uses a higher-grade steel than used for previous samplers, thus enabling a thinner profile that has an area ratio of 15%. Because the steel was thinner, the thread was changed to give greater strength. The tubes produced using this new design are more expensive than the original U100 tubes. Cheaper aluminium alternatives shortly followed, but having sufficiently thin walls to meet the area ratio criteria made them much weaker, even though they adopted the new square thread design.

The UT100 can provide a Class 1 sample from cohesive soils, which is generally of adequate quality to enable strength and deformation testing to be conducted with confidence. It is essential, however, that sampling process is meticulous, with the tube being clean, the cutting shoe being free from damage, without burrs, the drive equipment serviced, and the base of the borehole cleared of all debris prior to sampling. The use of a 'stubber' (see Chapter 1) should be avoided because it tends to leave cut marks in the soil at the base of the borehole. Similarly, if an SPT has been conducted just prior to taking a UT100, the hole should first be advanced to a depth equal to at least three times the diameter of the borehole below the base of the SPT.

In some soils the sample quality will be poorer than Class 1, particularly if the soils contain gravel. Gravel particles can be realigned as they enter the tube, disturbing the soil matrix as they rotate, and if a gravel particle becomes lodged on the cutting shoe, it may score the sample as it is driven.

2.3.6   Thick-walled open tube samplers

2.3.6.1 SPT split spoon samplers

This was probably the first borehole sampler used in the UK and arrived in around 1900 as a sampling tool. It was much later that it was adapted to provide some information on soil density/

compactness (see also Chapter 4). As a sampling tool, the substantial wall thickness, with an area ratio about 50%, and the fact that the sampler is driven, mean that it produces a very poor sample, Class 5, as shown in Figure 2.11.

The typical notation on logs is 'SPT (S)' for the test itself, although normally the sample is retained as a small disturbed sample, 'D'.

Often the tube will block if gravel is present, or recovery will be poor owing to the high internal wall friction. While the sample obtained provides some guidance on the material type present at the depth of the SPT, it is rarely of use for geotechnical laboratory testing. Although it may be suitable for plasticity index testing, the natural water content, which is conventionally measured alongside index testing, is likely to be unreliable.

Figure 2.11 A split spoon sampler open to show the highly disturbed sample inside (image courtesy of SOCOTEC)

Note the thick walls of the tube; also, the cutting shoe in the top left of the photograph that has been unscrewed to allow the tube to be split into two.

2.3.6.2   U100

The U100 is a thick-walled sample tube made from steel or aluminium, usually 100 mm in diameter (four inches) and 0.45 m long. The typical notation on logs is 'U', which sometimes will be followed by 100 to denote the diameter in millimetres.

Typically, the tube is fitted with a removable cutting shoe at its base, which has a slightly smaller inside diameter to the tube. If necessary, a core catcher can also be included, either as a plastic insert or built into the cutting shoe. The top of the sample tube is then fitted to a screwed adaptor connected to a slide link hammer to allow blows to be given to drive the tube into the soil at the base of the borehole. The connector has a one-way valve, which allows any air or water to be expelled that might be trapped above the sample as it passes up the tube. It is important to ensure that this valve is working correctly before beginning driving. The assembly is fixed to a swivel that can be attached by way of a shackle to the rig rope. The assembly is lowered to the base of the borehole, which should be prepared for the sampling process by removing any spoil in order to provide a clean flat surface. Specific tools to do this have been used in the past, although they are rarely seen on site today.

Once the borehole has been prepared the assembly is lowered so it is just touching the base of the hole. The tube is then driven its full length into the soil. It is normal practice to judge the required depth of drive by putting a mark on the rope 450 mm above the level of the casing.

Some practitioners attached a second tube to the sample tube by way of a threaded connector, as this prevents compressing the sample if the tube is overdriven, that is more than the length of the sample tube and cutting shoe.

Ideally, following driving, the sample should be left for a few minutes before winching it to the surface. Once at the surface, the tube is removed from the assembly, the cutting shoe removed from the tube and the ends trimmed and sealed with wax (see Section 2.2.2). The sample should be labelled clearly, as per Section 2.2.3. The cutting shoe material should be retained as a small disturbed (D) sample.

2.3.6.3 U100 using a liner system

This comprises a system which uses a thick-walled plastic tube, designed to replace the relatively expensive metal U100 system described above. The method became common practice during the 1980s because it was cheap and relatively simple to use. The system was used in a similar fashion to the U100; however, the plastic tube is held within a metal assembly to provide rigidity to the sample tube. The assembly incorporates a cutting shoe that is thicker than that for the UT100. Typical notation of such samples on logs is 'UL'.

The whole assembly is considerably thicker than the original U100 and, as such, has a very high area ratio, generally greater than 37%. The sample quality is therefore generally poor, with a quality no better than Class 3. It is therefore something of a misnomer for it to be called a 'U'100. This approach was a precursor to the development of the dynamic sampler that is in common use today.

The liners have the same diameter as plastic drainage pipe, the latter therefore being commonly selected as a cheap alternative and cut to length by the drilling operatives. However, this was a poor substitute for the purpose-made liner tubes that have a greater stiffness. In any event, the tubes are not always a tight fit in the assembly and may move as they are driven. In addition, the force of driving can deform the tube, causing a pressure wave to pass along its length. These effects can result in even more disturbance to the sample and may completely destroy any soil structure that might be present.

2.3.6.4 Field California Bearing Ratio (CBR) mould sample

The CBR mould, equipped with a cutting shoe, has been a normal method to obtain a field sample for testing to determine the CBR value in the laboratory. However, this form of sampling does provide a good quality sample that, if taken by pushing, can deliver a Class 1 sample.

The method was derived to obtain a sample from the field for CBR testing in the laboratory (see also Chapters 5 and 6). It uses a CBR mould with a cutting shoe screwed to its base and a collar screwed to the top of the mould. The mould is 152 mm (six inches) in diameter and 127 mm (five inches) in height. Typically, the sample is taken in a shallow trial pit where the surface can be prepared by hand, and the mould set vertically. A large block of wood, or purpose made dolly, is placed on top of the mould, and it is either hammered into the ground with steady blows or, as is more often the case, the bucket of an excavator with an extending back-actor can be used to push the mould to the required depth. Depending on the skill of the operator, the back-actor excavator is able to extend the arm in a smooth vertical direction providing a constant downward force on the mould. Once fully driven, the excavator bucket is then used to recover the mould by scooping below it and removing it, together with the surrounding soil. Once retrieved, the soil can be dug away from around the mould, the cutting shoe and collar removed, and the sample carefully prepared level with the top and base of the mould. Base and end plates can then be fitted to protect the sample during transportation. To preserve water content, the entire mould can be sealed with cling film and waxed.

The area ratio of the CBR mould is approximately 8% and when pushed in the manner described above may provide a good quality Class 1 sample, when used in cohesive soils. The method is much simpler and possibly more successful than conventional block sampling (see Section 2.3.3) when requiring a potential Class 1 sample. Safe access to the area in which the sample is being taken is, however, an important consideration, and the operation should only be undertaken where suitable precautions are in place (see Chapter 1) and the process has been fully risk assessed. Preparing for safe access is often time consuming and costly and, for this reason, this sampling method is not used routinely.

2.3.7 Large diameter samplers

2.3.7.1 General

Large diameter samplers generally provide good quality samples; however, the samplers are not common and are not normally used in conventional ground investigation. They are more often used for special situations or research where a high-quality sample is required to test for a specific design situation. For the most part, the samplers are quite complex and it can take a significant time to obtain the sample. They are included here for information and because some publications mention their use – for example, BS EN ISO 22475-1 (BSI, 2021).

2.3.7.2 Sherbrook sampler

The Sherbrook sampler (see Figure 2.12) is a mechanical means of obtaining a large diameter cylindrical 'block' sample. While the original design was manually operated and generally used at shallow depth, more recent versions, as shown in Figure 2.12, have been adapted for use in large diameter rotary boreholes and are rotated on drill rods operated at a slow speed. The Sherbrook sampler was originally developed in Canada (Lefebvre and Poulin, 1979); the original provided a sample of 250 mm diameter and 350 mm length. A smaller version has also been developed in Norway at 160 mm diameter (Emdal *et al.*, 2016).

The rate of rotation needs to be relatively slow and is adjusted depending on the soil type, with the operator controlling the rotation speed and rate of penetration. Blades located around the periphery of the frame carve the block of soil using three fixed cutting tools or blades. As the body of the sampler is rotated, the blades cut an annulus around the sample. The borehole, which for the original design should be of 400 mm diameter, is filled with bentonite during the sampling exercise to maintain stability of the free-standing soil block and to aid removal of the cuttings. Once the annulus has been carved to the required depth, blades are activated at the base to cut underneath the sample as rotation is continued and penetration is stopped. The basal blades finally meet at the centre of the base and provide support of the sample as it is lifted from the borehole. The sampling process can take up to 30 min with a further 5 min to undercut the sample.

Figure 2.12 Showing the Sherbrook sampler and retrieved sample (image courtesy of T. Lunne, Norwegian Geotechnical Institute)

This sampler is designed to obtain a high-quality Class 1 sample from soft to firm sensitive clays and soils of low plasticity.

2.3.7.3   Laval sampler

The geotechnical group at Laval University, Canada (La Rochelle *et al.*, 1981) designed this large diameter sampler. It is a piston style sampler that is capable of pushing and rotating. The sampler is operated in a slurry filled borehole, the sample tube being fully retracted into the guide tube at the start of the sampling process. The top of the assembly is fitted with a one-way valve that permits the drilling fluid to flow freely out from inside the sample tube. The sample tube and assembly are lowered to the base of the borehole with the sample tube locked into the top of the assembly. When in position, the sample tube is unlocked and pushed in such a way as to minimise the pressure in the drill fluid above the sample to ensure additional stress is not applied to it. When the pressure in the fluid is seen to exceed a head of 50 mm, the sampler is locked to the assembly and is then rotated to drill the rest of the sample in a conventional way. Drill fluid under pressure passes down the drill rods and between the sample tube and the drill assembly to remove drill cuttings. The fluid is under pressure to ensure the flow is maintained. When the sample tube has been fully inserted, the tube is slowly rotated 90 degrees to shear the sample at the base, the sampler then being lifted slightly and locked back in position. The whole assembly is then retrieved from the borehole.

The assembly is designed with a constant internal diameter and a very fine cutting edge. The sampler is 208 mm in diameter and 660 mm long, providing a Class 1 sample and was designed particularly for sampling in soft clay soils.

2.3.7.4 Deltares sampler

The Deltares sampler obtains a sample of 400 mm diameter and a length of 500 mm. The sampler comprises a thin sample tube housed in an outer casing. The cutting shoe is 320 mm in length and has a cutting edge with a 10-degree angle.

The sampler assembly is lowered on the rig rope into the borehole, which should be prepared by cleaning the base; casing should extend to the base of the borehole. During this process, the valve at the top of the assembly is left open to allow fluid to pass through without putting pressure on the soil. Once the base of the assembly is in contact with the base of the borehole, clamps are activated that lock the sampler against the borehole casing. The sample tube is then pushed into the soil using rams located at the top of the assembly. Pushing is carried out as a smooth single action. Drill fluid sitting on top of the sample passes out through the open valve at the top of the assembly. Pushing may be paused to allow the pressure within the drilling fluid trapped above the sample not to exceed 50 mb. Once pushed to the full depth, the top valve is closed and blades are activated that cut into the base of the soil block. The blades meet at the centre of the sample base and remain in place to aid removal of the assembly and sample. The assembly head is removed, and an end cap is secured to the top of the sample tube. The sample and the casing are then inverted, allowing their removal from the sample tube and freeing the sample to allow an end cap to be secured to its base. The sample is then returned upright and prepared for transportation or subsampling for the laboratory.

The resulting sample is generally a Class 1 sample, and the method is suited to soft soils that might collapse under their own weight, or peaty soils that might compress under normal sampling methods.

2.3.7.5 Ballast sampler – coarse-grained soil sampler

Soils that comprise coarse gravel-sized particles are particularly difficult to sample. This method was developed primarily to sample rail track ballast, and can provide a reasonably representative sample of the single-sized angular gravel, which is normally used as a track bed material (see Figure 2.13).

The sampler uses a robust steel tube 200 mm in diameter and 1 m in length, and a core catcher similar to that used for other sampling methods is incorporated at the base of the tube, behind a cutting shoe. The sampler is dynamically driven into the soil using a weight drawn up on a chain and ratchet mounted on a mast, as used on the drop weight-type dynamic sampling rig. Once driven to its full length, the tube is withdrawn from the ground with the core catcher retaining the soil within the tube. The sample at the top of the tube is discarded because this will be the part of the sample with greatest disturbance. The lower 0.5 m is retained for inspection and testing and is considered to be reasonably undisturbed, although it is probably a Class 2 sample at best. When testing is conducted, a note of gravel particles that have been broken in the sampling process should be assessed, as this will affect the accuracy of any particle size distribution tests undertaken.

Similar techniques have been developed in Europe. Figure 2.13 shows a sample of ballast obtained where a sonic drilling assembly mounted onto a rail track carriage drives the sampler.

2.4. Continuous sampling methods

2.4.1 General

There are many methods that sample by drilling, each attempting to recover a continuous core of soil or rock to the full depth of the borehole. This strategy has many advantages; in particular, it enables a full log to be made and strata boundaries to be defined that may dictate issues such as foundation depth and type, for example. Some methods are more successful than others at achieving this aim. Tables 2.2 and 2.6 provide a comparison of the methods generally used, their suitability in different soil types and the expected quality class that the methods might be able to achieve.

Although some methods attempt to obtain continuous cores in cohesionless soils, these are not always successful, and it should be considered the exception rather than the rule that samples of good quality might be obtained in such soils with any method.

In general, the larger the sample the better the quality and certainly the greater the detail that can be described during logging, features such as structure and discontinuities being clearer when large diameter samples are obtained. It is considered that, where possible, the sample should be at least 100 mm in diameter. When such samples are of sufficiently high quality, which typically means they have been obtained by rotary coring methods, this sample diameter enables subsamples to be taken for use in the 100 mm diameter triaxial apparatus for total and effective stress strength tests, as well as other advanced tests.

Because continuous sampling is the method of advancing the borehole as well as retrieving the material encountered, the recovered core is not seen as a sample in its own right. It is not targeted to a specific depth or material type, and indeed will cross any material boundary that is present. However, where subsamples are taken of selected sections of the recovered core, these should be given sample numbers and identified on logs by the notation relating to the sample taken, thus 'D' for small disturbed samples, 'B' for bulk samples and 'C' or 'CS' for core subsamples ('C' is the AGS agreed notation, although 'CS' is the abbreviation used in the Eurocodes).

2.4.2 Dynamic sampling methods

2.4.2.1 Introduction

Dynamic sampling is often referred to in some specifications and standards as windowless sampling, which is rather unfortunate and a complete misnomer. The detail of the method and appropriate

Table 2.6 Recommended sampling methods for different soil types and their achievable quality class (after Clayton *et al.*, with additions)

Soil type	Sampler/method	Achievable quality class
Predominantly gravel	Large diameter driven open tube sampler with catcher	3
	Sonic drilling	4
Medium to coarse sand	Thin-walled piston sample	2
Fine sand / silty sand	Thin-walled fixed piston sampler	2
Coarse silt	Thin-walled fixed piston sampler	1/2
Soft sensitive clay	Large diameter fixed piston sampler	1/2
Very soft clay	Thin-walled fixed piston sampler	1
Firm to stiff clay	Pushed thin wall sampler UT100	1
Very stiff clay; cohesive glacial till	Triple tube wire line coring	1
Weak rock (chalk, marl)	Triple tube wireline coring	1
Strong rock	Triple or double tube rotary coring	1

terminology is discussed in Chapter 1 and includes a description of all forms of dynamic sampling using both window sampling and open tube sampling (often referred to as windowless sampling) methods. These various sampling methods are also summarised in the following sections.

Correctly, the methods that use barrels without 'windows' should be called dynamic sampling (DS), (see also Chapter 1) but with a differentiation then being made between the driving methods – that is, either drop weight or hydraulic percussive. This differentiation is required because the two methods can provide samples of different quality classes, as discussed below.

The dynamic sampling method drives a 1 m long barrel, containing a rigid plastic liner, into the ground by applying repeated blows from a drop hammer. The sampler tube is thick walled, and the combination of these factors means that, at the very best, the method will provide a Class 3 sample when driven using a rotary drilling rig (DS-hydraulic percussive), or Class 4 when using a smaller 'dynamic sampling' rig (DS-drop weight) (see Table 5.2). The footnote to BS EN ISO 22475-1 (BSI, 2021) Table H4 states that dynamic sampling (windowless sampling) can only produce a disturbed sample Class 4 and will typically have a quality index of 2. In the authors' experience, this is only the case for dynamic sampling (drop weight), which is conducted using a small lightweight rig described above (see also Section 2.4.2.3). Slightly better samples tend to be recovered from the hydraulic percussive technique (see Section 2.4.2.4 and Chapter 1).

For both window sampling and the two dynamic sampling methods, it is normal practice to take subsamples of the material at specified depths for laboratory testing. For some tests, this will mean taking all the available material to be able to conduct tests on representative samples (remembering that subsamples should not be taken across strata boundaries, which might restrict the amount of material available for a sample even further). In some cases, the amount of material available will not provide a sample of sufficient mass to be representative of the materials in situ. This is particularly the case where the dynamic sampling drop weight method is used, and the tube diameters are reduced to minimise the friction on the outside of the sampler as the borehole depth increases.

2.4.2.2   Window sampling

This method is not normally differentiated on logs, usually given the notation 'DYS', meaning dynamic sample. The method of either window or windowless sampling is usually indicated in the borehole reference as 'WS', such as WS1, which only adds to the confusion between these methods (see Chapter 1 for terminology clarification). In any event, the material retrieved in the sampling barrel is subsampled in either small disturbed sample containers or bulk bags. Whichever method is used, it is important not to mix soils from different strata into the same container. Material from each individual stratum should be retained separately, and the sample container labelled accordingly.

This method of investigation was primarily derived to provide a rough estimate of where overburden meets a resource material such as sand or gravel. The tool was devised to enable the overburden strip volumes to be determined. The sampler tubes are designed as flow through, with a slot or 'window' cut into the side of the tube. The sampler is driven into the ground using a handheld jack hammer that delivers rapid blows to the rods. Much of the downward force to aid penetration is from the self-weight of the hammer and the weight of the operatives pulling down on the hammer.

Material can flow through from the base of the open tube and out of the top of the sampler. Material may be dragged into the sampler at random depths of the drive and often the tubes can be blocked by coarse particles. All of these factors result in a very unreliable sample being retained in the sampler, and for these reasons window sampling should not be considered as an appropriate geotechnical sampling method. The method is included here because of its persistent in use by some and is also described in Chapter 1.

The sample obtained using this method is highly disturbed such that even basic strata boundaries are difficult to determine. If carried out in cohesive soils, it might be possible to obtain samples adequate for liquid and plastic limit values; however, the natural water content cannot be relied upon. The method can be used where access is difficult and other methods cannot be carried out. At best, the method will enable the differentiation of material type and enable an estimate of where strata boundaries may lie.

Figure 2.14 A window sampler barrel showing access to the recovered soil through the longitudinal 'windows' (image courtesy of SOCOTEC)

The soil is subsampled by removal with a trowel, through the windows.

2.4.2.3   Dynamic sampling drop weight

The dynamic sampling tooling comprises an outer barrel that carries a 1–2 mm thick plastic liner; the liner has a retaining collar and is held in place by the removable cutting shoe. The assembly can accommodate a liner of 1 m length. The whole assembly is very thick walled and has an area ratio of about 46% at 100 mm diameter, which increases to in excess of 100% as the tubes are reduced in diameter; a 47 mm diameter sample tube, delivering a sample dimeter of 33 mm, has an area ratio of 103%.

Once recovered from the borehole, the sampler system is dismantled and the liner with sample inside is pulled from the tube with pliers. For this to work well, the barrel must be meticulously clean, otherwise the liner can jam making it difficult to retrieve. It is important that the plastic liner fits well in the sampler; poor fitting liners can become crumpled or torn, particularly where there is gravel present in the soil being sampled. Any damage will make extraction of the liner from the barrel very difficult and the whole sample may be rendered useless.

Because this system is much cheaper than alternative methods, the dynamic sampling drop weight method is much overused; the information and sample quality is generally very poor and often inadequate for any reliable assessment of the ground to be made. The samples obtained are highly disturbed and at best will only enable a sequence of strata to be determined. Some specifications will, however, ask for hand vane or pocket penetrometer testing along the core. This is of little value because the sample obtained is often highly disturbed, because it should not be assumed that such tests will necessarily provide a conservative indication of in situ shear strength. In some cases compression of the material during sampling can lead to strength tests overestimating in situ shear strength. Either way, results of such tests can be very misleading.

Dynamic sampling by this method can provide a general overview of the soils at a location, including information regarding the lateral and vertical extent of a particular stratum, as shown in Figure 2.15. When used in conjunction with SPTs it can also provide some useful data, which might relate to design parameters. For situations of Geotechnical *Category 1* (BS EN ISO 1997-1 (BSI, 2004)) this may be sufficient to satisfy the design requirements. For all other geotechnical categories, and where a rigorous design is required, other methods of investigation should be considered, either alongside or instead of dynamic sampling.

If the trip hammer of the sampling rig is set up in the DS super heavy configuration B (see BS EN ISO 22476-1, Table 1 (BSI, 2023a)) the energy provided by the drop weight is the same as that required to conduct an SPT. When SPTs are carried out, however, it should be noted that this will cause additional disturbance to the material in the top (potentially the entire top half) of the following sample.

2.4.2.4   Dynamic sampling (hydraulic percussive)

The same sample tube arrangement, as described above, can be used on a dual head rotary rig where one of the heads comprises a hydraulic hammer, thus enabling dynamic sampling to be carried out in superficial soils that are difficult to drill by rotary methods. DS-hydraulic percussive is carried out at a diameter of at least 116 mm, giving a sample dimeter of 100 mm. Larger diameters can also be used with a maximum of 145 mm with a sample diameter of 124 mm. These larger diameters are used to allow rotary drilling tools to continue in the borehole once competent soil or rock is met.

It is also possible to case the borehole to its full depth, enabling dynamic sampling to be progressed in granular soils. In this way, the borehole can often be sampled to depths of 15 m or more with near continuous sample recovery. Recently, this method has become more popular, enabling a single rig to complete the entire borehole rather than using the two-rig approach of cable percussive boring through superficial soils and installing casing to allow rotary continuation in more competent soil or rock.

Figure 2.15 Samples recovered by the dynamic sampling drop weight method (image courtesy of SOCOTEC)

Note sample diameter changes with successive runs. Selected sections of the sample are subsampled as D or B samples for laboratory testing.

In general, when used in this way the sampler achieves a sample of slightly better quality than the dynamic sampling drop weight method described in the above section, probably owing to the relatively rapid application of blows coupled with the generally larger diameter of the samples. The sample quality does enable good quality logging, and testing can be conducted to assess density and water content. These samples, however, are not generally suitable for strength and deformation testing, hence its inclusion in Table 5.2 as a Class 3 sample. Typically, the DS-hydraulic percussive method will provide samples of quality index 3.

2.4.3   Sonic sampling

Sonic drilling (sometimes referred to as 'resonance drilling' or 'rota-sonic') provides a method of sampling some soils, particularly granular soils, where other methods cannot achieve suitable results and may fail to obtain representative samples. The sampler is a continuous tube, usually of 100 mm internal diameter, which is used to drill a borehole and recover soils through which the hole is drilled. In normal use, the drilling is carried out dry. The sample tube is vibrated into the ground at a frequency of about 50 Hz. This action will enable the sampler to pass through concrete, wood, metal and other obstructions, as well as most soils and rocks. This ability can be very useful even although the vibration generated transmits through the soil and critically affects the pore water pressure and hence the effective stress. The sample quality can be poor in that the vibration can break up structure and disturb discontinuities that may be present. In finer-grained granular soils (such as silts and fine sands) the change in pore water pressure can cause liquefaction leading to significant disruption of the particles. Even so, the retrieved sample does comprise all the soil particles including the finer-grained particles that are often lost by more conventional methods. Rigs with rota sonic ability have the dual capability of operating in the same way as a standard rotary drilling rig in addition to sonic operation, and are able to swap to conventional coring methods if required.

Once the tube with sample is recovered from the borehole, the sample can be removed from the tube by tapping the tube with a hammer or applying a low resonance vibration to the sample tube and allowing the sample to slide into a gutter or a polythene tube (see Figure 2.16). The drilling tube needs to be held near vertical and hence the sample is further disturbed by this extraction process. The sample is often found to be perceptibly warm, and on cold days steam will be seen rising from the sample as it meets the air, thus the water content of this type of sample is unreliable. Cores obtained can appear to be of high quality, but under closer inspection they will be found to be highly disturbed. It is not uncommon for the sample to slip from the barrel as it is being lifted from the borehole, resulting in a loss of core.

The use of plastic liner systems with the sonic rig have been trialled, but for the most part these have been unsuccessful with the liner distorting and softening as the core barrel is advanced owing to the high frequency vibration inherent in the method, along with the heat generated by the high energy. As noted above, some sonic rigs have a dual head and can revert to conventional drilling methods that can greatly improve the quality of the sample recovered. In this way, Class 1 samples can be obtained at specific horizons.

Even although the sample is disturbed and not suitable for water content or density testing, the sample is generally representative from the perspective of composition because all particles, including fines, are recovered – as shown in Figures 2.17 and 2.18.

Figure 2.16 (a) Core being collected into a plastic sock (note the angle the sample is being turned, which invariably causes disturbance); (b) core showing distortion of clay sample (images courtesy of SOCOTEC)

(a)   (b)

Figure 2.17 Glacial till recovered by the sonic drilling technique (image courtesy of SOCOTEC)

Figure 2.18 Good recovery of gravelly sand by sonic drilling (image courtesy of SOCOTEC)

Note that although recovered sample is complete it is disturbed.

Other drilling methods that use water for drilling will often lose the finer end of the particle range and hence the samples are not representative. The retention of fines is an important factor, particularly when assessing the permeability of a soil, where permeability is a function of the smallest 10% of the particle sizes comprising the whole sample.

2.4.4   Rotary core sampling methods

2.4.4.1   Introduction

Rotary drilling is discussed in some detail in Chapter 1. The commonest reason to drill a rotary borehole is to obtain a core sample of the soil or rock. There are two main methods in use today: conventional coring, which requires a drilling rig with a head that can rotate drill rods with a coring barrel attached at the lower end; and wireline drilling, using a similar rig with a winch that can be used to lower the core barrel into a latching mechanism to enable the barrel and casing to be simultaneously drilled into the ground. A drilling fluid is used to remove cuttings and keep the drill bit cool. The fluid is usually water or air, a combination of water and air, or a mix of water with mud or polymer. The chosen fluid is used to circulate through the rods and down into the drilling bit, returning up the borehole outside the barrel, and may be collected at the surface for reuse.

The coring barrel design can be complex, but the basic principle is to rotate the core barrel that has a coring bit at its lower end, such that the soil or rock is cut away leaving a stick of core to enter it. The core is retained in the barrel and when at the required depth can be lifted from the borehole and retrieved as a complete continuous sample with little or no disturbance. Core barrels fall into three main types: single, double and triple tube core barrels, with the triple barrel being used either on drill rods or as part of a wireline system. Whichever type of barrel is used, the space above the sample is vented to prevent pressure build up in the fluid that will occupy this space, allowing it to be expelled as the sample moves up into the barrel.

There are many permutations of core barrel, drill fluid and bit design; the art is to find the combination best suited to the expected ground conditions to provide the highest quality core. It is highly likely that to drill a sequence of strata it will be necessary to use more than one combination. In addition, the drill speed and pressure will also influence factors such as core quality and bit wear. It is highly recommended that the advice of a suitably experienced drilling contractor is sought at an early stage in the borehole design process to ensure the adopted methods are suitable and successful. These factors are discussed in Chapter 1.

Whichever method of drilling is used, once recovered from the barrel the core should be sealed and stored in rigid core boxes suitable for protecting it and enabling efficient transportation (see Figure 2.19). It is essential that the boxes are clearly labelled identifying the core within the box. The core itself should be labelled with the depths at the top and bottom of each core run. Storage should be in an environment where the core is protected from freezing or high temperatures, and ideally held at a stable temperature and humidity (see Section 2.2.2).

Ideally, core should be cleaned of any drilling fluid on its surface and photographed and logged as quickly as possible after drilling and, if possible, at the drilling site (Reading and Martin, 2025). Subsampling (Section 2.2.4) should also be carried out at an early stage to preserve the natural water content of the sample. This process can be selective of only the better recovered core pieces; if this is the case, this should be recorded stating what percentage of the core is of a similar quality.

The core subsamples should be wrapped for protection, as described in Section 2.2.2. Once fully wrapped, each sample should be labelled securely and the top of the sample identified. The sample should be transported securely in a box capable of ensuring the sample is not affected by the transportation method.

2.4.4.2   Single barrel coring

Single barrel coring is normally used for coring concrete or masonry structures, road pavement and other man-made materials. It is generally used for short length holes, typically for structural surveys, or to create an access hole in surface concrete or blacktop layers ready for other forms of drilling – as shown in Figure 2.20(a).

The simplest form of core barrel is a single tube with cutting shoe; this design will invariably mean drill fluid will pass between the core and the barrel that in all but the most competent of materials will probably result in some loss or alteration of the sample.

Removal of the core from a single barrel is not easy, and either requires the barrel to be held in a near vertical position and the barrel tapped to encourage the material to fall from the barrel, or

Figure 2.19 A box of rock core ready for subsampling and description (image courtesy of SOCOTEC)

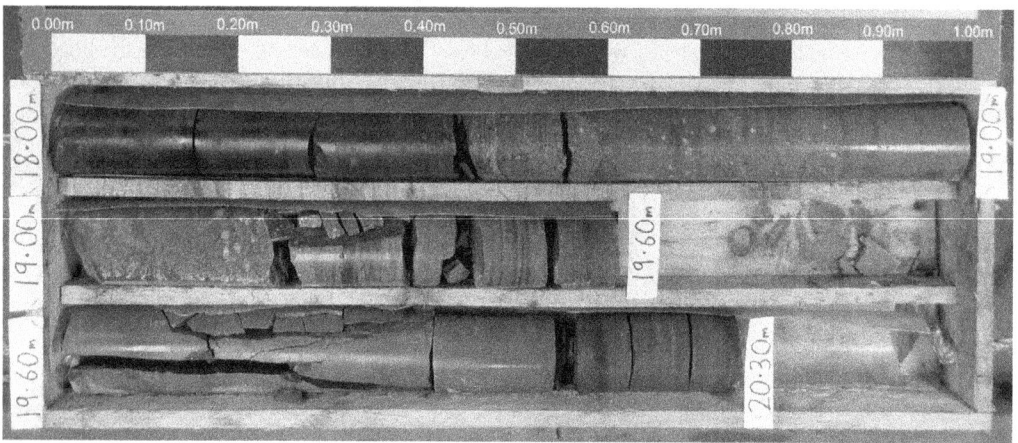

Figure 2.20 (a) Coring through macadam with a dynamic sampling rig and single core barrel; (b) typical core of road pavement recovered by this method (images courtesy of SOCOTEC)

(a)   (b)

alternatively the drill fluid under pressure can be directed down the barrel to push the core out. Either method will result in significant disturbance of the sample. The recovered sample will be at best Class 4 and probably poorer. The degree of disturbance should be taken into account when the materials recovered are inspected or tested– as shown in Figure 2.20(b).

2.4.4.3   Double tube coring

The double tube barrel incorporates an inner tube that is held on a swivel, which ensures the barrel with the core inside does not rotate as the outer barrel is turned together with the drill bit. Once at surface, the core bit and the catcher assembly are unscrewed from the end of the outer barrel, allowing the core to be removed from the inner barrel. If the core is not freely moving in the barrel, either the barrel is held near vertical and lightly tapped with a hammer or, with the assembly horizontal, a plate is placed at the top of the sampler and the assembly is then pressurised using the rig fluid pump to push the sample from the barrel.

The method of extrusion can cause serious disturbance to the core, which is the main disadvantage of this type of barrel. The double tube barrel is best used in competent rock where coring produces good intact core, which is less likely to be damaged on extraction, and can achieve Class 1 in these materials.

2.4.4.4   Triple tube coring

The triple tube core barrel design is similar to the double tube barrel, but with a rigid plastic sleeve, generally referred to as 'coreliner', forming a third tube within the inner steel barrel. The plastic sleeve is used to protect the core from contact with drilling fluid as drilling proceeds, and it also enables the core to be removed easily from the barrel (typically being pulled out of the barrel in a horizontal position, using a pair of pliers). The plastic liner also acts as the sample protection sleeve during transportation, meaning the core is less likely to be damaged or disturbed during the retrieval process or during onward transportation. Other systems use a metal split barrel, which again makes retrieval of the sample simpler.

The triple tube barrel is further enhanced using a wireline system (see Chapter 1).

The wireline system using a triple tube barrel enables much weaker materials to be drilled. It is now commonly used to obtain high-quality samples of stiff, fine-grained (cohesive) soils such as glacial till or the London Clay Formation. In combination with synthetic or natural polymers, drill muds can be designed to enable drilling in sands and silts with some success with respect to sample recovery. Successfully sampling soils such as fine sand or silt is difficult using any system (see Section 2.4), although some success has been seen using the wireline system with a combination of mud and drill bit design, as well as a highly skilled drilling operative.

In general, the triple tube barrel will provide a high-quality sample often of Class 1.

2.4.5   Wash bore sampling

Wash boring produces a flush return that carries chippings of the rock through which the borehole is drilled. By collecting the returns and sieving out the larger chippings, it is possible to obtain a sample that can be used to provide an indication of the materials being drilled; the sample is greatly altered and is not suitable for geotechnical testing. It is possible to detect changes of strata, but only a rough depth at which the change takes place can be determined owing to the uncertain delay as the cuttings flow up the borehole.

The method is commonly used in the mineral extraction industry and offshore, when drilling for oil and gas reserves. It allows the borehole drilling to continue without stopping and a high rate of penetration.

2.4.6   Hollow stem auger sampling

A continuous hollow stem flight auger can be used to obtain a continuous sample while drilling a borehole. The sample is taken where the central stem of the auger is a hollow tube, thus allowing the soil encountered to enter the stem as the auger turns is rotated into the ground. The sample is highly disturbed because there is nothing stopping the sample turning as the auger rotates. In addition, the hollow stem may be blocked preventing further recovery if gravel is present or the sample can be compressed as the auger advances. To remove the sample from the tube once drilling is completed it is necessary to hold the auger vertically and tap the flight to encourage the sample to fall from the tube. It is possible to incorporate a liner within the hollow stem, which will improve the sample quality and aid removal from the drill flight. In any event, the sample obtained can only provide a rough guide to the strata through which the hole has been advanced and is of limited geotechnical use.

2.5.   Sampling in soft and fine-grained granular soils
2.5.1   General

As discussed above, there is no sure way of sampling fine granular soils such as silts and sands. Adaptations of other methods have, however, resulted in a number of methods that do have some success in these materials. All are elaborate sampling barrels and are not simple to set up and use. The sampling process is usually slow, and handling of the barrel and sample recovered may result in the loss of a sample that has taken some time to remove from the ground. Invariably, these methods are time consuming and therefore expensive, and would only be used under special circumstances – for example, where a geotechnical design warrants complex testing, requiring the soil characteristics to be as close as possible to that in the field. Fine sand and silt are particularly susceptible to disturbance if they are allowed to freeze, and adequate protection must be provided if this is likely to occur.

2.5.2   Bishop sand sampler

This method of sampling would only be used in exceptional circumstances and would never be part of a routine ground investigation. It is included here, however, for the sake of completeness, and as it is mentioned in earlier texts (Clayton *et al.*, 1984).

The Bishop sand sampler (Bishop, 1948) uses compressed air to attempt to prevent a loss of sand during the sampling process. Using conventional sampling methods, there is normally a significant stress relief and lack of pressure balance that will often result in a loss of the sample. The Bishop sand sampler is deployed inside a compressed air bell and into the borehole that is full of water to ensure a positive water balance. The compressed air displaces water from around the sampler as it is pushed into the soil at the base of the borehole. The compressed air is then applied to the base of the sampler and around the sample to expel the water. The method relies on arching in the sand and capillary suctions that develop at the sample surface because air is present around the sample periphery. In this way, sample losses are reduced; however, the sample is still considered to be quite disturbed.

2.5.3   Gel sampler

This method was developed in Taiwan and has been successfully used to obtain high-quality samples for testing.

Gel samplers are either fixed or free piston samplers with a gel reservoir. The samplers use a fluid gel to protect the sample from losses. The method is relatively complex and messy in that the sample is coated with a gel that consists of a water and polymer mix. The main disadvantage is the gel does seep into the sample and fill the periphery pores. This can affect permeability and pore water movement, which may be a key factor in subsequent tests conducted on the samples. This technology is relatively new and may improve in the future to provide a more practical sampling method in loose, fine granular soils.

Samples were taken using this method to analyse the effects of the 2008 earthquake in Christchurch, New Zealand, and produced good quality results (Stringer *et al.*, 2015).

2.5.4   Foil sampler
Often referred to as the Swedish foil sampler, owing to its development being advanced by several Swedish researchers, the foil sampler is similar in concept to the gel method above. The sampler uses tin foil, which rolls onto the surface of the sample as the sampler tube is pushed into the ground. The foil provides a barrier to the drilling fluid as well as providing support to the sample as it is extracted from the sampling tube. This method performs best in very soft to soft soils. If larger particles such as gravel are present, the foil can be disrupted as it is deployed, resulting in significant disturbance and potential loss of the sample. The sample tubes can be particularly long, up to 20 m in length being achieved in suitable strata.

2.5.5   Stocking sampler
The use of a 'stockinette' was originally designed as an integral part of the drilling sample barrel known as the Delft stocking sampler (Begemann, 1974). The operation of the stocking sampler is similar to the foil sampler whereby a stockinette is rolled out as the sample enters the sample tube. The stockinette provides support to the sample and aids recovery from the sampler tube. Typically, the tube is pushed from surface and will continuously sample to depths of up to 18 m using the standard CPT equipment, and with some modification increased thrust depths of 30 m have been achieved. The open end of the tube is fitted with a cap that moves up the sample tube as the sample is pushed. The sampler was developed to obtain soil samples in very soft sediments.

A similar system, Mostap sampling, is again deployed using the CPT rig; the cone is pushed to the required depth and has a sample tube and stockinette system behind the cone. When the cone is at the desired sampling depth, the cone is hooked, using a line and hook, and is withdrawn into the cone body. This leaves an open-ended tube that is then pushed into the ground, and the sample enters it as the stockinette unrolls. Mostap samplers can recover samples up to 2 m in length. When used at 65 mm diameter, the sample can be of Class 1, and hence may be used for consolidation and undrained shear strength testing.

2.6.   Hand-driven samplers
There are several handheld sampling methods, including the Mackintosh probe, the Russian Core sampler and various small diameter tube samplers that use a lightweight slide link hammer to drive them into the soil. Typically, these samplers are used in soft or firm cohesive soils, in conjunction with manual methods of advancing a hole, such as hand digging or hand auguring. These techniques can be particularly useful where access is restricted, where other methods are not available. In most cases, the samples obtained are of poor quality (Class 3 or worse), and the depth is limited by the strength of the operative.

Tube samples can be obtained from a hand-augured hole, or trial pit, using a slide link hammer in conjunction with U38 or U50 sample tubes. The slide link hammer is joined to the rods (usually three-quarter BSP hollow tubes with threaded connectors) and lowered to the base of the hole where it can be driven to its full length by blows from the hammer. The rods are withdrawn with the driven sample tube. Although in most cases the tube area ratio is less than 15%, the method of driving is not controlled and the verticality cannot be determined, thus the quality class is generally no better than Class 2. The spoil from the hand auger also provides a means of obtaining a disturbed sample of the soil that can be used for classification testing. These would normally be placed in either tubs or bulk bags for onward transportation.

A similar sample can be obtained using a core-cutter sampler and drive hammer. The core cutter tube, which is 100 mm in diameter and approximately 200–250 mm in length, has a bevelled taper at the base that provides a cutting edge, and a flat upper edge that fits snugly into a dolly attached to the drop weight drive assembly. Samples are driven by blows from the drop hammer until the top is flush with the ground. The drop hammer is removed, and the sampler dug out using hand tools to free it from the ground. The base should be parred flat with the sampler and the whole sample wrapped in cling film and then waxed. The technique, which can be readily conducted by one person, is used to obtain a sample for field density testing (see also Chapter 5). The sample diameter enables the soil to be extruded in the laboratory using a conventional U100 extruder.

Most of these sample types do not have common notations. The samples, however, are usually retained either in small disturbed sample tubs or bulk bags. Tube samples are given the notation 'U' followed by the sample diameter. It should be noted, however, that although U represents 'undisturbed', many tube samples are, in fact, disturbed to varying degrees.

2.7.   Choice of sampling methods

The choice of sampling method is dependent on the required use of the sample and the quality class necessary for that intended use. Typically, the type of sample required for the laboratory testing needed to provide the parameters for design should dictate the method of investigation adopted. However, this is not always the case, and the ground investigation generally adopts a method of forming a hole that is driven by cost rather than the quality of the sample and the outcomes of in situ or laboratory testing.

The choice of methodology should be driven by the best achievable sample that can be obtained in the soils that are expected to be encountered. In any event, it is unlikely that a single method will be suitable for the strata encountered on a site, and a flexible approach should be adopted based on the known geology. Table 2.6 provides sampling methods that achieve the best results in different soil types.

Eurocodes use an abbreviated notation to describe various types of sample and sampler, although these are not universally adopted by industry. For clarity they are listed here, with a couple of examples of how these descriptors might be applied to the more conventional sample type abbreviations used throughout this book, to aid cross referencing.

- OS – Open tube sampler
- PS – Piston sampler
- LS – Large sampler
- BS – Block sample
- GS – Grab sample

These may be further described as follows.

- T/W – Thin walled
- TK/W – Thick walled
- PU – Pushed
- PE – Percussion

Core samples are given the following notations.

- CS – Core sample
- ST – Single tube
- DT – Double tube
- TT – Triple tube

Other samples are given the following notations.

- AS – Auger sample
- HSAS – Hollow stem Auger sample
- S-TP – Sampling from trial pit
- S-BB – Sampling from borehole bottom
- SN – Resonant or sonic drilling

Combinations of these are used to denote specific types – for example, a thick-walled open tube sampler would be given the notation OS-TK/W-PE where percussive methods are used; this notation would be applicable to the U100 sample.

emerald PUBLISHING   ice

Peter Reading and Miles Martin
ISBN 978-1-83549-891-0
https://doi.org/10.1108/978-1-83549-890-320251004
Emerald Publishing Limited: All rights reserved

Chapter 3
Down-hole logging: Geophysical and other methods

3.1.   Introduction

3.1.1   General

With the exception of Sections 3.4.6 and 3.6, this chapter has been written by Kim Beesley, formerly of European Geophysics Ltd.

This chapter describes the basic elements of down-hole geophysical measurements commonly used in geotechnical ground investigations, along with their applications and limitations and a selection of example logs.

3.1.2   Reasons to use geophysical logging methods

There are several reasons for geophysical measurements within boreholes as part of a geotechnical ground investigation.

- Information about the subsurface that may be unobtainable from surface methods or samples.
- Characterisation of the rocks in situ.
- Accurate repeatable measurements.
- Identification of lithology, minerals and aquifers.
- Determination/estimation of some engineering properties.
- Continuous data (most wire-line techniques) unlike sampling.
- Aid to reducing/optimising coring or sampling points.
- Aid to optimising the location of packers and piezometers.
- Enable the vertical and horizontal extrapolation of data derived from boreholes.
- Supplement geologist/drillers' logs.
- Provide control data for surface geophysical surveys.
- Check on borehole completion/construction/condition.
- Aid to correlation between boreholes.

Table 3.1 gives a guide to the selection of appropriate geophysical measurements in respect to the information sought. Often it is necessary to run a selection of logs dependent upon the type of rock and its anticipated properties. Limitations for each method are given in Table 3.2.

3.1.3   The logging system

The major elements of a geophysical logging (wireline) system are illustrated in Figure 3.1. Logging units are generally vehicle mounted. Portable units are also available but are subject to depth range and tool sizes. Figure 3.2 shows a typical vehicle-mounted unit used in geotechnical site investigations. The down-hole probes are commonly referred to as logging tools or sondes.

Table 3.1 Guide to the selection of geophysical logs

LOG PARAMETER	NATURAL GAMMA	GAMMA - GAMMA	NEUTRON NEUTRON	MAGNETIC RESONANCE	MAGNETIC SUSP/FIELD	SONIC SEISMIC	RESTIVITY CONDUCTIVITY	ACOUSTIC IMAGERS	OPTICAL IMAGERS	VERTICALITY DEVIATION	FLUID TEMP	FLUID EC	FLUID FLOW	CALIPER	VIDEO CAMERA	CAVITY SCANNER
Information sort																
Bed thickness	*	*	*			*	*	*	*							
Breakout								*	*					*		
Casing integrity								*	*					*		
Cavities														*	*	*
Clay content	*															
Core orientation								*	*							
Control data for other logs														*		
Density		*														
Diameter / volume								*						*		
Elastic properties						*										
Flow zones											*		*			
Fractures / fissures								*	*					*		
Hardness						*	*	*								
High resolution								*	*							
Lithological identification	*	*	*			*	*	*	*							
Magnetic metal presence					*											
Orientation of structural features								*	*							
Permeability				*												
Porosity		*	*	*												
Salinity							*					*				
Structural features								*	*							
Track of borehole										*						
Visual Identification of geological features									*						*	
Water quality												*				
Water content			*	*												
Washouts														*		

Table 3.2 Limitations of geophysical logs

GEOPHYSICAL LOGS USED IN GEOTECHNICAL SITE INVESTIGATIONS								
LIMITATIONS								
LOG	BOREHOLE CONDITIONS							
FORMATION	Fluid filled	Open hole	Air filled	Cased	Screened	Non-continuous	Good quality bore required	Clear water
Natural gamma	*	*	*	*	*			
Gamma -gamma (density)	*	*	Q	Q				
Neutron- neutron (porosity)	*	*	Q	Q				
Resistivity	*	*						
Induction (conductivity)	*	*		P	P			
Full wave sonic	*	*						
P & S micro seismic/suspension	*	*				*		
Acoustic imager	*	*					*	
Optical imager	*	*	*				*	*
Cross-hole seismcs	*	*		PG		*		
Down-hole seismics	*	*		PG		*		
CONSTRUCTION	*	*						
Verticality (magnetic)	*	*	*	P	P			
Verticality (gyro)	*	*	*	*	*	*		
Caliper	*	*	*					
Video camera	*	*	*	*	*			*
FLUID	*	*						
Flowmeter impellor	*	*			*			
Flowmeter heat-pulse	*	*			*	*		
Fluid temperature	*	*			*			
Electrical conductivity	*	*			*			
Q=qualitative only								
P=plastic only								
PG=plastic grouted								

Figure 3.1 Elements of a geophysical logging system (image courtesy of European Geophysics)

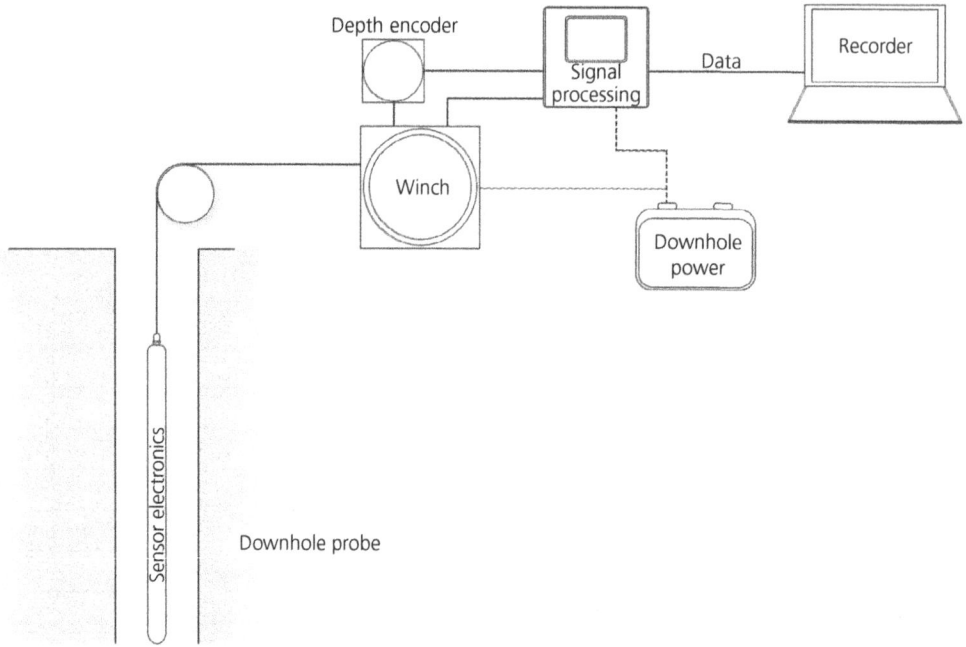

Figure 3.2 Example of a vehicle-mounted logging unit (image courtesy of European Geophysics)

3.2. Formation characterisation

3.2.1 General

The following methods use down-hole sondes to determine various geotechnical parameters and can be used to classify the various strata encountered.

3.2.2 Natural gamma

3.2.2.1 Property measured

The tool measures the naturally occurring gamma radiation found in rocks and sediments. The radiation is a result of the disintegration of the radioactive elements potassium, thorium and uranium present in certain rocks (see Table 3.3 below). The log is mainly used in sedimentary formations to distinguish clays, shales and marls (high gamma) from carbonates and sandstones (low gamma) but may also be used to identify certain minerals.

Table 3.3 Isotope content of common rocks

	Uranium	Thorium	Potassium K40
	g/ton	g/ton	g/ton
Shales/clays	1.2	10.1	324
Clean sand	1.2	6.1	132
Carbonates	1.3	1.1	32

3.2.2.2 Method of measurement

The most common detector of gamma rays in a down-hole logging tool is a combination of a sodium iodide crystal and photomultiplier. The rates of gamma ray emission are statistical in nature, therefore varying degrees of data filtering are required. Where a definitive 'sand' line (clean sand) and 'clay' line (i.e. 100% clay) are deemed to be present, then a percentage clay content may be calculated using the formula below.

$$\text{Volume of clay} = \frac{NG - NG_{min}}{NG_{max} - NG_{min}}$$

where NG : is the log reading at a given point

NG min : is the low value associated with the 'sand' line

NG max : is the high value associated with the 'clay' line

Measurement of gamma radiation at various energy levels allows identification of different isotopes (using a spectral gamma tool) and provides more detail on certain soils and rock types as well as aiding environmental monitoring around industrial sites.

3.2.2.3 Example natural gamma logs

In Figure 3.3, clear and distinct boundaries in the natural gamma response are seen particularly between the clays and the sands and gravel, as well as between the sands and the chalk (having made due allowance for fluid levels, casing and grout). Often in sedimentary sequences, the various formations may have distinct natural gamma responses, which subsequently prove useful in lithological identification and correlation.

Figure 3.3 Natural gamma log – lithology above the chalk – London area (image courtesy of European Geophysics)

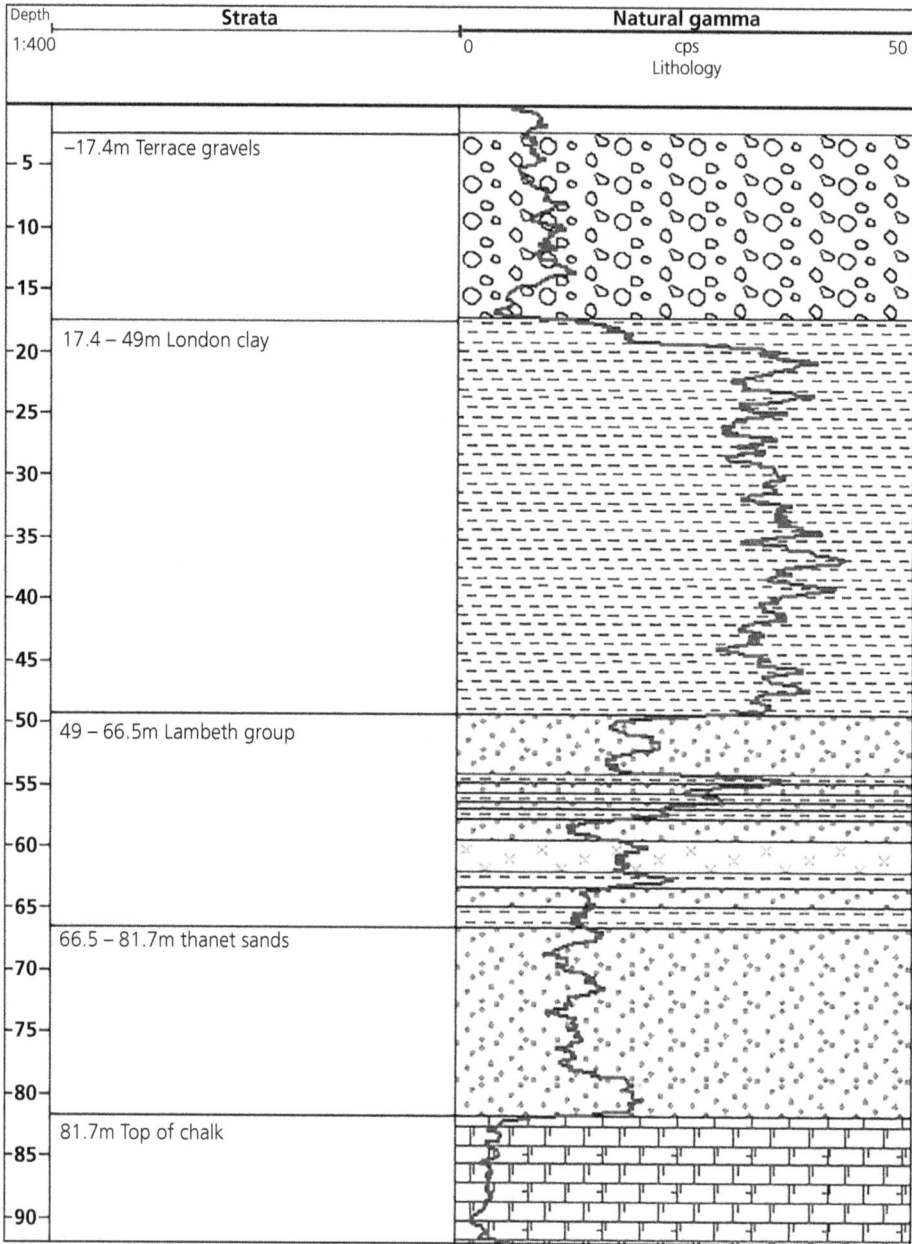

When interpreting natural gamma logs, it is important to use all available information such as the logs from drillers and site geologists as well as any borehole records of neighbouring sites. In fact, the natural gamma data may be used to fine tune the final geological log for more accurate depths (i.e. compensating for poor core recovery or cuttings returns) and aid in the standardisation of formation descriptions.

Where there are distinct natural gamma boundaries, contrasts or signatures, such as is often the case at the top of the chalk, then logs from several boreholes may aid correlation across the site, as illustrated in Figure 3.4.

Figure 3.4 Natural gamma logs – used for correlation (image courtesy of European Geophysics)

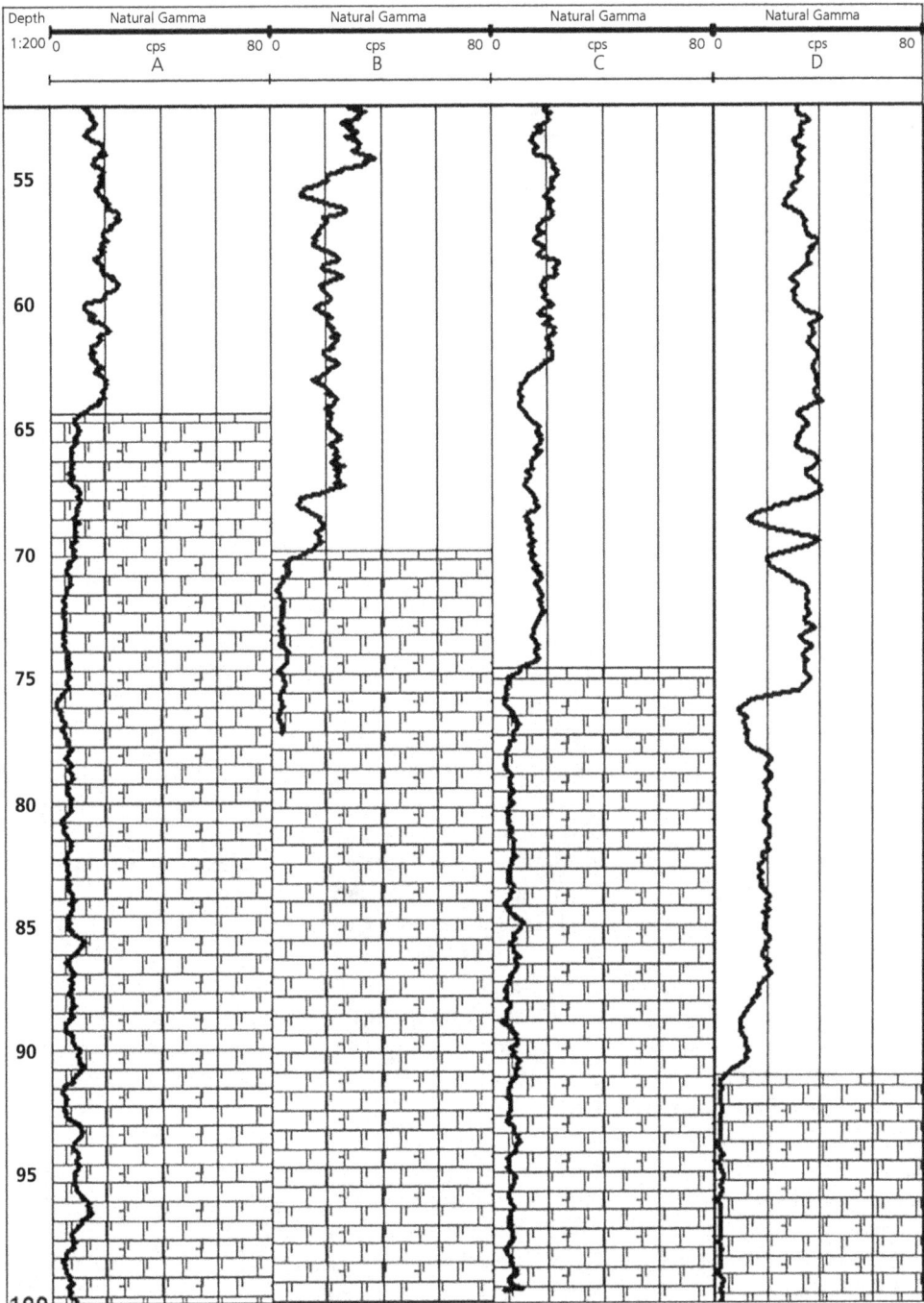

Further illustrations of the usefulness of natural gamma logging are given by Cripps and McCann (2000) and Mortimer (2012).

3.2.3 Gamma-gamma (density)

3.2.3.1 Property measured

Attenuation of back scattered gamma radiation as a function of the electron density of the rock surrounding the borehole. Typical densities of common strata are given in Table 3.4.

Table 3.4 Typical densities of common rocks

	$g.cm^{-3}$
Coal	1.2 - 1.6
Boulder Clay	1.8 - 2.2
Shale	2.2 - 2.7
Sandstone	2.0 - 2.6
Chalk	1.9 - 2.2
Limestone	2.4 - 2.7
Granite	2.5 - 2.7
Basalt	2.9 - 3.0

3.2.3.2 Method of measurement

The gamma-gamma (density) tool usually has at least two detectors at different spacing from a closed (sealed) source of gamma radiation, such as Caesium 137. The response from each detector indicates the apparent density of the material surrounding the tool at a radius of investigation related to each spacing. The long-spaced density (LSD) has typically a spacing of around 0.5 m and the short-spaced density (SSD) (sometimes referred to high-resolution density (HRD)), has a spacing typically of around 0.25 m. The detectors' outputs are initially in counts per second (cps), which are inversely related to the electronic density of material surrounding the tool. Gamma radiation is statistical in nature, therefore varying degrees of data filtering are required. The SSD/HRD has the smaller radius of investigation, up to around 10 cm under average/medium range of densities, and its response is also more affected by any borehole lining. The LSD has the greater radius of investigation, up to 15–20 cms under average conditions, but the least resolution. The tool is run side-walled up the borehole wall, typically in combination with a single-arm caliper and a natural gamma detector at a sufficient distance from the source to avoid its effect. A bulk density log (Db) in $g.cm^{-3}$ can be produced using the SSD log to compensate the LSD for borehole effects and mud invasion, and further corrections for borehole diameter changes may be made using data from the caliper log.

In lined sections of boreholes, the logs give qualitative information on the density of the material behind the linings and the logs may be expressed in terms of apparent density ($g.cm^{-3}$) or left in cps ('gamma- gamma' mode – common practice in cased coal exploration boreholes). Slim-line non-sidewall tools are often used in narrow boreholes, which are cased through worked coal measures/ bad ground or where safe access for the logging tool is problematic.

3.2.3.3 Example logs

The example in Figure 3.5 shows a definitive response over a band of coal: low densities and low natural gamma values. Note the SSD log has a sharper response over the coal – that is, better resolution and is therefore used to determine the bed thickness more accurately. The above-mentioned logs have been laid alongside an optical image by way of illustration.

As another example, Figure 3.6 shows logs run through a steel-cased borehole. In this situation, the density logs are left in gamma-gamma mode (i.e. cps), the peaks of high cps coinciding with low natural gamma values being the coals.

As well as aiding in formation identification and obtaining rock densities, the log may be used for determination of porosity. On the assumption that the formation is clay free, then the porosity (DPOR) may be calculated from the following:

Figure 3.5 Density logs (detailed) – responses over a coal seam (image courtesy of European Geophysics)

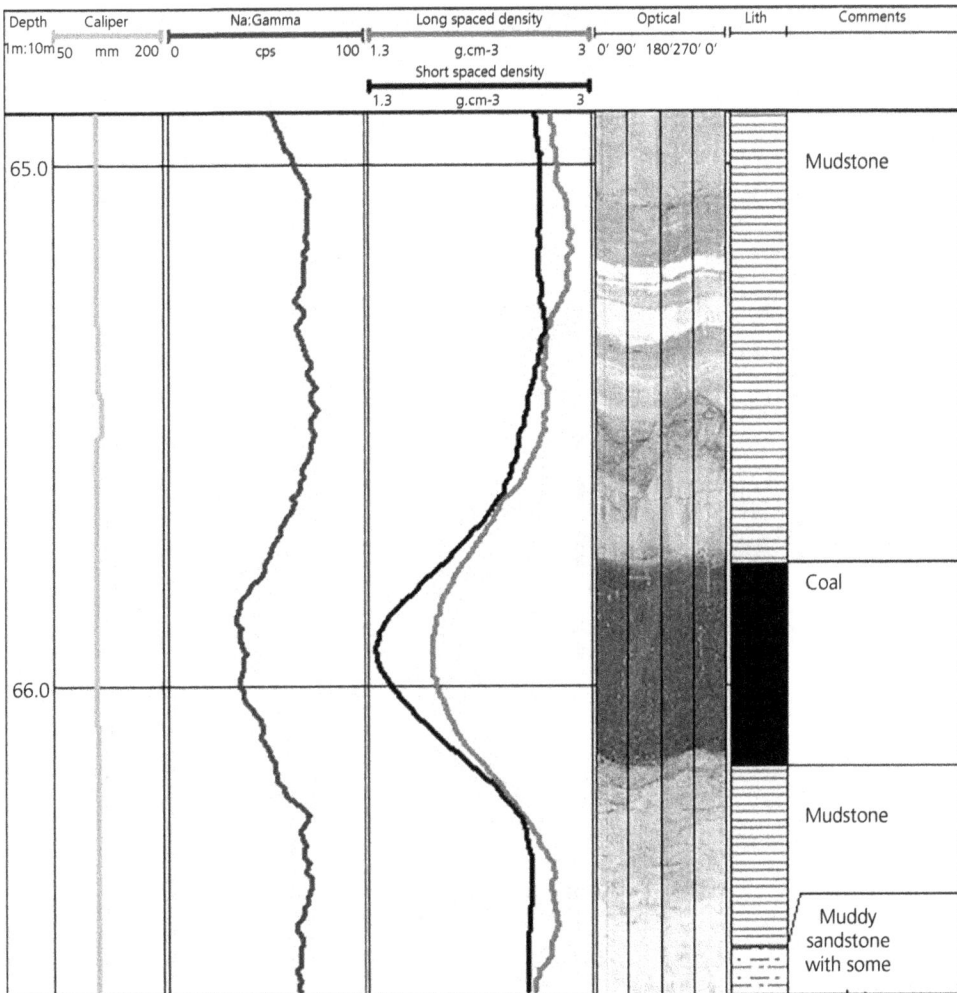

Figure 3.6 Gamma-gamma logs, within a cased borehole through coal measures (image courtesy of European Geophysics)

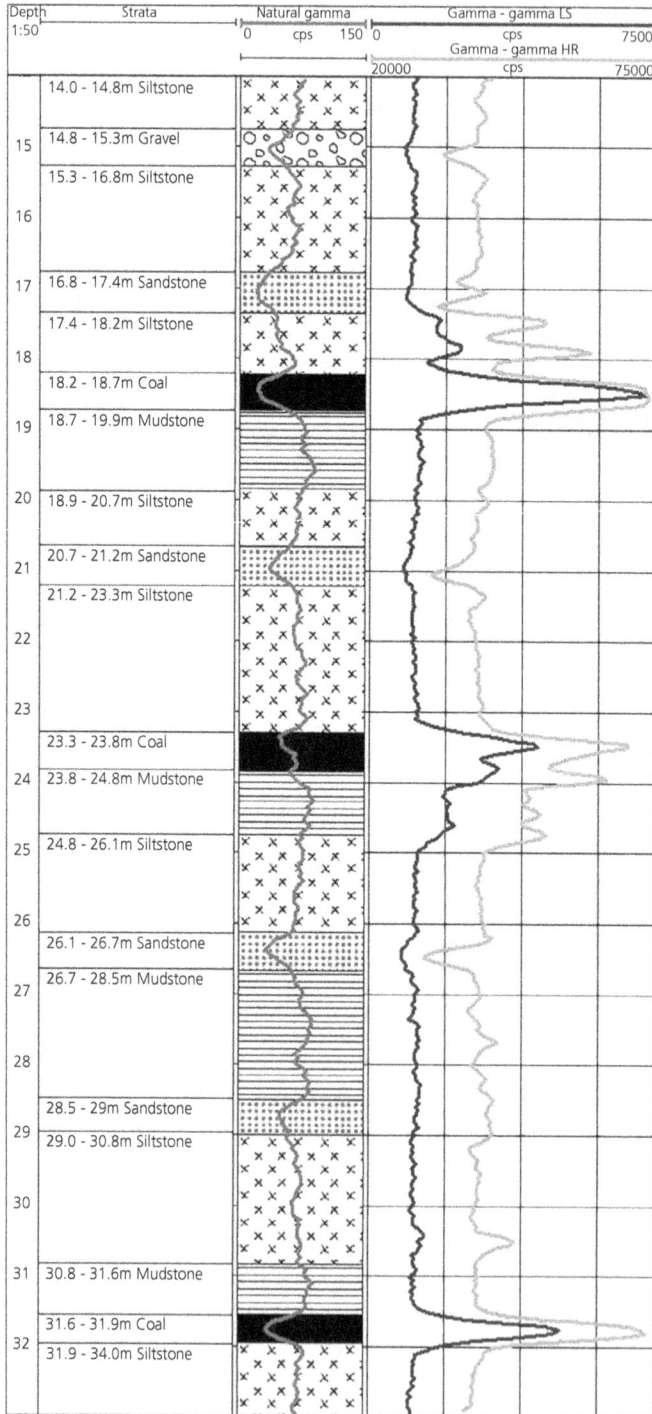

$$DPOR = (Dm - Db)/(Dm - Df)$$

where   Db : density from the log

Dm : sandstone (2.65)

Dm : limestone (2.71)

Df : water (1.00)

3.2.4   Neutron-neutron (porosity)

3.2.4.1   Property measured
Hydrogen content of formations.

3.2.4.2   Method of measurement
The neutron tool has detectors of thermalised neutrons at fixed spacing from a source of fast neu-trons, usually Americium Beryllium. The detectors' outputs are initially in cps, which are inversely related to the hydrogen content within the material surrounding the logging tool, as the presence of hydrogen atoms thermalise (de-energise) the neutrons. Below groundwater level, the pores in the rock will be filled with water and hence the neutron log can be used (in unlined hole conditions) to determine porosity. As the neutron log responds to hydrogen it also responds to water that is chemi-cally bound within the formation as in clays and certain minerals (e.g. hydrocarbons).

Where the borehole is lined or air-filled, it is possible to use the log in a qualitative way to identify the differing zones of relative hydrogen content. However, it should be noted that the presence of casings/grout will hold the tool away from the formation. Plastic lining contains hydrogen so the log can be affected by differing thicknesses of the lining or other construction features.

3.2.4.3   Example logs
In Figure 3.7, a comparison between neutron porosity (NPOR) and density derived porosity (DPOR) is made. Note that in the clay free sand zones both logs are very similar. However, in the

Figure 3.7 Neutron porosity log (NPOR) compared with density derived porosity log (DPOR) (image courtesy of European Geophysics)

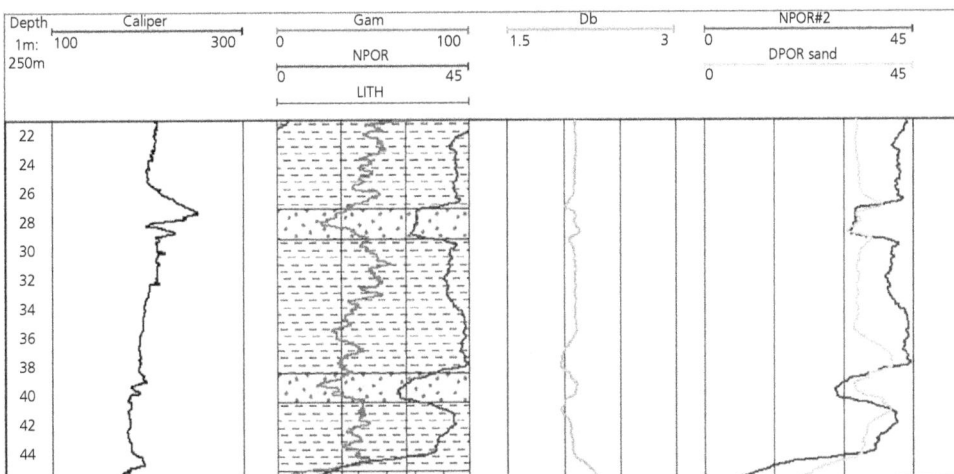

clay intervals the neutron log gives a higher apparent porosity owing to the bound water associated with clays.

3.2.5   Magnetic resonance

3.2.5.1   Property measured
Total porosity independent of lithology, with differentiation of bound and free water.

3.2.5.2   Method of measurement
Application of a strong magnetic field aligns the magnetic moment of protons (hydrogen nuclei). When this field is removed, they precess in the Earth's magnetic field and gradually return to their original state. Proton precession in free fluid produces a radio frequency signal, the amplitude of which is related to the amount of free fluid. The rate of decay of the precession signal depends on the interactions with neighbouring atoms and so the molecule of which the proton is a part. Analysis of the data can also produce information on bound water, grain size distribution and calculation of permeability. Applications of this method are more fully described by Kreici *et al.* (2018) and Dlubac *et al.* (2013).

3.2.5.3   Example log
An example of a typical magnetic resonance log is shown in Figure 3.8.

Figure 3.8 Example of results from magnetic resonance logging (image courtesy of European Geophysics)

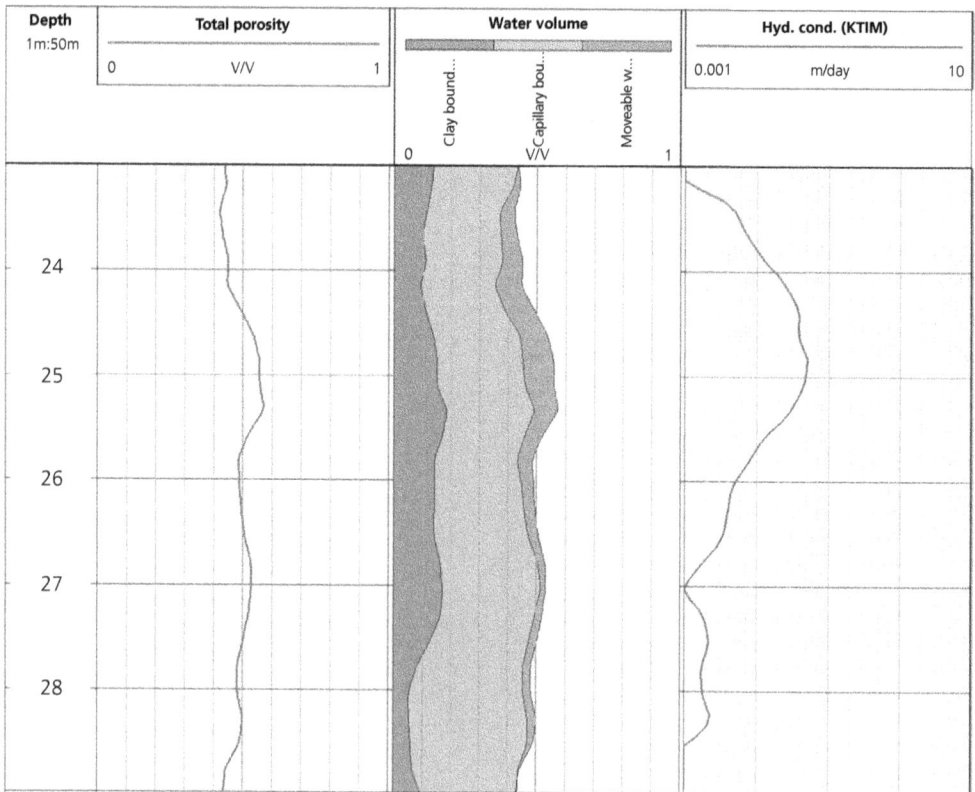

3.2.6   Magnetic susceptibility

3.2.6.1   Property measured

Magnetic susceptibility is a quantitative measure of how much a material may be magnetised in relation to an applied magnetic field.

3.2.6.2   Method of measurement

The tool uses electromagnetic techniques with a variety of coil arrangements.

3.2.6.3   Example log

In Figure 3.9, several thin ironstone bands have been identified clearly by the peaks in the susceptibility log. On the gamma-gamma logs (long-spaced and high-resolution densities), the responses are not as clear, as the gamma-gamma method is not as responsive at high densities, although very good at lower values as over the coal band.

Figure 3.9 Magnetic susceptibility and other logs in a borehole that penetrates iron-stone bands associated with coal measures (image courtesy of European Geophysics)

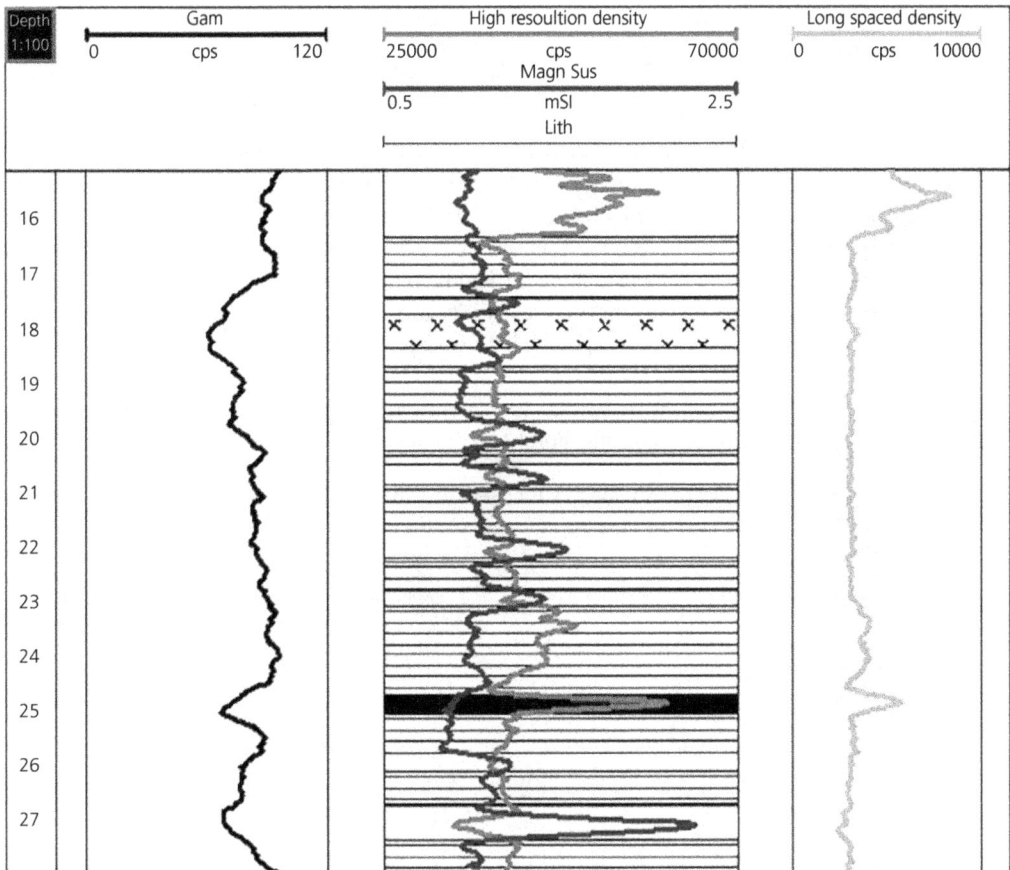

3.2.7   Resistivity

3.2.7.1   Property measured

The electrical resistivity of the formation, which is a function of mineralogy, porosity and pore-water salinity. Figure 3.10 illustrates the wide range of values encountered in common rocks.

Figure 3.10 Resistivities of common rocks (image courtesy of European Geophysics)

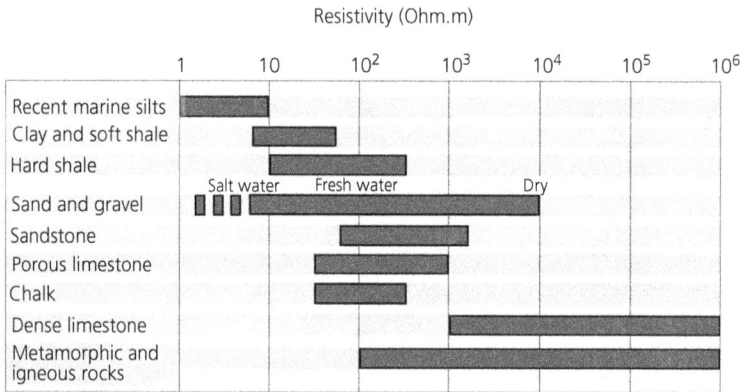

3.2.7.2   Method of measurement

Resistivity tools use a combination of electrodes to measure electrical resistivity of the ground. Early arrays were the 'Normal' configuration, with current/voltage electrode spacing of 0.4 m and 1.6 m being popular. This simple tool had several disadvantages in certain borehole/formation conditions and is now often replaced by a 'focused' tool that directs more of the electrical current into the formation. These tools give excellent vertical resolution and good penetration, especially in high-resistivity formations where a normal resistivity tool would not be as effective. Most tools have two electrode spacings to allow deep and shallow depths of investigation.

A comparison of the two types of resistivity tool mentioned above, run in boreholes that penetrated identical geology in neighbouring boreholes, is shown in Figure 3.11. The focused logs

Figure 3.11 1.6 m normal and long-spaced focused resistivity logs (image courtesy of European Geophysics)

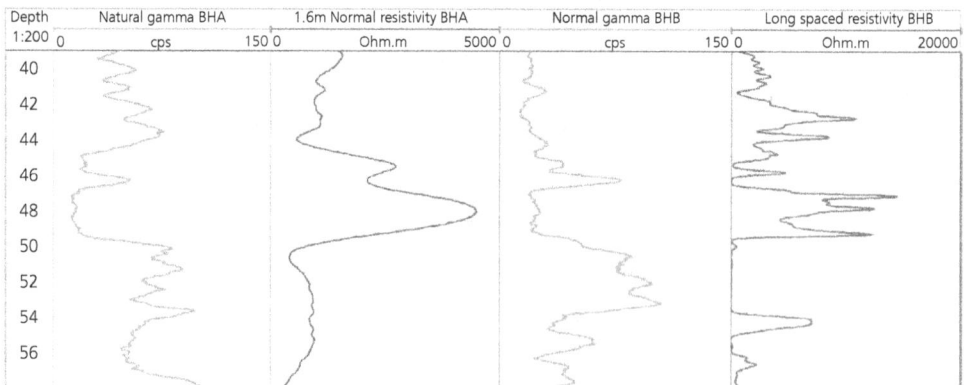

have greater detail and a sharper response, with higher values in the limestones (truer formation resistivity) compared with the rounder, smoother response of the normal resistivity logs.

3.2.7.3 Example logs

Resistivity logs may be used to identify flints and harder bands, although they are not as sensitive as an imager. The example in Figure 3.12 shows the peaks in the resistivity values at significant bands of flint within chalk. The logs have been plotted alongside an optical image by way of illustration.

Figure 3.12 Detection of flint bands within the chalk (image courtesy of European Geophysics)

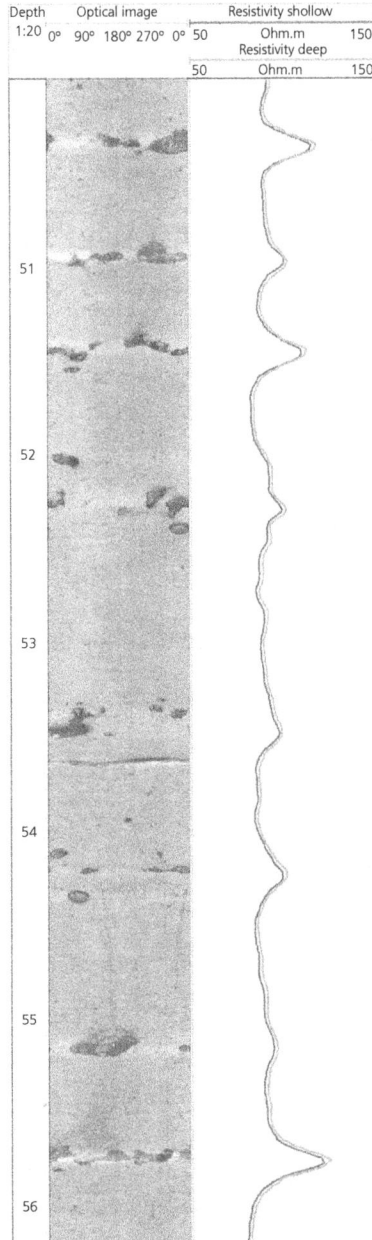

3.2.8   Conductivity (induction)

3.2.8.1   Property measured

Electrical conductivity (EC) (reciprocal of resistivity) of the formation, which is a function of mineralogy, porosity and pore-water salinity. The measurement is suited to low resistivity environments, typically less than 100–200 Ohm.m.

3.2.8.2   Method of measurement

The induction tool generates an electromagnetic field in the vicinity of the borehole and measures the response of the formations to this applied field from which the EC is determined, normally at two different coil spacings (deep and shallow). This technique has an advantage over resistivity logs in that it works through plastic casing.

3.2.8.3   Example logs

In Figure 3.13, the induction logs have a sharper response to features compared with the resistivity logs. This is attributed to the shorter spacing used in the particular induction tool.

Figure 3.13 Induction conductivity logs – comparisons with natural gamma and normal resistivity (image courtesy of European Geophysics)

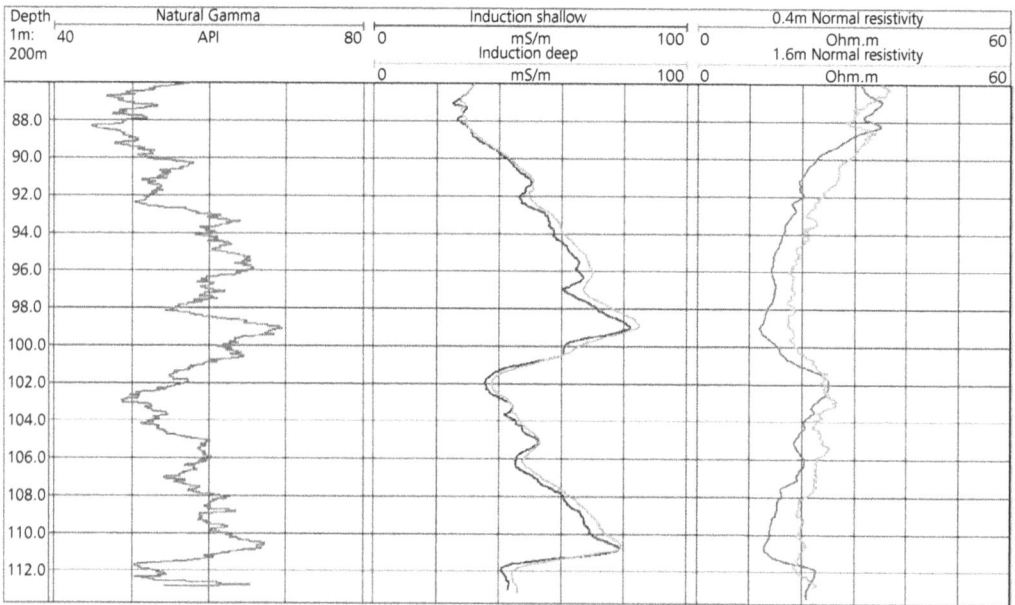

3.2.9   Sonic/seismic

3.2.9.1   Property measured

The seismic (sonic) velocities (Compressional Waves, P, and Shear Waves, S) of rock, and thus Poisson's ratio and other elastic moduli such as Bulk Modulus and Shear Modulus, may be calculated where density information is available. Figure 3.14 shows the range of compressional wave velocities encountered in common rocks and gives a guide to their hardness and rippability.

Figure 3.14 Range of compressional wave velocities in rocks (image courtesy of European Geophysics)

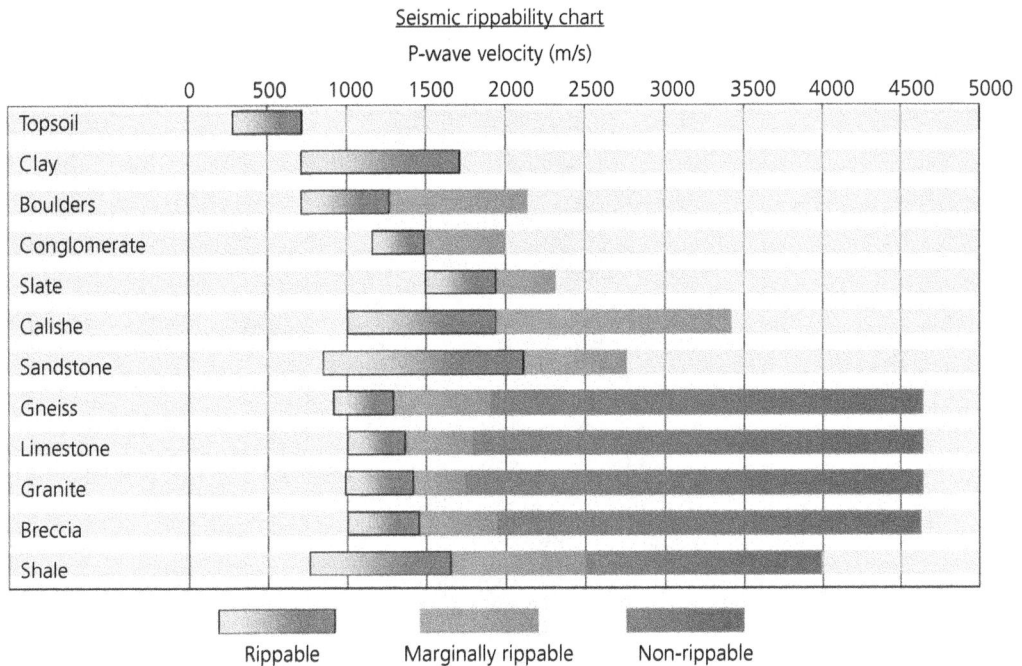

Seismic rippability chart

P-wave velocity (m/s)

3.2.9.2   Method of measurement

FULL-WAVE – RECORDING CONTINUOUSLY WITH DEPTH

This sonic tool provides a full waveform recording of sonic signals and uses fixed spaced transmitter – receivers. The time data are processed for Compressional Wave (P) velocity (or transit time). Estimates of Shear Wave (S) velocity may only be obtained under suitable conditions where transmodal conversion (P to S to P) occurs. They are waves that have travelled across the borehole fluid to the rock as P-waves and have undergone P to S conversion. Shear waves that refract at the fluid/rock boundary at the S-wave critical angle travel through the rock at V_s and if conversion back to P wave occurs, the waves can be received by the tool. These waves are normally identified by higher amplitudes and phase changes after the P wave arrivals (see Figure 3.16).

Results can be affected by the competency of the rock material, low velocity zones, irregular boundary conditions and complex interactions of non-direct P-waves and other fast waves. This last factor can be the main limiter on shear wave identification using this type of tool. By using a combination of two or more receivers, compensation for borehole diameter and fluid properties is obtained.

EXAMPLE LOGS The example in Figure 3.15 shows a full waveform recording in a variable density format where the higher amplitude waves are coloured red, magenta, yellow, blue, and the low amplitudes are green and turquoise. Initially, the travel times for each wave type are picked and

Figure 3.15 Seismic – full waveform (image courtesy of European Geophysics)

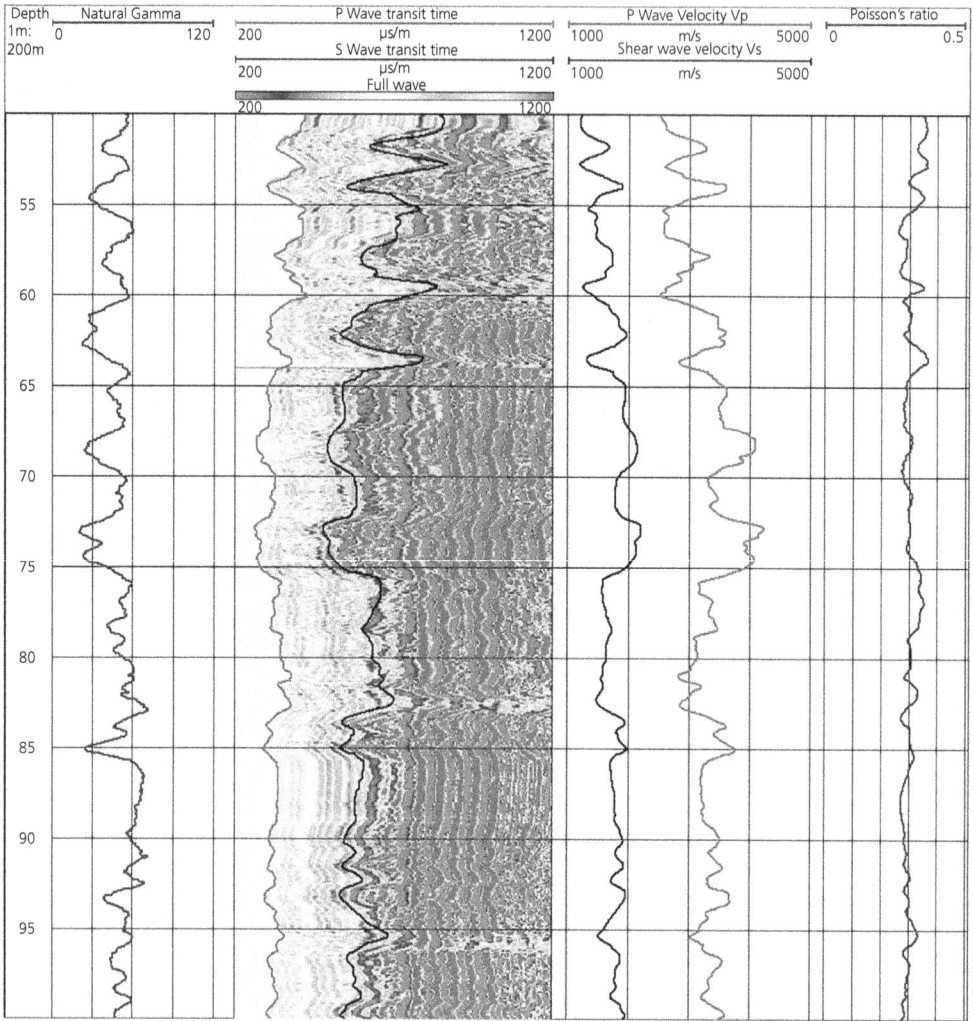

converted to velocities. Poisson's ratio and various moduli in terms of velocities and density may then be calculated using the following formulae:

$$\text{Poisson's Ratio } \sigma = \frac{0.5 * (Vp/Vs)^2 - 1}{(Vp/Vs)^2 - 1}$$

$$\text{Shear Modulus } \mu = D * Vs^2$$

$$\text{Bulk Modulus } K = D * Vp^2 - \frac{4Vs^2}{3}$$

$$\text{Young's Modulus } E = \frac{9 * K * D * Vs^2}{3K + D * Vs^2}$$

where:

D = density

Vp = Compressional Wave Velocity

Vs = Shear Wave Velocity

A SUSPENDED OR MICRO-SEISMIC ARRAY – AT SELECTED POINTS DOWN THE BOREHOLE

The suspended or micro-seismic array has been designed to obtain improved shear wave velocities from within a single borehole, particularly in shallower lower velocity formations (Kaneko *et al.*, 1990). The device consists of multiple detectors specifically chosen for their wave type response characteristics – for example, piezo-electric for P wave, geophone/hydrophone for S wave – and the down-hole transmitter of seismic energy is orientated perpendicular to the borehole wall for maximum effect. Longer spacing between transmitter and receivers are used (typically around 2 m and 3 m) to enable better detection of slower velocities. An example of a recorded waveform is given in Figure 3.16. The tool is therefore several metres in length and due consideration should be made with respect to required data points for a given borehole depth and / or casing position.

The example in Figure 3.17 shows a comparison between velocity data derived from a conventional full-wave sonic tool and that obtained from a suspended microseismic tool. The P wave velocities (Vp) from both tools overlay well, although the full wave sonic (FWS) Vp log has far more detail, being effectively a continuous recording against depth (typically at 100–200 mm intervals, compared with every 2–5 m on the suspended microseismic tool).

Also, in this example, sonic data were obtained in the borehole both prior to and after the installation of grouted plastic casing. In this case, identical P wave data were obtained. Where there is reasonably competent formation – that is, P wave velocities are generally greater than 2000 m/s – then logging through a grouted plastic lining has proved possible and sonic logs under either borehole conditions have proved useful to seismologists running cross-hole and up/down-hole surveys.

Figure 3.16 An example of P and S waveforms recorded by a microseismic/suspended tool (image courtesy of European Geophysics)

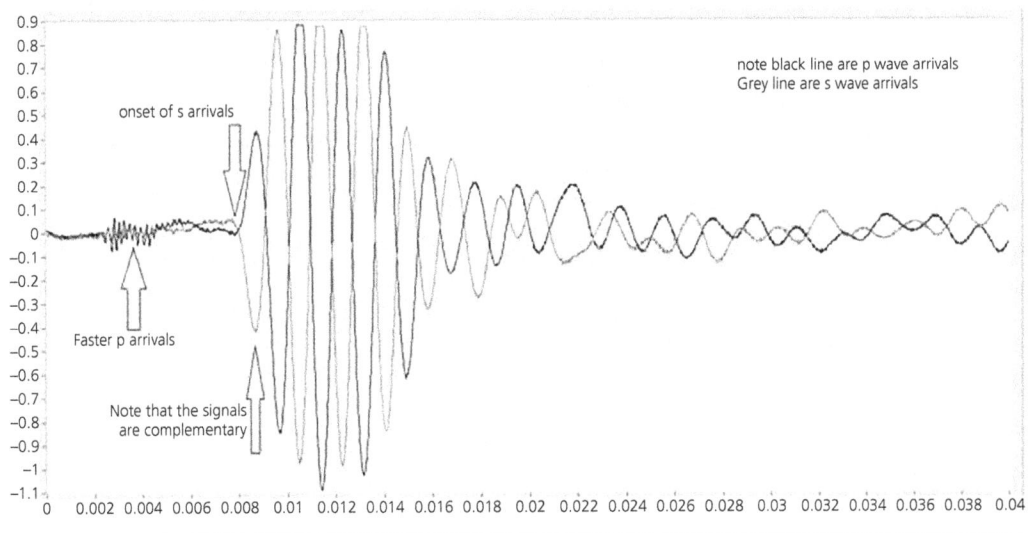

Figure 3.17 Suspended/micro-seismic results alongside full wave sonic (FWS) logs **P (-o—o-)** and **S (-o—o-)** waves in open hole using micro seismic/suspended tool (image courtesy of European Geophysics)

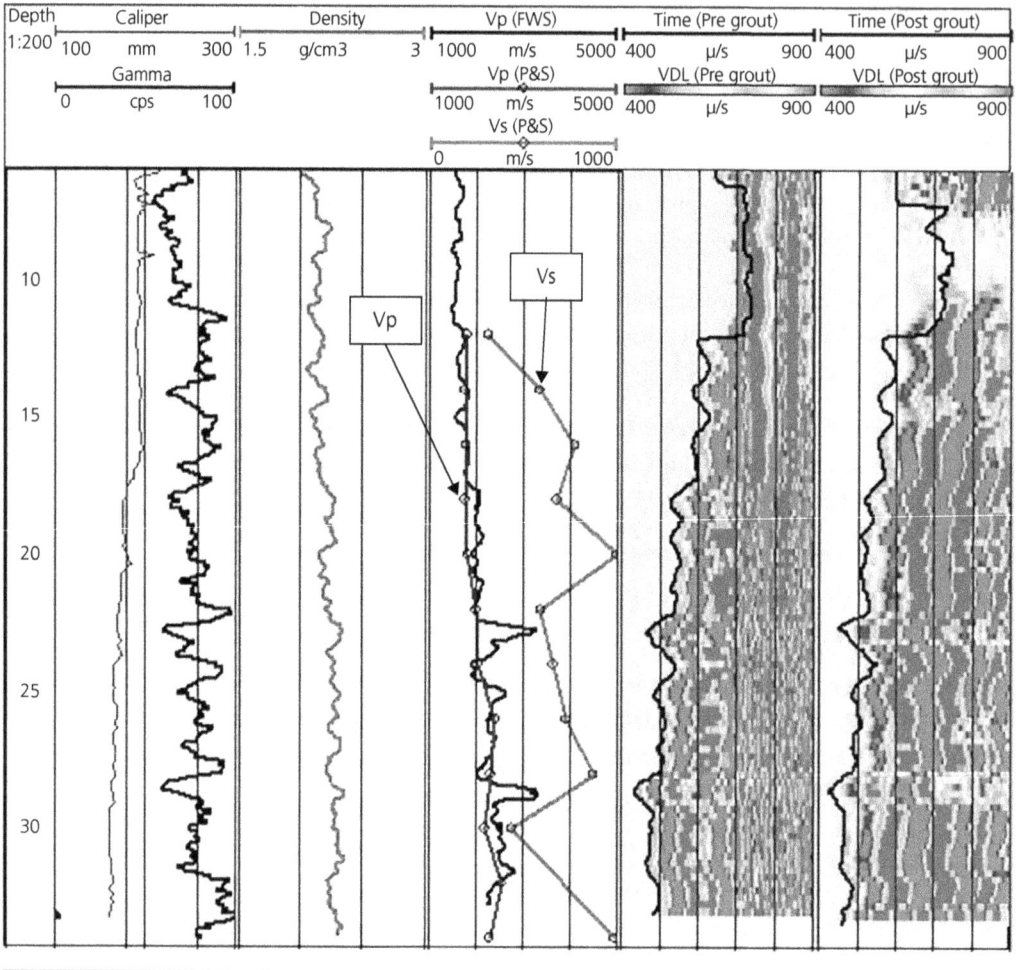

Although not a wire-line technique as such, the following methods are included here for the sake of total awareness of seismic measurements in geotechnical boreholes. These methods are further discussed in Section 3.6.

DOWN-HOLE SEISMICS (SEE ALSO SECTION 3.6.2 UP-HOLE SEISMICS) – MEASUREMENTS AT SELECTED POINTS DOWN THE BOREHOLE

In the down-hole method, seismic sources of compressional (P) and shear (S) waves are deployed on the surface of the ground near to the borehole and a down-hole geophone clamped to the bore-hole wall at selected intervals down the borehole.

The up-hole method is effectively the reverse of the above, suitable sources being deployed at selected intervals down the borehole with a suitable geophone placed on the ground surface.

This involves placing seismic sources of compressional (P) and shear (S) waves in one borehole and geophones in one or two neighbouring boreholes that are typically up to 1.5–3 m apart. The sources and geophones are placed at regular depth intervals and the P and S velocities obtained for the formation between the boreholes.

Boreholes drilled for such methods generally have grouted plastic lining installed for ease and surety of down-hole source and geophone clamping to ensure good seismic coupling through to the formations.

3.2.10   Acoustic imager (televiewer)

3.2.10.1   Property measured
Acoustic reflection from the borehole wall.

3.2.10.2   Method of measurement
This tool scans the borehole wall through 360° at a very fine depth sampling interval (typically 1–2 mm, hence the high resolution) and records the acoustic reflection of the resulting signal in terms of amplitude and transit time (the travel time from the tool to and from the borehole wall). This sensitive technique responds to small diameter changes, rugosity and the acoustic nature of the borehole wall. An orientation system is incorporated in the tool allowing resultant images to be orientated to Magnetic North. The 'unwrapped' image of the borehole wall is displayed from 0° through 90°, 180° and 270° back to 0°, examples of which are shown in Figure 3.18. The orientation system employs a flux gate magnetometer and therefore data near to steel casing are unorientated.

EXAMPLE LOGS
It is possible to obtain images through thin plastic casing under certain conditions, although the images may not be as good as in the open hole being dependent upon whether the casing is sufficiently central-ised within the borehole, and the nature and size of the annulus between the casing and formation. An example of such an obtained log is shown in Figure 3.18. Additionally, there is the option to obtain an electronic caliper from the acoustic signal, an example of which is shown Figure 3.22.

3.2.11   Optical imager

3.2.11.1   Property measured
High-resolution continuous optical image of the borehole wall.

3.2.11.2   Method of measurement
A scanned high-resolution optical image of the borehole wall is captured by a high-resolution digi-tal camera at a very fine depth interval, typically 1 mm. An orientation system is incorporated in the tool, allowing resultant images to be orientated to Magnetic North. The 'unwrapped' image of the borehole wall is displayed from 0° through 90°, 180° and 270° back to 0°, examples of which are shown in Figure 3.19. The orientation system employs a flux gate magnetometer and therefore data near to steel casing are unorientated.

EXAMPLE LOGS
Both the acoustic and optical imagers give mainly similar information but, as illustrated in the example in Figure 3.20, some finer features are detected by the acoustic imager at around 27.5 m, 28.6 m and between 29–30 m depth, as the acoustic imager responds to the hardness of the formation.

Figure 3.18 Acoustic images – unwrapped (images courtesy of European Geophysics)

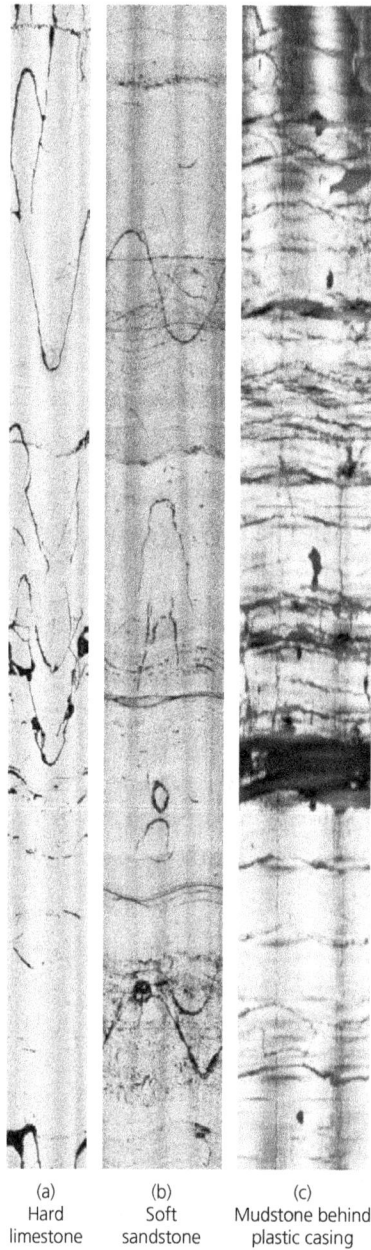

(a)
Hard
limestone

(b)
Soft
sandstone

(c)
Mudstone behind
plastic casing

From both the optical and acoustic imager logs structural features such as fractures, bedding planes, and so forth, are picked and corrected to true orientation, the results of which may be displayed in a variety of forms such as tadpole plots, stereo nets and rose diagrams (see example in Figure 3.21). The structural data may also be formatted for ready import into other geotechnical or engineering software.

Figure 3.19 Optical images – unwrapped (images courtesy of European Geophysics)

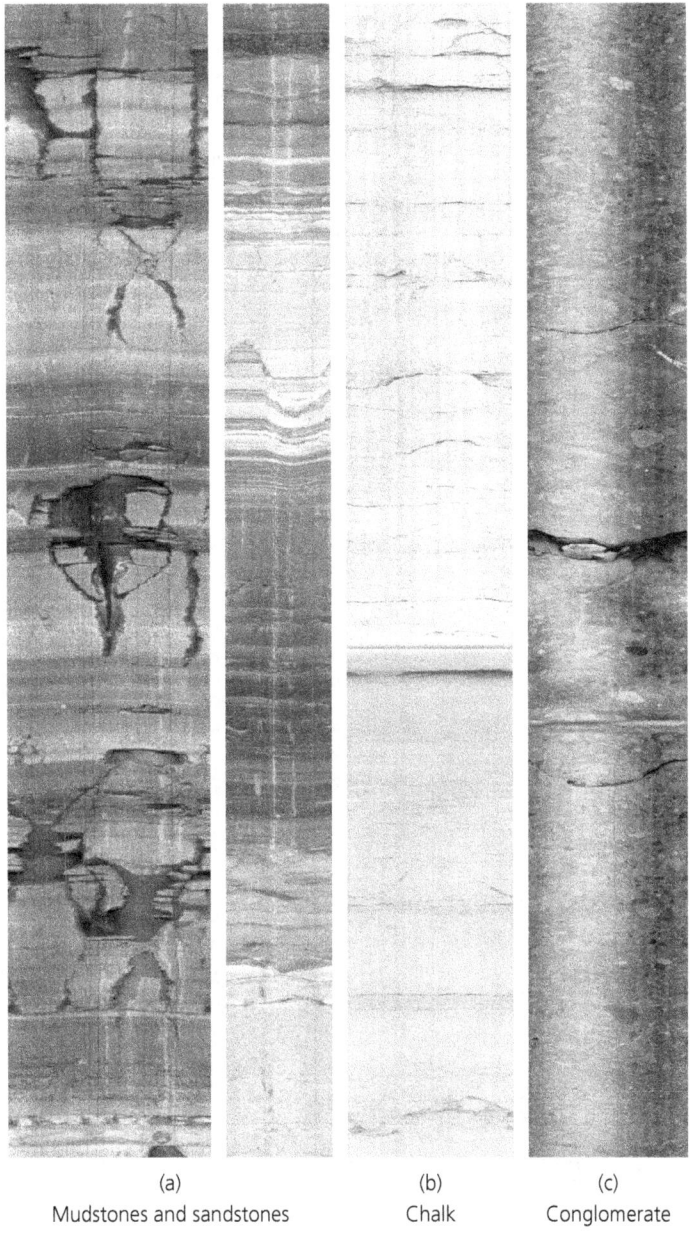

(a)
Mudstones and sandstones

(b)
Chalk

(c)
Conglomerate

Figure 3.20 Comparison of acoustic and optical images. Some features are seen on the acoustic and not on the optical (images courtesy of European Geophysics)

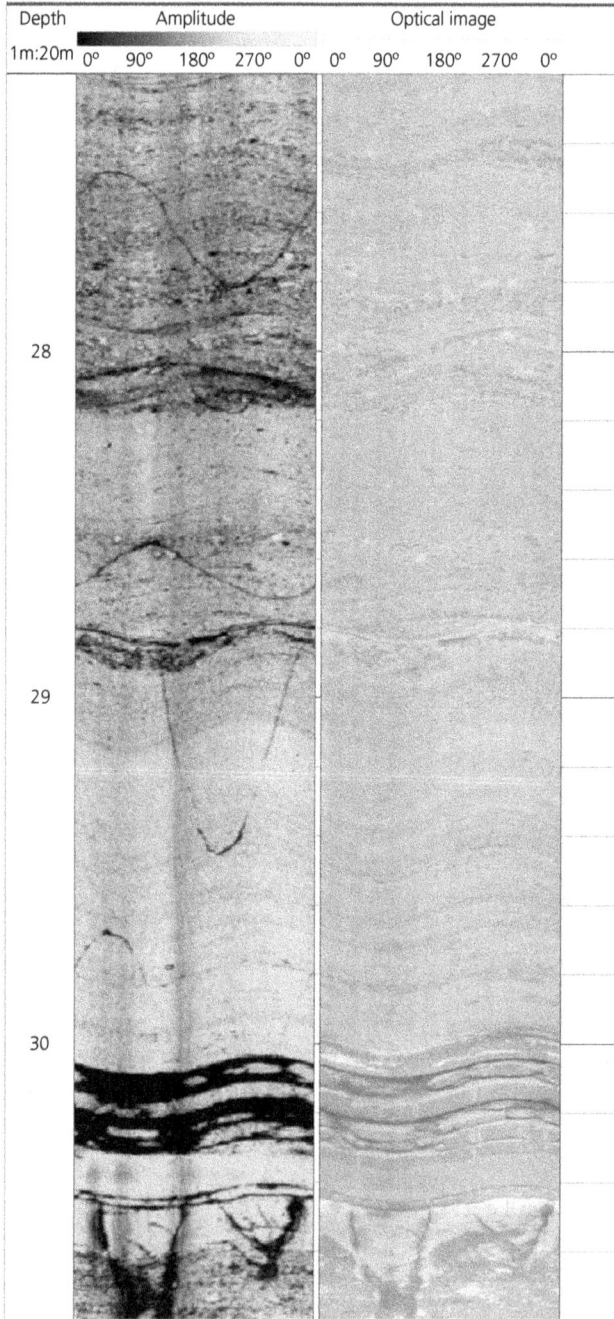

Figure 3.21 Example: processed and interpreted acoustic imager data (images courtesy of European Geophysics)

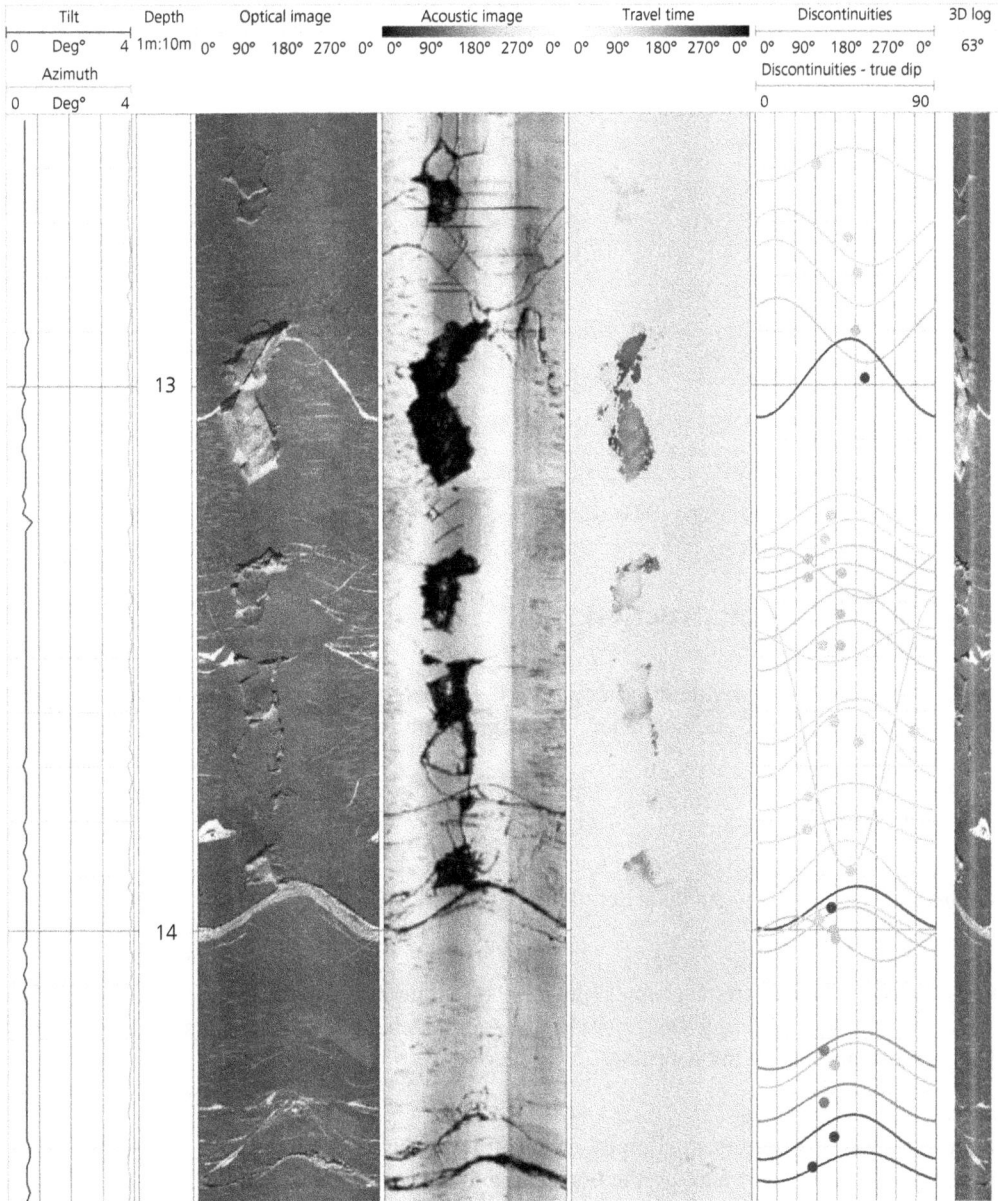

Figure 3.21 Continued

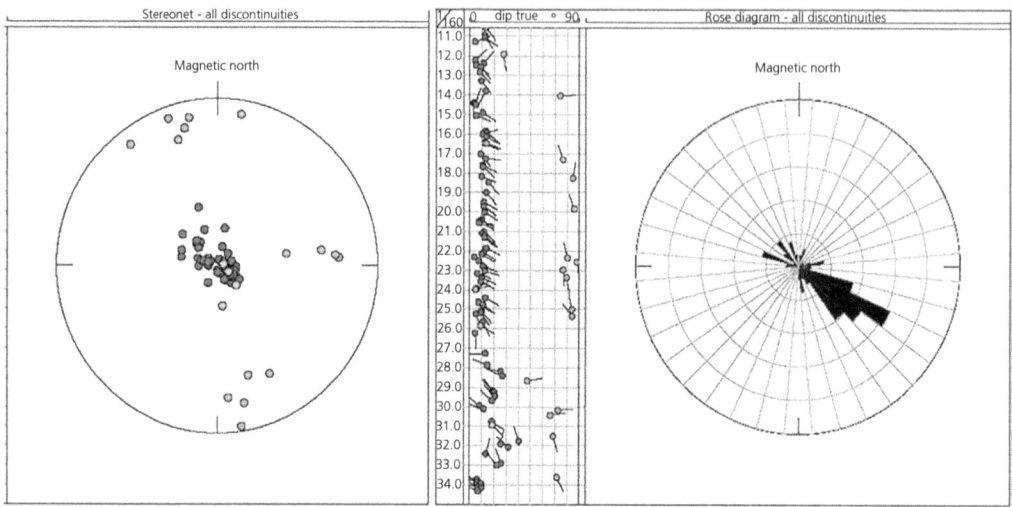

Note that the image on the right-hand side is a virtual core.

3.3. Borehole construction

3.3.1 General

The following methods provide information on the borehole and its construction, recording the width of the borehole and its inclination. These properties are often a function of the materials being drilled.

3.3.2 Caliper

3.3.2.1 Property measured

Continuous record of the mean diameter of boreholes.

3.3.2.2 Method of measurement

The majority of calipers are mechanical devices consisting of one or more arms that are pressed against the borehole wall; any movement in these arms being transduced to a suitable electronic signal for transmission up the wire-line.

EXAMPLE LOGS

Caliper logs have been presented on other examples, such as in Figures 3.7 and 3.17. Certain features of the rock penetrated by the borehole may be ascertained from the log such as identifying zones of washout, cavitation, breakout and fissuring. Also, an electronic caliper can be obtained from acoustic imager scanners at selected depths, an example of which is shown in Figure 3.22.

Caliper data may be used to provide control and correction data for other logs as well as calculating borehole volumes to aid gravel pack and grout requirements.

Figure 3.22 Acoustic imager data with derived caliper section (images courtesy of European Geophysics)

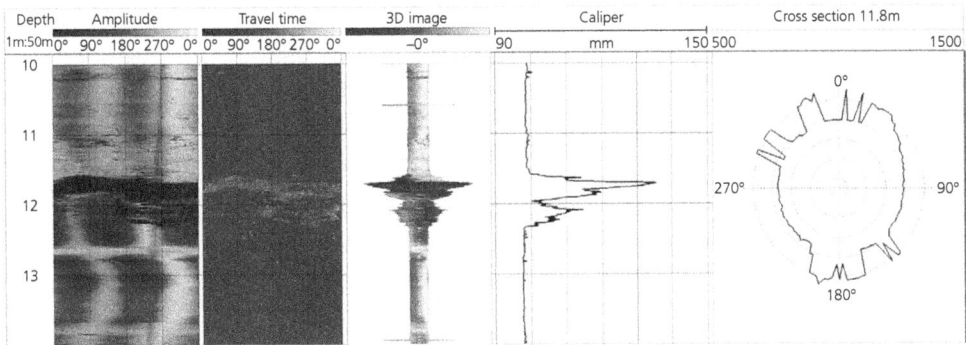

3.3.3   Borehole trajectory

3.3.3.1   Property measured

Inclination and direction of the borehole.

3.3.3.2   Methods of measurement

MAGNETIC DEVICE

Uses a combination of magnetometers and inclinometers.

GYROSCOPIC DEVICE

A North-seeking gyro provides full multishot orientation and deviation surveys of boreholes in any environment, including steel casing and drill pipe. It incorporates a gyro that is unaffected by steel or areas of magnetic anomalies, unlike the above-mentioned magnetic device.

The position of the borehole at any depth with respect to the log origin may be calculated in terms of X (West–East) and Y (South–North) co-ordinates with the z axis showing the true depth from origin. Readings are taken while the tool is stationary at selected positions down the borehole.

3.4.   Fluid properties

3.4.1   Fluid temperature

3.4.1.1   Property measured

Temperature of the borehole fluid. There is a natural geothermal gradient of increasing temperature with depth. This gradient varies with the thermal conductivity of the geological formation and is modified by water flowing in, out or vertically though the borehole.

3.4.4.2   Method of measurement

A sensitive temperature sensor located at the down-hole end of the logging tool so that it encounters the undisturbed fluid on a down run. This parameter is often run in combination with other fluid logs – for example, electrical conductivity.

Figure 3.23 Fluid logs in a borehole penetrating a fractured limestone (images courtesy of European Geophysics)

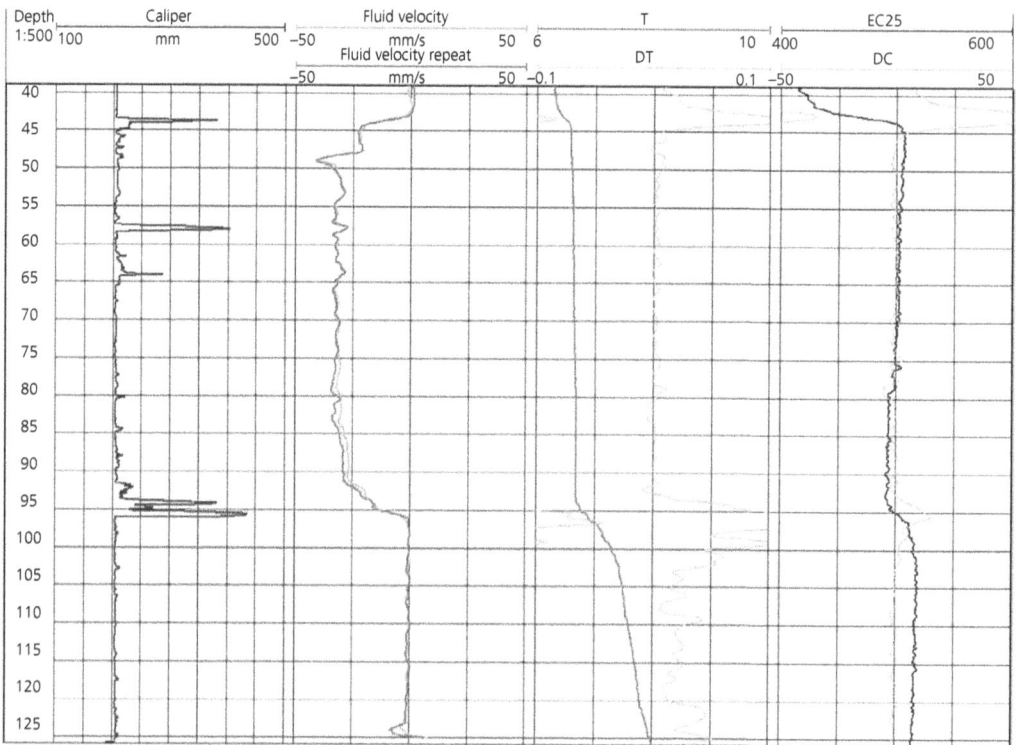

3.4.2 Fluid electrical conductivity (EC)

3.4.2.1 Property measured
EC of the borehole water. The EC of the water is related to its salinity and dissolved solids, and it is therefore a measure of the ionic quality of the borehole water. The shape of the log trace can indicate zones of inflow.

3.4.2.2 Method of measurement
A conductivity cell comprising several electrodes is normally fitted at the down-hole end of the logging tool, often in combination with a temperature sensor so that it encounters the undisturbed fluid on a down run. Using data from the temperature log, the EC is corrected to a standard temperature.

3.4.2.3 Example logs
In Figure 3.23, the steps/gradient changes in the fluid logs identify points of flow through fractures in a limestone. Towards the base of the borehole, the temperature increases with increased depth (a geothermal gradient) and suggests a lack of or little vertical fluid movement.

3.4.3 Flow

3.4.3.1 Property measured
Fluid velocity within the borehole from which, when combined with borehole diameter information, flow may be calculated.

3.4.3.2 Method of measurement
IMPELLOR METHOD

The tool uses an impeller and is normally run at a constant logging speed against the anticipated flow for the best response. The data are corrected for logging speed and a fluid velocity log is produced. Flow in l/s may then be derived from the fluid velocity and caliper (borehole diameter) data. This tool has a velocity threshold of typically 20–30 mm/s owing to friction effects of bearings, and therefore is not responsive in low flow situations. Best results are obtained in combination with pumping, where practicable.

EXAMPLE LOGS A fluid velocity log is depicted in Figure 3.23, showing water entering the borehole through fissures at 44 m and 48 m moving down and out through fissures at 94 m to 96 m.

SPECIAL CONSIDERATIONS The presence of suspended material in the borehole fluid can affect the impellor function.

HEAT-PULSE METHOD

This tool consists of two very sensitive temperature sensors that are positioned above and below a small heating element. The tool is positioned at a particular depth and left for a few minutes for the temperature sensors to stabilise. A heat pulse is then generated, and then the temperature sensors monitor for its movement. The time taken for the heat pulse to reach a sensor is recorded and is used to calculate the velocity of the fluid movement. Several readings are often taken at each position, suitably spaced to avoid convection loops being set up, and an average taken.

SPECIAL CONSIDERATIONS This technique can be very sensitive to logging line cable movement and thermal effects within the borehole column.

3.4.4 Fluid sampling

3.4.4.1 Property measured
Collection of a fluid sample from a given depth for laboratory analysis, environmental studies, groundwater quality monitoring run in conjunction with a fluid log.

3.4.4.2 Method of measurement
Although there are a number of sampling techniques such as pumped systems and disposable bailers available to run on a wire-line logging system, specially constructed tools, which have sealed chambers with activated valves to allow fluid access and containment, are also available. These sampling tools are lowered to a specific interval within the borehole and a valve is opened. Once the sample chamber is filled, the valve is closed, and the tool is brought to the surface. At the surface a valve is opened, and the sample drained into a suitable container. To optimise the choice of sample points, fluid logs are run prior to sampling to identify flow points and water quality changes. For example, sample points are often taken either side of a major flow point.

3.4.5 Measuring while drilling (drilling parameter recording)
Commonly known as drilling parameter recording (DPR), this process can be extremely useful to provide additional data regarding the drilling performance as a rotary drilled borehole is progressed. The technique is covered by BS EN ISO 22476-15 (BSI, 2016): 'Measuring while drilling'.

When drilling, three categories of parameters will influence the drilling operation. There are parameters that are imposed by the method of drilling adopted. These are primarily related to the type of rig, its mechanical condition and the performance limits these impose.

Some parameters are 'set' by the operator – for example, downward thrust pressure, rotation speed, and flush pressure.

There are also parameters that are essentially responses owing to the interaction of the rig with the ground, such as rate of penetration, torque, restraining (holdback) pressure and injection pressure.

The less intervention from the operator the more the data obtained will directly reflect the ground response. The recording of the drilling data in ground investigations is a relatively recent development. It has shown, however, that recording information regarding the depth and rate of penetration, the flush pressure, fluid return, drill bit pressure, rotation speed and torque can add significant value to the drilling process – as shown in Figure 3.24. Typically, the results have been used to provide accurate determination of soil boundaries, particularly where core recovery has been very poor or is generally difficult – for example, in strata such as sand channel deposits or other granular soils.

The results of the data are usually compiled into a series of graphs to provide a visualisation of the borehole performance. The data can then be compared with the borehole log to confirm important features, as shown in Figure 3.25. This shows a DPR record through the London Clay Formation

Figure 3.24 Results of DPR data clearly showing changes of strata (image courtesy of SOCOTEC)

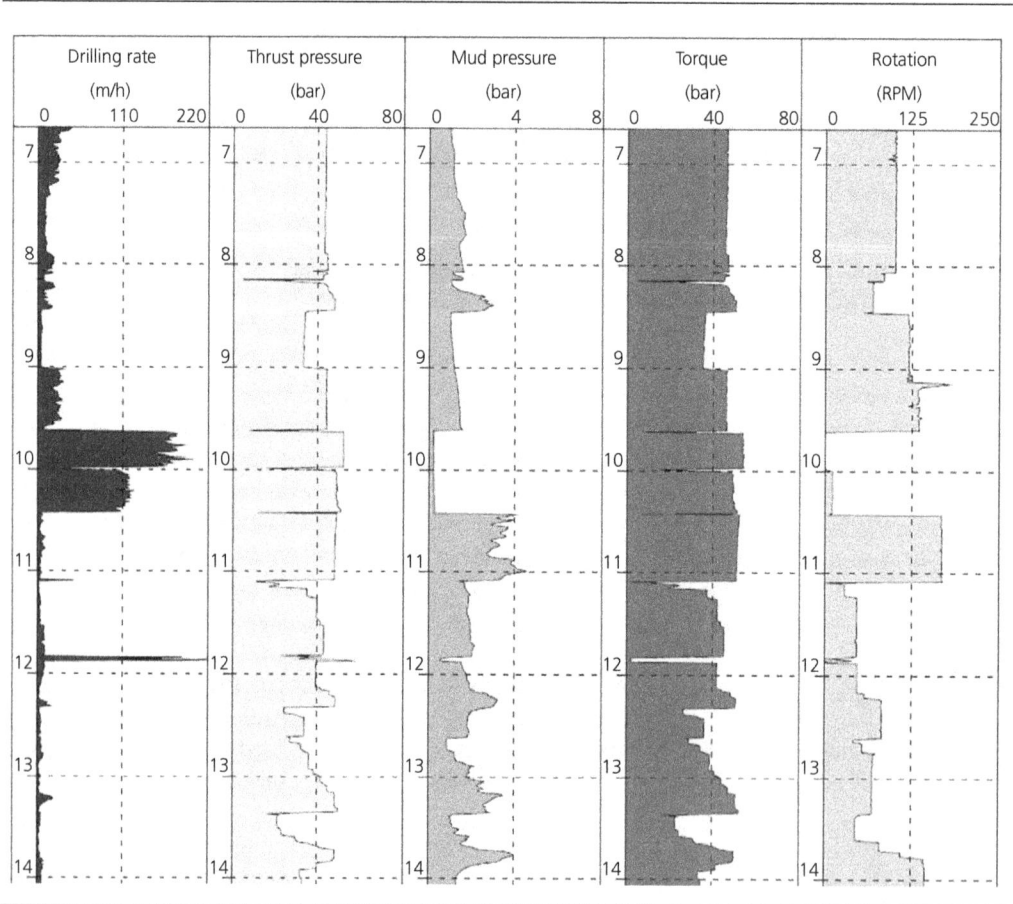

Figure 3.25 Graph showing DPR data for a borehole drilled in London (image courtesy of Soil Engineering)

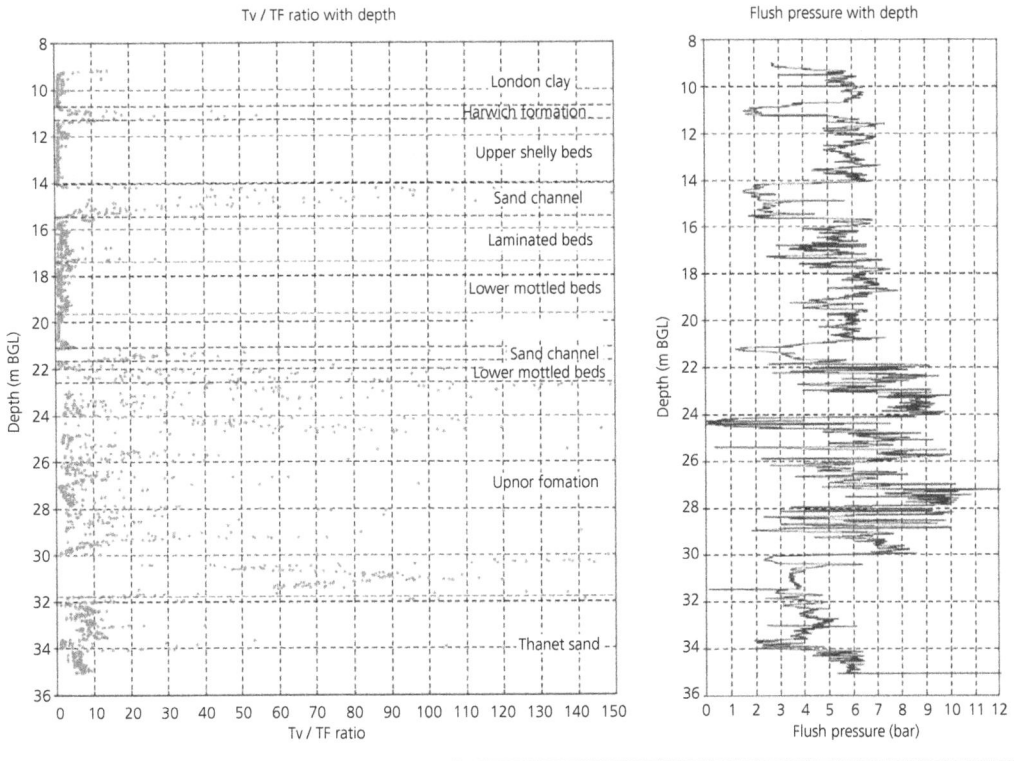

and into the Lambeth Group. The borehole encountered very poor recovery in the sequence marked 'sand channel'. However, the extent of the feature is clearly shown by the DPR record, thus enabling the location of the channel to be accurately determined.

3.5   Other techniques

3.5.1   Casing collar locator (CCL)

3.5.1.1   Property measured

Changes in the magnetic flux owing to the presence of magnetic material (such as steel casings).

3.5.1.2   Method of measurement

Using a magnet and coil arrangement in a moving tool generates an induced electro-motive force where there are mass changes in magnetic material (e.g. casing strings, collars and certain joins). This device is often run in combination with other parameters to confirm the position of steel casings that affect the measurements.

3.5.1.3   Example logs

An example log is shown in Figure 3.26.

Figure 3.26 Magnetic and electro-magnetic responses to a sheet pile (image courtesy of European Geophysics)

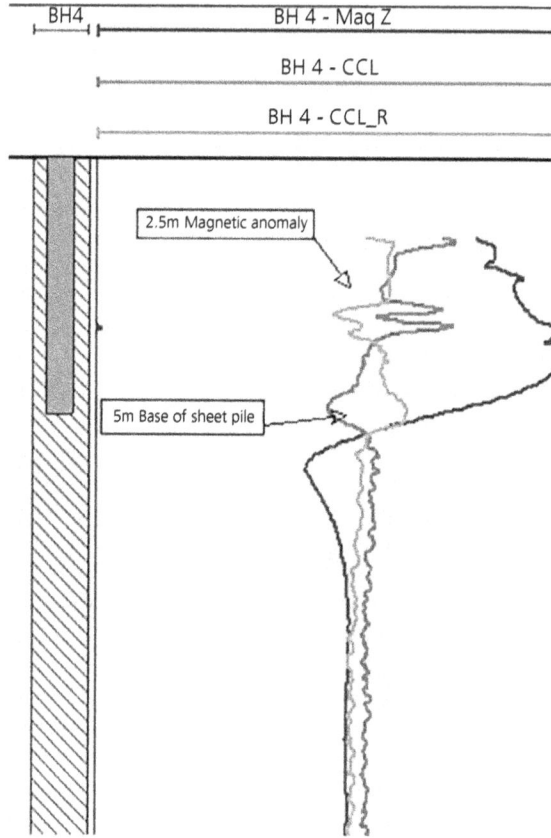

3.5.2 Magnetic field

3.5.2.1 Property measured
The strength of the local magnetic field in three axes (X-Y-Z) with a view to detecting magnetic anomalies near to the borehole.

3.5.2.2 Method of measurement
Using magnetometers that may measure the total field and/or its components in three (X-Y-Z) axes.

Figure 3.27 shows a response in the vertical magnetic field (Mag Z) to the presence of a sheet pile. This log has been plotted alongside results from an electro-magnetic device (CCL – see above) which may also be used for location or variations of magnetic material. The CCL was also run in the opposite direction and gives a mirror image repeat. For any of these methods to be successful, they need to be run in boreholes located close to the target features (typically within a metre).

Figure 3.27 Parallel seismic testing to determine the length of a pile (image courtesy of SOCOTEC)

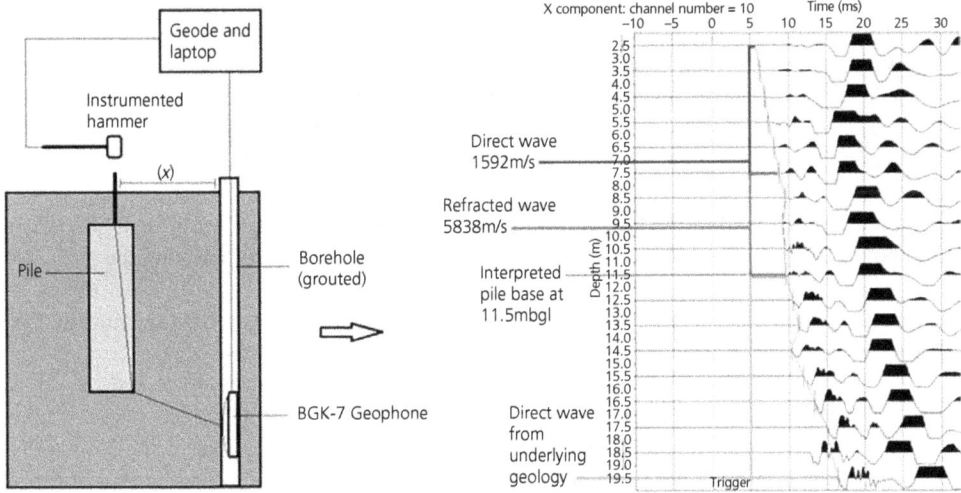

3.5.3   Video cameras

Video cameras are available with a variety of lenses, lighting and viewing aspects, such as forward, side viewing and pan and tilt.

3.5.4   Cavity scanners

3.5.4.1   Property measured

Three-dimensional scans of underground voids, cavities, tunnels and mines, using lasers (dry voids) or sonar (fluid filled voids).

3.5.4.2   Method of measurement

Narrow diameter probes are lowered using rods or wire-line (dependent upon the design) through an access borehole into the void. Once in the void, the laser-scanning or ultra-sonic head is rotated, measuring the three-dimensional shape of the void with full 360° coverage. Alignment of the probe may be achieved through the use of rods (mechanical alignment) or the use of a compass or 3-axis gyro that monitors the probe's heading. Accelerometers are also incorporated to determine the inclination. Some probes may also include a nosecone camera to give full visibility of the borehole during deployment so as to avoid obstructions and confirm entry into the void.

Sonar scanners operate by emitting an ultra-sonic pulse that is reflected by a solid object. Timing the delay between emission and the arrival of the reflected sound wave back at the tool allows for a calculation of distance. The sonar head rotates making radial measurements through a 360° arc; orientation of the beam is provided by a compass system.

The surveys are generally referenced to the local or National Grid system.

3.6   Seismic testing

3.6.1   General

Seismic testing using a borehole is an adaptation of the seismic testing carried out at the surface, as described in the *ICE Handbook of Ground Investigation – Planning and Reporting, Geoenvironmental*

157

and Non-Intrusive Techniques (Reading and Martin, 2025). The equipment is similar, with the deployment of receivers, consisting of geophones or transducers, housed in a container that can be held firmly against the borehole wall – for example, using air bladders. The receivers consist of three transducers arranged orthogonally to form a triaxial array, with one oriented vertically and two horizontally at right angles to each other. The receivers are arranged such that one transducer is perpendicular to the direction of the wave propagation. The energy source is chosen to provide the required energy type being rich in P and/or S waves. The recording system or seismograph amplifies the signal from each transducer, timing the arrival of the energy from the source with an accuracy of 0.1 ms.

A liner should be installed in the borehole and should be grouted in place ensuring a good contact with the borehole wall. The grout should be allowed to set fully before attempting the test.

Two forms of seismic test can be performed, down-hole seismic and cross-hole seismic, and these are discussed in the following sections.

3.6.2   Down-hole seismic testing

Down-hole seismic, or well shooting, is conducted once the borehole has been drilled to depth and accurately logged. The borehole is usually lined to prevent collapse and potential loss of the geophone. The geophone is lowered to the logged base level of the stratum of interest and an energy pulse is then generated at the surface, close to the borehole. From this the average, P- and S-wave velocities for the strata above the location of the geophone are determined. The geophone can then be raised to the top of the stratum and the test repeated. The difference between the times gives the average velocity of the P- and S-waves through the stratum (V_p and V_s, respectively).

The results may be used to derive the stiffness parameters for the stratum.

A variation of the down-hole seismic test can also be used to determine the length of piles, and when used in this way the method may be referred to as parallel seismic testing.

To conduct the test a borehole is required, ideally located within 2 m of the pile. The borehole is constructed with a liner. The pile head needs to be exposed to enable a hammer blow to be made on the pile. The blow should ideally be in a downward direction, which may require a notch to be chiselled into the pile concrete.

The geophone is lowered down the borehole in 1 m intervals and a hammer blow to the pile is made at each interval. The resultant profile is able to detect when the velocity of the energy pulse changes, which happens as the energy passes from the pile into the soil at the pile toe level. Figure 3.27 shows the results of one such test and indicated that the pile toe is at 11.5 m depth, below the datum point.

3.6.3   Cross-hole seismic testing

Cross-hole seismic testing is used to determine soil and rock stratigraphy and structure, as well as stiffness parameters. The method works by the construction of two or more boreholes in an array where the boreholes are spaced no more than 10 times the size of the target. For example, if a target stratum is thought to be 0.5 m in thickness, the distance between boreholes should not exceed 5 m. The distance is assessed from the borehole log.

A borehole deviation survey should be conducted to allow for accurate determination of the horizontal distance between the boreholes, with the horizontal separation of the boreholes being determined to an accuracy of 50 mm.

To conduct cross-hole testing, it is normal practice to use a string of equally spaced transducers. The transducers are housed in a container that is held in close contact with the borehole wall by an air bladder. An energy pulse is made in the adjacent borehole and the resulting P- and S-wave arrivals are detected by the geophones (see Figure 3.28). The technique is able to locate variations in physical characteristics, such as voids or channels within strata, or igneous intrusions for example. Structural features such as faults may also be detected.

Data processing involves calculating the wave velocities from the measured travel times for the travel path distances from source to receivers, determined from horizontal surveying. The processing follows methods described by the ASTM.

The following equation determines the straight-line distance, l, from source to geophone:

$$l = \sqrt{[(E_s - D_s) - (E_G - D_G)]^2 + (L\cos\phi + x_G + x_s)^2 + (L\sin\phi + y_G - y_s)^2}$$

where:
E_s = elevation of the top of the source hole
E_G = elevation of the top of the receiver hole
D_S = depth of the seismic source
D_G = depth of the geophone
L = horizontal distance between the top of the source hole and the receiver hole
$\emptyset$ = azimuth with respect to north from the top of the source hole to the geophone hole
x_S = the north deviation of the source borehole at the source depth
y_S = the east deviation of the source borehole at the source depth
x_G = the north deviation of the geophone borehole at the geophone depth
y_G = the east deviation of the geophone borehole at the geophone depth.

Figure 3.28 Cross-hole seismic configuration (image courtesy of J Milner, SOCOTEC)

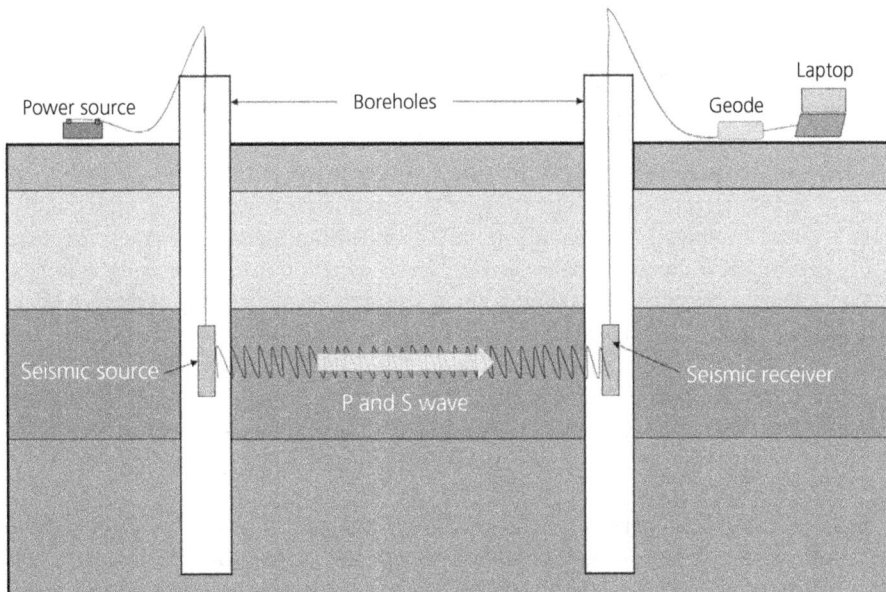

The P-wave arrival is identified on the horizontally orientated geophone as the first departure (first 'break') of the static horizontal receiver trace after time $T = 0$.

The S-wave arrival is identified on the seismic signature as a sudden increase in amplitude of at least two times that of the P-wave train, and an abrupt change in frequency coinciding with the amplitude change, which results in a period increase of at least two times that of the characteristic period of the P-wave.

The identified P- and S-wave first arrivals are picked and recorded to produce a series of travel times with depth, these being identified as:

T_p = P-wave travel time, and T_s = S-wave travel time.

The travel times of the P- and S-waves are used to calculate the P- and S-wave velocities, Vp and Vs, using the following equations:

$$Vp = l / T_p \qquad\qquad Vs = l / T_s$$

Given the P- and S-wave velocities (Vp and Vs), and the bulk density of the materials (ρ), it is then possible to derive other geotechnical parameters such as small strain shear modulus (G_{max}), Poisson's ratio (v) and Young's modulus (E) using the following equations:

$$G_{max} = \rho Vs^2$$
$$v = (Vp \cdot Vs)\, 2 - 2\{2\,[(Vp \cdot Vs)^2 - 1]\}$$
$$E = \rho Vs^2 (3Vp^2 - 4Vs^2)Vp^2 - Vs^2$$

Values for bulk density may be obtained from laboratory testing or typical values for the material.

3.7. Logging procedure

3.7.1 Site requirements
Clear line of sight access within 5–7 m of the top of the borehole is preferential for vehicle-mounted units. Space for the set-up of a tripod at the borehole should be available or provision to attach a pulley block onto the drilling rig.

3.7.2 Borehole quality – preparations for logging
Good quality boreholes are paramount to good down-hole data acquisition, particularly when deploying acoustic or optical imagers (diamond bit drilled holes are best). Also, for these techniques it is essential that the borehole is thoroughly cleaned out, to remove putty coatings, mud-cake or slime from the borehole walls. Failure to do this may result in a lack of data and images, as in the example in Figure 3.29, which shows part of the borehole wall covered in slime.

Where images through the borehole water are required, then enough time after flushing should be allowed to let any sediments settle out. The fluid column should remain undisturbed prior to logging, with no activities such as dipping the borehole being carried just before logging.

Where logging of fluid temperature is to be carried out, the fluid column should have been left for as long as possible to enable equilibrium with the groundwater system to be attained. In commercial practice, however, the available time for the borehole to settle is a compromise, typically 0.5–2 days.

Smearing on a borehole wall – optical image of sandstone where lower section has not been sufficiently cleaned out, resulting in a smeared borehole wall (image courtesy of European Geophysics)

Obstructions within a borehole should be removed, as the presence of pumps, risers, pipes and other hardware will give erroneous results as well as presenting a risk of tool entrapment.

Several logging tools need to be centralised in the borehole and therefore the internal diameters of casing and the open hole should be relatively close (±20 mm) for best transition through the borehole. In most cases, the quality and response of formation log measurements decrease with increased borehole diameter, generally best results being obtained in boreholes less than 300 mm in diameter.

The 'detection' point on most logging tools is often some way up the tool and, additionally, several different parameters may be measured on a given tool string. As such, data will inevitably not all begin/start at the very base of the borehole, and due allowance should be made for this in selecting the final drilled depth of the borehole. Consultation of the logging operator's tool specifications should be undertaken to obtain these measuring points. In the case of some tools, such as the P- and S-wave microseismic suspended tool, the start and finishing point of the data may be a few metres above the base of the borehole or below the casing.

161

3.7.3   Selection of logs

To aid in the selection of the appropriate logs for the information required by the ground investigation, a guide is given in the form of Table 3.1. 'Log selection'. The logs chosen will also be dependent upon their respective limitations imposed by the borehole construction and anticipated fluid levels (see Table 3.2. 'Log limitations').

Additionally, consideration of what depth resolution is required will govern the choice of log parameter and sampling interval of the data. For example, high-resolution imager data are typically sampled at 1 mm, HRD data at 10 mm or 20 mm, sonic at 150 mm or 200 mm, and P- and S-waves at 2 m or 5 m intervals.

The example in Figure 3.30 illustrates the respective resolutions of natural gamma, resistivity and imager logs. The gamma log picks up the main changes but lacks the detail over thin beds. The focused resistivity has a better response picking outer some additional features, while the high-resolution imager has the maximum definition and detail.

Several logs may be run in the same tool stack, which saves time and is more cost-effective – for example, natural gamma is commonly run with a number of combinations, thus providing geological control on each logging run.

3.7.4   Order of logging

It is important to consider the order of logging, particularly if fluid logs or imaging is to be carried out.

Where imagers are to be used these are normally run first and recorded downwards to obtain the best images before the borehole fluid is too disturbed. Similarly, if fluid temperature/conductivity is to be run this should be run first or as early as possible – for example, after the imager, and downwards for best response.

Where logging tools with radio-active sources (gamma-gamma, density and neutron) are deployed, it is prudent to run these after an imager, caliper or wide diameter nonnuclear tool has been run so that hole stability may be evaluated for safety reasons.

A typical suite of formation logs for ground investigations would be carried out in the following order:

(a) optical imager (recorded downwards)
(b) acoustic imager
(c) caliper/natural gamma/formation density
(d) full-wave sonic
(e) P- and S-wave suspension.

Logging may be carried out at several stages during a borehole's construction. For example, just prior to packer testing, caliper or imager data may prove useful in determining safe and optimum points for packer positions, while fluid profiles along with formation log data will help in the final design of piezometer or monitoring points.

Figure 3.30 Comparison of natural gamma, resistivity and acoustic imager over interbedded thin bands of mudstone and limestone (images courtesy of European Geophysics)

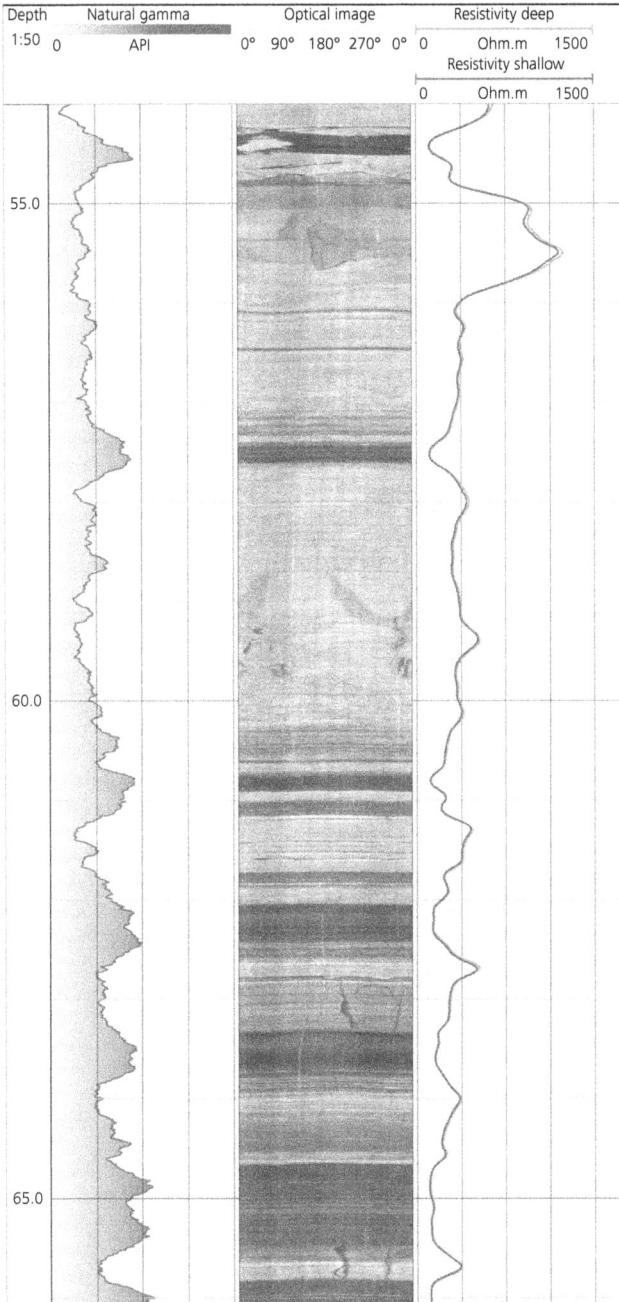

3.7.5   Data presentation

Initial field logs are produced on site at a convenient depth scale to allow ongoing site decisions to be made. Often, several log runs may be combined on one plot in a composite form, some examples of which have been illustrated above. Each log should have a header with relevant borehole details, datum used and date.

To aid subsequent interpretations, it is also important to view the geophysical data with other data – for example, borehole logs and key information on each borehole should also be recorded with the geophysical log data at the time of survey, including

- last circulated
- any drilling problems
- logging datum
- fluid level
- position and diameters of casings
- borehole diameters
- repeat intervals/runs.

A certain amount of filtering of the data is required, depending on the technique, particularly the nuclear logs that, by their nature, are statistically variant. Also, the amount of filtering applied is a function of the vertical scale – that is, generally a short filter on the higher-resolution detailed scale outputs. Imager logs are typically produced at detailed vertical scales of 1:20 or 1:10.

Most data recorded are in a raw form, which allows for further processing or reprocessing, including applications of different filters and vertical scales.

Final outputs may be produced graphically on paper, but more commonly as a PDF (portable document format) file. Most log data – that is, log curves – are also presentable in in a digital format, such as CSV (comma separated value), LAS (Logging ASCII Standard – Canadian Well Logging Society) or AGS (Association of Geotechnical and Geo-environmental Specialists), for ease of import to other software. Images may be converted to BMP (bitmap) and seismic/sonic waveform data may be presented in SEG (Society of Exploration Geophysicists) format.

Peter Reading and Miles Martin
ISBN 978-1-83549-891-0
https://doi.org/10.1108/978-1-83549-890-320251007

Chapter 4

Instrumentation and monitoring

4.1. Introduction

Owing to significant advances made over recent decades, in areas such as electronics, remote sensing and digital data transfer, it is now possible to monitor many parameters, record the data and transfer it to online monitoring software (OMS), or view the details almost in real time anywhere in the world. This ability has radically changed the nature of instrumentation. While every effort has been made to provide up-to-date information in this chapter, the pace of change within this field will almost certainly mean that methods discussed here will be updated and probably deliver greater accuracy than suggested in the following sections. The general principles, however, do not change and, as always, attention to detail is paramount to providing success and useful information from the instrumentation installed.

In order to obtain long-term information on properties such as groundwater level or stability, it is necessary to install instrumentation on the site. However, installation is just the first part of a three-part process that must also include monitoring and interpretation. A common rule of thumb is that the cost of installation will be one-third of the total cost, with monitoring and interpretation each comprising a further third of the total cost. It is commonly found, however, that although instrumentation might be installed during an investigation, neither of the other two requirements are carried out, making the installation a costly waste.

> For every pound spent on installation a further pound should be spent on monitoring and then a further pound should be spent on interpretation. (Bromhead, 2008)

While the commonest measurements are those of groundwater level, monitoring can also be made of ground movement, settlement and slope failures as well as stresses in the ground. The following is a list of information that can be obtained from monitoring that is commonly conducted in a civil engineering environment. Although some of the parameters listed are not directly related to ground investigation, they are all elements that might be found on site and may well be included within a ground investigation package.

Section 4.5 of this chapter deals briefly with the other forms of monitoring that are less commonly included in an investigation; these are shown in italics in the following list. This section also includes a table of the most commonly used methods for monitoring.

Measurements of the ground

- water depth
- ground gas concentrations (*ICE Handbook of Ground Investigation – Planning and Reporting, Geoenvironmental and Non-Intrusive Techniques* (Reading and Martin, 2025))

- soil pore water pressures
- water flow
- slope stability/movement/failure
- soil pressure
- expansion/contraction (shrink/swell)
- settlement/heave (profiles and direct)
- displacement.

Measurements of structures

- bending (deflection)
- convergence (ovalisation of a tunnel)
- tilt (rotation)
- cant/twist – railway measurement
- crack propagation
- strain (ratio of change in dimension to original dimension).

Measurements of the environment

- noise levels
- wind parameters (speed and direction)
- temperature
- barometric pressure
- air quality – dust (PM10, PM2.5 and PM1)
- vibration.

Monitoring can be conducted prior to and during construction, and in some cases well beyond construction and throughout the design life of the structure. It is therefore important to ensure the equipment installed is robust, sufficient and fit for purpose to allow for the duration of monitoring required along with the data output to allow for correct data analysis.

Measurements can be made to add value to a ground investigation. They can also be used to monitor performance, to ensure specified criteria are not exceeded, and give warnings when excessive movement or levels occur, which then informs action to be taken to mitigate the risks.

Whatever is being measured, it is essential that the programme and the type of instrumentation is carefully planned to ensure that the installation will measure the intended parameter, will not influence the ground and the parameter being measured, and will provide data of sufficient accuracy. When considering what types of instrumentation to use, it is essential to understand what is needed to inform the investigation, how often readings are required, and what actions or decisions need to be taken once the readings have been gathered. When making these plans, it is useful to consider the unexpected and how unusual events might influence decisions.

Such events might include rapid rises or falls in water levels or sudden increases or decreases in pore water pressure, perhaps significant movement signalled by an inclinometer, or rapid changes in ground level. Whatever the possible event, the instruments used should be suitable to measure the changes and the plan should include trigger levels whereby if a parameter is exceeded there is a clear action plan.

The location of instruments should be chosen to ensure that they can be read over the intended period. Many instruments need to remain accessible to enable readings to be taken, or for future maintenance of an automated system. They should therefore be positioned so they are not covered or blocked by vehicles or site activities, while also being able to capture the intended parameter changes. There are times where meeting both of these objectives is not possible, and under those circumstances the correct instrument and logging choice is even more important. It is increasingly common now for instruments to be automated, whether this be through the latest wired or wireless dataloggers or a standalone logger with manual download. This automation, and the step away from periodic or ad-hoc readings, enables a more detailed picture of the site conditions to be established.

Almost all instrumentation, be it periodically manually read or data logged, will also require protection from vandals, site traffic and other activities that might damage the installation.

Before deciding what instrumentation might be installed, the following questions (based on Dunnicliff, 1999) should be answered.

- What are the project conditions?
- What parameters requires monitoring?
- Why does the parameter need to be monitored?
- What are the mechanisms that control the parameter behaviour?
- What instruments are there that will measure the parameter?
- How much is the parameter likely to change? (What is the range?)
- What external factors might influence the parameter and the recorded results?
- What spatial distribution of instruments is required to provide the level of confidence in the monitoring?
- Are the locations practical over the timeframe of the monitoring period in relation to the changing site activities?
- What accuracy/repeatability is required?
- Does the parameter vary rapidly?
- At what frequency should readings be taken?
- What is the instrument response time?
- How long should monitoring be conducted? (Is it for initial ground investigation and design, or will it form part of a long-term monitoring strategy for the asset?)
- How are results to be recorded?
- Are measures for consistency, repeatability and back up required?
- How often are results to be reported?
- Who needs to see the report?
- Is the responsible person likely to change as the site activities change?
- What are the trigger levels?
- What actions should be taken at the trigger level?
- What is the budget for monitoring?

Armed with the answers to the above, the installation and monitoring processes should be programmed and planned, incorporating commissioning, initial calibration, ongoing maintenance, data collection, processing and reporting, together with details of what actions might be needed if a trigger level is met. A method statement drawn up for each installation suite and subsequent monitoring procedures should accompany the programme.

BS5930 (BSI, 2015) provides information and guidance on some field instrumentation, although this is confined to instruments installed within boreholes.

The following Sections 4.2 to 4.4 deal with the installation and monitoring methods that provide information of the ground and groundwater and would regularly be included within a ground investigation package. Sections 4.5 and 4.6 provide descriptions of other methods that may be of use or included within a ground investigation to provide addition site information. These include developing technologies that will invariably become more widely adopted.

4.2. Groundwater monitoring

4.2.1 General

Groundwater monitoring is essentially the measurement of water pressures in the ground, or of the groundwater level (piezometric surface) where the water pressure is equal to atmospheric pressure. Typically, a series of measurements are taken over a period of time to assess natural variations.

Good knowledge of the groundwater level and behaviour is essential, and a poor understanding is often the cause of problems within civil engineering projects, resulting in many serious events, which have disrupted and delayed construction programmes. An example is the development of 'blowing' conditions in the base of an excavation supported by sheet piles (i.e. base failure), whereby the pressure exerted by the groundwater within a cofferdam exceeds the stabilising downward force of the soil within the cofferdam owing to insufficient penetration of sheet piles.

Under such circumstances, an understanding of the groundwater level and pressure inside and outside the cofferdam could ensure that this is avoided, allowing an adequate depth of soil to be maintained within the cofferdam to balance the hydrostatic conditions. This is just one example of where inadequate monitoring might result in a difficult situation. Knowledge of the groundwater level, and possible variations over time, is essential to enable appropriate measures to be taken.

Groundwater monitoring is facilitated either by the installation of a water level observation tube (generally referred to as a standpipe), to monitor the water level directly by measuring its depth below a datum, or alternatively by using a 'piezometer', which measures the water pressure at a particular point. When the latter is used, the pressure can then be translated into an equivalent height of groundwater above the monitoring point.

Groundwater can be found in isolated pockets or zones of permeable material within low permeability soils, as found in glacially derived materials, for instance. It can also occupy deposits of granular soils that are underlain by low permeability materials, resulting in a 'perched aquifer', as is often seen in river terrace deposits. At depth, more substantial aquifers are found within extensive porous and permeable materials, such as sandstone and chalk.

In the UK, water extraction is carried out from both perched and deeper aquifers to provide potable water for domestic and industrial use. Where this is the case, these aquifers are often designated protection zones and care must be taken to ensure they are not exposed to contamination. In most cases, water extraction in significant quantities from these sources – as, for example, might be needed for a pump test – will require a licence.

For many construction processes, it is as useful to know that groundwater is not present as it is to know where it might be met. However, in a ground investigation there are several factors that can hinder the simple observation of groundwater. The groundwater level may be masked by the drilling process, which can block pathways through which water might seep into the borehole, as with the use of casing, or by the addition of water or other drilling fluids. This can make the assessment of where to place a standpipe or piezometer difficult. It is therefore essential that some forward planning is undertaken before starting the investigation.

Decisions such as the probable required depth of a standpipe or piezometer should be estimated prior to beginning drilling, to ensure that sufficient pipework and materials are available once the borehole has been completed (Reading and Martin, 2025). This requires a knowledge of the strata likely to be encountered, and the approximate depth of strata boundaries, as well as a knowledge of the likely depths at which groundwater might be met. This is particularly relevant when installing vibrating wire piezometers (see Section 4.2.8.2) as these are supplied with connected wires that need to be preordered to the required length.

Backfill materials are used to define a specific section, or sections, of a borehole for which information on groundwater is required. Free-draining filter material, such as sand or gravel, is used as backfill around the slotted section of a standpipe, filter tip or piezometer, situated within the water bearing strata. This provides continuity between groundwater outside the borehole and inside a standpipe or, in the case of piezometers, a link between the water pressures in the ground and in the body of the piezometer. The top and bottom of this zone, known as the 'response zone', is sealed with layers of bentonite clay. Where finer-grained soils are present, a prewrapped perforated section of standpipe can be deployed, these having a filter sock covering the slotted zone of the pipe, protecting the standpipe from siltation. The filter sock may be prefitted by the supplier or may be fitted by the site operatives.

Sufficient time must be allowed for the installation process. Many installations fail owing to the process being rushed by pressures to complete work within a fixed preconceived timeframe.

Where an installation does not extend to the base of a borehole, the section of borehole below the installation should ideally be grouted, using a tremie, a pipe extending to a specific depth – usually the base of the borehole – through which grout may be pumped. Time should be allowed for the grout to cure before starting with the installation above. If the borehole is backfilled with arisings below the response zone, which is not recommended, the fill must be adequately compacted as it is placed to avoid arching and potential subsequent collapse. Sufficient time must be allowed for bentonite used in the seal below the response zone to cure, a process that will involve a significant volume change. If the sand filter zone is placed too soon on top of a bentonite seal, the latter can swell into the sand and clog the instrument being installed in the filter. Time must also be allowed for the sand or gravel filter material to settle around the instrument tip, taking regular measurements of the depth to the top of the material. It is essential that the instrument tip, or slotted section of a standpipe, is completely encased within the filter material. This whole process will regularly take more than a single shift, and the hole may need to be left overnight to allow sufficient time for certain materials, such as grout, to set before completing the installation.

Once installed, the water level in an installation may take some time to reach equilibrium, and piezometers can take several days for pressures to settle at an equilibrium level.

Changes in groundwater level may be directly related to variations caused by rainfall, nearby water bodies or tidal effects. It is worth measuring these to see how influences such as rainfall affect groundwater, and compare variations seen in groundwater levels to tide times or changes in river water levels.

It is not uncommon for the specification of both a water level observation tube and a piezometer to measure water levels in two different strata in the same borehole. However, the readings obtained may not be relied upon with confidence because it is highly likely that the two instruments cannot be sufficiently isolated to ensure they are reading independently of each other. It is recommended that, wherever possible, only one water table should be measured in any one borehole.

4.2.2   Direct measurement of water level

Direct measurement of water levels is achieved by the installation of an observation tube, generally referred to as a standpipe, which is designed for the purpose and provides direct access to the groundwater. The water level is determined using a dip meter, described below, and this type of instrumentation therefore requires the location to be accessible for subsequent visits to the site.

A dip meter comprises a tape measure with insulated wires passing down each edge to a brass or stainless steel probe. One wire is connected to the body of the probe the other to the tip. A plastic spacer insulates the tip and body from each other. The tape is wound onto a reel, which typically houses a 9-volt battery and a buzzer that sounds when the circuit is complete. The probe is lowered down the pipe to the water, and as the probe enters the water it connects the tip of the probe to the body, completing the circuit and signalling that the probe is at the water level in the pipe. The audible signal is often accompanied by a light that illuminates at the same time, which is useful in noisy environments where it might be difficult to hear the buzzer.

The operator may then record the depth to groundwater by reading the measuring tape against a fixed datum point, such as the upper edge of the pipe or headworks. Should it be required to reference water levels to ground level, and should the datum being used be above or below ground level, a suitable adjustment may be applied to the reading. As such devices are battery operated, it is advisable to carry a spare battery on each site visit.

Figure 4.1 A dip meter reel showing the probe and measuring tape (image courtesy of MGS)

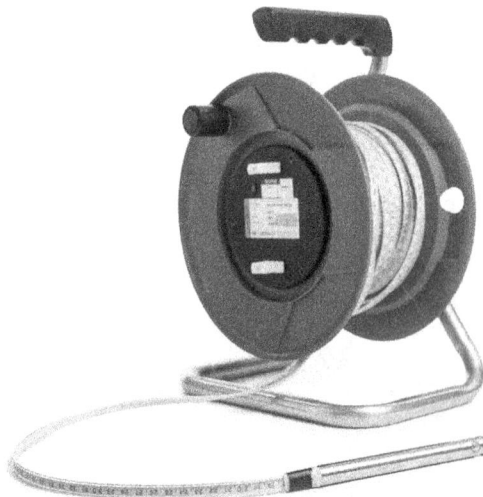

The probe should be raised and lowered a few times to ensure repeatability of the measurement, and that a false reading has not been obtained from water condensed on the wall of the observation tube. False readings are particularly likely in brackish or saltwater conditions, owing to increased conductivity; under such circumstances, consideration should be given to the use of a dip meter that can be adjusted for sensitivity (typically by turning a knob by hand or using a screwdriver).

In some cases, there may be lighter oils (a light nonaqueous phase liquid, or LNAPL) sitting on top of the water. If this is suspected then an interface probe can be used to determine the top of the oil and the interface between the oil and water. This type of probe is designed to detect the difference in conductivity between oil and water, and causes the sound emitted by the probe to be of a different frequency, thus indicating the top of each fluid.

The observation tubing is generally installed to a depth below the water table that has been encountered in the borehole. When installed correctly, the water level in the ground will be represented by the same level in the tube. The tubes are usually of high-density polyethylene (HDPE) and are typically supplied in 1 m or 3 m lengths, which connect by male and female threads at either end of the tube, or by using connectors to join the lengths. The lower sections of tubing that extend into the water table may either be slotted or fitted with a piezometer filter tip (see Section 4.2.3), to allow water into the pipe. The sections above are 'plain' – that is, with solid walls. When slotted, the slots are cut horizontally and are 1.5 mm wide. The smallest pipe in general use is 19 mm in diameter; this diameter ensures the dip meter used to measure the groundwater level is able to pass freely down the tube. Larger diameter tubes may be selected (typically 50 mm), which are particularly useful should water sampling be required at periods after installation has been completed, or if it is perceived that other types of instrumentation, such as pressure transducers (see Section 4.2.8.4), may be required at the location.

A 50 mm dia. tube is usually selected for monitoring water levels in granular soils that, being free draining, respond quickly to changes in groundwater level. When monitoring water level in soils of lower permeability, a smaller diameter pipe and smaller filter zone will enable a quicker response to water level changes.

When gas monitoring is required as well as water level monitoring, the slotted section of standpipe needs to extend through unsaturated ground, to allow the gases to pass into the tube, and a gas tap, mounted in a sealing cap or bung, is typically used to cap the tubing and prevent the gas venting to the atmosphere (Reading and Martin, 2025).

The tubing should be central in the borehole throughout the installation length. The use of spacers is recommended to achieve this, although in the authors' experience these are rarely used in practice. Without spacers, however, the tubing will invariably rest against the side of the borehole, which may result in inaccurate readings and possibly the malfunction of the installation.

The annular space between the standpipe and the borehole walls is filled with various layers of permeable and impermeable materials. The permeable materials are used to form a 'response zone', encasing the slotted section of the standpipe, or the piezometer tip, and allowing groundwater (and in unsaturated zones, ground gas) to pass into the pipe. The impermeable materials (typically bentonite clay) are used to seal off other sections of the pipe, thus delineating the response zone. See also Section 4.2.1.

The permeable filter materials may consist of single-sized fine gravel if the stratum of interest is predominantly gravel, or a single-sized sand when the stratum is finer grained. Ideally, the filter material should have a permeability at least one order of magnitude higher than the strata that the response zone is designed to target. This is particularly important if permeability testing is to be conducted in the pipe, although this method of testing should still be used with caution primarily owing to the potential influence the filter may have on the test result. The primary function of the filter medium is to prevent the observation tube becoming clogged with silt. Should strata have a significant silt content, a prefitted filter wrap is recommended (as shown in Figure 4.2), although this may not be suitable should monitoring for possible LNAPL be envisaged, as filter wrap can exclude LNAPL as well as silt.

The design of the installation will be dependent on the strata encountered and the water level within the strata. The response zone or filter tip should be positioned within the zone of interest. This will require an understanding of the behaviour of the groundwater. A number of typical arrangements for a water level installation are shown in Figure 4.3. These are for guidance only because every situation will be different, and a specific design may be necessary to suit the particular conditions seen in the borehole.

Typically, the bottom of the response zone will be positioned near the base of the water bearing stratum, or at least two metres below the water table. If the water table is likely to be affected by tidal influences, a greater range may need to be accommodated. Where the borehole extends some depth into less permeable soil, the lower section of the hole may be grouted using a cement/bentonite mix and then a seal formed using bentonite clay pellets up to the base of the zone of interest. The filter material of clean pea shingle or sand should be placed extending between 0.1–0.25 m above the top of the bentonite seal. At this stage, the fully assembled pipe with the slotted section at the base and plain pipe above may then be lowered into the borehole to rest onto the filter

Figure 4.2 Various sizes of slotted standpipe fitted with a filter sock, held in place by a layer of open plastic netting (image courtesy of MGS)

Figure 4.3 Some typical diagrams showing installation configurations (authors' images)

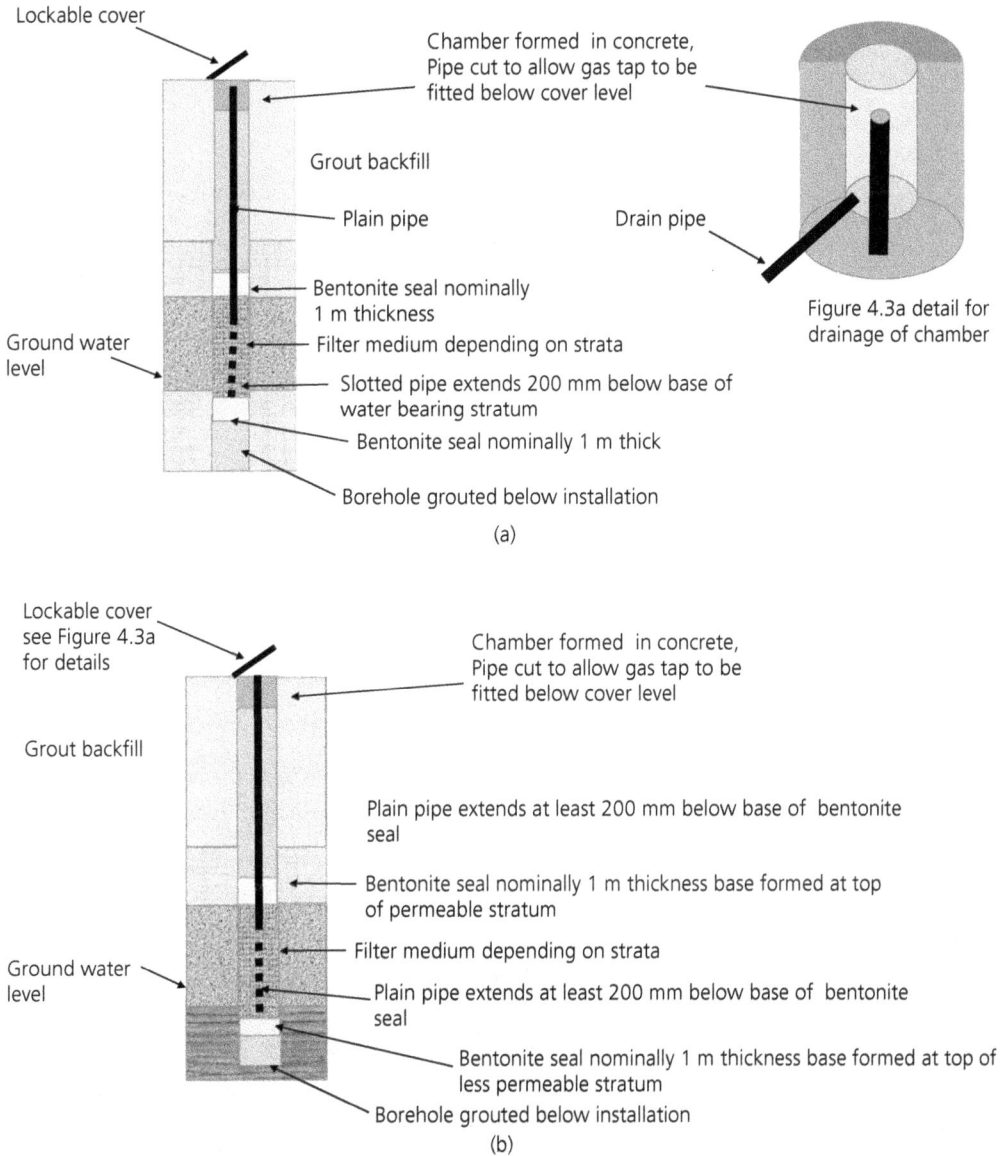

Lockable cover

Chamber formed in concrete, Pipe cut to allow gas tap to be fitted below cover level

Grout backfill

Plain pipe

Drain pipe

Bentonite seal nominally 1 m thickness

Ground water level

Filter medium depending on strata

Slotted pipe extends 200 mm below base of water bearing stratum

Bentonite seal nominally 1 m thick

Borehole grouted below installation

Figure 4.3a detail for drainage of chamber

(a)

Lockable cover see Figure 4.3a for details

Chamber formed in concrete, Pipe cut to allow gas tap to be fitted below cover level

Grout backfill

Plain pipe extends at least 200 mm below base of bentonite seal

Bentonite seal nominally 1 m thickness base formed at top of permeable stratum

Filter medium depending on strata

Ground water level

Plain pipe extends at least 200 mm below base of bentonite seal

Bentonite seal nominally 1 m thickness base formed at top of less permeable stratum

Borehole grouted below installation

(b)

medium, ensuring at all times that the pipe is central in the hole and ideally held in position by spacers. Once the pipe has been positioned and is resting on the filter medium, further filter material can be placed around the pipe. The filter medium should extend approximately 0.5 m above the top of the slotted section of pipe, as this will reduce the risk of bentonite seeping from above and into the slotted pipe. Keeping the borehole topped up with water while placing the filter medium will reduce the risk of arching. Another bentonite seal should be formed on top of the response zone, which should be between 0.5–1 m in thickness. The remaining section of the borehole may then be grouted up toward the surface. An additional bentonite seal, again at least 0.5 m thick,

173

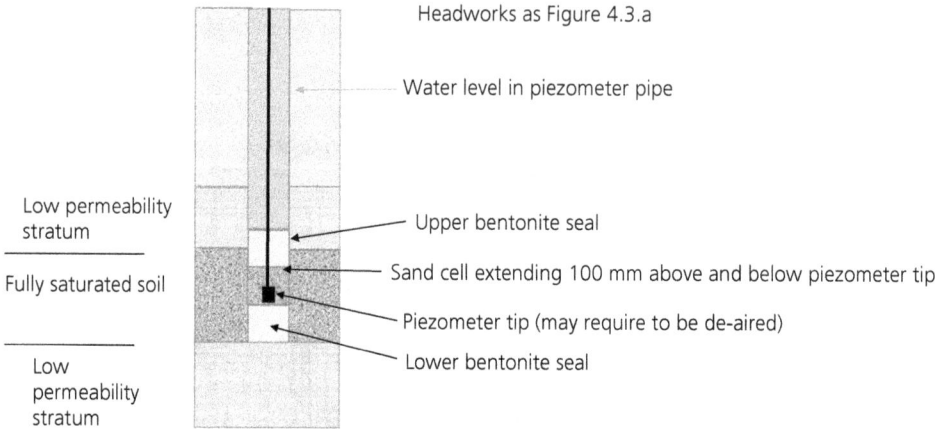

Headworks as Figure 4.3.a

Water level in piezometer pipe

Low permeability stratum
_____
Fully saturated soil
_____
Low permeability stratum

Upper bentonite seal

Sand cell extending 100 mm above and below piezometer tip

Piezometer tip (may require to be de-aired)

Lower bentonite seal

(c)

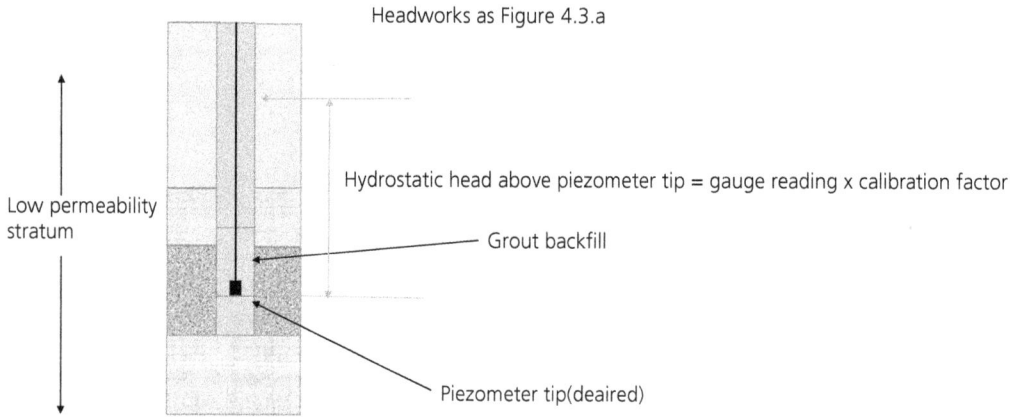

Headworks as Figure 4.3.a

Hydrostatic head above piezometer tip = gauge reading x calibration factor

Low permeability stratum

Grout backfill

Piezometer tip(deaired)

(d)

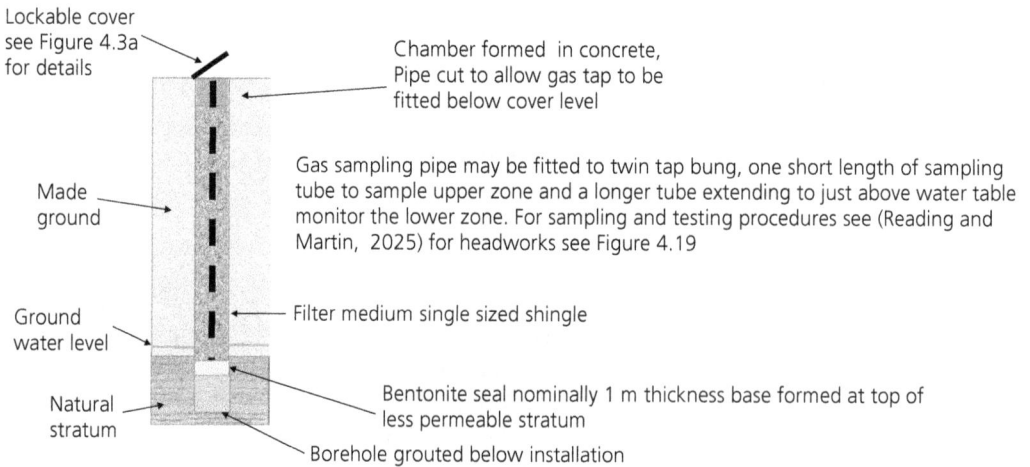

Lockable cover see Figure 4.3a for details

Chamber formed in concrete, Pipe cut to allow gas tap to be fitted below cover level

Gas sampling pipe may be fitted to twin tap bung, one short length of sampling tube to sample upper zone and a longer tube extending to just above water table monitor the lower zone. For sampling and testing procedures see (Reading and Martin, 2025) for headworks see Figure 4.19

Made ground

Ground water level

Filter medium single sized shingle

Natural stratum

Bentonite seal nominally 1 m thickness base formed at top of less permeable stratum

Borehole grouted below installation

(e)

should be placed at the top of the grouted section, to restrict the seepage of surface water into the backfill materials. Above this should be a gravel drainage layer, to disperse any surface water, and then at least 0.5 m of concrete, which is used to fix the headworks in place. Completion of the installation headworks is discussed in Section 4.4.3.

As liquid grout needs to be left a number of hours to set, and as the next layer of the backfill materials cannot be placed until the grout is solid, its use can result in a full installation taking two or three days to complete. This might not be problematic on some sites, but under certain circumstances may be inconvenient. In practice, bentonite pellets are often used in place of grout because, although a more expensive material, they tend to be cleaner and less time consuming to use, enabling completion of the installation in one operation. Where pellets are used, care must be taken to avoid clogging and arching. Pellets should be introduced slowly into the borehole, particularly where water is present within the bore. Time should be allowed for the pellets to hydrate, a process that will cause them to swell. Pellets will swell by 500% over 2 days; this increase in volume is largely accommodated by closing the void spaces between the pellets, but can also result in the bentonite rising into the filter zone (see Section 4.4). The top of the standpipe should be cut to a level such that the cover can be concreted in, making sure the cover closes. The cover should be lockable to prevent vandalism (see also Section 4.4.3: Headworks).

It is recommended to take the position of the borehole using a GPS (global positioning system) device and clearly mark the underside of the installation cover with the borehole identification number. This may prove to be particularly useful if a long period of monitoring is required, during which some borehole locations become increasingly difficult to find if site conditions deteriorate. A wooden marker post or wooden fencing around the location can be used, although on some sites it is preferable not to draw attention to the locations.

Apart from installation issues such as clogging of the filter pack, the main problem with installations of this type, where site visits to carry out monitoring are made periodically, is that water level behaviour, such as fluctuations, that may occur between such visits will remain unknown. It is only with instruments that are data logged, and that are set up to undertake readings at frequent intervals, will a full picture of water level behaviour be obtained. To this end, standpipes, as described above, can be retro fitted with logging devices such as divers (see Section 4.2.8.4).

4.2.3   Casagrande piezometer

Piezometers measure water pressure at the point they are installed. The Casagrande piezometer consists of a porous element that can be installed into a small sand cell (i.e. a short response zone), with bentonite seals placed above and below the sand response zone at the required position within a borehole. Piezometers are normally installed in soils such as sand, silt and clay or mixtures of these three soil types.

The simplest form of Casagrande piezometer is a porous tip attached to 19 mm dia. observation tubing, similar to the configuration described as a standpipe in Section 4.2.2 above, and is sometimes referred to as a 'standpipe piezometer'. The tip is usually plastic with a porous expanded foam insert to reduce siltation. The tip replaces the slotted section of pipe described in Section 4.2.2. Water enters the tip and will rise to an equilibrium level within the tubing, equal to the water pressure in the ground at the point of installation. The depth to this water level may be measured using a dip meter, as described in Section 4.2.2, and when it is required to automate the water level readings a submerged data logger, or 'diver', as described in Section 4.2.8.4, will provide a more complete picture of what is happening to the water level.

When the Casagrande-type piezometer is installed, the tip is usually placed in a small sand cell between 0.5–1 m in length, isolated by bentonite seals above and below the cell, as shown in Figure 4.3(a) and Figure 4.4. This will enable the piezometric pressure at the location of the sand cell to be recorded, which will be equal to the head of water in the access tube.

The response time for the standpipe piezometer can be slow and sufficient time should be allowed for the instrument to reach equilibrium after installation and to respond to changes in the pore water pressure. For this reason, the water level recorded from this type of installation may lag behind rapid changes in water level – for instance, when it is tidally influenced.

It is possible to conduct permeability tests in this type of installation, although the permeability of the cell and piezometer tip will need to be at least two orders of magnitude greater than the surrounding soil in order to provide meaningful results.

Figure 4.4 A typical Casagrande piezometer installation (authors' image)

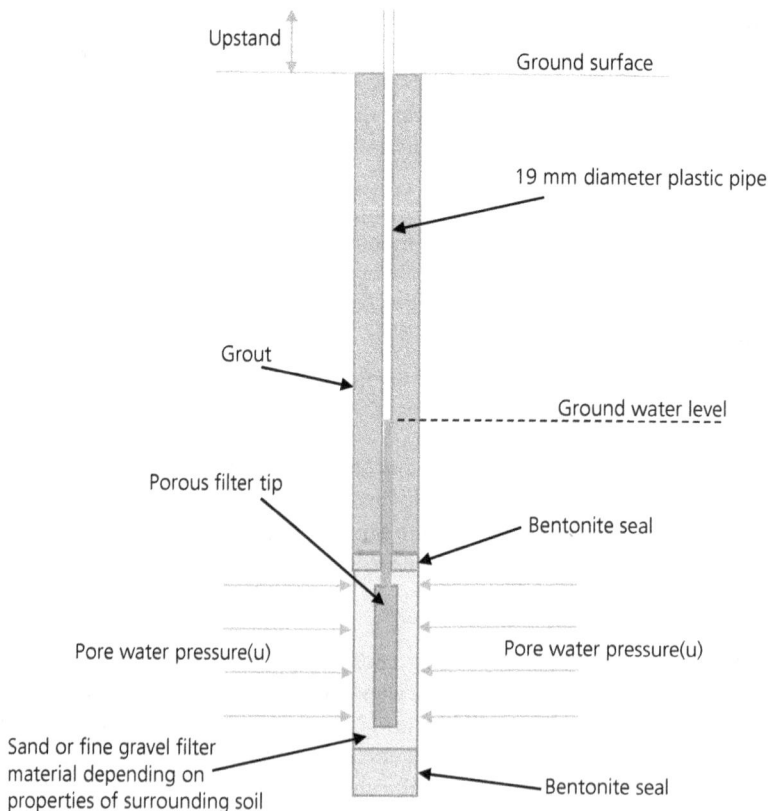

4.2.4   Driven piezometer

The driven piezometer comprises a steel tube with closely spaced access ports in its side and a 30-degree point at the base. The tube is 19 mm in diameter (3/4 inch) and approximately 300 mm long. It couples to steel 19 mm tubing, usually in 1 m lengths, with steel threaded connectors. The equipment is sufficiently robust to enable it to be driven into the ground to a few metres with a hammer. The tip has an expanded foam plastic insert filter.

This type of piezometer is limited by the depth it can be driven or pushed, which in some cases might result in the instrument being installed above the level of the water table. In general, these instruments will be installed using a dynamic sampling or dynamic probing rig. However, they can also be driven by hand and are therefore of some use where only limited access by foot is afforded. The main disadvantage of this equipment is that water levels may not be reliable because the ports, which allow water into the tube, can become blocked during the driving process, particularly if the piezometer is driven through cohesive soils. Measurement of the water level is made using a dip meter or, if the vibrating wire versions of push-in piezometers (see Figure 4.5) are used, data may be obtained using a data logger or remote monitoring systems. These are discussed in detail in Section 4.2.8 and because they give many more readings will provide a fuller understanding of the groundwater behaviour.

Figure 4.5 Push-in version of the vibrating wire piezometer (image courtesy of MGS)

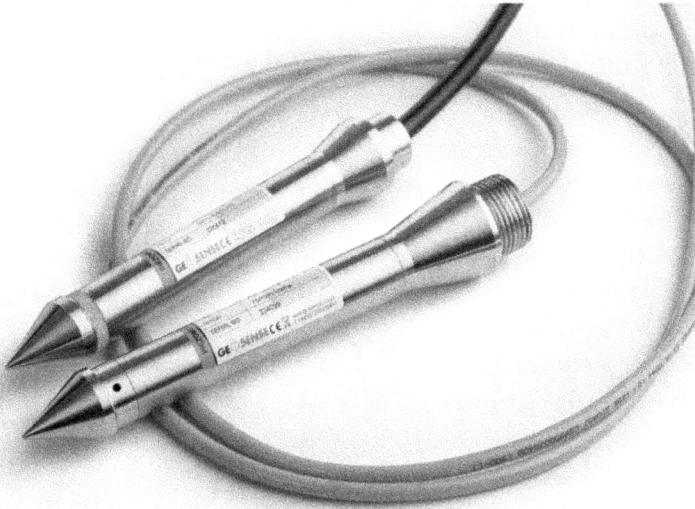

4.2.5   Pneumatic piezometer

As with the Casagrande piezometer, the pneumatic piezometer consists of a ceramic porous tip that allows groundwater to enter a chamber in the device where the pressure at the point of installation is measured. While in the Casagrande-type piezometer the water entering the tip then rises through a standpipe, in the pneumatic piezometer water is held within the tip, the water pressure acting on a diaphragm. Nitrogen gas, under pressure, is then used to operate a valve within the piezometer tip and this process determines the groundwater pressure. The piezometer is connected to the surface by two small-bore flexible tubes that introduce the pressurised gas to the piezometer tip. The diaphragm, which is sealed by the water pressure onto a plate, is lifted off the plate as the gas pressure from one of the tubes increases. This allows the gas to escape up the second tube to the monitoring box at surface. The pressure at which the diaphragm lifts off the plate is equal to or greater than the pore water pressure, and this is registered at the monitoring box or by using a digital readout unit. The pressure is then allowed to drop and the point at which the diaphragm reseals as it reaches the pore water pressure is recorded.

By knowing the precise depth at which the tip has been installed, and the pore water pressure at that point, the depth to groundwater may be calculated, if needed.

The porous ceramic tips (see Figure 4.6) should be saturated in deaired water prior to use and should be kept saturated for as long as possible until the point of insertion in the borehole. The lower bentonite seal, beneath the response zone, and the filter section up to the depth the piezometer is to be installed will need to be in place before the piezometer tip and the tubing are inserted. Assembly of the piezometer should be carried out on site, ensuring the tip remains fully saturated. The tip needs to remain saturated because any air bubbles within the it would compress under pressure and therefore affect the accuracy of the readings.

The borehole, which should still be cased to the level of the installation, should be filled with water at least above the expected level of the groundwater. The attachment of the tubing to the piezometer tip uses compression joints and therefore requires the fitting of compression olives and a retaining nut on site while keeping the unit saturated. This can be particularly difficult if the weather is cold. The piezometer unit has a reverse threaded connecting port to which a thin rod can be attached.

Figure 4.6 A pneumatic piezometer (image courtesy of MGS)

The rod is supplied in sections to be long enough to reach the desired depth of installation, thus avoiding dangling the tip in the borehole from the flexible tubing. Once the tip has been seated onto the sand filter, placed earlier, the rest of the filter material can be added around the tip. With the tip secured at the correct depth, the rod is then unscrewed and removed. After the sand has settled and its depth to the top of the filter zone measured, the upper bentonite seal can be placed; this should be at least 0.5 m thick. The rest of the hole is then grouted as the casing is removed. Great care is required when doing this to avoid pulling the tubing from the tip or pulling the tip from its location. The twin tube is carefully passed through each length of the casing as it is removed. This process is aided by ensuring there is sufficient length of tubing to be threaded through the casing. It is also recommended to incorporate loops into the tubing both in the hole and, if the monitoring point is to be some distance from the borehole, within the cable trench running from the top of the borehole to the monitoring point. The loops will provide sufficient slack should the borehole grout settle or if the trench is subjected to traffic that might cause settlement of the trench fill.

Provided the piezometer tip is free of air, these piezometers provide an accurate measurement of the water pressure at the point of installation. For this pressure to be meaningful, it is important to know the exact depth to which it applies. Care is therefore needed to measure the depth of installation accurately. The water level can then be determined knowing the level at which the piezometer tip was installed, and the pressure recorded as the diaphragm reseals. Where groundwater level fluctuations are anticipated, it is important to install the tip below the lowest point at which the groundwater may fall; otherwise, there is a risk that air will enter the tip, which will reduce the accuracy of the readings.

One of the main advantages of this type of piezometer is that, owing to the flexibility of the twin tubes, the point of measurement can be remote from the borehole. This can be useful if, for example, pore water pressures need to be measured beneath a new embankment, as it allows the boreholes to be located directly beneath the embankment, but the monitoring point to be situated at the edge of the construction. This requires the twin tubes to be run in a shallow trench to the monitoring point. Where convenient, a single monitoring point can be used to make observations of several instruments in an array of boreholes in close proximity, such as those used as observation wells for a pump test. However, with the tubing of several boreholes being terminated at a single access point, care is needed to ensure each piezometer tube is clearly identified.

4.2.6   Hydraulic piezometer

The hydraulic piezometer is generally used in situations such as earth dams where the soil is not fully saturated, such that the pore spaces are filled with both air and water, and where the pore air pressure may be much higher than the pore water pressure. To measure the pore water pressure will typically require a 'high air entry' value for the ceramic tip.

Ceramic tips can be either high or low air entry – denoted by HAE and LAE, respectively. The tip features are defined below, and typical specification parameters are shown in Table 4.1.

High air entry (HAE) – describes a piezometer ceramic tip where the pores are very small and thus it has a high resistance to air entry into the chamber. These are normally used to monitor pore pressure in clay core dams and embankments that may not be fully saturated. The surface tension of the water in the very fine pores of the filter prevents the entry of air. HAE filters must be saturated before installation by means of a saturation device along with the use of deaired water within the system. These piezometers may be supplied fully saturated and sealed in a plastic sleeve, which should be removed just before installation.

Table 4.1 Typical values for LAE and HAE piezometers (air entry value used by suppliers to define the parameters above)

	Low air entry (LAE)	High air entry (HAE)
Pore size (μ)	60	1
Permeability k (m/s)	3×10^{-4}	2×10^{-8}
Air entry value	5	100

Low air entry (LAE) – means air can easily enter the tip chamber because these have larger pore spaces that will allow air to pass at lower pressure. They are often used in permeable saturated soils such as sand and silt. They do not need to be saturated by a special device because the porosity of the filter permits water entry with simple immersion in water for a few hours. Typically, LAE tips are used where there is little or no risk of the water level dropping below the tip, ensuring the tip remains saturated at all times. This type of piezometer should be used if the installation is to be fully grouted; alternatively, flushable devices may be considered.

Hydraulic piezometers have twin plastic tubing connected to the piezometer tip, which consists of a ceramic pot (either HAE or LAE) that allows water into the chamber and is directly connected to the twin tubing by two ports. A typical installation is shown in Figure 4.7. The twin tubing allows the piezometer tip to be flushed with water to remove air that will influence the pressure reading.

The twin tubing is run to the monitoring point and the pore water pressure can be measured in one of four ways:

(a) manometer system
(b) Bourdon gauge

Figure 4.7 Schematic for the installation of a hydraulic piezometer (authors' image)

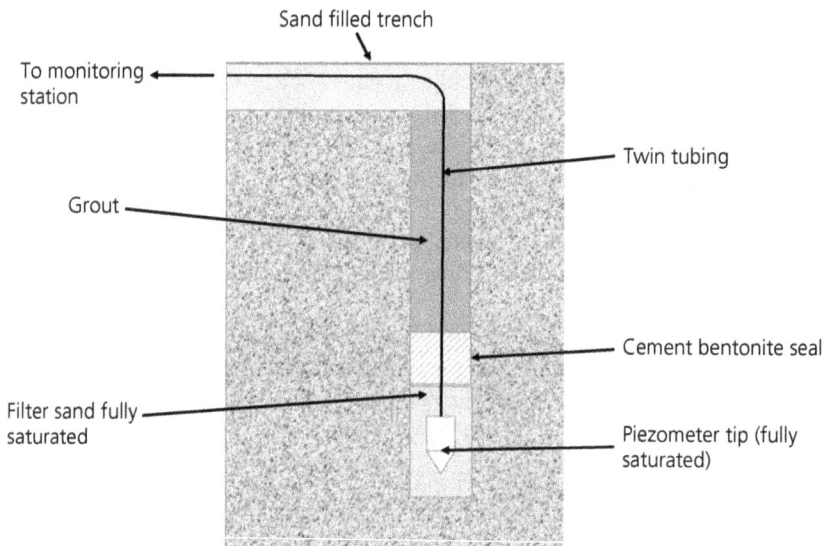

(c) terminal panel and pressure transducer
(d) scanner and data logger.

Methods *(a)* and *(b)* require manual reading while methods *(c)* and *(d)* may be logged electronically.

Typically, the cabling for several instruments will be run to the monitoring station where a manifold is used to connect to each instrument, allowing to be read in turn. This type of set up is easier to operate using the manual options.

If the pressure in the system drops below atmospheric pressure, there is a risk that air will form in the system. It is important to keep the system free of air to avoid erroneous readings, and therefore if this occurs deaired water can be flushed through the twin tubes and around the chamber in the tip. This should be done at a low pressure so as not to increase the water pressure in the materials surrounding the tip. During colder months, antifreeze may need to be added to the system and, if this is the case, the potential environmental impact should be considered because this may introduce antifreeze into the surrounding ground.

Although hydraulic piezometers are capable of recording negative pore pressures, the range – which is dependent on the depth of installation – can be limited. It is advisable to adopt the flushable type of piezometer if this is required (see Section 4.2.7).

4.2.7   Flushable piezometers
The flushable piezometer has been designed for use in partially saturated soils (see Figure 4.8). It is similar to the hydraulic piezometer in its construction, with twin tubes that can be used to flush the pot with de-aired water to expel air. The tip is of HAE low permeability ceramic. A pressure

Figure 4.8 Detail of a flushable piezometer installation and tip (image courtesy of Geo Observations)

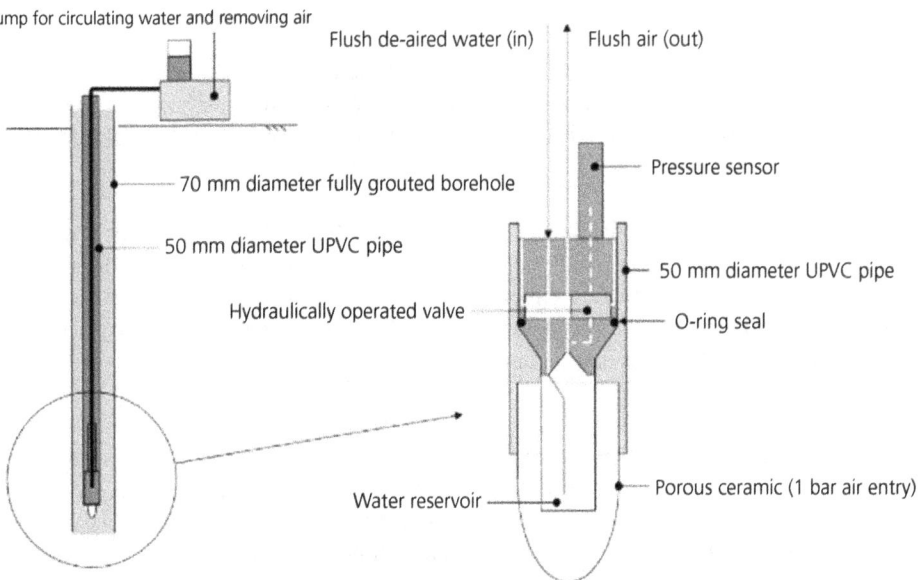

Picture courtesy of Geotechnical Observations Limited

transducer, fitted to the top of the piezometer cell, measures the pressure in the piezometer tip. Readings can be collected by way of a data logger at surface.

This type of piezometer can measure pore suctions and is particularly useful in environments where shrinkage and swelling might be experienced from seasonal moisture content changes, such as embankments. They may also be used to conduct permeability testing.

The piezometer is easily deployed using most rig types, although in confined positions the drop weight dynamic sample rig is commonly used. The piezometer tip is usually fully grouted. As with hydraulic piezometers, provision of antifreeze may be required in cold weather.

4.2.8   Electric piezometers

4.2.8.1   General

There are three types of electric piezometer: the vibrating wire piezometer (see Section 4.2.8.2), the strain gauge piezometer (see Section 4.2.8.3) and pressure transducers (see Section 4.2.8.4). The main difference is that the strain gauge piezometer cannot be used with long connecting cables because the reading is affected by the cable length. They all react very quickly to variations in pressure.

The vibrating wire piezometer is not always stable in chemically aggressive ground conditions or where there is a strong electromagnetic interference (EMI).

Vibrating wire piezometers are not affected by the length of the cable, meaning the monitoring point can be some distance away from the point of installation, even several kilometres.

Electric piezometers can be used to provide electronic logging with either a logger situated at the top of the borehole, within the cover, or using a wireless node similarly situated within the cover.

4.2.8.2   Vibrating wire piezometers

The vibrating wire piezometer, as shown in Figures 4.9 and 4.10, is commonly used to measure water pressure at the point of installation. It can be installed with most methods of forming a borehole and does not necessarily require access to the borehole position once the installation has been completed.

Typically, a piezometer will be installed in a small sand cell (see Figure 4.11), or it can be fully grouted, in which case the LAE tip may be installed using a bentonite rich grout. It is considered that because bentonite grout has a very high water content, close to 100%, the water within the grout is in close contact with the piezometer, and will be in hydraulic continuity with the surrounding groundwater. This ensures a rapid measurement of water pressure/level fluctuations is possible.

Each individual piezometer has four connecting wires, two of which are used for recording the frequency while two are connected to a thermistor for temperature measurement. The latter is a function that, in the authors' experience, is rarely used, although it does provide very useful information. Typically, the temperature within a borehole will be very stable, and so a fluctuation may indicate that there is an issue with the general reliability of the piezometer.

Figure 4.9 A vibrating wire piezometer note calibration details on body (image courtesy of MGS)

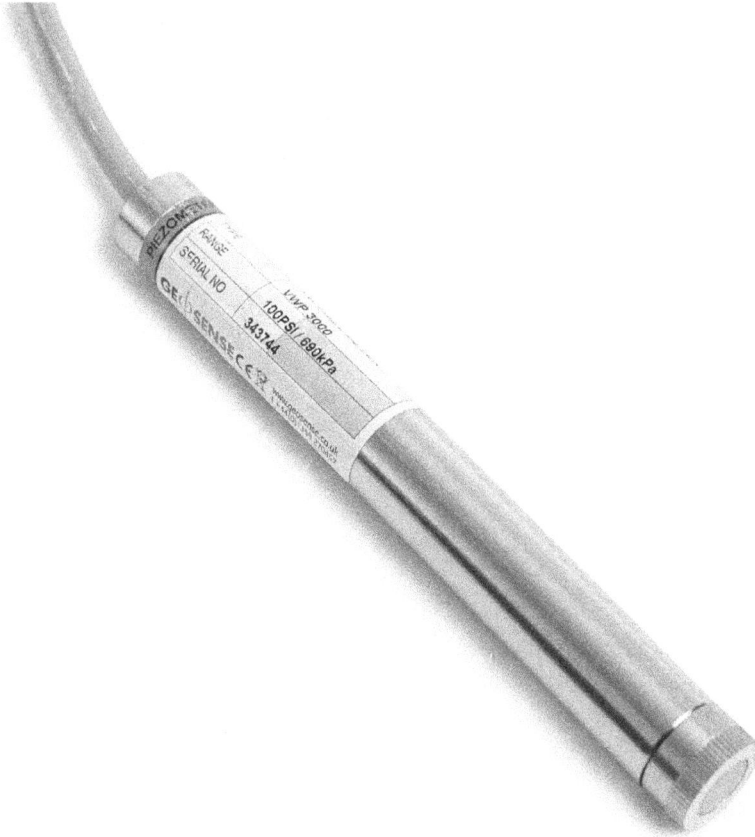

Figure 4.10 Diagrammatic section through body of a vibrating wire piezometer showing key component parts (authors' image)

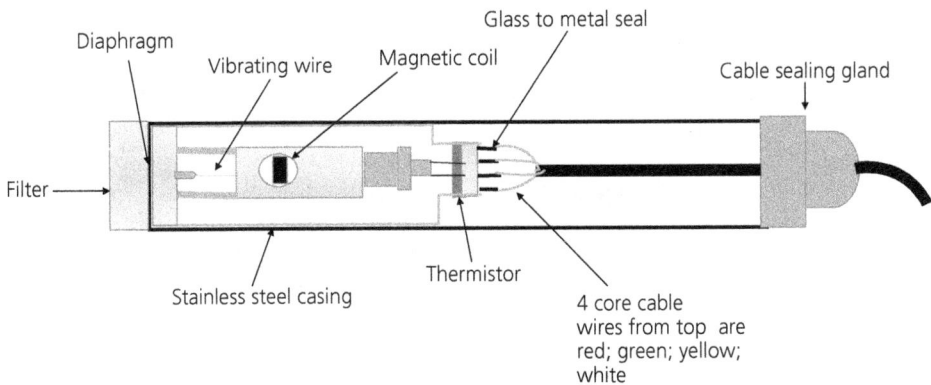

183

Figure 4.11 Typical installation detail for a vibrating wire piezometer (authors' image)

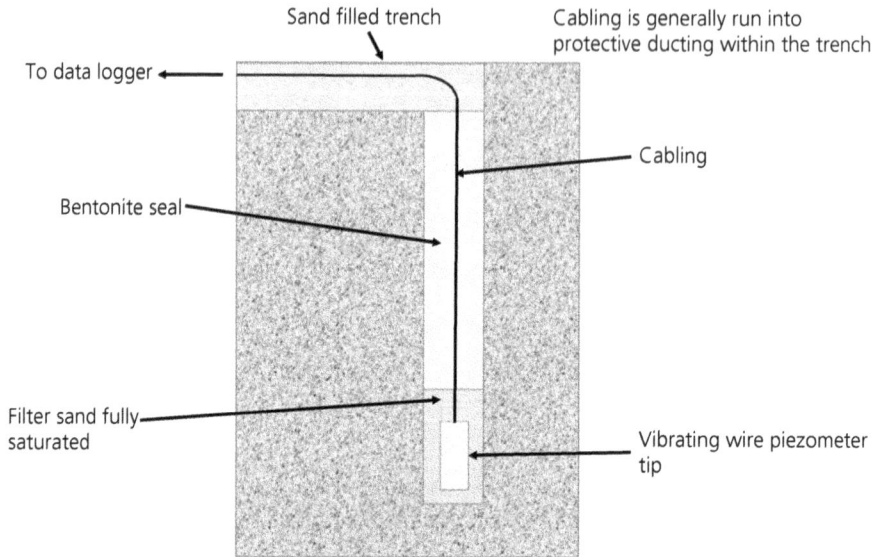

The vibrating wire piezometer may be used in situations where the pore water pressure can be negative, although the flushable piezometer has a more reliable and greater range in these situations. Very high suctions can irreparably damage the diaphragm of the vibrating wire piezometer. If exposed to long periods of negative pore water pressure, the vibrating wire piezometer chamber will tend to fill with air, which will significantly reduce the accuracy of the readings.

Vibrating wire piezometers are simple devices and, in some configurations, have no electronics. Therefore, they do not suffer from any form of long-term degradation, and some instruments installed in the 1960s are still functioning today. This enables the logging devices to be upgraded in line with modern technology but still read existing instrumentation. It also means that with the use of an isolation barrier they can be used in intrinsically safe (exclusion) zones such as fuel storage sites.

Within the piezometer body a filter, which is usually of LAE ceramic, allows water into a small chamber with a diaphragm at the opposite side to the filter. A taut wire extends from the diaphragm to a fixing point. The pressure in the chamber will equal the pore water pressure in the ground and this will deflect the diaphragm, thus changing the tension in the wire. When a measurement is taken a brief voltage excitement is applied to two electromagnetic coils on either side of the wire which induces a magnetic field. The wire oscillates (vibrates), which generates an alternating current. The change in tension of the wire causes the resonant frequency of oscillation of the wire to change. The frequency of oscillation is directly proportional to the root of the applied pressure:

$$\text{frequency } f = \{[\sigma g / \rho]^{1/2} / 2l\} \text{ Hz}$$

where σ = tension of the wire; g = gravitational constant; ρ = density of wire; l = length of wire.

The instrument calibration enables the frequency to be converted to a pressure that relates to the water pressure at the point of installation. Each piezometer will have different characteristics and thus each piezometer must be individually calibrated. The relationship between frequency and the tension in the wire, and hence pressure, is not linear, requiring a calibration curve to be used to provide the pressure reading for each instrument.

The calibration curve is a polynomial of the form:

$$y = ax^2 - bx + c$$

where a, b and c are factors unique to a particular piezometer; y = pressure at the point of installation; x = current reading from the piezometer.

Because each unit has a unique calibration factor, it is important to be able to identify each piezometer's wiring at the monitoring point. While long cable lengths are possible, the wires should not be joined or cut because this will change the calibration.

The calibration and operation of the unit should be checked immediately prior to installation, with the piezometer at surface. It should again be checked by taking readings at measured intervals as the instrument is lowered into the borehole. For this operation, the borehole should be either full of water or at least with the water level just above the water table. The precise depth of installation should be recorded.

It should be noted that in Figure 4.12, which shows multiple vibrating wire devices, the piezometers have been installed with the porous filter end facing upwards, which helps ensure that the unit remains de-aired.

Figure 4.12 A multilevel vibrating wire piezometer installation (authors' image)

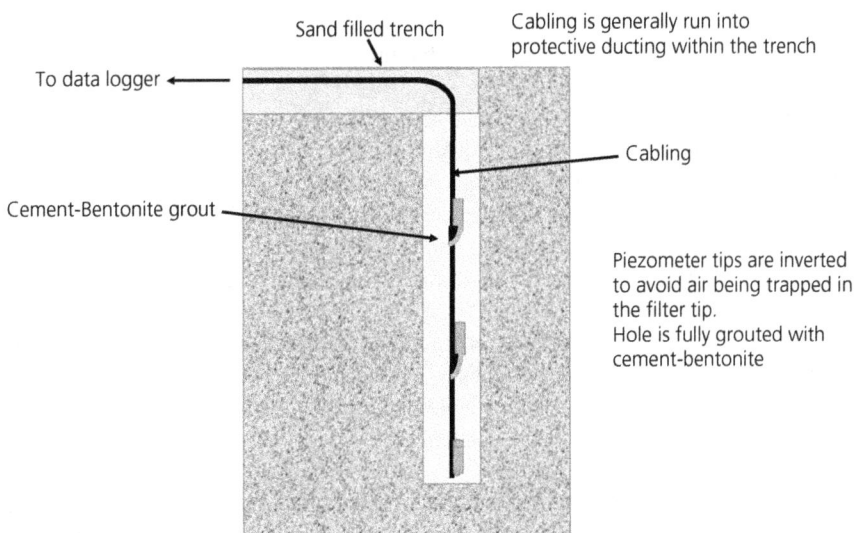

4.2.8.3   Strain gauge piezometer

Similar to the vibrating wire piezometer, the strain gauge piezometer uses the same type of body to house the gauge, and a filter tip with either HAE or LAE characteristics. The pressure measurement is achieved with internal strain gauges, which convert the displacement of a diaphragm, as it reacts to variations in water pressure, into an electronic signal. This uses the principle that the electrical resistance of a wire under tension is proportional to its cross-sectional area and when the tension in the wire is altered by a variation in water pressure the cross-sectional area of the wire also changes.

As with vibrating wire piezometers, each piezometer requires a calibration to convert the readings, measured in milliamps, to a gauge reading. Each instrument will have a unique calibration; it is therefore very important to be able to identify individual piezometers.

Strain gauge piezometers have rapid reaction times and are highly accurate, so they are particularly suited for dynamic measurements where influxes of water or changes in pressure are expected.

The main limitations of this type of piezometer are that they cannot be used where cable lengths need to be increased, as any changes in cable length will affect the calibration. They are also affected if moisture enters the cable sheathing or the piezometer's sealed body.

4.2.8.4   Pressure transducers

The pressure transducer, commonly known as a diver, measures and records pressure and can be installed in a standpipe or standpipe-piezometer. This is achieved by lowering the electronic piezometer into the standpipe below the water level. In situations of variable water levels, obtaining a continuous set of monitoring data depends on the diver remaining submerged at all times. However, as there may be only a basic understanding of likely water levels prior to undertaking the monitoring, a conservative approach is often taken, suspending the diver at a relatively deep level in the standpipe. Where this approach is taken, or where a large range in water levels is expected, care should be taken to use a diver that is designed to operate under the range of pressures anticipated.

As well as being designed for different pressure ratings, divers are also of different physical sizes, and if a 19 mm dia. standpipe is being used, a smaller, 'micro' style diver will be needed. Advice on these factors should be sought from the manufacturer.

These transducers measure the water pressure above the point of installation using a sealed air entrained diaphragm, the pressure exerted on the diaphragm being converted into an electrical signal that can be calibrated to measure changes in pressure. The transducer is also able to measure temperature and fluid density, and these data are then used to determine the pressure at the point of measurement. The transducer is programmed and houses a logger capable of taking readings and storing them over several years, typically resulting in some 12,000 readings.

The standard type of diver is self-contained, incorporating a battery and logger within the body of the instrument, and therefore does not need to be linked to the surface by tubes or wires. It is simply suspended in the water column on, for example, a stainless steel cable. The unit is periodically retrieved from the borehole so the data can be downloaded, enabling graphing and display of the readings obtained.

One complication with a diver is that the transducer records total pressure – that is, the barometric pressure as well as the head of water above the diver. As such, in order to assess the water pressure, the transducer reading needs to be 'compensated' by subtracting the relevant barometric pressure at the time of the reading. As barometric pressure typically changes relatively slowly, in many instances it would be sufficient to rely on values based on local weather station data. However, best practice is to take a site-specific measurement of barometric pressure that coincides with each diver reading. This can be achieved by using a separate diver, generally referred to as a 'baro diver', placed above the water level in one of the boreholes being monitored. This diver can be programmed to take a reading of barometric pressure at the same time as the water monitoring divers are taking a total pressure reading, making 'compensation' a simple task. The baro diver must not become submerged and so is often suspended at a high level in one of the standpipes. If very shallow water levels are anticipated, an alternative is to keep the baro driver above ground – for instance, stored in a site office.

There are some variations to the standard diver, including a 'vented diver', which has a small-diameter vent tube linking the diver to the surface. This maintains the device at barometric pressure, meaning that the diaphragm reacts only to pressures exceeding the barometric pressure – that is, the groundwater pressure. This design therefore negates the need for compensation using separate barometric pressure readings. With such devices, however, care needs to be taken to ensure the vent pipe is not kinked or becomes blocked. For routine applications, the standard diver is still favoured, for its cost effectiveness and relative simplicity.

Other variations include divers that are linked to the surface by a data cable. These enable live readings to be taken, data to be downloaded or devices reprogrammed while they are in situ. Such data cables can also be linked to telemetry systems that allow data to be accessed remotely using various communication methods such as a radio link, mobile phone network or even satellite link.

An advantage of a diver placed in a standpipe-piezometer over other piezometer types that are permanently installed in the backfilled borehole is that water levels can be checked manually using a dip meter (Section 4.2.2). Periodic manual dips not only provide an independent validation of water levels, but also are commonly used to define the depth of the diver, which is essential if the water pressures it is measuring are to be meaningful. As the manual dip determines the depth from ground level to water level, and the head of water measured by the diver determines the depth of the diver below water level, the sum of these two measurements can be used to define the diver depth below ground level. Plotting water levels ascertained from manual dips and diver data together generally provides good continuity, particularly if the diver depth is defined using the head of water measured by the diver after it has been in place for some time. If diver data are to be downloaded on a regular basis, which is recommended, it is good practice to take a manual dip of the water level just before the diver is removed from the borehole and use this with the diver's corresponding water pressure reading to define the diver's depth over the monitoring period just completed.

Another benefit is that, once the period of monitoring is complete, divers can be removed and reused elsewhere, making them relatively cost-effective.

4.2.9   Multilevel water monitoring systems

4.2.9.1   General

There are a number of proprietary systems that have been developed to enable pore water pressure to be measured at multiple levels, along with testing and sampling, within a single borehole. While the systems are generally more expensive when compared with the systems discussed in the preceding sections, it can be cost effectively installed into a single borehole rather than several different boreholes, thus possibly reducing the need to drill numerous boreholes. These systems are particularly useful for sites with limited space, or where monitoring during construction might be difficult owing to restricted access.

In general, all multilevel installations involve pipework that has access ports at intervals along its length and are able to seal each section of interest from other sections, usually with packers (inflatable sleeves), as seen with the Westbay System, or by using a segmental interior to the access pipe as seen with the Waterloo System. Most multilevel systems are able to measure pore water pressure and sample groundwater. Some systems, such as the Westbay System, can also be used to conduct certain tests, such as for permeability, at different levels within a single installation, by installing a testing and sampling port rather than a monitoring port at the desired test location.

In general, all require a borehole of at least 100 mm diameter and careful installation to ensure the port locations are positioned correctly, together with any seals or packers, to ensure each zone is isolated from others and is located within the zone of interest.

4.2.9.2   The Waterloo System

This comprises a single tube where the interior is divided into up to six segments. A port can be made at different levels in individual segments thus allowing water levels to be determined at six separate locations within the borehole. Water level can then be determined using a micro dip meter. This system might also be used to sample water from various levels within the borehole.

4.2.9.3   The Westbay System

In this, the pipe is supplied in diameters from 76–160 mm and uses packers, consisting of expandable polyurethane sleeves, to isolate each zone for measurement. Between the packers and in the zones of interest smaller diameter plastic sleeves allow for the measuring tools to access any particular zone. The tools are designed to lock into ports within the sleeve. There are two types of port: one for measurement and another for pumping. The system can be installed to depths in excess of 300 m. In aggressive ground conditions, the system can be supplied in stainless steel.

The measurement ports are located by the measuring probe, which then latches to the port and opens it to enable pore pressure measurement and sampling.

The pumping port enables access for injection or removal of fluid. This permits various processes to be undertaken, such as hydraulic conductivity/permeability testing, purging, sampling and the injection of tracers that can be used to assess flow direction.

This system is very flexible and can be readily decommissioned by deflating the packers, allowing easy removal from the borehole.

4.2.9.4   Vibrating wire piezometers

These may also be used to measure water pressures at several different depths in a borehole and there are proprietary systems that string piezometers together using an access tube and cabling. The piezometers are usually fully grouted in the borehole, which makes installation much simpler. A typical arrangement is shown in Figure 4.12.

The multipoint vibrating wire piezometer is often deployed to monitor water pressure variations within a dam or embankment and can be used to determine the flow net through the dam to enable comparison with graphical flow net models.

4.2.10 Fibre optic piezometer

The use of fibre optic cables is becoming more common; they offer much smaller instruments on long flexible cables that are capable of extremely accurate measurement and fast transmittal of data. In addition, unlike most other means of measurement, electromagnetic forces or harsh conditions do not influence them.

The fibre optic piezometer is significantly smaller than other electric piezometers, although similar in appearance. The piezometer uses a miniature micro-optical mechanical system (MOMS) to measure pressure. The device is less than 5 mm in diameter and 54 mm in length.

The pressure transducer has a flexible diaphragm assembled on top of a sealed vacuumed cavity, and the pressure measurement is based on Fabry-Pérot white-light interferometry. Pressure creates a variation in the length of the Fabry-Pérot cavity, which is the distance between the inner surface of the flexible diaphragm on one side and a reference optical surface attached to the lead optical fibre on the other side.

These piezometers can be embedded within a structure or grouted in position. The small size of these piezometers, along with their very stable nature and high degree of accuracy, enables them to be used within structures such as dams, retaining walls and tunnels, as well as in harsh environments. They are very accurate and provide data stability over long distances. Unlike vibrating wire piezometers, the fibre optic piezometer is unaffected by EMI.

The readout units for these devices are expensive, and use a considerable amount of power, which makes them less popular than other types of piezometer.

4.2.11 Monitoring direction of flow

The direction of flow of groundwater may be determined by using a tracer, but for this to be successful it is necessary to have some indication of the potential direction of flow. That will be dependent on the hydraulic head between two points, which might be influenced by dipping strata, or large elevated bodies of water such as lakes, seas or rivers. Tidal influences may also impose a hydraulic gradient on the groundwater, which will induce flow.

The use of tracers is controlled, and it may be necessary to obtain permission from the Environment Agency to conduct a groundwater tracer survey. The test usually involves adding a dye or chemical to the groundwater within one 'dosing well', and then using a series of other wells, positioned at various distances and directions from the well, as observation points.

Typically, concentrations of sodium chloride or red dyes are used. Testing can be lengthy and is dependent on the permeability of the strata within which water is flowing. By monitoring the concentration of the additive at each observation well, the direction of flow may be determined.

4.2.12 Barometric pressure

In some situations, it is necessary to compensate the water pressure measured by a piezometer for the effect of barometric pressure to give a correct value of water pressure and therefore water level. The 'baro logger' is a pressure sensor that is used to measure the barometric pressure at the monitoring location (see also Section 4.1.8.4).

Barometric pressure is generally assumed to be relatively constant at any one point in time within a radius of about 20 miles and within 300 m of elevation. The barometric pressure is a function of the air pressure that is affected by weather systems. In general, high pressure is associated with periods of calm weather while low pressure is associated with stormy weather. This is particularly relevant to recording gas levels where higher levels of gas concentrations in the air are generally associated with barometric pressures less than 1000 mb. This is because gases come out of solution as the atmospheric pressure drops. It is recommended that, where readings of gas concentrations are required, at least one set of readings should be obtained when the barometric pressure is below 1000 mb. This may require readings being taken at different times to those required by a specification, which would often be taken at regular intervals. For further information on measuring gases, see Reading and Martin (2025).

The following formula is used to correct the water level readings from piezometers:

$$H = (P - Pr) / pg$$

where H = water level; P = water pressure from piezometer; Pr = atmospheric pressure; p = density of water; g = gravity.

Most vibrating wire wireless dataloggers are equipped for barometric measurement, and barometric sensors can be added to any datalogger setup to enable this correction to be automated.

4.3. Ground movement

4.3.1 General

There are many situations where it is useful to make observations of ground movement. It may be required, for example, to verify the stability of man-made slopes as they are being constructed or assess natural slopes that are either currently unstable or form part of a development where they may become unstable in time.

During construction, settlement may require monitoring, either to provide warnings when settlement may become excessive or to control timings for staged construction that may be required when building an embankment on poor ground. In cases where a water table is reduced by a construction process, changes to the effective vertical stress may induce settlement of the ground, causing distress to any structure or infrastructure in the vicinity.

There are several methods that may be used to monitor the ground for lateral movement or vertical settlement. These methods range from simple visual observation to physical measurement, using a range of techniques from instruments installed in the ground to satellite radar imagery. The methods adopted depend on how important the risk is perceived, and hence the budget the client is prepared to make available.

The following sections provide some detail for the most commonly used methods.

4.3.2 Visual observation

For most ground movement situations, a visual inspection is a good indicator that movement is occurring. This is simple and may be catalogued with a series of photographs taken from fixed locations at each inspection visit. Observations should be checked against a baseline record, ideally made before any other work at the site starts. The timing of visual inspection visits should take place at regular intervals, as well as at the start and end of each construction phase.

Lateral movement of a slope may be monitored by surveying a line of wooden stakes or metal pins driven into the ground across the slope. The stakes should initially be positioned in a straight line, and on larger slopes several lines might be deployed. Once positioned, it is relatively easy to see if movement is occurring by sighting along the line. If any pegs are seen to be off the original line, this will be an indicator that movement has occurred. As such observations are only relative between the pegs, surveying the points from a stable datum position will enable the absolute movement to be quantified.

The disadvantage of using wooden pegs is that they can be moved by other events not related to the slope movement itself. To avoid this, datum markers or permanent ground markers may be installed. These are steel pins of between 300–500 mm in length, some proprietary pins being provided with a brightly coloured plastic head caps that enables easy recognition. These caps are also recessed to allow them to hold a survey target for use with a total station survey. If a total station survey is conducted, each pin can be individually located, and any movement occurring between surveys may be readily identified and quantified in three dimensions. Pins may be concreted in place for extra security.

When using pegs or pins, it is essential to check for services prior to installation.

4.3.3   Slip indicators

The simplest borehole installation to determine if a slope is moving is that of a slip indictor. This involves the installation of a small-bore access tube and two brass or stainless-steel rods which will pass freely up and down the tube (see Figure 4.13). A 19 mm dia. standpipe is used to guide

Figure 4.13 Slip indicator equipment including 9 mm dia. standpipe guide tubing, flexible small-bore tube, brass rod, securing line and base stop (image courtesy of MGS)

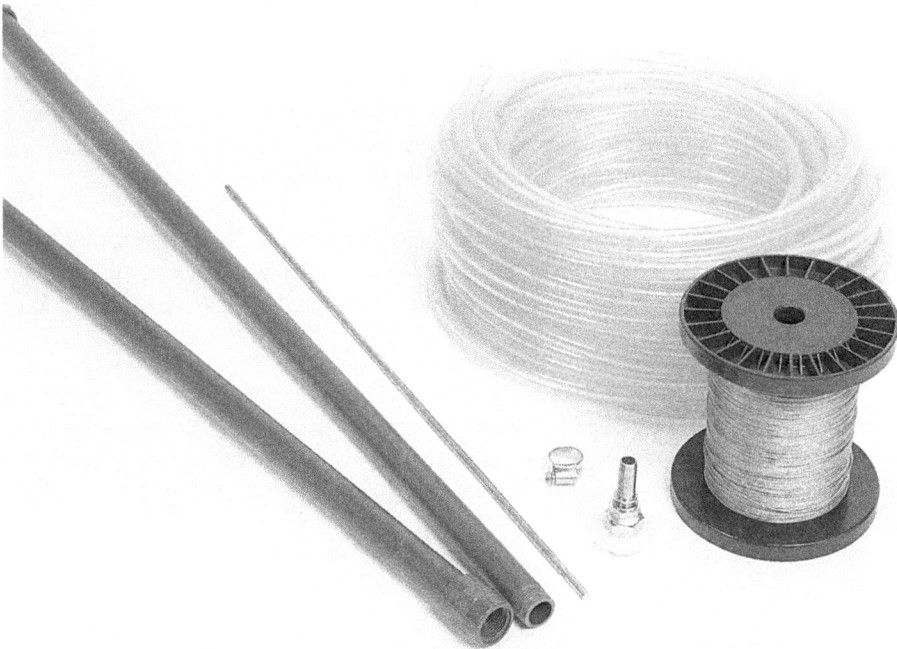

the flexible tubing and keep it both straight and central in the borehole as the borehole is filled with sand. The flexible tubing is securely connected to a base stop that is of sufficient size to prevent it from passing up the standpipe. This enables the flexible tubing to be held taut and straight in the standpipe while the sand is placed by way of a tremie. The standpipe is slowly withdrawn as the sand fills the borehole. Once in place, one of the rods is lowered to the bottom of the flexible tube secured on a thin cord (usually fishing line), the line being secured at the top of the hole ready for use at each monitoring visit. On installation of the rod, the line should be held taunt without raising the lower rod and a datum position marked on the line. Crimping a fishing weight onto the line is a useful way to mark this upper datum point. The second rod is similarly fixed to a line and is deployed from the surface if required.

At each monitoring visit, the lower rod is drawn up the tubing, and if it passes easily to the surface, it can be deemed that no significant deformation of the tube, and therefore ground movement, has occurred. Should the rod jam at a certain point, the length of line removed from the hole to the position where the jam occurs can be measured to determine the depth. The second rod is then lowered down the tubing from the ground surface and the position it jams is again measured. The two measurements will provide upper and lower positions in the borehole and indicate that a slip surface is potentially located between the two points.

Such information is very useful when analysing a slope failure using a slope stability programme as it helps define a position of the slip surface, which would otherwise be unknown. Often, only the position of the slip surface at the crest of the slope will be known, as this may be indicated by the formation of a tension crack. Most slope stability programmes analyse numerous possible positions of the slip surface to obtain the most critical alignment within the slope. By locating known positions of movement, a more accurate analysis of the slip surface is obtained.

4.3.4 Inclinometers

4.3.4.1 Probe type inclinometers

The inclinometer has been used for many decades as the primary method to determine lateral movement, mainly for slopes but also in other situations such as monitoring the lateral movement of retaining walls. An inclinometer is a torpedo-like probe, designed to record its inclination from the vertical, that is lowered down an access tube, sometimes called 'inclinometer casing', to map the profile of the tube and determine how that profile might change over time.

The installation of the access tube requires meticulous planning. It is important that the tube extends below the expected level of ground movement into the stable strata, a minimum depth of at least 2 m below the slip surface being recommended.

The tubing is made from extruded acrylonitrile butadiene styrene (ABS), which is a combination of two plastics and rubber making it strong and flexible. The tubing, which is manufactured with tight tolerances for spirality, is supplied in 1 m, 1.5 m and 3 m lengths, and at diameters of 70 mm or 83 mm. Each tube has four grooves, or 'keyways', cut into its inside wall, forming two opposing pairs running the full length of the tube (see Figure 4.14(b)).

Tubing is generally supplied with quick fit connections, as shown in Figure 4.14(a), or using connecting sleeves. Tubes cannot be screwed together as the keyways need to align precisely between each section. The connecting sleeves allow the tubing to be cut to length and joined. However, the sleeves require gluing and riveting to the tube, which is particularly challenging if, owing to its overall length, the tubing needs to be assembled as it is inserted down the borehole. Once joined,

Figure 4.14 Inclinometer tubing showing (a) connections and (b) internal keyways (images courtesy of MGS)

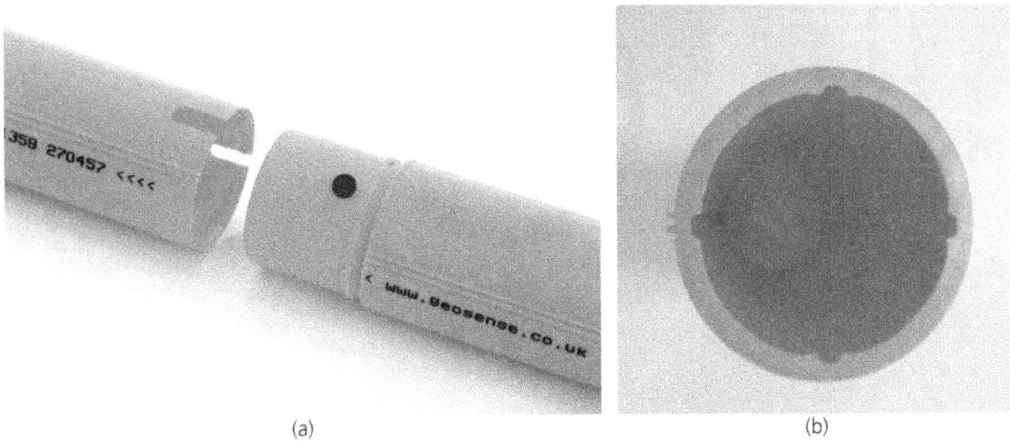

(a)   (b)

the joints are usually wrapped in a fabric-based waterproofing tape to protect them from ingress of grout. A bottom cap is fitted to the base of the lowermost tube.

The tube should be lowered to the bottom of the borehole and then grouted in place with a bentonite cement grout. Great care will be required if the tubing is assembled as it is inserted into the borehole to ensure is does not slip down the hole. For short sections, the tubing can be assembled next to the borehole, prior to lowering it into the hole. Where the tubing is placed below the water table, it will be necessary to fill the tube with water to counter the hydrostatic pressure in the borehole and enable the tubing to be inserted to the required depth. The tubing should be orientated, where possible, such that the keyways are aligned in the optimum direction to detect movement, with a selected keyway (generally referred to as the 'A0' keyway) corresponding to the direction of anticipated movement (for example, down-slope), as shown in Figure 4.15. However, an attempt to correct the tube orientation once grouting has started should not be made as this will distort the tube.

The grout should be designed to match the strength of the surrounding ground. The design mix will vary in respect of the proportions of water, bentonite and cement used, and laboratory strength trials are recommended to determine the most comparable mix. Guidance on grout mixes can be found in the BDA *Manual of Rotary Drilling* (Tomlinson, 2021). The tubing should be held centrally in the borehole and the grout placed using a tremie pipe.

As soon as possible after the tubing has been installed, a dummy probe – which is simply a wheeled probe attached to a nylon cord – should be run to the base of the tube in both orthogonal orientations of the keyways, to ensure the tubing is functioning correctly and the keyways are free from debris. This should be done before the grout has set, and care should be taken to avoid wobbling the tubing and causing a void between the tube and grout. However, if the probe is used as soon as the grouting has been completed and the tube does move slightly, the grout should reform around the tube.

Once the grout has set and the borehole has stabilised, typically after a week or so, the first set of inclinometer readings may be taken. These will be the 'datum readings' or 'base readings', against which all future profiles will be compared. Typically, at least three sets of readings are taken using the

Figure 4.15 Keyway alignment to the down slope direction (authors' image)

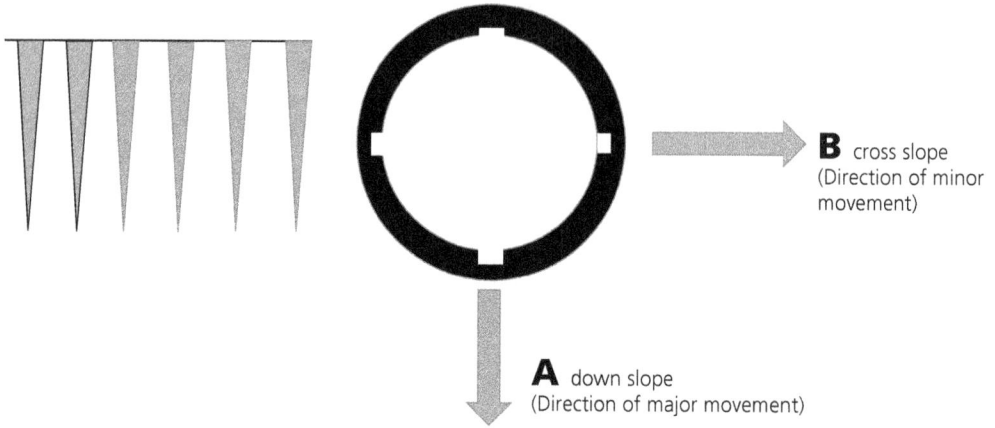

B cross slope (Direction of minor movement)

A down slope (Direction of major movement)

inclinometer probe, the objective being repeatability in the profiles it produces. Occasionally, a reading error may occur, and this will affect the whole profile. The reason for carrying out three sets of readings is that if one of them contains a reading error, the other two may still show a high degree of repeatability, giving confidence that they are representative of the current profile of the inclinometer tube. If only two profiles are taken and they differ, it may not be possible to determine which is the representative profile. Of the three sets, one that is considered representative is adopted as the datum set.

Readings should be taken from the bottom of the tubing upwards. The base plate should be sufficiently deep to remain in the same location, thus providing a fixed datum point. The top of the installation cannot be used as a datum because it will move if there is movement of the slope or wall being monitored. In some cases, an optical survey prism is fitted to the top of the tube to allow measurement of changes in its level.

The inclinometer probe is generally between 0.5–0.7 m long, depending on the manufacturer, and is attached to a cable with markers every 0.5 m; the inclinometer is held securely in the keyways by a pair of sprung wheels, usually 0.5 m apart. This ensures the probe has exactly the same orientation in the tubing for each set of readings. After lowering the inclinometer probe to the base of the tubing, it should be left for a short period to acclimatise (predominantly to the temperature within the tube).

The probe is equipped with a means of measuring the tilt of the unit in the vertical plane as it moves up the tube. Measurement is made using one of several methods, such as a vibrating wire with force balance accelerometers, pendulums attached to rotary electrical potentiometers, strain gauge cantilevers, and – in more recent equipment – micro–electro-mechanical systems (MEMS) accelerometer devices.

Some inclinometers, referred to as 'uniaxial', take measurements in a single plane, which is aligned with the plane of the guide wheels. These devices need to be run in the borehole twice, once in the orientation generally coincident with the direction of anticipated movement, known as the 'A-plane', and then at right angles to this, in the 'B-plane'.

More sophisticated inclinometers, including those using the MEMS accelerometer technology, measure in both A and B planes simultaneously and are therefore referred to as 'biaxial'. In theory, these would only require the probe to be run in the tubing once. In practice, however, it is normal to run the device twice, turning it through 180 degrees before the second run. This enables the inclinometer to assess the same inclinations twice, and an average of the two readings is then reported. This method will reveal any discrepancies in the readings. It is conventional to label the axis of the wheels as the A-plane and that perpendicular to the wheels the B-plane.

Considering the above discussion, running inclinometers up the borehole twice for one set of readings (turning a uniaxial inclinometer 90 degrees between runs and a biaxial inclinometer 180 degrees), and then requiring three profiles to be taken, to provide confidence in which may be considered as representative, means that six passes of the inclinometer is generally needed at each location on each visit.

Older equipment requires the cable and inclinometer to be calibrated together. Once calibrated, the two parts should not be interchanged with other devices without recalibration. The MEMS system, however, has an internal calibration and may be interchanged. These systems are also available with Bluetooth, enabling wireless communication with a PDA (personal digital assistant) or mobile phone.

Readings are made at a distance equal to the guide wheel spacing, usually every 0.5 m, as the probe is withdrawn from the tubing. It is important to take readings at a fixed point each time, starting with the fixed point closest to the base of the tube. The depth for each reading is usually marked on the cable with a metal collar that can be hooked onto a cable guide, or 'gate', temporarily connected to the top of the tube. This ensures that readings are always taken at exactly the same depth. The depth of the lowermost reading in the tube is conventionally found by lowering the probe down to the very base of the tube, and then lifting it until the first depth marker can be hooked onto the cable gate.

The results are processed using the datum set of reading to normalise further readings; this enables a series of different plots to be drawn of any movement of the inclinometer pipe with respect to the base profile. This may be represented as

- absolute movement against depth
- incremental movement against depth
- cumulative movement against depth.

The results are usually presented as incremental and cumulative movement plots, the incremental plots showing spikes in displacement data at depths where movement has occurred, and cumulative plots adding each measurement of displacement to the previous location to give a more visual impression of how the tube is deforming, as seen in Figure 4.16. The plots enable the depth at which movement is occurring to be determined which, in the example of slope instability, can be used to back analyse the potential failure. Figure 4.16 shows that readings were obtained more frequently as the ground movement accelerated.

Many inclinometers are designed only to operate within a range of 30 degrees from the vertical. Other inclinometer systems, however, may be used either in an inclined position or horizontally. The MEMS inclinometer is particularly suited to these variations.

Inclined inclinometers are particularly useful to look at deformation along the sides of dams while horizontal inclinometers can be used to measure displacement or settlement of embankments, the inclinometer casing being installed transversely across the proposed embankment alignment. This is particularly useful where significant settlement from loading is expected, the monitoring

Figure 4.16 Incremental (left pair) and cumulative (right pair) inclinometer plots obtained from a moving slope over a period of about nine years, showing movement in both the A and B directions (image courtesy of SOCOTEC)

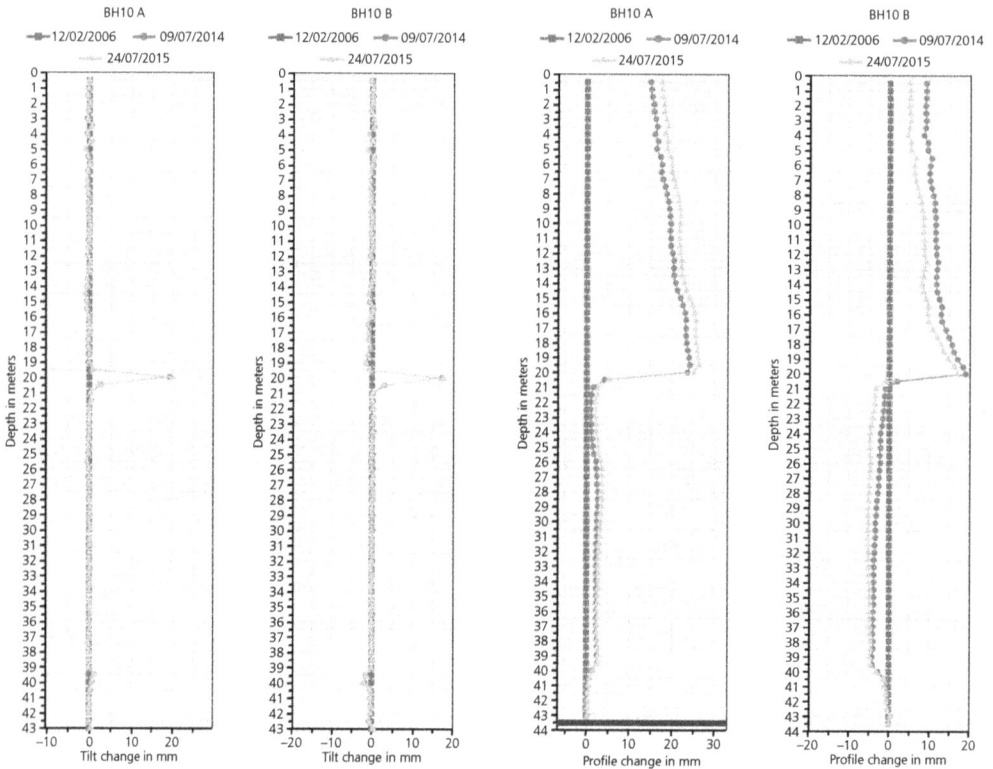

allowing the planned construction on top of the embankment to begin once the majority of settlement has taken place.

When used horizontally to measure settlement, the inclinometer forms a potential alternative method to the hydrostatic profile gauge described in Section 4.3.5. The two approaches have some similarities: the access tubing through which the inclinometer will pass is installed prior to the embankment construction, typically in a shallow trench on a bed of compacted sand. Once the tubing is in place the trench is also backfilled with sand. Datum concrete slabs are constructed at each side of the section being monitored these should be approximately at the same level as each other and above the rest of the tubing. They should be installed such that the slope at either end allows the inclinometer to pass freely through. A cord must be run through the casing, from one end to the other, typically fed into the pipe as it is assembled. The cord is used to draw the inclinometer along the tubing and to return it to its start position. Where one end is closed, or will not be accessible, a pulley wheel can be included around which the cord is passed, allowing the inclinometer to be drawn in both directions from one end. The datum slabs usually incorporate survey pins that can be surveyed at intervals to check if they have been subjected to movement. This allows the inclinometer readings to be compared with a fixed datum outside the zone of influence of loading.

The monitoring process is carried out in a similar way to that of vertical inclinometers, described above, including the need to start with a set of datum readings, which should be taken as soon as installation has been completed, and the need for readings to be taken at fixed intervals as the inclinometer is drawn from one end of the tubing to the other. Monitoring will generally take place both during and after construction, and for some time after the required embankment height has been reached, in order to determine when the majority of settlement is complete.

4.3.4.2   In-place inclinometers (IPI)

IPIs use strings of measuring devices, connected by way of a series of short rods installed within a grooved access tube similar to that shown in Figure 4.13. Orientation is relayed using electromechanical signals that can be compared with a set of datum readings taken at the time of installation. Typically, readings are collected at intervals by an inbuilt data logger that can be remotely interrogated to provide real time readings of movement. The logger also measures temperature variations that are used to compensate the effect on the readings

The IPI uses MEMS sensors, either of uniaxial or biaxial type, housed within a stainless steel body, the gauge lengths ranging between 0.5–3 m. The system is extremely flexible and can be installed vertically, inclined or horizontally. The base unit has a rigid joint that is used as the datum from which readings to the units above are referenced.

An IPI can either monitor the entire length of the borehole, using an interconnected rigid system, or if there is a specific depth of interest the IPI can be suspended on a wire rope at the required location.

Other proprietary systems use similar MEMS sensors deployed in strings that comprise sectional lengths of between 200–600 mm, or in a rugged form with a length of 1 m. Each rigid sensorised section contains three tiltmeters, temperature measurement device and a microprocessor. The string is installed within a 27 mm dia. access tube. Because the string fits loosely into the access tubing, it is possible to remove the string and potentially recover them for reuse once the monitoring period has been completed. The string can also be inserted into larger diameter tubing and so it may be retro fitted to pipework that has been installed earlier for water level reading, for example. Because the system does not require a specific casing type, it can also be installed around the circumference of a tunnel to monitor the tunnel response to loading.

The array sits in the access tube, resting against the wall, in a zigzag fashion down the tube. The three tiltmeters detect movement in any direction. By interrogating the readout from each set of sensors, and calculating the movement between sets of sensors, a three-dimensional visualisation of the movement taking place can be compiled. Readings can be taken and processed in real time and relayed to a remote monitoring point wirelessly or made available using cloud-based data capture systems.

Most in-place style inclinometers are supplied with a bespoke cover assembly. It is also possible to incorporate extensometer systems within the same in-place systems.

Typically, the results are shown with a series of curves showing the change over time, as discussed for probe type inclinometers.

4.3.5   Hydrostatic profile gauge

The hydrostatic profile gauge has similarities to a horizontal inclinometer, in that a torpedo-like probe is run along a preinstalled access pipe. In this case, however, the gauge uses a closed system of hydraulic fluid, and a pressure sensor, to determine the elevation of the probe.

The horizontal access pipe through which the probe is run is 50 mm in diameter and supplied in a single length, on a reel, so that there are no joints that could shear as the pipe deforms or on which the probe might snag. Planning is needed to select the required length of access pipe. As this method is typically used to determine settlement beneath a new embankment, the length of pipe needs to be a few metres greater than the width of the base of the embankment being monitored. The pipe is laid into a trench with 150 mm of sand in the base and then covered with a further 300 mm of sand, as shown in Figure 4.17. A nylon cord is threaded through the pipe, either pushed through with drain rods or piezometer pipe, or blown through with compressed air, and left in place. This is used to pull the probe from one end of the pipe to the other. A concrete plinth is constructed at each end of the pipe, to hold the pipe in place and provide a location to situate the profile gauge when in use. A datum pin is incorporated into the concrete plinth, to act as a temporary benchmark.

The steel probe of the profile gauge is attached to a reel of flexible nylon tubing, filled with a hydraulic fluid. The end of the tubing is terminated at a hydrostatic pressure transducer that measures the elevation of the probe relative to the reel.

A profile is measured by attaching the probe to the cord in the access pipe and pulling it through the full length of the pipe, the hydraulic tubing being taken off the reel as it is extended. The probe is then drawn back along the access pipe, gradually winding the hydraulic tubing back onto the reel. This process is paused at metre intervals, which are marked on the hydraulic tubing, and an elevation reading taken at each of these intervals. Once the probe has reached the near end of the pipe, a reading is also taken of the datum pin. The reel is usually linked to a PDA by way of Bluetooth, allowing all readings to be stored digitally.

The readings are plotted to provide an elevation at each interval and therefore a profile of the pipe. A datum set of readings should be obtained as soon as the access pipe has been laid. Further profiles are then compared with the datum readings, both during and after construction of the embankment, to assess the settlement magnitude and rate induced by the loading of the ground. Because the settlement of the embankment may affect ground levels immediately adjacent to the structure, the datum pin should be surveyed at the time it is installed and checked periodically for movement, thus allowing the results to be related to an absolute datum.

Figure 4.17 Hydrostatic profile gauge installation details (authors' image)

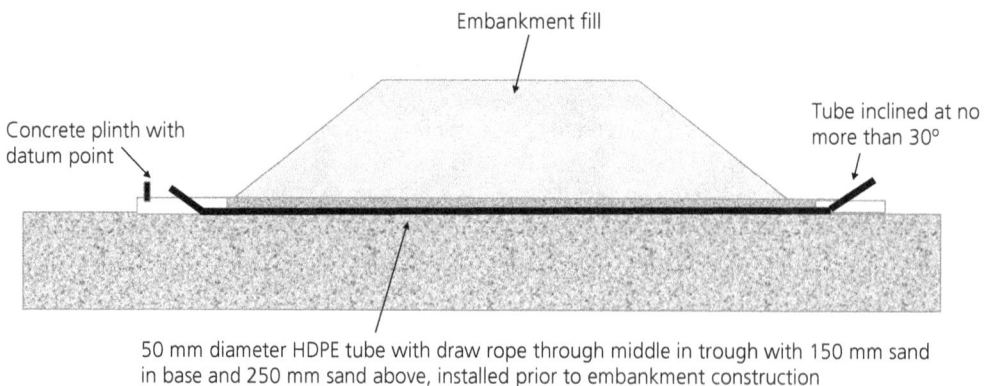

50 mm diameter HDPE tube with draw rope through middle in trough with 150 mm sand in base and 250 mm sand above, installed prior to embankment construction

4.3.6 Extensometers

4.3.6.1 General

Extensometers measure vertical movement, the most widely used methods either measuring changes in distance between a fixed base datum and ring magnets that are installed at intervals down a borehole (see Section 4.3.6.2), or using a head assembly and a system of rods similarly fixed at anchor points in a borehole (Section 4.3.6.3). In-place extensometers (IPX) can measure the positions of the magnets and enable the automation of the monitoring of existing magnet extensometers.

IPIX are IPI that also have the ability to measure the magnets installed as part of an inclinometer/ extensometer combination. This enables the monitoring of a manual inclinometer/extensometer combination to be automated and assess borehole movements in three dimensions. While the three extensometer methods are able to measure settlement and heave, the IPIX method is able to measure angular rotation of the borehole in addition to settlement and heave.

4.3.6.2 Magnetic extensometer

The magnetic extensometer is generally installed into a borehole that is either specifically drilled for the installation or has been drilled for other geotechnical purposes. The installation requires a guide pipe, usually of 19 mm diameter, to be installed to the required depth. This pipe must be sealed to exclude grout. Usually, the base of the installation is placed at a depth below the level where the effects of movement are negligible. For large embankments or structures, this can be considerable and often several tens of metres. A base datum, which consists of a ring magnet, is fixed about 250 mm above the base of the guide pipe (which passes through the centre of the ring magnet). This is to ensure that when the monitoring probe is lowered down the guide pipe it will clearly pass the magnet, ensuring the internal reed switch will be activated. All readings are taken relative to the depth of the base magnet datum, which is designed to remain in the same position and not be influenced by ground movements in the soils above.

Further magnets are then placed at various levels in the borehole, throughout the zone where movement is expected. These magnets have strip metal legs (and are therefore commonly known as 'spider magnets'), as shown in Figure 4.18. The legs can be tied, using fishing line, so they close around the access tubing, allowing them to be lowered down the borehole. The line is passed through a pneumatic cutter, so that when the magnet is at the desired depth in the borehole a foot pump can be used to activate the cutter and severe the fishing line. This causes the legs to spring outward, becoming embedded into the wall of the borehole. Other systems use a wire loop that is wound round the legs and fixed in position with a release pin. The release pin is attached to a cord that runs to the surface and can be pulled to release the wire loop allowing the legs to spring open.

A three-legged spider, with the legs pointing in just one direction, can be inserted without tying the legs together: with the legs uppermost they can be pushed down the borehole to the required depth. Care should be taken to ensure the magnet is set at its desired position and not pushed beyond this point; if the device is pushed too far it will not be possible to pull it back. Once at the desired position, the spider is pulled slightly upwards, which causes the legs to dig into the borehole wall, thus fixing its position.

Ideally, the extensometer would be installed in an uncased borehole, although this is not always possible. Where casing is required to maintain the borehole stability, the extensometer tubing will need to be fed through the casing as it is withdrawn from the borehole. While this is being carried

Figure 4.18 Two configurations of the 'spider magnet' and the detection probe (image courtesy of Soil Instruments Ltd.)

out, the pipe should be held in position and not allowed to move, particularly upwards. Such operations should never be rushed to ensure the installation is completed successfully. It should be noted that the legs of the spider magnet should not be released until the casing has been drawn above the location where the installation is to be made. If the legs are released too soon, inside the casing, the whole installation will need to be removed and reinstalled correctly.

Where the expected movement is likely to be large, telescopic lengths of the access pipe can be incorporated into the pipe. These are particularly useful where the ground surrounding the installation is highly compressive, such as Alluvium or poorly compacted fill, or where the ground will be made up around the location of the instrument.

Once the installation of all the magnets has been completed the borehole may be backfilled using a cement-bentonite grout. As the borehole needs to deform, to allow the magnets to move, the grout mix needs to be designed such that the grout has a similar consistency to the surrounding ground. This is related to the relative proportion of cement, bentonite and water in the mix, and guidance on this is given in Table 4.2 and also in Tomlinson, BDA *Manual of Rotary Drilling* (Tomlinson, 2022).

When the borehole has been grouted, a datum set of readings should then be made by measuring the distance between each of the magnets referenced to the base magnet. The measurements are taken with a monitoring probe situated at the end of a measuring tape, similar to a dip meter for water levels. As the probe switch passes each magnet either a light will come on or a buzzer will sound. It is good practice to conduct two sets of measurements, while lowering the probe down the borehole and again when the probe is withdrawn. It is often found that these measurements are marginally different, either because of the operator's reaction time or from variations in the probe position as it is drawn past the magnet. It is recommended that three passes are carried out and the average used for this base set of readings. It is normal to take further readings at specified time intervals, depending on the activities that might be carried out. For example, if an embankment is being formed on top of the monitoring position, then monitoring may be timed to coincide with the completion of a stage in the construction. The procedure is the same as described above for each set of readings.

Where an embankment is being built over the installation, the access tubing will need to be extended upwards to ensure it is always above the surface of the fill material. Additional plate magnets may be introduced at various levels as the embankment height is increased, thus also allowing the consolidation of the embankment materials to be assessed. The advantage of the spider magnet system is that many magnets can be positioned in the borehole, while other methods are limited by the space available in the borehole.

In addition to the assessment of the ground beneath a loaded area, described above, magnetic plates can be used solely within the embankment, where the consolidation of the embankment materials is of interest. Where this is required, a datum magnet can be positioned approximately 0.5 m from the bottom of the access pipe. This lower section may then be inserted into a preformed hole, of at least 50 mm diameter, penetrating about 0.6 m into the ground below the base of the proposed embankment. The tubing should be grouted into place. As the embankment fill is built up, additional lengths of access pipe should be added to the original length to keep the top of the tube above the level of the fill. Further plate magnets can be added as required, each magnet being slipped over the tubing and not fixed. If required, the plate magnets may be replaced with cross-arm magnets. These are magnetic rings set on a steel member that covers a larger area and is considered to be more representative of the embankment performance away from the installation position.

When readings are taken to determine the location of the plate magnets, it is essential to determine the level of the top of the tube, either against a site datum or Ordnance Datum.

The magnetic extensometer can measure changes of up to 50 mm/m with an accuracy of 3–5 mm.

4.3.6.3 Rod extensometer
The rod extensometer is installed into a borehole and enables measurements of the distance between anchor points and the referencing head, which is situated at the top of the borehole. Systems generally allow up to eight positions in a single borehole to be monitored. The device is very accurate and is used to measure movements between 50 mm in extension and 30 mm in shortening.

The ends of the rods are anchored by grouting or using packers at their required locations down the borehole, and a protective sleeve or tube ensures the rods can move without restriction. Each rod is located into the head assembly where it can move freely up or down, the relative movement between the base grouted location and the head being measured either with dial gauges, digital depth gauges or displacement transducers.

The depths recorded are relative to the head position at the time of measurement, and it is therefore essential to determine the head location relative to a fixed datum point to provide absolute displacement measurements. This is normally carried out using conventional surveying methods.

4.3.7   Pressure measurement

There are several types of pressure measuring devices, the one most commonly used as part of a ground investigation being the total earth pressure cell. These can measure changes in the ground pressure arising from various construction activities. They are thin circular plates, welded together around their circumference. The space between two plates is filled with hydraulic fluid. The cell is connected to a pressure sensor such as a vibrating wire unit that converts the changes in pressure to an electronic signal. The vibrating wire sensor generates a frequency signal that is not affected by changes in resistance, meaning the signal wire can be extended, giving reliable readings up to 1 km from the point of installation. Typically, these devices would be used to measure the pressure exerted by soil on the back of a retaining wall or some other structure. They may also be used to measure earth pressures generated by an earthworks operation, such as an embankment construction.

A push-in spade cell uses a similar construction, with a deaired glycol fluid between the steel plates that form the blade. These are robust spade-like instruments that are pushed into the ground to the desired depth. Typically, these would be installed down the back of an existing retaining wall to measure the pressure exerted onto the wall by construction activity. They may be used during a ground investigation to assess the in situ horizontal stress. The spade cells can be inserted either in a borehole or from the surface. The main potential problem is penetration resistance owing to stiff ground or gravel. Where this might be a risk, a pilot hole may be drilled to about 1 m above the position of installation and then the cell may be pushed to its final depth. The cell may also include an integrated pore water pressure measurement capability, by way of a separate port, which can be similar in principle to any of the electric piezometers discussed above.

The spade cell is a very useful means of monitoring the changes in both horizontal and vertical stresses that occur within cliffs that are prone to degradation and collapse. There are significant stress changes prior to a collapse occurring, and these cells can provide an early warning of an imminent failure. These instruments provide a safer means of monitoring because the readout can be installed a safe distance from the point of installation.

4.4.   Backfilling instrumented boreholes

4.4.1   Backfill materials

The backfill used around an installation needs to be formed in such a way that it does not affect the parameter being monitored. Typically, bentonite clay or a grout consisting of a cement and bentonite mix is used to form low permeability seals or to provide properties of stiffness that match the soils within which they are places. Single-sized sand and fine or medium gravel are used to provide a permeable zone from within which the water level or pore water pressure may be measured.

Bentonite is a clay mineral that is thixotropic, meaning if mixed with water and left undisturbed, it will form a thick almost solid material, but when disturbed it will become more fluid. Typically, bentonite is supplied dry and either as a powder, chips or pellets. When supplied as a powder, it will require hydrating and mixing before introduction into the borehole. This requires a mixing pan and suitable pump and tremie pipe. The tremie pipe is used to deliver the pumped mix to the required depth in the borehole (typically from the base of the hole, to avoid the formation of air pockets). Alternatively, the bentonite may be formed by hand into balls, which are then dropped into the borehole. Chips or pellets may be introduced to the borehole in a dry state, although care is needed to ensure they do not clog and arch. They will require hydrating with the addition of water, and the hydration process may take some time. Bentonite on its own may not be volumetrically stable, potentially increasing its volume during hydration by up to 500%. This can be problematic when placing a seal below a permeable filter zone, as a change in the intended boundary depth between the two materials can be detrimental to the design. Under dry conditions the reverse may result, allowing filter material to fall into the lower bentonite seal, or bentonite above the filter zone to contaminate the filter material.

Cement bentonite mixes are essentially manufactured grout mixes and can be obtained as a powder or in both pellet and chip form. The addition of even small amounts of cement significantly reduces the expansion during hydration, while not having a substantial effect on the permeability. The cement produces a solid backfill that is less susceptible to change in the borehole and will go some way to prevent collapse of the backfill materials and arching. These mixes have a permeability of the order of 10^{-8} m/s or less.

Bentonite fly ash mixes produce a softer grout mix but act in the same way as the addition of cement, with the pozzolanic property of the fly ash creating a chemical bond with the bentonite. These are particularly use when trying to simulate low strength soils with a relatively low strength grout. They exhibit similar permeability to the cement bentonite mixes.

Sand and gravel are used to provide a free draining surround, or 'filter zone', around an installation, particularly standpipes and piezometers. The choice of grading is dependent on the permeability of the material within which the instrument is required to function. An estimate of the permeability from a uniformly graded sand or gravel can be obtained using Hazens formula (see Chapter 6). As well as providing a free draining surround to the instrument, which will allow groundwater and ground gas to migrate towards the instrument, the material is required to filter out finer particles being drawn in, which might over time block the instrument and render it inoperable.

4.4.2   Backfill process

For water level monitoring in standpipes or piezometers, it is generally required to isolate the monitoring zone of interest (known as the 'response zone') from the influence of soils and water above and below the zone being monitored. This ensures that any readings obtained are relevant to the stratum under observation and not influenced by water for other sections of the borehole. These upper and lower seals would normally comprise pure bentonite to form a near impermeable zone ($k < 10^{-10}$ m/s) of between 0.5–1 m thickness both above and below the response zone.

Owing to the volumetric expansion that occurs during hydration, the use of pure bentonite should be kept to a minimum to avoid subsequent issues and potential infiltration of bentonite into the filter zone. Where this zone is some distance above the base of the borehole, a bentonite cement mix might be used to fill the space. Some practitioners will use the arisings from the borehole to fill this space, although this practice is to be avoided because it is difficult to obtain adequate compaction of the material, particularly at depth. Consequently, this may result in collapse of the fill, with a loss of the effectiveness of the response zone and hence the integrity of the instrumentation and readings obtained.

Cement-bentonite mixes can be tailored to provide similar strength properties to the surrounding soils, this being particularly useful where they surround inclinometer or extensometer installations. Table 4.2 provides a guide to the strength that might be achieved from various proportions of cement, bentonite and water. Where the strength needs to be accurately determined it is recommended that once the raw materials have been sourced laboratory trials are undertaken to assess the

Table 4.2 Typical design ratios (by weight) for mixing grout to a specified strength at 28 days

Grout composition	Soft to firm consistency soil	Stiff consistency soil	Very stiff consistency soil	Rock
OPC	1	1	1	1
Bentonite	0.4	0.35	0.3	0.05
Water	6.6	3	2.5	1
UCS at 28 days (kPa)	30	300	700	1000

OPC = Ordinary Portland cement; UCS = unconfined compressive strength.

proportions required for the specific location. Further guidance may be found in Tomlinson (2021) and Mikkelsen (2002).

The backfilling process must be conducted in a careful and methodical way. Each stage of introduction of materials requires time to be given to allow the materials to settle in the borehole, and each stage requires the depth of the fill to be repeatedly measured until it has settled. This process must not be rushed. If materials are introduced too quickly into the borehole and mixing occurs, or the depth of the seal or filter medium is wrong, then the only recourse will be to remove everything from the borehole and allow any silt, which will invariably be in suspension, to settle before starting again.

4.4.3 Headworks
Forming a neat housing or headworks for the installed instrumentation is a necessity for its longevity, particularly where it has been installed in an area that might be trafficked by vehicles, pedestrians or livestock.

The finishing or headworks of the installation in a borehole is as important as the installation itself. The headworks is the only visible part of the installation and, as such, is vulnerable to vandalism or accidental damage. If the headworks is damaged, it is generally unlikely that the instrumentation can be retrieved for ongoing use in the monitoring programme. It is therefore important that the type of headworks and the workmanship used in its construction are planned and conducted in a meticulous manner. Some typical examples of headworks are shown in Figure 4.19.

Generally, for water level monitoring a small chamber will need to be formed in concrete to house the pipe and possibly a gas valve where this is specified, together with any data logging devices that are to be installed. It is essential that the chamber is of sufficient size to readily accommodate all the necessary equipment, is secure and protects the installation from surface water ingress. To control water ingress, a drain should be provided and may be achieved using a short piece of pipe in the base of the chamber, providing drainage into the surrounding soil. Without a drain, rainwater can collect in the chamber and find its way into the standpipe, in which case it could potentially give a false reading of groundwater level. Any installation pipework should extend at least 50 mm above the floor of the chamber to avoid water entry. This may become irrelevant should the local area be prone to flooding. Where this is the case, the fact should be recorded along with any other readings obtained.

Figure 4.19 Examples of headworks for installations: (a) de-airable piezometer in chamber, (b) flush cover, (c) raised/barrel covers, open, (d) closed and (e) surrounded by stock proof fence (images courtesy of SOCOTEC)

(a)   (b)   (c)

(d)   (e)

Where the headworks consists of a chamber, it is generally set down into the ground with a cover flush with ground level – for example, where pedestrian or vehicular access needs to be maintained at the location. The chamber should be concreted in place, and its cover should have an appropriate rating commensurate with the traffic loading that might be expected.

Alternatively, where the location demands, the headworks can be raised above ground level, and these typically take the form of cylindrical steel barrels of various sizes. This is preferable where the location needs to be visible from a distance – for example, in a field – or perhaps where there is a rough surface that could bury a flush cover, as on a landfill site.

Whichever type of headworks is selected, sufficient space needs to be provided for access to the top of the installation tube or wires, as relevant, so that the technician is able to conduct connections, or freely use dip meters and probes as necessary. Where a standpipe has been installed for gas monitoring, there needs to be enough room for a gas tap to be positioned without hindering closure of the lid, and the tap will need to be accessible to allow it to be opened. If the lid of the cover bears onto the gas tap, it could be forced down into the top of the pipe and become difficult to retrieve.

Where a raised cover is used, any pipework for piezometers, inclinometers or extensometers should be extended above ground level such that the end of the tube can be readily accessed from the top of the cover (see Figure 4.19). In any event, the installation should be such that the location is accessible for maintenance, taking readings or purging and sampling, as required.

It may also be necessary to accommodate coils of wire or tubing from some piezometers, and potentially data loggers' telemetry equipment as required. In some cases, where these additions are sizable, it may be prudent to form a manhole using manhole rings and a suitable cover. The cover should be lockable to prevent unwanted intrusion and access to the instrumentation.

Where there is vehicular access to the area of the installation, or where livestock is likely to be present, the installation may need to be protected by a wooden fence. Alternatively, manhole rings set above ground level can be used to provide a more robust form of protection, which may be necessary on an active building site, for example.

It is essential that each location can be readily identified; it is therefore useful to tag each installation with details of the borehole, including identification number and the depth and type of installation. This can be marked on the underside of the lid or provided on a tag kept inside the cover. Each position should also be surveyed to determine grid reference and ground level. Where the ground around an installation is uneven, it is often beneficial to use the cover of the headworks as a stable datum against which depths, such as depth to groundwater, may be referenced. For this reason, it is good practice to survey the level of the headworks cover as well as the surrounding ground.

For location, some practitioners are also using the 'what3words' phone application to locate installations. This uses three unique words to identify a 3 m square grid and can be used anywhere in the UK. This can be extremely useful in remote locations, although it does require a mobile phone signal to work.

4.4.4 Reinstatement and decommissioning

Once monitoring has been completed, it is often necessary to remove part or all of the installed materials. There are several reasons why this would be required.

- To remove the potential for contamination being introduced into the borehole, which may pose a risk to the groundwater, or enable contamination to migrate around the site and potentially into surrounding areas.
- To remove obstructions that might cause a hazard to farming equipment, such as ploughs.
- To avoid interfering with proposed or further works – for example, an excavation for a basement where there might be a potential to create a pathway for water to enter the excavation, or where the borehole might be intersected by the path of a tunnel, possibly resulting in ground conditions encountered by the tunnel changing unexpectedly.
- Where the installation might become an obstruction to the proposed construction works.

Where instruments are removable, such as divers and IPI, these should be retrieved.

For headworks and pipework, the simplest process is to excavate by hand to the required depth, allowing the removal of all steel covers and concrete chambers. The depth of excavation would usually be the base of the concrete, or to 1 m, whichever is deeper. The exposed pipework may be filled with cement bentonite granules. These should be introduced slowly and allowed to settle to the base of the pipe. Once filled, the pipe and any wires or tubes may be cut off to the base of the excavation. Liquid grout may be used, although this should ideally be placed using a tremie, which may not be practical in a small diameter pipe. Should grout be used, the possibility of it seeping through the filter material and into the surrounding ground should be considered.

If full removal of the installation and backfill materials is required, it will need to be 'over-drilled' by drilling a slightly larger diameter borehole to the original. This process may be quite arduous. Most pipework is made from very durable HDPE or PVC, which is difficult to cut with drilling tools, particularly where the instruments extend to greater depths. Often this will force the drilling equipment off the drill line and the pipework cannot be retrieved. Open hole rotary drilling, using a rock roller, might be chosen to force the hole to depth. This method will completely destroy the instrumentation but will not guarantee removing the pipework, which is likely to be forced into the borehole wall.

Once the pipework or other form of instrumentation has been removed, the borehole can be backfilled with low permeability materials such as bentonite, or cement-bentonite grout. Alternatively, if required, the backfill materials can be designed to mimic the surrounding ground conditions (i.e. low permeability backfill materials in zones of low permeability ground, and higher permeability materials in zones of higher permeability ground).

The Environment Agency recommends that the upper 2 m of the borehole be backfilled with an impermeable plug (of cement, concrete or bentonite grout) to prevent seepage of potentially contaminated surface waters into the borehole. A concrete slab, extending at least 0.5 m beyond the edge of the borehole, should then be installed, either at the ground surface or, in agricultural land, at least 2 meters below plough depth.

4.5. Monitoring using imagery

4.5.1 Photogrammetry and surveying

There are several methods of topographic surveying that can be adapted to provide information on the amount and rate of movement taking place, including the following.

- Precise levelling; uncertainty ranges of 2–0.5 mm.
- Total station surveying; measurement uncertainty ranges of 1–5 mm.
- Electronic distance measurement; uncertainty ranges of 3–0.5 mm.
- GPS; measurement uncertainty ranges of 20–50 mm.

The accuracy of these systems is affected by the skill of the operator, the terrain and external influences such as weather. In the case of GPS, the signal reception at any one location may reduce the effectiveness and accuracy of the method. The accuracy of all the above is improving as technology advances and our ability to remove 'noise' is enhanced.

As with many methods deployed during a ground investigation, it is practical to employ more than one method of measurement to sense check the main method selected.

The use of photographic and radar imagery to monitor movements is a relatively efficient method of determining if movement is ongoing or has taken place. The value of such records relies on images being recorded from the same position over time, and when combined with more traditional survey methods, such as a GPS device (see also Section 4.5.6), a high degree of accuracy can generally be achieved. GPS devices are generally very accurate, and can provide absolute grid references and elevation, although in built up areas the location accuracy can be compromised. It is always prudent to use a secondary positioning method or even visually locate positions on a survey drawing as a sense check.

4.5.2   Time lapse photography/photogrammetry

With the use of a digital SLR camera, it is possible to track the progress of a failure. To ensure accuracy, it is essential to set the camera at the same location and height each time the photographs are taken. The lens aperture and focus also needs to be at exactly the same setting for each photograph. The camera should be mounted on a tripod placed on a stable, sound foundation, ideally a concrete plinth with sockets to receive each leg of the tripod and with the tripod remaining in place for the duration of the monitoring. Where the tripod is left in situ, consideration should be given to its protection form damage or vandalism. For greater accuracy, a colour chart and scale should be included in each photograph. This can be readily achieved using a photo board, which allows the location to be recorded and the date and time of the photograph. Most digital cameras will automatically record such data, although if this method is being relied upon it is prudent to check the camera has been set with the correct time and date before starting.

This method has been used to document crack development in buildings with photographs being taken at regular intervals over a prolonged period of time. Depending on the situation that requires monitoring, at least one of the methods discussed below will invariably provide additional useful information at relatively low cost.

More sophisticated photogrammetry can use pixel and pattern recognition by comparison of images taken at different times from the same position.

4.5.3   Total station

The total station was developed from the theodolite used in traditional topographical survey methods. In addition to measuring horizontal and vertical angles, as with a theodolite, a total station incorporates an electromagnetic distance measurement (EDM) device, and can therefore provide distance, direction and elevation of fixed target points. It requires line of sight to each location.

The positioning and height of the tripod upon which the station is mounted should be fixed. By taking readings at intervals of time any movement of the ground or structure may be recorded. Typically, targets can be fixed to structures and monitoring points positioned using steel datum pins. These are nail-like, and generally between 25–100 mm in length and are driven to be flush with the ground. The pin has a locating dimple to ensure accurate positioning of the target. Alternatively, a permanent ground marker can be formed, concreting the pin into a small foundation to ensure the location is not disrupted by site activities. When monitoring the pins, a target is used mounted on a 1 m long survey rod and is held on position by the survey technician.

When deciding on the location for the survey station it is important to choose a location that will not be disrupted by site activities and the line of sight is maintained for the duration of the monitoring period.

The total station can be a robotic unit, which is totally automatic and programmed to take sightings onto fixed targets at chosen intervals, giving precise direction elevation and distance. Should readings be found to deviate from earlier readings, alarms can be programmed into the system to alert if change is outside predetermined limits.

4.5.4   Laser imagery

Laser scanning, along with 360-degree high-definition photography, is a fast and reliable way to provide a three-dimensional image of a site and structures. The scanner will determine the distance from the scanner to millions of points (a point cloud). These data can then be processed, together with high-definition photography, to provide a three-dimensional image. This type of survey is extremely useful when the location of interest is not readily accessible, such as high rock faces, unstable slopes and structures. With a good quality processing programme, fly-through images can be viewed, allowing details such as rock joint sets to be measured. By repeating the exercise over a period of time, images may be overlain to show where changes have taken place.

4.5.5   Drone imagery

Where the site is not readily accessible, the use of drone imagery may be a useful method to assess where movement is taking place. Drones fitted with high-definition cameras can obtain useful three-dimensional imagery. The flight path is determined at a fixed height and pattern to provide overlapping images of the site. Surveyed datum points provide fixed locations beyond the expected zone of movement, which enables the flightpath to be positioned.

By conducting several flights over a period of time, the imagery can be accurately superimposed to show where movement has occurred. Approximate measurements of movement can be made when the imagery is combined with scale features on the ground and traditional topographic survey techniques. This method has been used to document large scale slope failures as well as to enable cliff regression rates to be measured.

Drone surveys can also be used to determine geological features and information on the orientation of joints where access is very difficult, such as cliff faces and in quarries. This can be particularly useful if the rock face is unstable and presents a hazard for closer inspection.

In order to fly a drone, the operator needs to have undergone specific training and be registered with the Civil Aviation Authority (CAA). There are some areas where drone flights are not permitted, such as close to airfields and sensitive facilities such as military establishments. Before conducting a drone survey, checks should be made to ensure the area lies outside a restricted zone.

4.5.6   Global navigation satellite system (GNSS)

GNSS are a developing science. The system requires a network or constellation of satellites which work together to provide positioning navigation and time. The system works on a global scale; the more commonly known GPS is a small part of the umbrella system GNSS. GPS is run by the USA and is not as reliable as GNSS, the signal often becoming distorted or blocked, giving false data. When using GPS, it is essential to sense check the information produced.

The GPNSS systems are run by individual countries and give partial global coverage; only the Russian system GLOSNASS is a fully global system. The main system used in the UK is the Galileo system, which is run by the European Union.

4.5.7   Satellite imagery

The use of satellite imagery in the UK is an emerging science that has developed following the launch by the European Space Agency (ESA) of the ERS1 and ERS2 satellites in 1991 and 1995, respectively. More recently were the Sentinel missions, consisting of a series of six satellites including Sentinel 1 A and 11 B (to be replaced by 1 C) launched in 2014 and 2016, respectively. These two satellites have been orbiting and scanning the UK using Interferometric Synthetic Aperture Radar (InSAR). The satellites, which each orbit along the same trajectory every 12 days, give full UK coverage and can be used to monitor movements of large-scale areas with an accuracy of a millimetre per year. The concept uses the data from two satellites in opposing orbits, one north–south the other east–west. This removes terrain issues and improves accuracy.

The radar signal is able to measure the distance from a point on the ground to the satellite. Images are compared, to give an indication of movements both in terms of settlement and swelling, between subsequent passes of the satellites.

The signal requires significant processing to provide a series of spot points from which the attitude of the point can be compared with that of previous images of the same point, enabling the change in distance to be measured. By collecting data from numerous passes of the satellites, a graph of height change can be obtained over time. The number and location of points varies and is dependent on topography and available natural reflectors, potentially including roof tops, posts and other

Figure 4.20 Examples of measurement using satellite imagery: (a) the Folkestone Warren landslip area; (b) observations of a motorway bridge (images courtesy of SatSense)

(a)

(b)

fixed features. In less developed areas, the number of reflectors is significantly reduced. Alternatively, corner reflectors can be placed on the ground in areas of particular interest, although where this method is used it may take some time to accumulate sufficient data.

The use of satellite imagery, such as the InSAR data, can provide very accurate long-term information, often covering a period significantly in advance of any engineering works being undertaken. These methods are particularly useful to inform the desk study or to provide data of ground movements in areas of large landslide complexes often seen in coastal environments, where access and the terrain make traditional methods of monitoring and measurement difficult. The area covered by the survey can extend well beyond the site perimeter and enables global movements to be visualised allowing the risk to the development site to be determined. The database is large, with the Sentinel mission retaining imagery from 2014, thus much can be gleaned about a site well before any planned development begins.

Other areas where this method can inform future investigation are related to mining and the movement associated with the gradual collapse of mine workings, or to tunnelling where it can be used to track the associated settlement trough, as shown in Figure 4.20.

Care should be taken to consider the reasons for the detected movement and to understand the actual risk. For instance, the method is able to detect movement from vegetation, which may induce seasonal ground swelling and shrinking of a few millimetres. While the resulting graph of movement against time may appear to show a dramatic fluctuation, this might only be a few millimetres in total, which for many situations is acceptable. Any decision made using these methods should be confirmed by other observations and measurements.

The software required to analyse this data is complex, and several agents currently process the available data and make the results commercially available either through downloadable packages or by request for specific locations.

4.6.   Other types of monitoring

4.6.1   General

The following instrumentation is not directly linked to ground investigation. However, sometimes the concern about a structure, or a need to monitor the impact on the environment, means that the monitoring scheme will be extended to measure other factors related to these elements. These are briefly described in the following sections.

4.6.2   Measurements of structures

4.6.2.1   Bending (deflection)

Bending or deflection of structural members can be measured using strain gauges. These come in many forms and are general vibrating wire instruments, similar to vibrating wire piezometers except the ends of the wire are attached to blocks that can move independently of each other. The tensioned wire vibrates at its resonant frequency when plucked, and the square of this frequency is proportional to the strain in the wire. The vibrating wire readout or data logger interface activates a magnetic coil around the wire, which plucks the wire and measures the resultant resonant frequency of vibration. Deformation of the structural member will cause the end blocks to move relative to each other and hence will alter the tension in the wire, changing the resonant frequency. Strain gauges may be cast within concrete or attached to the surface of the structure or member.

4.6.2.2   Tilt (rotation)

Tiltmeters and tilt beams use MEMS systems to monitor movement in either two or three dimensions. These are similar to those used in the MEMS inclinometer system (Section 4.3.4.2). The tiltmeter is housed in a small protective box and is fixed to a mounting plate that can be attached to the structure of interest. The device is read using either a PDA or datalogger, and several monitoring devices may be strung together on a single cable.

A mobile tiltmeter can be used to monitor several locations, a locating plate being secured to each position to be monitored. The plate has a locating mechanism that ensures the tiltmeter is connected securely and at the same position each time a reading is taken. In this way, a single device can be used to monitor numerous locations. Where this method is used, the location will need to be accessed over the monitoring period.

4.6.2.3   Cant/twist

These measurements relate to the monitoring of railways, assessing the distortion of the rails and track and, typically, they are conducted using tilt sensors and tilt beams. Often sensors will be set up across a series of sleepers to provide a mesh of points. The data are wireless, allowing remote monitoring. A MEMS inclinometer system can also be used in these scenarios to measure the longitudinal settlement between sleepers. Typically, these measurements would be taken when a tunnel is being constructed beneath a railway track, or where there are changes to the traffic pattern at a particular location. MEMS devices are also used to determine tilt, which can be related to the cant and twist of the rail.

4.6.2.4   Crack propagation

Monitoring of cracking within a structure can enable the cause of strain to be identified. Methods include demountable mechanical (DEMEC) strain gauges, which use steel studs, set either side of the crack, and a vernier caliper to measure the distance between selected studs. The caliper can measure to 0.1 mm, and if several pairs of studs are used along a crack it is possible to see if the crack is wider at a higher or lower level, which will indicate how the movement is being generated. Gauges may also consist of vibrating wire devices that allow automatic logging and remote monitoring.

4.6.2.5   Strain

There are several applications where strain might need to be measured, including on structural members, piles, retaining walls and ground anchors. Typically, strain gauges use vibrating wire methods as described above.

4.6.3   Measurement of the environment

4.6.3.1   Noise levels

Noise is measured to protect site workers and to ensure people local to where the noise is generated are not inconvenienced. Individual noise dosimeters can offer protection for workers who carry out noise generating activities by providing a warning if specified levels are exceeded. As a rule of thumb, a noise is approaching a threshold level if operatives must raise their voices to be heard. The lower exposure action level is 80 decibels. Noise exposure is a combination of how loud and for how long the noise is generated. Most construction sites are required to record noise generation and will need to act if levels exceed those dictated as excessive by legislation. Noise meters can be handheld or fixed to prominent features around a site. The devices download wirelessly, and proprietary computer programmes will enable the data to be presented, usually as a graph of noise level against time of day.

4.6.3.2 Air quality and dust (PM10, PM2.5 and PM1)

Air quality and dust monitoring covers a wide range of different particulates and gases. Measurement techniques to be used to determine the concentrations of specific measurands are defined by the UK's Automatic Urban and Rural Network (AURN) and are show in Table 4.3.

Table 4.3 The AURN recommended methods of measurement of air quality

Measurand	Method of measurement
Ozone (O_3)	Ultraviolet absorption
Nitrogen dioxide NO_2 and nitrous oxide NO)	Chemiluminescence
Sulfur dioxide (SO_2)	Ultraviolet florescence
Carbon monoxide (CO)	Infra-red absorption
Particulate matter 10 µg/m³/particulate matter 2.5 µg/m³	Taper element oscillating microbalance Beta attenuation monitor Gravimetric monitor Filter dynamics measuring system Optical light scattering Fine dust analysis system

In the UK, air quality is determined by measuring the daily concentration of ozone, nitrogen dioxide, sulphur dioxide and particulate matter at both 2.5 and 10 µg/m³. Depending on individual concentrations, the contaminant is given an index between 1 and 10, which are further divided into four groupings of low, medium, high and very high. For construction sites, it is incumbent on the company to minimise or eradicate air pollution as well as to provide suitable protective equipment for their workforce. The UK Clean Air Directive provides guidance. Air quality should be considered as part of an Environmental Impact Assessment (EIA) at the planning stage.

4.6.3.3 Vibration

Vibration produces three characteristics: acceleration, displacement and velocity. Vibration monitoring devices use accelerometers, displacement transducers or velocity transducers to measure vibration at fixed locations. Typically, triaxial geophones or accelerometers are used to measure movement in the x, y and z directions.

Sensors can either record vibration over time or individual occurrences where limits are exceeded. Typically, a log of the measurements over time is provided. Often, these devices are coupled to a warning system to provide an alert when movements exceed a given threshold.

4.6.3.4 Weather

It is sometimes necessary to measure rainfall, temperature, barometric pressure, wind direction and speed. This is because weather can have a direct effect on the measurements obtained from site instrumentation, with factors such as precipitation, temperature and barometric pressure potentially influencing the readings obtained. For example, significant precipitation or prolonged dry periods are likely to affect the groundwater level.

Weather factors might also influence how work is conducted, or if work on a site needs to be halted. This is typically the case when working overwater, where high winds might create an unsafe work environment, or if conducting earthworks where working in excessively wet weather will result in detrimental soil conditions, making the placement and compaction of fill very difficult, if not impossible. The provision of a weather station where parameters can be recorded electronically can provide an early warning of poor conditions, an understanding of how weather might influence the monitoring data, and an irrefutable basis for claims. Weather events that are recorded only anecdotally can lead to disputes, and such a situation can be clarified by monitoring data.

4.7. Data management

Much time, effort and money are used to put instruments onto and into the ground, and this can be wasted if readings are then either not taken, are poorly reported or are just not acted upon. It is essential that instrumentation and monitoring is properly planned and implemented, as mentioned at the start of this chapter.

It is now possible to link instrumentation to data loggers or to post results in real time to cloud-based data visualisation software. The data can be manipulated to show trends and warnings can be set such that actions can be taken when certain limits are exceeded. While electronic data logging will provide many more observation points and can give a fuller picture of the behaviour of the measurands, they can also provide false readings, or fail completely. Issues such as power supply failure, instrument malfunction and external influences, such as lightning strikes, are more likely to affect results collected electronically than those collected manually. It is therefore essential that backups are considered, and instrument checks are carried out to ensure the data being collected are valid, an example being the use of the temperature measurement capability within a vibrating wire piezometer to check it is functioning correctly.

It is essential to choose instruments that can measure to the required accuracy and ensure that readings are obtained at a frequency that will detect change sensibly. Where systems are intended to provide a warning, such warning levels should be set at realistic and measurable limits, and someone must be appointed to take responsibility to take appropriate action should warnings be triggered.

In general, most monitoring is conducted over a fixed period of time and thus the measurements are most usefully plotted against time. With modern systems using cloud-based data visualisation software, almost real-time measurements can be obtained, thereby giving all interested parties the ability to see changes as they occur.

emerald PUBLISHING   ice Publishing

Peter Reading and Miles Martin
ISBN 978-1-83549-891-0
https://doi.org/10.1108/978-1-83549-890-320251005
Emerald Publishing Limited: All rights reserved

Chapter 5
Geotechnical field testing

5.1.   Introduction

Field testing, sometimes referred to as in situ testing, provides a means of testing the soil while avoiding the complexity of sampling, handling and preservation of soil samples in order to conduct tests in a laboratory. Field testing takes several forms, and this chapter deals with tests for various geotechnical parameters that are conducted at the ground surface or in exploratory holes, generally by a technician or drill crew. Down-hole geophysical and borehole properties testing, which are conducted in a borehole using sondes or electronic devices to measure specific soil or borehole parameters, are discussed in Chapter 3.

The decision to carry out testing in the field should be made considering all the practical limitations that may have an impact on the results obtained. There are strong arguments for performing tests in the field rather than in the laboratory, which are shown in Table 5.1.

The choice of test method will depend on the parameter required and to some extent the accuracy needed. In any event, care should be taken to ensure that all limitations of the test and testing method are fully understood before embarking on a testing campaign. For most applications, if designs are to be verified or sought from the data obtained from field testing, it is recommended that more than one source or test type is used to independently verify the results obtained. It is rare that a single method of testing will provide all the information required or provide the confidence in the results sufficient for a design application.

In general, field testing is quicker and cheaper than laboratory testing and can often provide results shortly after completing the test. Test locations can be chosen based on site features and the topography and additional testing can be conducted if results are not showing the expected condition. This introduces the ability to identify zones where the soils are either better or worse than expected, which may then be used to optimise the design.

While many field tests are able to provide parameters as the test is conducted, such as the standard penetration test (SPT), others require samples to be taken to a laboratory to complete the parameter determination – for example, field density tests. Where this is the case, it may be a few days before the results are available. Other tests require significant computation and derivation of results, such as the cone penetrometer (CPT) and pressuremeter, and for these it will be some time before results are made available.

Field tests have the advantage of being able to provide test parameters of the material in a natural state, the degree of disturbance often being minimal, allowing the measurement of parameters that are prone to change when sampled. However, in order to achieve high quality, it is necessary to ensure the test equipment is in good condition and is calibrated. To this end, it is essential that the equipment is transported carefully and in a way such that it is not damaged, or its calibration

Table 5.1 Advantages and disadvantages with field tests

Advantages	Disadvantages
Test soil is intact and generally undisturbed	Penetration tests may be affected by unseen obstructions such as cobbles and boulders
Drainage conditions and soil structure effects are present that cannot be replicated in laboratory testing	Equipment may fail resulting in an abortive site visit *(carry plenty of spares)*, and such failures can mean significant loss of time
If anomalous results are recorded, further testing can be carried out relatively easily	Weather can influence results. Testing conducted within 1–2 m of surface may be significantly affected by precipitation and temperature fluctuations
In most cases, results can be reported as test is conducted	Can require significant labour and setting up – for example, large plate tests
Provides a means to obtain parameters directly	Can be expensive and time consuming
Useful where soils are not easily sampled or where sampling method might be complex or expensive	Limited depth of influence for test zone can make interpretation difficult (although volume of ground tested is typically greater than in an equivalent laboratory test)
Useful to obtain parameters for coarse granular soils that cannot be sampled and are difficult to assemble into laboratory test equipment in a way that can represent field conditions	Requires meticulous attention to detail. The simplest things can have significant influence on the results

is affected. Checks of the equipment condition and confirmation of calibration status should be conducted before use at the site.

When considering whether to use field testing that requires driving or pushing a tool into the ground, it is essential to check that the area for test is free of services in the same way that locations are checked prior to drilling a borehole.

Most field test procedures are provided in either BS 1377 (BSI, 1990) or have been replaced by the Eurocode series of British standards. Table 5.2 provides the references to each test method discussed in the following sections. Some methods are not covered by either code, and where this is the case the generally accepted standard is listed.

5.2. Soil density and water content

5.2.1 General

A number of field tests can be used to determine soil density, most of which require a good knowledge of the test procedure and the pitfalls that can be encountered. Traditional direct methods of testing are based on measurement of the mass of soil excavated from a known volume. This sounds relatively simple but in practice can prove difficult in a field environment with all the associated constraints including the issue of the weather. Unfortunately, ensuring controlled conditions is often a challenge. Indirect methods are less sensitive to this difficulty but are themselves based on empirical correlations between density and measurement of some other soil property that may not always be reliable.

Table 5.2 Reference standards for specific field test methods (continued on next page)

Test	British Standard	Eurocode (BS EN)	Other
Soil density			
Core cutter	BS 1377-9 test 2.4		
Sand replacement	BS 1377-9 test 2.1/2.2		
Water replacement	BS 1377-9 test 2.3		
Nuclear density	BS 1377-9 test 2.5		
Nonnuclear density			none
Moisture condition value	BS1377-4 test 5.4		
Rapid water content methods			none
Load testing			
California bearing ration (CBR) (field method)	BS 1377-9 test 4.3		
Static plate test (k_s)			none
Plate loading test	BS 1377-9 test 4.1		
Ev1 and Ev2			DIN 18134
Skip test			none
High loading plate test	BS 1377-9 test 4.1		
Zone test	BS 1377-9 test 4.2		
Lightweight falling weight deflectometer (LWFWD)			ASTM D 4695.140
Heavyweight FWD			ASTM D 4695.140
Probing tests			
Standard penetration test (SPT)	BS 1377-9 test 3.3	BS EN ISO 22476-2	
Dynamic probe	BS 1377-9 test 3.2	BS EN ISO 22476-3	
Transport Research Laboratory (TRL) cone pressuremeter (CPT)			
Mackintosh probe			Manufacturer's instructions
Hand penetrometer			Manufacturer's instructions
Weight sounding		BS EN ISO 22476-10	

Table 5.2 Continued

Test	British Standard	Eurocode (BS EN)	Other
MEXE cone			Manufacturer's instructions
Static cone testing	BS 1377-9 test 3.1	BS EN ISO 22476-1	
Field vane testing	BS 1377-9 test 4.4		
Pressuremeter tests			
Menard		BS EN ISO 22476-4	
Self-boring		BS EN ISO 22476-6	
HPD		BS EN ISO 22476-5	
Flat plate dilatometer		BS EN ISO 22476-11	
CPT			
Field shear box test			None/ASTM for rock
Permeability and related tests			
Infiltration test		BS EN ISO 22282-5	BRE digest 365
Borehole variable head		BS EN ISO 22282-2	
Borehole constant head		BS EN ISO 22282-2	
Slug test		BS EN ISO 22282-2	
Water pressure (packer) test		BS EN ISO 22282-6	
Pumping test		BS EN ISO 22282-4	

Measurement of the soil mass and the occupied volume enable the bulk density (or wet density) to be calculated:

Bulk or Wet Density ρ_b = mass/volume Mg/m^3

It is useful to measure the water content of the soil removed to enable the dry density to be calculated:

Water content w = $[(\text{mass of wet soil} - \text{mass of dry soil})/\text{mass of dry soil}] \times 100\%$

Dry density ρ_d = wet density / $[1 + (w/100)]$ Mg/m^3

For example if a soil has a wet density of of 2.00Mg/m^3 and a water content of 20% the dry density is given by 2.00 /(1 + 0.2) = 1.67Mg/m^3.

To determine density in any soil is challenging and in granular soils is not truly possible. This is because in order to determine density an undisturbed sample is required, or soil needs to be removed without disturbance or effect on the remaining soil. While some methods get close to undisturbed, no method is truly undisturbed. It is therefore essential to be meticulous when trying to determine density in order to approach as close as possible the undisturbed condition.

There are several other soil parameters that can be derived from the basic parameters of dry density, water content and particle density (as discussed in Chapter 6), including voids ratio, porosity, air voids ratio and degree of saturation.

Degree of Saturation, S_r = (volume of water / volume of voids) . 100 (%)

$S_r = (V_w/V_v)$. 100 (%)

Where the volume of voids, V_v = volume occupied by both water and air

Porosity, n = volume of voids / total volume (dimensionless)

$n = V_v/V$

Voids ratio, e = (volume of voids / volume of solids) . 100 (%)

$e = (V_v/V_s)$. 100 (%)

Air Voids Ratio, n_a = (volume of air / volume of voids) . 100 (%)

AVR, $n_a = (V_A/V_V)$. 100 (%)

Degree of saturation, $S_r = 100 -$ Air voids ratio (%)

Unit weight or weight density, $\gamma = \gamma_w (S_r . e + G_s) / (1+ e)$ (kN/m³)

Where both S_r and e are expressed as decimals

G_s = specific gravity

γ_w = unit weight of water (10kN/m³); Density of water = 1Mg/m³

Bulk or wet density ρ_b = mass of sample / volume of sample

The following commonly used methods of obtaining soil density are discussed in more detail in the following sections.

- core cutter
- sand replacement and water replacement
- nuclear and radar.

Methods for the site determination of water content are also described.

5.2.2   Core cutter method

The method is applicable to fine soils, ideally cohesive, and represents one of the simplest forms of density test, using a core cutter that is driven or pushed into the ground. The cutter comprises a short steel tube, 100 mm in diameter and 200 mm long, with a tapered cutting edge at one end. The tube is driven into the ground using a sliding drive hammer that fits over the top of the core cutter. Blows are applied until the full depth of the cutter has been driven. The sample may then be dug out of the ground and ideally should be weighed immediately. At this stage, it is possible to determine the soil bulk density provided the actual length of the sample can be measured. The sample should be sealed either using wax or cling film and end caps and transported to the laboratory where the test may be completed. At the laboratory the dimensions of the sample, height and diameter, are measured and the volume determined. The mass of the sample is measured to enable the bulk density to be calculated, using the formula given in Section 5.2.1.

Should the sample be considered fragile and likely to break up if it is removed from the core cutter, then the mass of the sample and core cutter can be measured before removing the sample. Once this is done the sample may be removed and the mass of the empty core cutter determined allowing the mass of the sample to be determined and the bulk density to be calculated.

The use of the slide hammer as a drive mechanism has fallen from common use and the driving is now done using a variety of methods. These include placing a block of wood on top of the core cutter that may then be pushed into the ground using the bucket of a backhoe excavator or driven using a wooden block and a sledgehammer. While these methods can be effective, they lack the control the slide hammer provides, and can lead to issues of verticality during driving that may cause sample disturbance resulting in an incorrect measurement of the density. This can, to some extent, be avoided if an extending backhoe excavator is used where the backhoe can be extended vertically downwards whilst pushing the core cutter. Where this method is used care should be taken not to over drive the core cutter which may result in compression of the sample.

This method will give good results in cohesive soils, but unreliable results may be obtained where such soils contain gravel. It is unusable in gravel and sand and driving the core cutter in such soils will be difficult and often impossible. Where only occasional gravel is present the gravel can cause grooving and gouging, which will give erroneous results and may damage the sample tube.

5.2.3   Sand replacement method

The sand replacement test uses a fine, uniformly sized dry sand to determine the volume of a hole dug into the soil to be measured. The test can be conducted in all soil types although it does become problematic in loose soils prone to collapse. The test is generally carried out at surface or in a shallow pit. It is often used to assess the effectiveness of soil compaction during earthworks construction to ensure that the required compaction is achieving the necessary density in the placed soil.

The 'replacement' sand will require calibrating in the laboratory, to obtain its density. This is achieved using a calibration mould (commonly referred to as a 'top hat' owing to its shape). The volume of the 'top hat' is calculated by linear measurement. A known mass of the sand is placed into the pouring cylinder sufficient to fill the 'top hat' and the cone of the pouring cylinder. The cylinder is placed centrally on the top hat and the valve is then opened allowing the sand to flow into the cylinder and the cone. When all flow has ceased the valve is closed. The mass of the sand in the top hat and cone is calculated by weighing the mass of sand remaining in the cylinder.

To calculate the mass of sand in the cone, the procedure is repeated but this time the cylinder is placed on a sheet of glass. The following measurements and calculations are carried out:

Original mass of sand kg = M_o

Mass of sand remaining in cylinder after pouring into top hat kg = M_1

Mass of sand in top hat and cone kg = $M_2 = M_0 - M_1$

Mass of sand in cone kg = M_3

Mass of sand in top hat only M_4 kg $= M_2 - M_3$

Internal volume of top hat m$^3 = V$

Density of sand is ρ_{sand} kg/m$^3 = M_4/V$

The test requires preparation before going to site, bags of the calibrated clean dry sand need to be pre-weighed and labelled with the weight noted and the bags sealed. The batch should be prepared from the same batch as that used for the calibration and the identification for the batch should be noted on each prepared test bag. It is always prudent to prepare a few more bags than expected, to allow for additional testing that may be required. This is a lesson quickly learnt when conducting testing in the field; it is both time consuming and costly to find that there is something missing that is integral to any test being conducted on arrival at the site!

To conduct the test, any vegetation and topsoil present should be removed to provide a clean level surface sufficient for the template to lay flat. The template tray is placed on the ground. Soil is removed through the hole in the template tray to form a hole as a right cylinder, this should be the same diameter as the template hole and approximately 100–150 mm deep. It is essential that all the material removed is retained, as shown in Figure 5.1. Great care must be exercised while removing soil to avoid compressing the soil around the sides of the hole because this will artificially make the hole larger and hence the volume greater than that which the removed soil occupied.

Once the required depth has been reached and the hole is clean, all the soil that has been removed should be placed in a strong bag labelled and sealed. The cylinder is then placed over the hole on

Figure 5.1 A sand replacement test being conducted in the field using the large pouring cylinder (image courtesy of K4 Soils Laboratory)

the template and filled with sufficient sand to fill the hole and the cylinder cone. The valve is then released, and sand will flow into the hole and cone and will fill both. When the sand stops flowing, the valve should be closed, the cylinder removed and the sand remaining in the cylinder put back into the original bag of sand with any remaining unused sand. Both the bags containing the soil removed and the sand should be labelled with the location identification. It is recommended that the bags are taped together and returned to the laboratory. The following calculations can be completed to calculate the soil density.

The weight of the soil removed, and the weight of the sand remaining is measured, and a sample of the soil taken for moisture content:

Mass of soil Mg = M_s

Original mass of sand Mg = M_{so}

Mass of sand remaining Mg = M_{sr}

Water content of soil % = w_s

Density of sand Mg/m^3 = ρ_{sand}

Volume of hole = V_h = $(M_o\text{-}M_{sr} - M_3) / \rho_{sand}$

Bulk density of soil $\rho_b = M_s /V_h$ Mg/m^3

Dry density $\rho_d = \rho_b / 1+(w_s/100)$ Mg/m^3

While the procedure sounds relatively easy to carry out, this test often produces inconsistent results. The test requires meticulous measurement and procedures. A common error is to assume that all batches of the density sand will have the same density; this is rarely the case and should be determined for each batch. The sand must be kept dry, which makes the test challenging in the UK's climate. Careful identification of the samples related to a particular test is also essential. When conducting several tests at a site, it is easy to mismatch samples of sand and soil. The bags must be handled with care to ensure they are not damaged, and bags should be sealed to prevent moisture loss from the soil or the sand becoming wet.

The most critical element for this test is ensuring that the soil is removed from the ground in such a way that the hole size is not larger than the space occupied by the soil being removed. This is a greater risk in softer soils whereby the digging tool can compress the soil in the wall of the hole, thus making the hole larger. Where this does occur, it will have the effect of increasing the volume and hence giving a false low value of the density calculated. The test is often used to assess quality control of earthworks compaction; thus if an unreliably low density value is reported, this may result in a dispute with the earthworks contractor and the controlling engineer. It is always useful to check sand replacement density test results using an alternative test method.

5.2.4   Water replacement test
This test is rarely used but it does provide a relatively quick and easy way to determine soil density. The test requires a steel ring to be set into the ground using Plaster of Paris. The ring must be level

and set to the ground level of the soil to be tested. A pointer that fits onto the ring is positioned to be in contact with the ground surface at the centre of the ring. The pointer is locked into position and then removed. A hole is dug through the ring into the soil. Ideally, the hole should be dug as a right cylinder; this will make measurement easier. The area chosen for the test needs to be completely level. Great care must be taken to ensure the ring is not moved while digging and removing the soil. The hole should then be lined with a thin plastic sheet. It is useful if the hole is of sufficient size that the sheet can be manipulated to fold the corners or base such that the plastic takes up no more space than is necessary. All the soil from the hole should be retained and weighed, but if this is not done at site then the soil will need to be bagged and sealed to be returned to the laboratory for weighing.

The pointer is repositioned onto the ring and water is then poured into the plastic-lined hole from containers that enable the volume of water to be measured accurately. It is important that the volume of water added is accurately measured. The lined hole should be filled to the tip of the pointer. The volume of water used represents the volume of soil removed and is used to calculate the soil density. An approximate check on the volume is provided by the measurements of the hole dimensions.

The advantage of this test is that if the hole is relatively large then the effect of creases in the plastic sheet on the volume will be reduced. Unlike other tests, this test can be used in an irregular hole and where the material is granular. When the test has been completed the plastic sheet can be removed.

If the soil taken for weighing has been removed from site it will be necessary to have other suitable material available to backfill the resulting hole.

The soil density is calculated using the formula:

Bulk density of soil, Mg/m^3 = ρ_b = M_s / V_h

Where M_s = mass of soil removed from hole Mg

V_h = volume of hole = volume of water used m^3

5.2.5 Nuclear density testing

The nuclear density test is commonly used to carry out a soil density test for earthworks assessments. The test is relatively simple and quick. Each device is different, so it is essential to calibrate the device for each material type it will be used on. The test can determine both density and water content. The results are not direct and are derived following calibration, which is key to obtaining accurate results from this test. Usually, a test block will be formed at the laboratory, compacted blocks using different materials at different densities. Ideally, the test block should be from the same material as that being tested on site. The soil is compacted into a box with dimensions of 500 mm × 400 mm × 350 mm. Several test readings need to be taken to give a good statistical average for any one material type. Readings for both density and water content should be taken and compared with alternative test methods for comparison. The instrument should be adjusted for the test calibration before being used in the field, and recalibration is usually carried out every three months.

The test is performed using a small nuclear source held in a protective casing while not in use. The case needs to be robust and both water and dust resistant (as shown in Figure 5.2). Two methods of testing are used: direct measurement for density measurement and backscatter for both density and

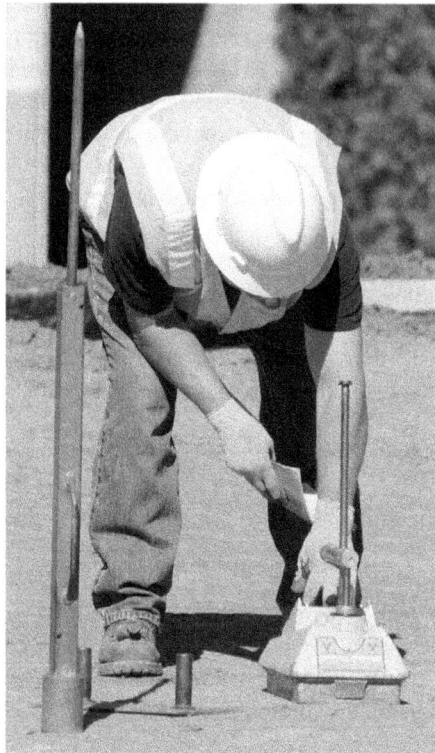

water content determination. There are several different manufacturers of these devices, and they all differ slightly but the general operation is the same.

The direct measurement is made by forming a test hole in the soil of interest by driving a pin to form a hole of sufficient depth to accommodate the apparatus probe and with a diameter at least 3 mm larger than the probe. The probe, which contains the radioactive source, is held within the device until required. The instrument is set up over the preformed spike hole and the probe is then extended into the hole and brought into contact with the wall of the spike hole. The radioactive source is then exposed by opening an automatic shutter that remains closed until the probe is in place.

The backscatter method uses a source of both neutron and gamma radiation. The device is set in contact with the soil and the sources are exposed, allowing detectors to measure the radiation passing through the soil. This test is able to assess both water content and density in the near surface layer of the soil. Tests are repeated to provide a statistical average of values.

The backscatter method tests the near surface soils that are likely to have been disturbed and affected by weather. The results obtained tend to be approximate. While density determinations are generally quite reliable, water content measurements are less reliable using this device because they detect the moisture density that may vary depending on void distribution and this may not be comparable to water content. Calibration with site determinations of water content using a different method is strongly recommended.

The use of the nuclear device requires stringent controls both during storage and transportation of the device. Storage requires a suitable sealed lead lined container and control documentation that is registered and regularly inspected by UK customs.

The operator requires training on the stringent health and safety measures that must be followed and the vehicle used to transport the device must be clearly labelled with the radioactive symbol. In the event that the device is damaged on site, the site must be closed, and a specialist team will be required to deal with the device. For these reasons, the use of a nonnuclear device has become more common.

5.2.6   Nonnuclear density testing

The test device was developed in order to mitigate the risks of the nuclear device and its incumbent control measures. The device uses an electromagnetic wave generator to deliver a radiated wave into the ground and measures the reflectance of the pulse on the materials encountered. This is proportionate to the soil density and hence the device can be calibrated to provide density determinations. To calibrate the instrument soil properties such as plasticity index values, particle size distribution and compaction test results are used. The results also need checking using the compaction test block discussed in Section 5.2.5. This level of calibration is sometimes seen as a hindrance and calls into question the validity of the results. The data display provides density, water content and temperature. Most devices are able to store the property details of 20 or so different materials. Care needs to be taken to choose the appropriate material properties for the analysis. The devices incorporate a full data logger that can store many results and generally are able to send the results electronically to interested parties.

This device was originally adopted to test bituminous road pavement surfaces, although in recent times it has also been used with some success on earthworks projects (shown in Figure 5.3). Its use in this sector has generally been limited to larger projects, where the calibration can be used for a reasonable duration, rather than on smaller sites where there may be a need to recalibrate frequently. The instrument is particularly useful detecting variations in material and poor compaction.

The base plate requires a good contact with the soil to ensure the pulse is evenly passed. This device can typically measure density of the soil to a depth of 300 mm and it can also measure water content. Most devices incorporate a GPS locator to pinpoint the test location. The results are displayed within 3 s and derived values of density, both wet and dry, are given along with water content and temperature. Although both the nuclear and nonnuclear devices provide a water content, it is considered that this measurement cannot be relied upon from either device.

The operation of the nonnuclear device is simple and does not require the stringent controls of health and safety, which the nuclear device requires. This method, however, does still require the calibration controls with checks using another method of density determination.

The take up of the use of this device has been slow in the UK mainly owing to the variability of the results obtained. However, provided the material type can be defined by the soil properties, the standard deviation is generally no worse that the nuclear device, and research has shown that, in general, the scatter is less than that found when using the nuclear device. It also has the advantage of being considerably cheaper particularly over the long term and, of course, it does not have the added restrictions of needing to be stored in a safe place and its movement being rigorously tracked.

Figure 5.3 Typical nonnuclear density device (image courtesy of Transtec)

Figure 5.3 Typical nonnuclear density device (image courtesy of Transtec)

5.2.7 Rapid assessment of water content

Several devices are available based around the reaction of calcium carbide with water, which produces acetylene gas, the volume of gas being proportional to the water available. The method uses a container with a valve, which opens to a gauge that will indicate the volume of acetylene gas given off. The soil and calcium carbide are vigorously shaken to mix them, the reaction of the calcium carbide with the free water in the soil causing gas to be generated. The result obtained is not reliable, particularly in cohesive soils, because the water is held in the pore spaces and is not readily accessed by the calcium carbide powder. This method is not recommended for design or compliance testing and should be seen as a technique of providing only an approximate water content value.

There are also a range of devices that are used to determine water content of materials such as wood and masonry, these devices generally working by passing a low voltage current through the material and measuring the resistance. The analysis assumes that if water is present then there will be less resistance. For a predictable uniform material, this method works in a satisfactory fashion. Soil, however, is a complex substance with many variables that will influence the resistance measured over a few centimetres, and such devices are therefore unlikely to provide reliable results.

In conclusion, currently there is no quick assessment method that has proved sufficiently reliable for the assessment of water content in soils.

5.2.8   Moisture condition value (MCV)

It is evident from the above that most tests to determine water content are neither reliable nor quick. This poses a problem to earthworks projects where soil is required to be placed at a particular water content value during compaction. Compliance testing is required to ensure the soil condition is appropriate. If the soil is too wet it will not compact adequately, plant manoeuvrability can be hampered, and the resultant density may lie outside the specification. Conversely, if soil is too dry then compaction is difficult and there is a risk of collapse mechanisms causing failure at a later date should the placed soil become inundated with water. The MCV test was devised to provide a rapid assessment of water content in the field to resolve some of these issues. Figure 5.4 shows the test device in use in the field. It should be noted that, although the device is intended to be portable, it is heavy and will require two people to lift it to the location where it is to be used.

For the test to be meaningful, it is necessary to carry out a calibration test in the laboratory on the same material that is to be compacted in the field as part of the proposed earthworks. The calibration test is discussed in Chapter 6: Geotechnical laboratory testing. The laboratory test provides a calibration curve of MCV against water content. For the test to be reliable, it is essential that the calibration test is performed on the same material to that being tested in the field, as most discrepancies occur when soils are mixed in the field, or the material is variable. The Atterberg limits should be determined to confirm the material being tested in the field has similar properties to that calibrated in the laboratory.

Figure 5.4 MCV test device with height to top of soil being measured (images courtesy of K4 Soils Laboratory)

(a)

(b)

The main advantage of the test is that the result can be obtained quickly. Therefore, if there is doubt regarding the suitability of the water content of the soil, either immediately prior to or shortly after placing, the results can help in the decision whether to exclude the material until it improves or to carry out remediation if the material has already been placed. Both decisions are much easier to make with early reliable information.

The field test is a single point test conducted on the material being used for the earthworks. The sample is taken either from the stockpile or from the earthworks surface following compaction. A sample of approximately 1500 g passing the 20 mm sieve is loosely placed into the compaction mould and a separating disc placed on top of the soil. The rammer is positioned 250 mm above the

Figure 5.5 (a) graph showing penetration against number of blows and MCV and (b) a typical calibration plot for MCV against water content (authors' image)

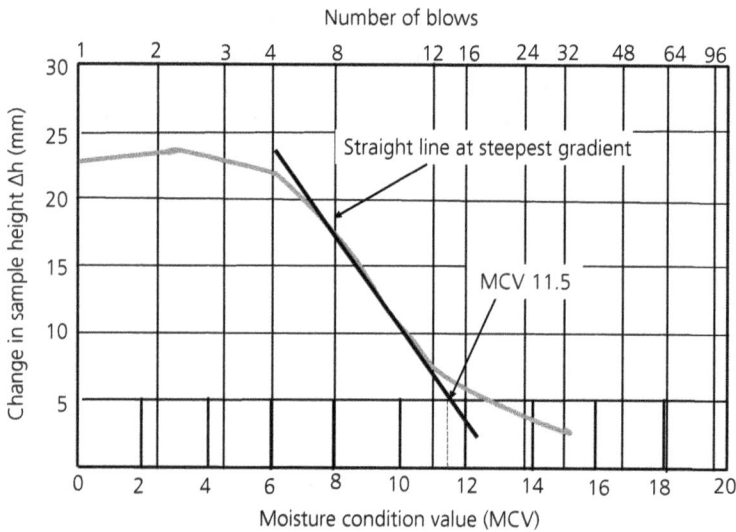

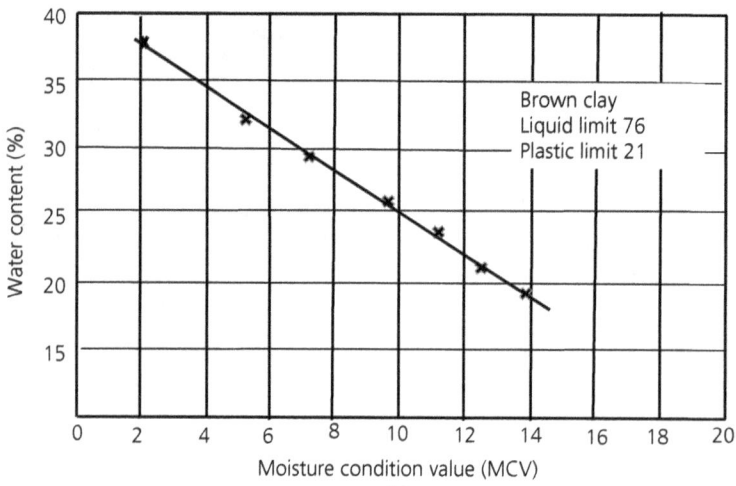

sample. A single blow of the rammer is allowed to freely fall onto the sample. The change in height of the cylinder of soil is measured and the process is repeated with further blows being applied. The change in penetration of the rammer is calculated between any given number of blows and four times that number of blows – for example, between 1–4, 2–8 and 3–12 blows. A graph of change in penetration is plotted on a linear scale. The steepest line possible is then drawn before the line meets the 5 mm penetration point, as shown in Figure 5.5(a). The number of blows where the line meets the 5 mm penetration point is taken as B, which is used in the following formula to calculate the MCV value:

MCV = 10 Log B where B is the number of blows corresponding to 5 mm penetration

The calibration graph obtained from the multi point test in the laboratory (Figure 5.5(b)) can be used to determine the water content in the field. This should be accompanied by the plasticity index for the material tested in the laboratory, which will assist in ensuring that the calibration is used for similar material. A typical relationship set of graphs is shown in Chapter 6 (Figure 6.19). As an example, if the MCV value for the material shown on the graph in Figure 5.5(b) is 10 this would indicate that the natural water content is 25%.

In general, the water content at which 10% air voids or less would be achieved is roughly equivalent to a maximum MCV of 12.5 for cohesive soils and 5% air voids or less will be achieved at MCV of 11.5 or less. For granular soils, the equivalent MCV will be higher – for example, for a well-graded sand an MCV of 14.5 will achieve 10% air voids or less.

As a check of the field assessment, it is recommended that samples are taken for subsequent laboratory testing for water content and, where materials are cohesive, plasticity index testing (see Chapter 6).

5.3.   Static load testing

5.3.1   General

This group of tests all apply a static load directly to the ground and measure the ground's response to loading. The test results are generally plotted as stress–strain curves and either peak values are taken at failure or the relationship between load and a given settlement is reported as a modulus value. Often high loads are required, which can present health and safety issues particularly during inclement weather or when working on uneven ground. It is therefore important to risk assess each situation prior to starting the proposed test.

The test methods discussed in this section are: (a) CBR test, (b) plate loading test – with low-, medium- and high-static loads and (c) zone test.

5.3.2   California bearing ratio (CBR) testing

5.3.2.1   Introduction

The CBR test was devised by the California State Highway Department to evaluate sub grade strength and determine the required thickness of the road pavement construction depending on the intended final use. CBR is empirical and is the ratio of the strength obtained for the soil expressed as a percentage of the value obtained from a sample of California Limestone subjected to the same test. This test has been widely adopted in the UK and around the world because there is a large database of information that directly relates the CBR value to the performance of road pavements. However, the test can be unreliable in some situations and in recent times dynamic testing to determine the stiffness modulus has started to replace the CBR. The use of the stiffness modulus is now the preferred design method recommended in The National Highways' *Design Manual for Roads and Bridges* (*DMRB*). A comparison of the various methods has been documented by Zohrabi and Scott (2003).

The CBR test may be conducted in the laboratory (see Chapter 6) and in the field. The field test has some advantages over the laboratory test because it measures the CBR value of the soil under field drainage conditions and can take account of conditions that cannot be replicated in the laboratory. The test should not be conducted if the ground is frozen and, as a rule of thumb, if the ground temperature is below four degrees any test used to determine a CBR value should be viewed with caution. Although not part of the test procedure, it is useful to record the ground temperature at the time of testing.

The test uses a similar procedure in both the laboratory and field and measures the force required to insert a 63.5 mm dia. plunger into the ground at a constant rate of penetration. In the field, the rate of penetration is achieved by using a screw jack, which is manually turned at a constant rate, and a reaction normally provided by a vehicle such as a backhoe excavator or a four-wheel drive truck. Where a four-wheel drive truck is used, additional reaction loading can be provided using an intermediate bulk container (IBC) that can be filled with water once on site and drained once testing has been completed. Care should be taken to ensure the vehicle is not overloaded.

The load applied is measured using a load cell or proving ring and the penetration of the plunger is measured either using a dial gauge or displacement transducer measuring against a datum beam.

Calculation of the CBR value is made by plotting the load applied against penetration of the plunger.

The CBR value is obtained by comparison of the load values measured at 2.5 and 5 mm penetration to the normalising values obtained when testing crushed California Limestone.

A variation of the test can also be made using the light plate load test as described in Section 5.3.3, the result providing a modified CBR value that is derived using the relationship shown below. In addition to providing an estimate of the CBR value, the test also enables the Modulus of Subgrade Reaction (k) to be determined. The test is conducted in a similar way but using a plate rather than a plunger, and the load, which is applied in several increments, is recorded against the penetration of the plate. The calculation to estimate equivalent CBR requires that at least 1.25 mm penetration of the plate is obtained. Critical measurements should not be taken outside the calibration range of the force measuring equipment. Note that even if the 1.25 mm penetration figure is reached after the first few increments of load, the test must continue until at least five increments have been applied. The equivalent CBR value is normalised to the 762 mm dia. plate relative to the plate size used. Plate sizes can range from 300–1000 mm and are chosen depending on the soil strength.

The equivalent CBR value is calculated from the load required to achieve 1.25 mm of penetration, using the equation below.

First the results are normalised to a 762 mm dia. plate using the following formula to determine the equivalent modulus value for the 762 mm dia. plate (i.e. k_{762}):

$k_{762} = p/y \times F$

Where p = bearing pressure kPa   y = settlement of plate = 0.00125 m

k = equivalent modulus value kN/m^2/m

and F = 0.079 + (0.001209 x plate dia.)

Then the CBR (%) is estimated from the following relationship:

CBR = 6.1 x 10^{-8} x (k$_{762}$ x 1000)$^{1.733}$

The plate method is more suited to use on granular soils than the smaller cross-sectional area of the plunger method. The method described in BS 1377 (BSI, 1990) is the definitive method, as described above. The plate method is empirical and therefore only indicative. Other methods of determining or deriving the CBR value are discussed within this chapter; see Section 5.5.4: TRL dynamic cone penetrometer test and Section 5.4.2: Falling weight deflectometer testing.

In addition to the field and laboratory CBR tests and the derived value from the plate test, the CBR value may also be obtained using the plasticity index together with a knowledge of the depth to ground water. With such a range of methods, it can sometimes be difficult to know which method to use.

The following sections (5.3.2.2 to 5.3.2.8) provide an overview of the various methods that are currently in use to determine the CBR value and provides a comparison of the relative reliability of each method.

5.3.2.2   CBR test in the field using the 50 mm plunger method
This test is the most appropriate and reliable method. It is the original method used to determine the CBR value. It is important to take a sample of the material after testing in the zone influenced by the test. Tests may be conducted on cohesive, granular and mixed soils. Results may be unreliable if large particles, coarse gravel or larger, are present within a finer matrix. The test result may also be significantly influenced by particularly wet or freezing weather, and testing in these conditions should be avoided.

5.3.2.3   CBR test in the field using the plate method
This is a modified method using a relationship to obtain the CBR value. It is particularly useful when testing coarse soils that may give less reliable results using the plunger method and cone methods described below.

5.3.2.4   CBR test in the laboratory
See also Chapter 6, Section 6.4.5. The preferred method is by sampling using a mould and cutting shoe to obtain a sample in the field, which can then be taken to the laboratory for testing. Care needs to be taken to avoid disturbance during the sampling process. This method requires the sample to be cohesive and ideally without gravel. Water content must be preserved. Other methods may be used to form the test sample. such as using the soil compaction methods as described in Chapter 6. Results are generally reliable but may not be representative of site conditions.

5.3.2.5   CBR using the dynamic cone penetration (DCP) test
See also Section 5.5.4. This test was designed by the Transport and Road Research Laboratory (Jones and Rolt, 1986) and is used in soils where gravel is not present. The test can provide an

approximate CBR profile with depth and is often used to determine the depth of competent soil for pavement design. It also enables a greater number of tests to be carried out relatively quickly. Results can be obtained as the test is conducted, making this a very useful method.

5.3.2.6 CBR from the falling weight deflectometer (FWD) and lightweight falling weight deflectometer (LWFWD)

See also Sections 5.4.2 and 5.4.3. The CBR may be determined from the results of the LWFWD, a dynamic plate load test. The deflectometer primarily provides an indication of the Subgrade Stiffness Modulus and can be directly related to the CBR value. The relationship $E = 17.6\,(CBR)^{0.64}$ MPa is applied.

5.3.2.7 CBR derived from plastic index values

Table C1 in TRRL Report LR1132 (Powell *et al.*, 1984) provides a rapid assessment of the CBR value using the plasticity index and the known depth to the water table in relation to the road base. This method should be used in conjunction with one of the other methods described above. By calibrating the site CBR values to the plasticity index values, the CBR value can be assigned across a site without the need for numerous field tests. This method assumes that soils with the same plasticity will exhibit the same behaviour and hence the same CBR value. The method is only applicable to soils that exhibit plasticity, that is, predominantly cohesive in nature. If this method is used, it is recommended that it is used in conjunction with one of the field tests described above to verify the relationship.

5.3.2.8 CBR using the MEXE probe

Also known as the soil assessment cone penetrometer, or MEXE cone. This test is only included here because the test kit is still available. While this apparatus appears to give rapid assessment of the CBR value, the loading control method for the test is too erratic to provide worthwhile results. As such, the authors recommend that it should not be applied to any design function (see also Section 5.7.3).

5.3.3 Plate load testing
5.3.3.1 Introduction

Plate load tests consist of the application of a load to a steel plate in order to monitor the reaction of the ground beneath the plate. The majority of tests are performed on horizontal surfaces, using vertical loads, although it is also possible to test inclined surfaces if required. The objective of most plate load tests is to determine one or more of the following properties.

- **Modulus of subgrade reaction**: Either to be used in its own right, or to estimate CBR values in coarser soils (see Section 5.3.2). This test measures the pressure required to cause a small deflection (1.25 mm) in the plate, is typically of short duration and often requires only relatively low pressures.
- **Soil stiffness/settlement characteristics**: Generally used for materials that are difficult to sample for laboratory analysis, such as granular soil, or where the assessment needs to include in situ features such as fissuring in clay or fracturing in weak rock. Larger diameter plates are usually beneficial for such tests, and relatively high loads may be required. These tests generally have multiple stages in which a range of pressures are applied and maintained until full settlement has been achieved. These tests are therefore often of long duration.
- **Ultimate bearing pressure**: Plate tests can also be used to load the ground to failure. These generally use a steadily increasing load, with the plate deflection being recorded until the

point that the material yields. Should a clear point of yield not be observed, BS 1377 (BSI, 1990) recommends taking the ultimate stress as that required to achieve a penetration equivalent of 15% of the plate width (e.g. 90 mm for a 600 mm dia. plate). Such tests can require the use of high loads.

In all cases, the deflection, or settlement, of a loaded plate is recorded against time, and is measured as a function of an increase in stress within the ground induced by applying a load to the plate. While this concept can seem simple, in practice doing these tests can be very challenging, particularly in inclement weather. The test requires meticulous adherence to procedures for setting up the equipment and a high degree of accuracy in measurement of the resultant movements in order to avoid numerous sources of error that can invalidate the data obtained. In any event, it is essential to inspect the ground beneath the test position once the test has been completed to understand the natural variations in the soil present within the stress zone, which may have influenced the results.

The majority of tests are carried out at or close to the ground surface, although it is possible to undertake a plate loading test at the base of a large diameter borehole (see BS 5930 (BSI, 2015)).

The preparation of the test surface requires particular care to ensure the ground directly beneath the plate is not disturbed in any way. This is essential because what might appear to be a small amount of disturbance could influence the settlement of the plate and possibly result in the plate tilting, which will induce a variable bearing pressure on the soil and will give rise to significant errors in the settlement calculations. It is perhaps useful to understand that settlement of the plate will be measured at between 0.01–0.001 mm accuracy. In general, settlement of the plate may only be of the order of a few millimetres in total, and therefore very minor disturbance could have a significant effect. Similarly, temperature changes during the day can significantly influence the final result causing distortion of the datum bar or fluctuation in digital readout units.

Most tests involve placing a rigid steel plate, which is usually circular, onto the soil. Typically, the plates are either 300 mm, 450 mm or 600 mm in diameter and thick enough to be sufficiently stiff that they will not distort under the applied load. A load is applied to the plate by the use of a hydraulic jack situated between the plate and some form of kentledge (to provide a reaction load). The kentledge is often in the form of a heavy vehicle, which is useful as it can easily be moved between test locations. A practical consideration, however, is that the vehicle needs to be of sufficient weight to prevent it being lifted off the ground by the jack. The type of vehicle appropriate depends on the loads envisaged for the test and, while the use of a pickup truck, loaded dumper, roller or wheeled excavator may be possible for lighter loads, more often a 13 T or 20 T tracked excavator will be required (see Section 5.3.3.2).

Once the load is applied, the ground response or settlement of the plate is measured. Figure 5.6 shows a general arrangement for the test. Loading is generally incremental with a number of loading stages. Ideally, the load for each stage should be held until the plate has stopped settling. It is necessary to ensure the load does not decay as the plate settles, which generally will involve adjusting the pressure at intervals throughout the loading stage.

To obtain meaningful results great care must be applied to setting the plate to ensure it is in full contact with the ground and that it is level. Smaller plates are usually placed on a bed of uniformly graded sand spread in a thin layer to take out any undulations in the prepared ground surface. For larger plates, Plaster of Paris or quick drying cement may be used.

Figure 5.6 Schematic diagram for the general set up of a plate load test (authors' image)

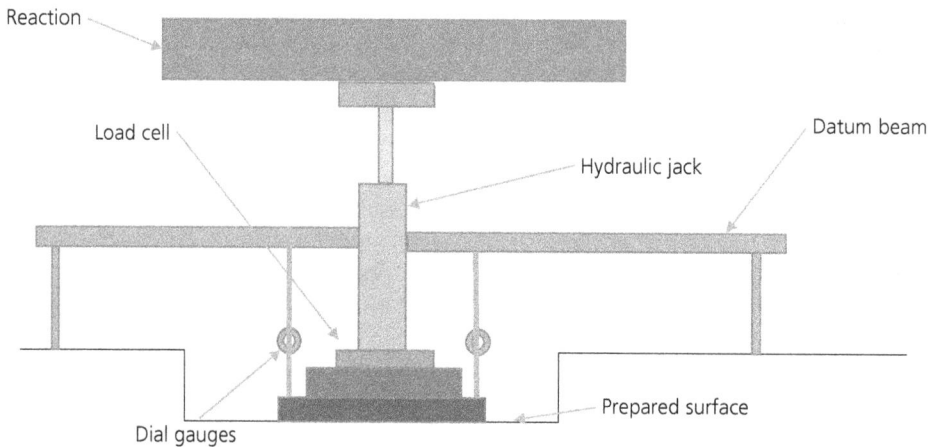

Similarly, care is needed in the placing of dial gauges or transducers, which are to be used for measuring the movement of the plate. These are normally set around the edge of the plate at three or four equally spaced points close to the plate perimeter. However, the ideal location to measure movement is at the centre of the plate. This is not possible using the standard method adopted in the UK because the load column acts at the centre of the plate. The DIN 18134 2012-04 (DIN Media, 2012) uses a load column with a slot cut into it, which enables the deflection to be measured at the centre of the plate (see Section 5.3.6).

The datum beam that carries the gauges must be positioned such that the supports are outside the zone of influence of the plate and the reaction that is to be used to deliver the required loading. For quick tests, load is applied rapidly, and the test is conducted within an hour or two. For larger tests, the testing may be conducted over a period of days, possibly weeks, with loading applied in stages. More sophisticated tests may also incorporate unloading and reloading stages. Full settlement of the plate, particularly on cohesive soils, can take a long time, often several hours or even days. Many tests are curtailed after 1 h; however, such short durations may not provide sufficient data for a realistic analysis to be undertaken. Where tests need to be conducted over a long period, it will be necessary to provide protection from weather and to measure changes of temperature over the duration of the test. To reduce temperature effects, the datum bar should be of Invar steel and it should be supported outside the zone that will be influenced by the plate. Should the supports of the datum bar be too close to the test area, the deflection of the plate and the ground around the plate may result in a degree of settlement of the datum bar itself, and consequently an underestimation of the plate deflection. Similarly, the supports for the kentledge, such as the tracks of an excavator if this is used to provide the reaction, should lie outside any zone of influence of the plate. Using an excavator, however, this is difficult to achieve as the jack will generally need to be placed beneath the belly plate of the machine to generate sufficient reaction load, and this is naturally close to the tracks. Ground conditions at the time of test must be noted. Testing may need to be delayed if, for instance, the ground is frozen or waterlogged, unless the tests are being conducted to investigate behaviour under these conditions.

When larger scale tests are conducted, it is recommended that the whole test set up is shielded from weather by either using a pop-up gazebo or, if something more substantial and longer term is needed, a scaffold frame covered with tarpaulins. The authors have been involved with larger scale tests that were run for a full month and involved a scaffold/tarpaulin structure, as well as full welfare and support facilities. Testing of this scale requires an attention to detail, otherwise a great deal of organisation and cost may be futile.

When deciding the type of test to conduct some thought should be given to the material being tested. Typically, granular soils will settle quickly and as such the duration of the test will be relatively short taking an hour or two to complete. Tests on cohesive soils are generally carried out quickly assuming undrained conditions however if the drained condition is sought the test duration may take a considerable time often taking several days or weeks to provide useful answers.

Another point to consider is that if a small area is tested using a small diameter plate, the results may not be applicable to a larger area, such as a full-size foundation loading. The stress zone forms a bulb shaped zone of increased stress in the soil, which results in a change in the effective stress and subsequent settlement as pore pressures dissipate. A rough rule of thumb is that the plate will stress to a depth of the order of two to three times the diameter of the plate. Therefore, a 300 mm diameter plate will stress to a depth of about 1 m. In comparison, a foundation that is 1 m square, for example, will stress a zone that extends to about 3 m below the base of the foundation. While in some cases this would mean a conservative design from the plate test, this is by no means guaranteed.

Plate load tests generally take two forms, which depend on the method of loading and the load required. Testing can be conducted relatively rapidly with single or incremental loading using relatively light reaction loads provided by a vehicle of some form; these are discussed in Section 5.3.3.2. Larger, more complex tests, using much higher reaction loads, will require designing and are discussed in Section 5.3.3.3.

Further information on plate load test procedures is provided in BS 5930 (BSI, 2015) and BS 1377 (BSI, 1990).

5.3.3.2   Plate load tests – quick tests with low to medium reaction loads

These simple tests use vehicles to provide the reaction and would generally be run for a few hours. Figure 5.7 shows a general arrangement for the test. Typically, the reaction is provided by a pickup truck suitably ballasted, an excavator or other available type of plant. When a vehicle such as a pickup truck is used, the vehicle should be jacked up off the suspension and safely supported on stands. Often, the loading column is attached to the vehicle towing arrangement to enable the loading assembly to be attached or to provide a convenient place to position the jack. It is essential the jack is not placed on a structural part of the vehicle that is not adequate to carry the intended loading. On a tracked excavator, for example, the belly plate of the machine is usually used as a suitable jacking point. The plate can then be set on level ground beneath the jacking point and the jack and load measuring cell can then be placed in position.

The measurement of the applied load can either be recorded using a calibrated load cell with a readout unit or by using a gauge pressure reading on the jack pressure unit. Load cells provide a more accurate measure of the applied load but do introduce another link in the load column. In any event, the cell must be calibrated. It is important that the load column is vertical and central on the plate. If during loading the column tilts, the test should be stopped and the column reset, and the test rerun.

Figure 5.7 (a) Test arrangement using a backhoe excavator and (b) 360-degree excavator (authors' images)

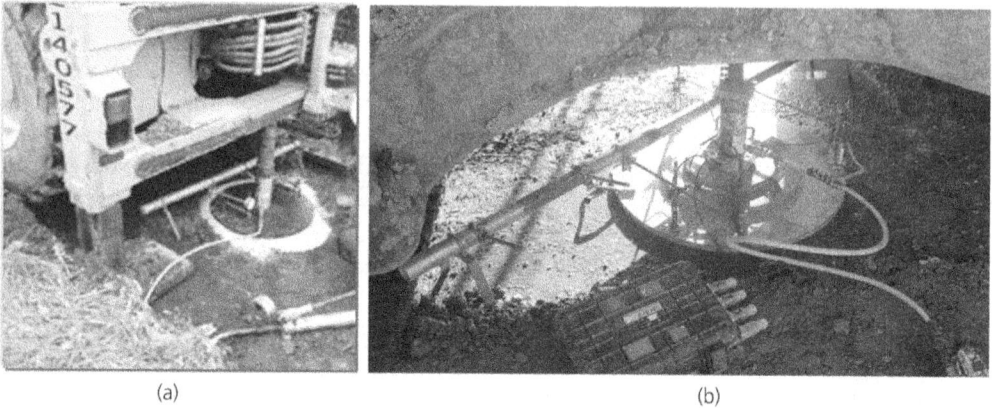

(a)   (b)

Note that on both, the datum beam is likely to be influenced by the machine.

Testing may be conducted in a shallow pit at the intended founding level. This may require the loading column to be quite tall. If this is the case, there is a much greater risk of the column tilting, particularly if the plate does not settle uniformly under loading. In reality, the depth of test is rarely more than 0.5 m without significant preplanning. Should a test be required at a greater depth, this may be possible so long as the reaction load, such as an excavator, can be taken down to the required level. This typically means an excavator digging a large enough excavation for it to track into. Consequently, as the sides of such an excavation will need to be battered such that access and testing can be carried out safely, this is only feasible on sites where sufficient space is available.

Safety warning

Work a safe distance from the load column and never work under the vehicle once loading has started. As load is applied the column can become unstable particularly if settlement is not uniform across the plate which can cause the load column to topple without warning. For this reason, the load column should be kept as short as possible.

An initial reading should be taken as the zero point prior to loading beginning. Readings for settlement at each of the gauges should be taken and plotted on a graph of settlement against time; this is normally carried out using a computerised system and data logger that will generate the load settlement plot as the test progresses. This will make deciding the termination of the test much easier to recognise.

Some specifications suggest that the test should be run for 1 h using a single loading stage. As a guide, however, this should generally be taken as the minimum length of time to allow for a test. For multistage tests undertaken to determine settlement characteristics, for example, the termination of an individual loading stage should only be made when the settlement is demonstrated to be 90% complete. In such tests, each loading stage may take several hours to complete. The settlement-time curve is plotted against square root time or Log_{10} time where time is in minutes.

The loading sequence can take several forms

- single load application
- incremental loading in stages – for example, 50 kPa; 100 kPa; 150 kPa; 200 kPa
- incremental loading and unloading
- constant rate of strain.

As noted above, plate loading tests are often used to determine the modulus of subgrade reaction (k_s) – see also Section 5.3.2. At first glance, this seems to provide a very quick and simple way of designing a ground bearing slab. Nevertheless, it would be a misconception to consider this as a simple parameter. Terzaghi (1955) presented the test and calculation of the subgrade reaction, and clearly recognised that this is not a fundamental soil property; rather, it is a lump constant that requires adjustment. The factors by which the adjustments need to be made are as follows.

- The elastic properties of the soil – due consideration must be paid to both initial and long-term response that occurs when the structure is put under a consolidation load.
- The load intensity.
- The area over which the load is applied and the shape of the bulb of pressure induced by the loading.
- The stiffness of the foundation.

The key elements of these factors are that the value k_s is not a true modulus and should more correctly be referred to as the subgrade reaction. This value is not a constant and should be modified for each loading situation. There are also some situations where the analysis of the foundation performance should not be carried out using this method.

In addition to the points above, there is a significant scale effect that must also be considered.

5.3.3.3 Plate loading tests – with high applied loads

The plate load tests described above apply low to medium loads to conduct the test. If higher loads are required, testing can be conducted using a structure designed to carry larger loadings from either steel beams or concrete blocks. Testing is conducted in a similar fashion to the above, but loadings are much higher and generally testing is run over a longer period of time, often several days or weeks.

Tests of this type require considerable organisation and adequate plant to handle the kentledge blocks. Kentledge can comprise concrete or steel blocks each weighing several tonnes. The blocks need lifting eyes, and the operation requires planning and to be carried out by trained and competent persons taking due consideration of the health and safety requirements during the movement and placing of the kentledge. The ground around the area that will be carrying the weight of the kentledge must be suitable. If the ground is soft then this type of test configuration cannot be used unless deeper foundations are adopted, placed on more competent soil.

The general arrangement is to start with two steel I-beams placed parallel to each other either side of the intended plate location. The plate will need to be positioned first and set using sand, Plaster of Paris or cement. The beams should be sufficient to provide access beneath the construction to enable the load column to be manually set up. This may require casting reinforced concrete foundation blocks or placing blocks to raise the height of the beams. A suitably heavy beam should be positioned above the plate to provide the jacking point for the test. The kentledge may then be built up in a planned fashion. Once constructed, the datum beam and load column can be positioned on and around the plate. The datum beam should extend beyond the zone that might be affected by the zone stressed by the kentledge supports and placed where they do not hinder observation of the testing progress and load column. The plate should also be positioned far enough away from the kentledge supports so that the test area is not affected by the stress zone developed at these positions. A typical plate load test configuration using kentledge is shown in Figure 5.8.

The measurement of plate deflections is usually conducted using linear variable differential transformers (LVDTs) connected to a data logger. It is normal to use staged loading for such tests, as described earlier, the only difference being that much greater loads can be delivered to the plate. The plate size is often larger, commonly up to 1 m in diameter. Testing is carried out over longer periods, often lasting several days or even weeks.

Where this is the case, the whole test area might be protected from weather effects by erecting a cover over the set up. It is normal to provide 24 h attendance to ensure the test is being continuously recorded and to protect the work from vandalism. Additional facilities, such as fencing and accommodation with welfare facilities, might be required. If the tests are to be run for a long duration, then some form of site accommodation may be preferable so that staff can work shifts in comfort.

An alternative to the use of kentledge blocks is to install tension piles on either side of the test area that can support the beams and provide the reaction load. Reactions of up to 200 tonnes can be developed using threaded bar grouted into short boreholes constructed either side of the test location, and using a steel beam bolted to the threaded bar. This method does have the advantage of being simpler to set up and not requiring both large loads to be delivered to site and the need to carry out a major construction to support the kentledge. Consideration must be given to the ground disturbance and zone of influence of construction of the piles and the zone of soil affected when

Figure 5.8 Large-scale plate load test using kentledge to provide the reaction (image courtesy of SOCOTEC)

the piles are put into tension by the loading process, to ensure that neither affects the zone that is being tested. It is recommended that calculations are carried out to ensure the test zone influenced by the plate loading is not influenced by the piles. Similar calculations should also be applied if the tests are conducted with kentledge.

5.3.3.4   Ev1 and Ev2 DIN plate testing

This configuration was initially derived in Germany to provide pavement parameters and avoid the use of the CBR test, which is a comparative design parameter. As mentioned previously, the test measures the deflection/settlement of the plate at its centre and is therefore deemed to be more representative than measuring deflection around the perimeter. The test results are analysed to determine the bearing capacity modulus values Ev1 and Ev2 and the ratio Ev1/Ev2. Two load cycles are carried out: the first provides Ev1 and the second Ev2, in kN/m^2. From Ev2, the strain modulus is derived and is related to the settlement of the plate as $kN/m^2/m$. Ev2 is termed the subgrade reaction.

The test method is described in the German standard DIN 18134 2012-04 (DIN Media, 2012), which is used extensively across mainland Europe. Similar to the method described above for other static plate tests, this test method uses relatively low reaction loads to carry out loading. Load is applied by way of a hydraulic jack system to the centre of the plate while measuring the plate settlement at the centre of the plate. The system uses a balanced datum beam to hold the displacement transducer that measures the settlement of the plate. This method ensures the datum beam is outside the loading zone of influence. Setting up the test is simpler and often quicker than the standard method. The arrangement ensures the operative is not in close proximity to the plate or reaction load and hence the operation is safer than other methods. The plates used are commonly of three sizes: 300 mm, 600 mm and 762 mm in diameter, as shown in Figures 5.9(a) and (b).

As with other plate loading tests, the applied load is plotted against the settlement to provide a stress–strain curve. Modulus values are determined using the average settlement across the whole test giving a bulk modulus, rather than the steepest section of the curve that represents the portion where elastic behaviour is exhibited.

The simplicity, accuracy and safety of this apparatus has resulted in its common use in recent times.

Figure 5.9 Showing the Ev2 DIN plate test in operation (image courtesy of SOCOTEC)

(a)   (b)

5.3.4   Zone test

The zone test, or shallow pad maintained load test (BS 1377 (BSI, 1990)), requires a suitable base to be provided. This can either be a steel plate or a trial concrete foundation. It is normal practice to provide a base of the same dimensions as the proposed foundation.

Loading is started by stacking steel or concrete blocks. These are preformed and usually of uniform size and weight such that the total applied load can be readily determined. The blocks are designed to stack neatly and safely, and each block has lifting eyes or similar to assist with placing. Loading is determined to provide the design bearing pressure for the intended foundation. A test configuration is seen in Figure 5.10. Settlement of the base plate is determined either using dial gauges, displacement transducers or by surveying using a precise levelling technique. Measurements are taken at each of the four corners. The average settlement is plotted against time to enable the total settlement characteristics to be determined.

The authors have successfully used this method to test foundations for a structure straddling natural chalk and engineered chalk fill. The test was used to assess the potential differential settlement between foundations placed on natural undisturbed soils and differing thicknesses of fill. This enabled the differential settlement to be predicted and the necessary structural accommodation to be incorporated in the final design.

Figure 5.10 Zone test being conducted using a steel plate, with deflections being measured by precise levelling. Kentledge applied to provide actual design loading to the plate (image courtesy of SOCOTEC)

5.3.5   Skip test

This is probably the simplest type of zone test to set up and, while often considered as a crude test, it can provide good data over a relatively large area of test. The area for test will require some preparation to ensure it is level and flat, with voids being filled with either dry sand or a lean cement and sand mix. The area should be surveyed with a precise level with datum points set into the corners. The skip may then be placed in position. Conveniently, most skips have lifting points on each corner that can be used as the survey points. The skip can then be loaded with a suitable material such as sand or soil. If a specific load is required, the weight of the skip and contents can be obtained using a weighbridge, either prior to or after the test has been completed.

This test has the advantage that sensitive equipment is not required except for a level and staff or survey targets. Equipment does not have to be left on site if the test is required to run for several days. If the test is to run for an extended period, or if the test is carried out during wet weather, the skip should be covered to prevent rain or materials being added through fly tipping increasing the intended applied load.

Precise level survey readings should be made at each corner of the skip to a datum located outside the zone of influence. The results should be plotted on a time-settlement graph for the movement on each corner and then averaged across the four corners if they are similar.

Because loading takes place over a relatively large area – that is, the base area of the skip – the scale effects are reduced when compared with most other static load tests, such as plate loading tests. If higher loads are required, a further loaded skip can be stacked on the first, although only in accordance with recommendations from a relevant risk assessment. Even with this additional loading, the applied pressures are generally quite low. The main errors occur during the taking of level readings, which need to be as accurate as possible to provide meaningful results.

5.4.   Dynamic load tests

5.4.1   General

This group of test methods use a dynamic impact, typically onto a steel plate, to assess the ground behaviour under a dynamic load, as experienced by the ground beneath rail and road structures.

The tests are divided between those that induce an energy wave through the ground, by impacting the ground with a heavy weight and measuring the ground response, or by inducing a shallow seismic wave in the ground and measuring the travel characteristics of the pulse of energy. The tests measure values of stiffness, Evd, where vd relates to vertical dynamic loading.

The impact load causes the ground to deflect. The method is primarily used to investigate the condition of roads and paved areas as well as subgrade suitability. The applied load pulse generates a deflection basin with the deflections becoming smaller further from the load plate. A series of sensors measure the level of pavement deflection. In addition to producing information on shallow ground conditions, the FWD can help engineers to design a robust road structure as well as helping to plan more efficient repairs by detecting areas of poor integrity and cracking within an existing pavement. In this way, the FWD can be used to verify patching locations and identify areas of potential future failure that would not be identified by visual assessment alone.

5.4.2   Falling weight deflectometer (FWD) testing

Deflectometer testing is a nonintrusive test that provides information on soil stiffness and changes in material density. The tests are conducted by causing a shallow Rayleigh wave in the near surface soils. This is achieved by dropping a weight onto a buffer that is located on a circular plate. The buffer shapes the energy pulse and transmits it into the plate. The drop weight has a trigger that switches on impact with the plate. The sensor, either a geophone located at the centre of the plate or an array of geophones set radiating out from the source, record the arrival time of the induced Rayleigh wave. The reflected signals from the strike are collected to determine soil density changes that are proportionate to the soil stiffness. A soil stiffness profile is obtained and a graph of soil stiffness against depth is produced. Some systems use more expensive force balance seismometers; these have built-in calibration and are less prone to external interference.

Deflectometers range in size and method of energy delivery. The lightweight version (Section 5.4.3) may be carried by hand. Larger versions are either towed as a trailer or can be integral in a vehicle (Section 5.4.4). For these tests, the impact can be delivered either as a single or double mass or a combined total mass. Other devices are towed use a rolling weight.

5.4.3   Lightweight falling weight deflectometer (LWFWD) testing

The handheld LWFWD enables the design thickness of the proposed pavement construction to be determined (an alternative to the CBR test) or changes in soil density in earthworks, thus providing an indication of changes in the construction make up. When testing a road or earthworks the test is usually used in conjunction with intrusive coring or density testing methods. For the most effective application, the LWFWD can be used to locate areas that show anomalies, which can then be investigated with intrusive works. This is particularly useful when the test is used on the subgrade, enabling areas of poor suitability to be detected and thus enabling remediation to be carried out before the pavement is constructed. Results are generally reported as a mean surface modulus (see Figure 5.11).

The test results record the Rayleigh Wave as the energy pulse is reflected from the soil, denser soils producing a stronger response. The return energy wave is detected by a geophone located at the centre of the plate. Analysis of the return signal strength and travel time enables variations of stiffness with depth to be calculated. The test can be interpreted to give a derived CBR profile or used to determine the stiffness modulus of the soil placed during earthworks. The authors have used the device to determine areas of compaction variation of rail track beds, as shown in Figure 5.12. The advantage of the LWFWD is that it is readily portable and can be operated by a single operative. The equipment incorporates a GPS system so that the location is readily determined, and connection to the internet means that test results can be rapidly transmitted to the engineer for analysis, thus giving a fast confirmation of compaction quality.

When conducting the test, it is essential that the plate is in good contact with the soil or pavement surface, and this can be achieved using a fine dry sand to fill surface voids if the soil matrix is granular.

5.4.4   Heavy falling weight deflectometer (Heavy FWD) testing

Heavy FWDs use a similar method with a weight dropped from a fixed height to impact a buffer connected to a circular plate in contact with the ground. The plate is typically 300 mm or 450 mm in diameter, the larger plate size achieving a greater depth of penetration. The resulting impact creates a shallow pulse of energy or Rayleigh wave through the soil. These devices use an array of geophones deployed to pick up the reflected energy pulse. Typically, these will include a geophone

Figure 5.11 Typical results from the LWFWD (image courtesy of SOCOTEC)

Drop No	Deflection µm	Peak load kN	Peak stress kPa	Pulse duration m8	E-modulus MPa	Status	Remarks
1	1278.9	6.9	97.6	15.9	20	Seating drop	
2	726.6	7.0	99.0	16.0	36	Seating drop	
3	699.2	7.3	103.3	15.6	39	Seating drop	
4	682.5	7.4	104.7	15.5	41	Accepted	
5	684.7	7.6	107.5	15.7	41	Accepted	
6	648.5	7.5	106.1	15.7	43	Accepted	

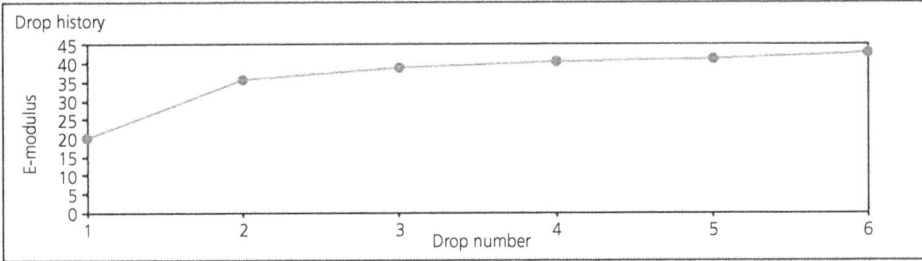

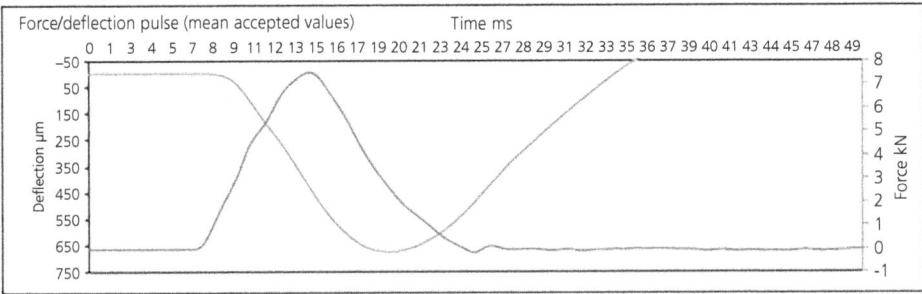

Mean surface modulus (E)   41.5 MPa   Estimated CBR : 4%

Figure 5.12 LWFWD being used in a rail environment (image courtesy of SOCOTEC)

at the centre of the plate and then radiating outwards at 200 mm, 300 mm, 450 mm, 600 mm, 900 mm, 1.2 m and 1.5 m. Because an array is used to receive the reflected pulse, the data obtained cover a larger area and include greater definition. The device is towed behind a vehicle, which makes the test particularly useful for assessing road pavement and subgrade condition without the need for intrusive works. The vehicle is positioned at the test location and the test is made before moving to the next location.

Single impact tests can overestimate the modulus values of softer soils. This issue is addressed by using a double impact test, this configuration using two weights each dropping onto a buffer and plate. This produces a 50 kN pulse of longer duration and is similar to a typical wheel loading. The results obtained are generally more accurate, particularly when conducted on softer soils. The double loading system can also be carried out by locking the two weights together to provide a single higher impact load.

Other deflectometer test configurations are conducted using larger vehicles that tow an array of geophones and use a rolling weight impactor. This configuration is used almost solely to assess pavement construction such as motorways. However, unlike the methods described above, the vehicle does not need to stop and can travel at speeds up to 55 m/h, thus avoiding the need for lane closures when performed on a live road.

5.4.5 Dynamic plate testing – dynamic impact

This dynamic load test method uses a mass of 120 kg that is dropped 500 mm onto a plate that has a spring shock absorber set onto it. The dynamic strain modulus, E_{dyn}, is determined by analysing the rebound force and the deflection of the ground owing to the impact. Figure 5.13 shows a test being undertaken on a chalk surface.

Figure 5.13 The measurement of the dynamic strain modulus E_{dyn} using the Dynaplaque 2 model (image courtesy of SOCOTEC)

Figure 5.14 Plot of the data set from comparative tests on the chalk using the E_{v2} static plate test and the dynamic modulus, E_{dyn}, Dynaplaque 2 (image courtesy of SOCOTEC)

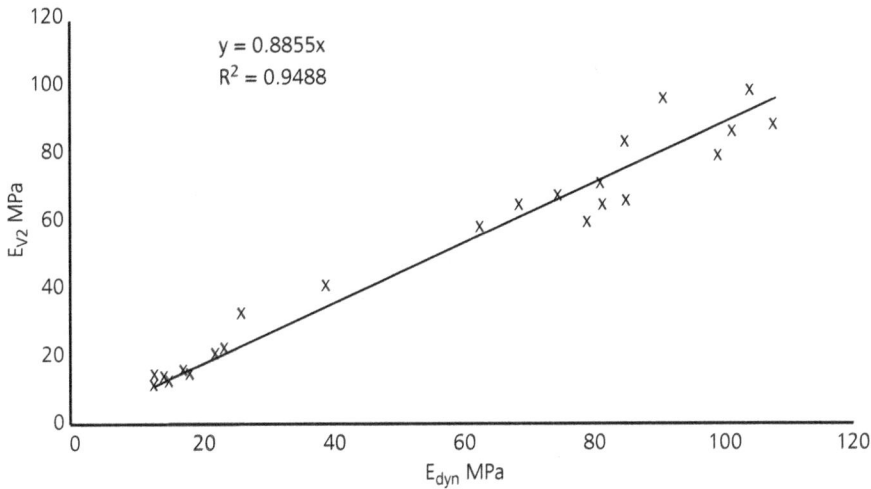

The impact loading simulates the working load of the subgrade, and provides good correlation against other loading tests (as shown in Figure 5.14, where E_{v2} strain modulus from static tests is plotted against the dynamic strain modulus (E_{dyn}), these specific results being measured in chalk).

As previously mentioned, the modulus value is the gradient of the stress–strain curve, either at a particular point on the curve or in the case of elastic modulus over the steepest part of the curve recorded from the test.

5.5. Dynamic probe and penetrometer tests

5.5.1 General

These tests fall into two categories, the simplest using a drop weight to drive into the ground, a series of rods tipped with a cone. More sophisticated methods use a pushed, instrumented cone able to measure forces at and around the cone tip. All tests are then interpreted, using various correlations, to provide an indication of geotechnical parameters such as density, shear strength and compressibility. None of these devices give direct measurement of parameters, and few provide a sample of the soil being tested, making interpretation subjective.

There are several devices using a cone or probe that is either pushed or hammered into the ground by hand. The key to the effectiveness of such tests is the method by which energy is applied. Any test where the energy is not controlled should be considered as providing general guidance only and should not be used for design. Such tests may be of some use in restricted access situations, where larger pieces of equipment cannot be used. Even so, the test will only provide comparative information regarding ease of penetration and can only be used if other tests are conducted in the same materials alongside more substantial and proven equipment.

5.5.2 Standard penetration test (SPT)

The SPT is one of the most commonly used field tests in the UK, although its value to provide high-quality information on soils is debatable. The origin of the choice of driving energy, now

245

standardised, is based on the height a man might be able to comfortably lift a mass of 63.5 kg using a single pulley block, which was deemed to be about 760 mm. These values are directly converted from earlier imperial measures.

The standard penetration test as we know it today was first described by Terzaghi and Peck (1948). At that time, the equipment had been in use for more than 30 years as a sampling tool, albeit a rather crude one. The sample is invariably highly disturbed, but does provide an indication of the material within which the test was performed. The original equipment was modelled around similar equipment used in the USA and attributed to Colonel Charles R Gow in 1902. The original sampler was driven at the bottom of the borehole using a 110 lb (50 kg) slide link hammer. In around 1927, part of the Raymond Piling Group, the Gow Company, began using a split spoon sampler of 2 inches (50.8 mm) diameter, and at around the same time a similar system was being used by Sprague and Henwood. In both systems, the drop weight was winched by hand. The split spoon sampling tool, which comprised a thick-walled tube with a cutting shoe on one end, was being used to recover a sample of the soil.

Powered winches were not reported until 1937. The equipment was not standardised, and it is recorded that the drop weight ranged from 110–140 lb (50–63.5 kg). It was not until Terzaghi (1947) and Terzaghi and Peck (1948), when describing the sampling system, suggested recording the number of blows required to drive the sampler by standardising the distance driven to 1 foot (304.8 mm) following seating the sampler 6 inches (152.4 mm) beyond the base of the borehole. Terzaghi suggested that, by recording the number of blows, valuable information might be obtained at little additional cost. Proposing that the information may be of some use in soils where little other information could be obtained, such as granular soils, he coined the term 'Standard Penetration Test'. Up to this point, there were few other tests that might indicate the potential competence of granular soils. Terzaghi proposed using a 60-degree cone SPT(C) rather than an open shoe SPT(S) because many UK soils contain gravel with can block or damage the shoe. The cone tip and the shoe are shown in Figure 5.15 (upper and lower examples, respectively). It was proposed that the number of blows should be recorded to drive the cone a distance of 1 foot (i.e. approximately 300 mm). This has been standardised into the test we use today, although the open shoe tends to be used by default, unless drilling in ground where it is considered likely to be damaged.

Figure 5.15 SPT split spoon sampler with solid cone tip (top), for SPT(C) tests and open shoe (bottom) for SPT(S) tests (authors' image)

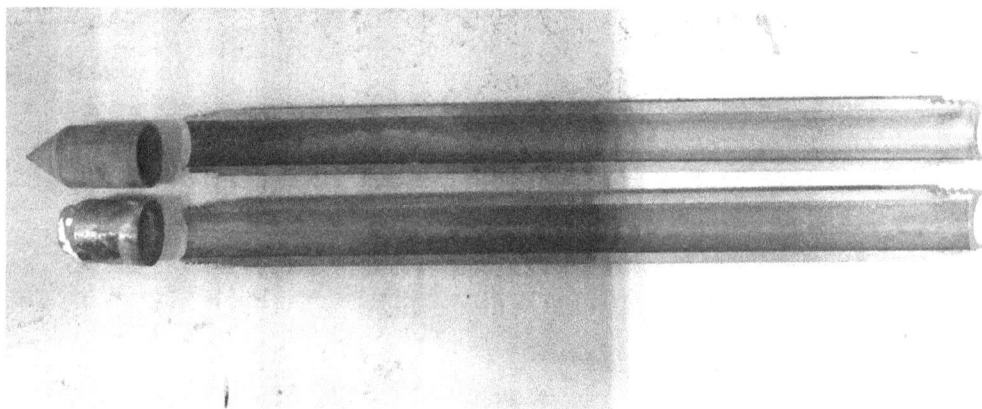

Figure 5.16 Diagram showing the elements of the SPT drop hammer assembly (authors' image)

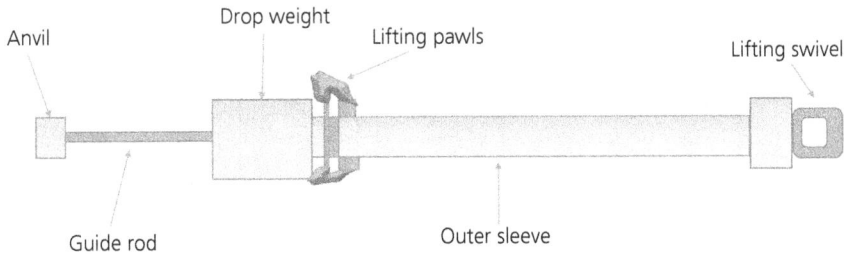

Current practice (BS EN ISO 22476-3 (BSI, 2005b)) requires the test to be conducted with a 63.5 kg weight falling 760 mm driving a 60-degree cone or hollow spilt sampler with a similar angled cutting edge, shown in Figure 5.15. The tool should be carried on rods. Blows are recorded for each 75 mm of penetration. The complete hammer assembly should weigh no more than 115 kg. Figure 5.16 shows the component parts of the drop hammer assembly. The assembly should be clean, well maintained and used without lubrication. The solid cone can either be a replaceable tip that screws into the base of the split spoon sampler instead of the open-end cutting shoe, or an integral solid steel unit. If this latter option is used, the tip will require re-pointing after several uses while the replaceable tip is much simpler to change.

Results obtained from using the open shoe and solid cone are treated the same, although little research has been conducted on the differences between values obtained using either form. Evidence would suggest that the use of the solid cone in sand may give significantly higher results; see Thorburn (1963) and Montague (1990). It is logical to think that there might be significant differences in some soil types. The use of a solid cone is only recommended in gravelly soils, this mainly owing to the risk of damage to the cutting edge of the open shoe if driven in gravelly soils rather than any influence on the test result.

To begin the test, the base of the borehole should be cleaned, removing any loose material. The SPT tool is connected to rods and lowered to the bottom of the borehole. The test is conducted as two stages: the 'seating drive' followed by the 'test drive'. Testing begins with the seating drive, which consists of driving the tool by 150 mm (two 75 mm increments) over which the number of blows should not exceed 25. These are defined as the seating blows. If the number of seating blows reaches 25 before driving the full 150 mm, the seating drive is stopped and the depth of penetration achieved by the 25 blows is recorded.

This is followed by the test drive, during which blows are counted for each of the next four 75 mm increments of penetration, giving a further 300 mm penetration (i.e. a total of 450 mm including the seating drive). If the number of blows over these four increments reaches 50 in total then the test is stopped. The SPT result is reported as the number of blows to achieve 300 mm penetration in the test drive – that is, the penetration resistance, known as the 'N-value'. Where the number of blows reaches 50, the result is reported as 50 blows for the penetration achieved (which should be recorded in mm). It is important that, in addition to the number of blows, ancillary details such as the casing depth, borehole depth and water level are recorded prior to starting the test. Table 5.3 shows some common result combinations and their reporting information for the SPT.

Table 5.3 Examples of test results for the SPT (courtesy of Equipe Group)

Sample Depth (m)		Standard penetration test											Casing	Water
Type	From	To	Self-penetration weight (mm)	75	75	PEN (mm)	75	75	75	75	PEN (mm)	N	Depth (m)	Level (m)
SPT	2.00	2.45	0	5	6	150	7	9	10	12	300	38	2.00	dry

Example showing full penetration of SPT with N value of 38.

Sample Depth (m)		Standard penetration test											Casing	Water
Type	From	To	Self-penetration weight (mm)	75	75	PEN (mm)	75	75	75	75	PEN (mm)	N	Depth (m)	Level (m)
SPT	3.00	3.555	105	1	1	150	1	0	1	1	300	3	3.00	1.00

Example showing self-penetration of 105 mm and N value of 3 for the full 300 mm penetration.

Sample Depth (m)		Standard penetration test											Casing	Water
Type	From	To	Self-penetration weight (mm)	75	75	PEN (mm)	75	75	75	75	PEN (mm)	N	Depth (m)	Level (m)
SPT	15.50	15.95	0	9	11	150	16	19	15	-	175	-	15.50	3.25

Example showing SPT refusal during the third test drive increment recorded as 15 for 25 mm giving 50 blows for 175 mm.

In some cases, the tool will sink into the base of the borehole under its own weight and should this happen the magnitude of this self-weight penetration should be measured and recorded. Where this occurs, consideration should be given to adopting other forms of testing, such as CPT or vane testing, to provide measurements of parameters that are more applicable to such soft or loose soils.

When the test is carried out below the water table, it is essential that a positive head, or water balance, is applied to prevent upward water flow in the borehole. In finer-grained granular soils, such as fine sand, this will cause loosening of the soil and SPT N-values will be significantly reduced. Water balance is seen as an important necessity when conducting the SPT, and provision should be made to ensure that water balance is maintained in the borehole before and during the test. Failure to do so will invariably change the soil density in proximity of the base of the borehole and results will be unreliable.

When using cable percussion drilling methods in granular soils, such as gravel or sand, the use of undersized tools is recommended to reduce the suction effects that may occur as the tools are lifted from the base of the borehole (see Chapter 1). There is a distinct possibility that if suctions are developed at the base of the borehole, the granular soils will be drawn into the borehole causing significant reduction in the density of the soils adjacent to and below the borehole base. Should this occur, then the recorded SPT N-values will be significantly reduced, thus underestimating the soil's relative density.

Great care needs to be exercised when conducting the test. The 75 mm penetration intervals are generally indicated by chalk marks on the drive rods, as shown in Figure 5.17. The driller needs to judge when each interval is reached and make a note of the number of blows required to achieve this. Many errors have been observed with this visual assessment part of the process. Some more recent rigs do have automatic recording of drive distance and blows, although this is an exception rather than the rule.

Some specifications and BS EN ISO 22476-1 (BSI, 2023a) state that the total number of blows can be increased to 100 before terminating the test; an approach that might be taken if testing in soft rock such as mudstone or chalk. In reality this would exceed the original test limits and could damage the equipment. On a more pragmatic note, 50 blows would define granular soils as being very dense or partially cemented. It is suggested that a more practical limitation would be to specify that the number of blows per 75 mm increment should not exceed 25, with the test being terminated

Figure 5.17 Sequence of standard penetration testing: (a) marking up the rods in 75 mm increments; (b) counting the blows; (c) results recorded on rig cross member (images courtesy of J. Winson and P. Horton, SOCOTEC)

(a) (b) (c)

if this condition is exceeded. Such changes to the test method, however, would need to be agreed with the investigation designer.

Where hard rock is met, which can include a cobble or boulder of hard rock within a generally weaker material, the SPT tool may bounce on the application of each blow, and where this is the case, the test should be halted and the condition noted as 'no penetration'.

It is common to conduct an SPT during rotary drilling. When the test is conducted using a rotary rig, the hammer can be operated using the rig winch rope, in the same way as when using a cable percussive rig. Alternatively, where the rotary rig is equipped with a hydraulic hammer, this may be used to conduct the test provided it complies with the required mass and drop height. Testing has shown that these hammers are much more reliable, with more consistent impact energy, and generally the energy ratio is often seen to be close to 90%. This is because the hammer impact tends to be more efficiently applied without the friction losses seen with the conventional trip hammer. The results are reported in the same way to that described above.

Little comparison work has been conducted to determine if the use of the hydraulic hammer provides comparable results to the conventional drop weight hammers. Although, anecdotal evidence would suggest that the values obtained give slightly lower N-values in the same soils when compared with those conducted using the trip hammer. This is because some of the errors inherent in the trip hammer are avoided when using the hydraulic hammer system, which would suggest that the tests performed using this method are more reliable.

The same basic weight and drop height configuration of the SPT are also used in the dynamic probe test when the apparatus is set up as Dynamic Probe Super Heavy, Configuration B (DPSH–B). However, there are significant other differences with this test configuration that are discussed in Section 5.5.3.

A key objective of the SPT is consistency between tests. However, even using a standardised test method and hammer design, the energy actually delivered by each blow will vary, depending on the efficiency of the individual hammer, and consequently different hammers will measure a range of N-values in the same material. Since 2005, the Eurocode BS EN ISO 22476-3 (BSI, 2005b) has required individual SPT hammers to be tested to determine their energy ratio on a regular basis. This determination, often referred to as the hammer calibration, means that the test results from different hammers can be normalised to provide a value that, in theory, might have been obtained from a hammer with 60% efficiency (i.e. a hammer that delivers energy representing 60% of the theoretical energy values assuming a frictionless device). This value of 60% was suggested by Clayton (1995) as being the average energy efficiency ratio of hammers he observed. In practice, the normalisation process involves a calculation that increases N-values measured by a hammer with an efficiency greater than 60%, the high efficiency from such a hammer having resulted in an underestimation of the number of blows required to drive the SPT tool 300 mm. Conversely, the process reduces the N-value from a hammer with an efficiency lower than 60%. This is shown in the examples below:

General expression: $N_{60} = E_r/60 \times N$ Where N is the N-value recorded in the field

Example of an efficient hammer with an energy ratio of 75%: $N = 20$; $N_{60} = 75/60 \times 20 = 25$

Example of a low efficiency hammer with an energy ratio of 55%: $N = 28$; $N_{60} = 55/60 \times 28 = 26$

The two examples above may have been from the same material, being tested in two adjacent bore-holes by rigs using SPT hammers with dissimilar energy ratios. The apparent discrepancy between the two field results ($N = 20$ and $N = 28$) is shown to be a factor of the test equipment rather than the relative density of the materials, the N_{60}-values being much more consistent.

To obtain the average energy ratio of an individual SPT hammer, the energy per blow is measured using accelerometers and strain gauges attached to a short section of rod coupled directly below the anvil. The test requires that the tool is in contact with the soil at the base of a borehole. Practice would suggest that the optimum test depth is of the order of 6–8 m. Annex B of BS EN ISO 22476-3 (BSI, 2005b) provides further details of this process. The hammer energy ratio is calculated by:

Energy ratio, E_r = (actual energy of individual hammer, E_{meas} / theoretical energy per blow, E_{theor}) %

In addition to the energy ratio and normalisation to N_{60}, other corrections have been suggested by researchers; these include a depth correction to account for the mass of the rods (Skempton, 1986) and an overburden correction, depending on the overconsolidation ratio, for the soils being tested. Skempton also suggests that where fills are placed an age factor is relevant. Additional potential correction factors are set out in Annex A of BS EN ISO 22476-3 (BSI, 2005b).

There are much wider questions that are frequently raised regarding the use of the SPT. It is clear that the test does not provide a true parameter for the soil tested. It should be borne in mind that the test drop height and drop weight are arbitrarily chosen by practical considerations rather than scientific reasoning. Nevertheless, these have ostensibly remained unchanged for over a century and thus there is a significant bank of data for comparison. It should be noted, however, that it is only since 2008, when the first energy analyser was introduced into the UK (Equipe Group), that the energy ratio could be determined on a regular basis. As discussed above, this at least enables SPT results to be normalised relative to the actual energy applied, and allows direct comparison between tests conducted using different test hammers.

The test may be influenced by many other factors that have a significant bearing on the actual measured value, and these will often influence results much more than any correction factors. These include the following.

The condition of the equipment

- A loosely fitting guide rod to the assembly may prevent the weight falling vertically, losing energy against the guide rod as it falls.
- The impact with the anvil should be clean and true – if the mechanism has too much 'play', the impact may be greater on one side of the anvil delivering an erratic transmission of energy down the rods.
- Repeated misaligned impact may damage the anvil.
- The cone or shoe must not be burred or damaged.
- Rods must be straight.
- Connectors should be tightened.

Poor operator control

▨ Jarring the weight into the pawls that retain the weigh can lift the SPT tool upwards.
▨ Towing the hammer behind the rig and dragging it on the ground will wear the weight, changing its mass (note that when the hammer is tested the weight is not measured).
▨ The drop height is not generally checked and can be altered if the safety pin location is reset.
▨ Miscounting the number of blows: very few systems have an automatic blow count (often the number of blows is noted using chalk on the rig leg (see Figure 5.17) or in a notebook).
▨ The depth driven is generally not measured automatically but judged by eye.

Borehole conditions

▨ The boreholes should be cleaned of debris prior to test.
▨ Water balance must be maintained during the test, this being particularly important when testing in finer-grained granular soils.
▨ The risk of developing suctions if undersized tools are not used in some soil types.

In any event, and in accordance with BS EN ISO 22476-3 (BSI, 2005b), the SPT hammer should be checked for wear and correct operation by the rig operator on a regular basis and any damaged or worn equipment should be replaced.

The standard recommends that, along with other equipment checks, the rod straightness is checked after every 20 tests. This is not common practice but can have a significant effect on hammer efficiency. In deep boreholes, it is recommended that rod centralisers are used to ensure the rods remain straight in the borehole. To prevent buckling of the rods during driving, it is recommended that AW rods (43.8 mm dia.) are used to a depth of 20 m and BW rods (54.2 mm dia.) are used at greater depths.

When conducted with care, using well maintained equipment, the SPT gives an indication of relative density that, although not being directly related to parameters such as strength or compressibility, can be used to estimate such parameters.

Many researchers have provided comparison data for specific sites to enable parameters such as strength and angle of friction to be deduced. The test may be used as a comparative measure that can inform the performance of foundations, the design of piles, structures and materials providing comparisons with the performance of similar results in similar soils for the same end use. Thus, on this basis, many empirical foundation design relationships have been developed. Such relationships give reasonable design outcomes provided an appropriate factor of safety is applied, commensurate with the considered level of reliance on the source, and provided that the test results can be confirmed.

Typically, the SPT N values are used to assess the peak effective angle of friction $\emptyset'_{max}$, that is, in turn, correlated to values of Nc, Nq and N_γ, which are used in Terzaghi's bearing capacity equation when assessing the bearing pressure of granular soils. In clay soils, relationships have been suggested that draw comparisons to the skin friction owing to remoulding that is developed in pile foundations – for example, Schmertmann (1979).

Clayton (1995) suggests that there is a relationship between N_{60} and c_u in cohesive soils such that $c_u/N_{60} = 11$ for unfissured cohesive soil. Terzaghi and Peck (1967) suggest a factor of 10 in similar soils. Stroud (1974) uses the relationship $c_u = f1.N$, where f1 is a factor (between about 4.5 and 7) dependant on the plasticity index. These factors must be used with care, and ideally the factor should be tested for each site by comparison between total stress triaxial tests on Class 1 samples and c_u values estimated by correlation with the normalised results of SPTs.

5.5.3   Dynamic probing

Dynamic probing was developed during the 1980s and was seen as a simple and cheap method of obtaining soil profile data. The method is particularly useful where access is difficult or restricted, or where the impact of investigation works needs to be minimised.

Traditional dynamic probing is used as a comparison tool providing useful information to relatively shallow depths (generally up to about 15–20 m). It is particularly useful for determining the depth of superficial deposits that rest on more competent soils or, for example, to determine overburden thicknesses for gravel extraction sites.

The dynamic probing rig has a chain carriage that is used to lift a weight to the top of the mast where it is released and falls to strike an anvil. The anvil is connected to a string of rods, each 1 m in length, linked together by couplers and with a 90-degree cone fitted to the bottom rod. This cone is a slightly larger diameter to the rods and is either fixed (referred to as 'Cone Type 1') or sacrificial ('Cone Type 2'), the latter being left at the base of the drive when the test has been completed. This is done so that the friction on the rods is reduced within the oversized hole formed by the cone, thus aiding recovery of the rods. Some practitioners weld a sacrificial cone to the bottom rod to enable its recovery and reuse, to reduce cost. The standard BS EN ISO 22476-2:2005 + A1:2011 (BSI, 2005a) requires that where cones are reused, they must be checked for wear and discarded if the tip is found to be blunt.

The cone is driven for increments of 100 mm, and the number of blows used to achieve this penetration is recorded. The standard requires the torque to be measured after driving each 1 m, to assess the effect of friction on the rod from material collapsing or swelling as it is driven into the ground. The results are plotted as number of blows per 100 mm penetration, as shown in Figure 5.18.

Tests conducted using lightweight rigs are described in BS EN ISO 22476-2:2005 + A1:2011 (BSI, 2005a). Within this document, five configurations of hammer weight and drop height are listed in Table 1, with their respective work per blow ranging from 50–238 kJ/m². These various configurations (light, medium, heavy and two classes of super heavy) allow the possibility of selecting an option that would be most suited to the ground being investigated. Ideally, the configuration chosen will have sufficient energy to penetrate the required strata, but not be excessively 'heavy' otherwise subtle variations in penetration resistance may be lost. In addition, the light and heavy configurations (respectively DPL and DPH) may be selected in order to correlate results with another parameter such as density index (see Annex G of BS EN 1997-2 (BSI, 2007a)).

The configuration delivering the heaviest blows is the DPSH-B, the hammer weight and drop height of which are comparable to the SPT, and in recent times this configuration has been used to drive a spilt spoon and open cone in the same way as the SPT. Nevertheless, while this configuration delivers the same energy per blow as the SPT, it uses smaller diameter rods and a 90-degree cone, and there is no conclusive evidence that the results are comparable to the SPT, which is carried out using a 60-degree cone.

Figure 5.18 Typical dynamic probe (DP) test result (image courtesy of SOCOTEC)

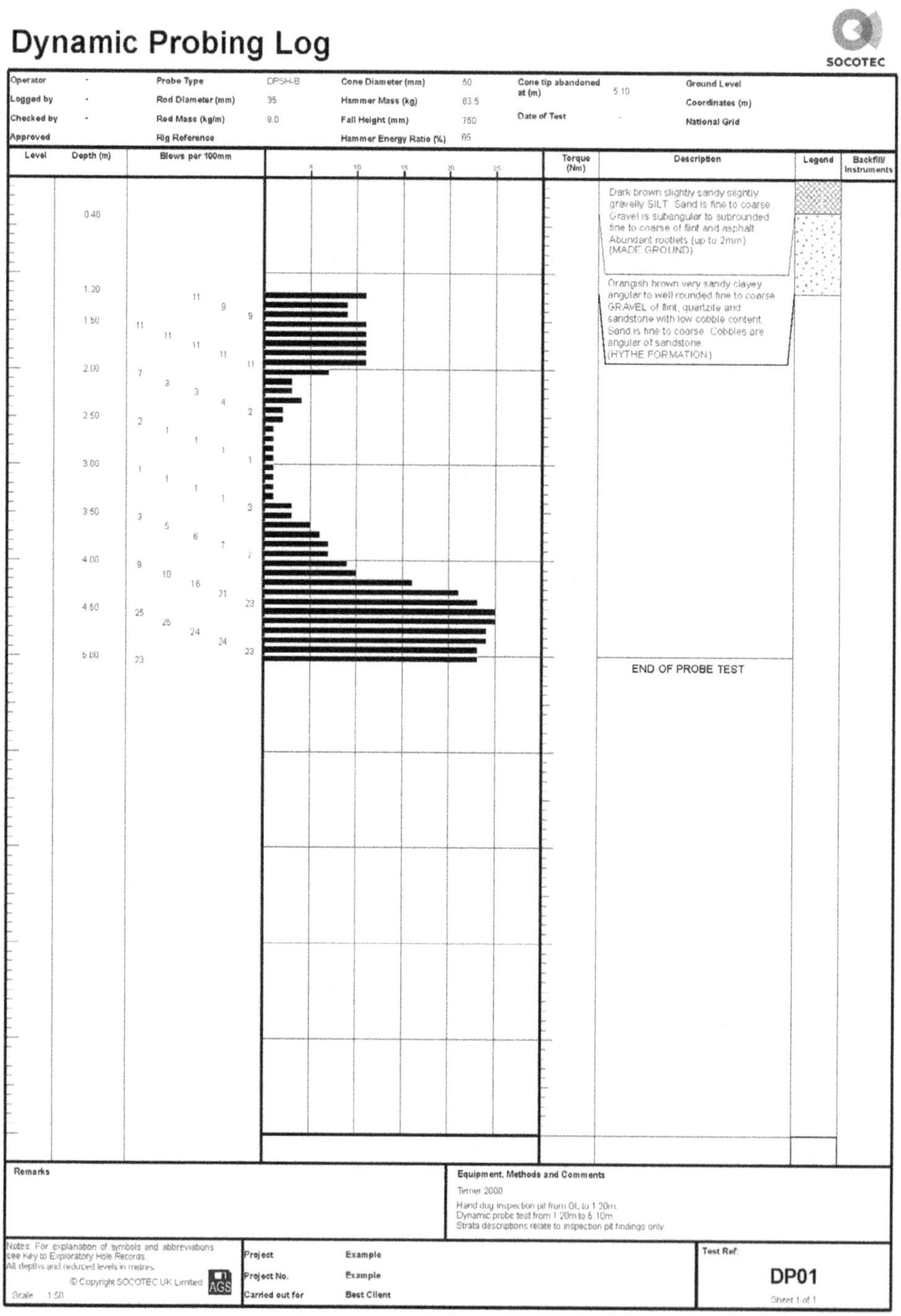

Dynamic Probing Log

SOCOTEC

Operator	-	Probe Type	DPSH-B	Cone Diameter (mm)	50	Cone tip abandoned at (m)	5.10	Ground Level	
Logged by	-	Rod Diameter (mm)	35	Hammer Mass (kg)	63.5			Coordinates (m)	
Checked by	-	Rod Mass (kg/m)	9.0	Fall Height (mm)	750	Date of Test	-	National Grid	
Approved		Rig Reference		Hammer Energy Ratio (%)	65				

Level	Depth (m)	Blows per 100mm		Torque (Nm)	Description	Legend	Backfill/ Instruments
	0.40				Dark brown slightly sandy slightly gravelly SILT. Sand is fine to coarse. Gravel is subangular to subrounded fine to coarse of flint and asphalt. Abundant rootlets (up to 2mm). (MADE GROUND)		
	1.20	11, 9	9		Orangish brown very sandy clayey angular to well rounded fine to coarse GRAVEL of flint, quartzite and sandstone with low cobble content. Sand is fine to coarse. Cobbles are angular of sandstone. (HYTHE FORMATION)		
	1.50	11, 11, 11	11				
	2.00	7, 3, 3	4				
	2.50	2, 1, 1	2				
	3.00	1, 1	1				
	3.50	3, 5, 6	3				
	4.00	9, 10, 16, 21	7				
	4.50	25, 26, 24	23				
	5.00	23	24		END OF PROBE TEST		

Remarks

Equipment, Methods and Comments

Terrier 2000
Hand dug inspection pit from GL to 1.20m
Dynamic probe test from 1.20m to 5.10m
Strata descriptions relate to inspection pit findings only

Notes: For explanation of symbols and abbreviations see Key to Exploratory Hole Records. All depths and reduced levels in metres.
© Copyright SOCOTEC UK Limited
Scale 1:50

Project	Example	Test Ref:
Project No.	Example	**DP01**
Carried out for	Best Client	Sheet 1 of 1

As set out in BS EN ISO 22476-2:2005 + A1:2011 (BSI, 2005a), the test is conducted by driving the cone vertically at a rate of 15–30 blows per minute. For light, medium and heavy configurations the number of blows should be recorded for every 100 mm of penetration, whereas for super heavy tests this can be either 100 mm or 200 mm intervals. The number of blows, 'N', should be quoted with a subscript relating to the interval in centimetres, typically N_{10}.

In materials with a particularly low penetration resistance ($N_{10} < 3$), it is generally more suitable to record the penetration per individual blow. Similarly, for materials with a high penetration resistance ($N_{10} > 50$) recording the penetration for a selected number of blows, such as millimetres penetration per 50 blows, is likely to be more useful.

At a maximum of 1 m intervals (for example, after each successive rod is driven) a torque wrench should be used to rotate the rods by 1.5 turns, or until maximum torque is reached. These torque measurements should be recorded along with the blow counts (see Figure 5.18).

Should a particularly high resistance to penetration be encountered, either $N_{10} > 100$, or $N_{10} > 50$ for at least 1 m of penetration, the test should be terminated.

Research has shown that the blow count per 100 mm of penetration (N_{10}) is affected by particle size and does not therefore provide comparable results between different soil types.

Work has been carried out to determine design parameters and in particular conversion factors to provide SPT N values from the Dynamic Probe results. A rough comparison is suggested by Card and Roche (1988), using the DPSH configuration, where $N_{30} = N_{(SPT)} \times 1.5$, although this relationship should be used with caution and will almost certainly vary within the same material and between sites. In this case, N_{30} is used so as to match the penetration of an SPT. Relationships that may be of use can be developed in particular soils and on specific sites, but general conversions should be viewed with caution. While such relationships have been proven of value, it cannot be recommended that these are applied without comparison with other forms of data, such as CPT or SPT testing at the same site. Essentially, if the project requires SPT N-values, it is recommended that SPTs be undertaken rather than relying on a conversion from dynamic probing.

5.5.4   TRL dynamic cone penetration test (DCP)

This test is not covered by a standard although it is discussed in the National Highways *Design Manual for Roads and Bridges* (*DRMB*) and will often be specified for road construction projects to assess and check that the design value of CBR has been achieved by the construction. BS 5930 (BSI, 2015) states that the test should be conducted in accordance with the guidance given in the DMRB document and the manufacturer's instructions, and results should be treated with caution.

The DCP test was devised by the Transport and Road Research Laboratory (later becoming the Transport Research Laboratory, or TRL) to provide a rapid assessment of the CBR value for pavement construction (as shown in Figure 5.19). The results of the test show good correlation with those obtained by the standard CBR test, as described in Section 5.3.2 (Zohrabi and Scott, 2003). The test is conducted by driving a 20 mm dia. 60-degree cone with blows from an 8 kg weigh that falls 575 mm onto an anvil. The penetration is recorded for a series of blows, the number of blows varying between 1 and 10. Alternatively, the number of blows needed to deliver approximately 10 mm penetration is counted. From these results, a graph of penetration against blows is plotted.

Figure 5.19 The DCP test being conducted on a construction site (image courtesy of SOCOTEC)

The CBR value is determined using the formula:

$$Log_{10} CBR = 2.48 - 1.057 Log_{10} P$$

Where P = penetration rate (mm/blow)

The test is relatively easy to perform. The instructions for use suggest three people are required: one to steady the equipment, another to operate the drop weight and a third to take the measurements of penetration per blow. This can be reduced if equipment with automated logging is used. Its lightweight nature means that numerous tests can be conducted during a shift. The test is particularly useful to detect variations of CBR value in the subgrade. For this reason, it is often used to ensure a prepared surface is adequate for paving. The test can produce spurious results when gravel is present in the tested material. Because the test can be carried out to depths of up to 2 m (using the principal 1 m long penetrometer plus a 1 m extension rod) it does allow a profile of CBR values to be assessed in near surface layers, which can be useful if the formation level has not yet been decided, or if it lies at some depth below the existing site level.

5.6.   Cone penetrometer tests (CPT)

5.6.1   General

The static cone test is sometimes known as Dutch Cone, acknowledging its country of origin, or referred to as cone penetrometer testing. There are two means of conducting the test: the mechanical cone and the electric cone. The latter can also be deployed with the ability to measure the pore water pressure where it is referred to as a Piezocone.

The mechanical cone is the forerunner of the electric cone. Mechanical cone testing is discussed in Section 6.7.1 and BS EN ISO 22476-12 (BSI, 2009c). Electric cones are discussed in Section 6.1.7.2 and in BS EN ISO 22476-1 (BSI, 2023a). Lunne *et al.* (1997) provide a detailed and practical discussion of the use and interpretation of CPT.

Early testing using a mechanical cone was conducted in the Netherlands from 1932 by Barentsen (1936). Development progressed in Belgium and the Netherlands experimenting with different shaped cones and methods of penetration. The first electric cone is attributed to German researchers during the Second World War. Development was continued and in between 1947 and 1957, Delft Soil Mechanics Laboratory developed a system that was very similar to that used today. This was further modified and by 1965 Fugro had developed a cone test that has become the international reference test procedure (ISRMFE 1977 and 1989).

The cone test, unlike other field tests, provides almost continuous information, recording the soil's resistance as the cone is pushed into the ground. In order to achieve this the rods need to be stiff and require a heavy reaction (15–20 tonnes) to develop sufficient force. Early tests using the mechanical cone were conducted using an A-frame and pulley system to push the cone into the ground. This was later replaced with a set of rams and a rod clamp.

The test commonly consists of pushing a 60-degree cone into the ground and calculating the forces acting on the cone and friction sleeve. The cone resistance q_c is given by the force on the cone Q_c divided by the cone area A_c and the sleeve friction f_s is given by the total force acting on the friction sleeve F_s divided by the surface area of the sleeve, A_s. The friction ratio, f_r, is obtained by dividing the cone resistance, q_c, by the sleeve friction, f_s, and is expressed as a percentage:

$$f_r = (q_c / f_s)\,\%$$

Interpretation of the sleeve and cone resistance, together with the friction ratio, enables the soil type to be deduced, along with many other parameters (see Figure 5.25). The interpretation is sometimes tenuous but can be made more reliable by comparison with other data from boreholes ideally from the same site and drilled in similar materials.

As with many field tests, a significant part of setting up the cone for testing includes meticulous calibration of the cone and checking that the systems are correctly recording the load cells and transducers. It is essential to record the cone dimensions, check zero values and ground temperature before each test.

5.6.2   Mechanical cone test (CPTM)

The mechanical cone is rarely used in the UK, having been almost entirely replaced by the electric cone penetrometer. The mechanical cone does not have internal load sensors, all measurements being made from surface. The original designs measure the applied forces using manometers whereas more recent variants use electric sensors to determine the pressures generated. The test measures forces in three distinct operations. The cone is mounted on an inner rod while the sleeve is attached to a hollow outer rod. Initially, the sleeve and cone are lowered to the test location, the cone is then pushed out from the sleeve and the force required is measured. The sleeve is then pushed until it is in contact with the cone and again the force required to do this is recorded. Then the sleeve and cone are pushed as one to the next test level the force is again measured. These three measurements may then be analysed, and in particular the ratio of the sleeve friction to cone resistance is calculated to give the Friction Ratio. During its development, the configuration of the cone and sleeve has been changed such that there several variants of the geometry and method, although the basic principles are similar.

5.6.3   Electric cone and piezocone test

5.6.3.1   The standard electric cone penetrometer (CPT) and the piezocone test (CPTu)
Modern test equipment measures the cone and sleeve resistance forces together using a 60-degree cone with a base area of either $10cm^2$ or $15cm^2$ and a friction sleeve with an area of $225cm^2$. The rods that are 1 m in length are hollow to carry the umbilical data cable. The cable is prethreaded through the rods, which are held in a rod cage while the data cable is connected into a computer system capable of storing the data and showing a visualisation of the information as the test progresses. While the above dimensions are commonly used, other sizes of cone are available.

With the advances being made in electronics, the conversion of the analogue data to the more robust digital data is a natural progression and provides a significant improvement in reliability and accuracy. This conversion can be performed at the cone by using a digital piezo static cone, whereby each cone has its own memory and identity stored within it. These cones incorporate independent measurement of the tip and sleeve friction by separate load cells, which greatly improves the accuracy of the device. They also incorporate temperature measurement and compensation.

The pore water pressure can be measured using a filter section in the cone. This can be placed either within the cone tip (U_1), behind the cone shoulder (U_2), as shown in Figure 5.20, or behind the friction sleeve (U_3), although this is rarely used in practice. Using the filter section, the pore water pressure can be monitored, and it may also be used to assess the rate at which it dissipates.

To conduct a dissipation test, the cone is pushed to the required test depth and held in position while the pore water pressure is recorded. Readings are taken at frequent intervals at the start of the test, increasing in time with each log time minutes cycle. The test should be run for a duration of at least the time for 50% pore pressure dissipation to have taken place. This is the time commonly used

Figure 5.20 (a) Diagram showing position of U_2 and the forces on the cone (authors' image); (b) a cone with outer sheath removed showing component parts (image courtesy of In Situ Site Investigation)

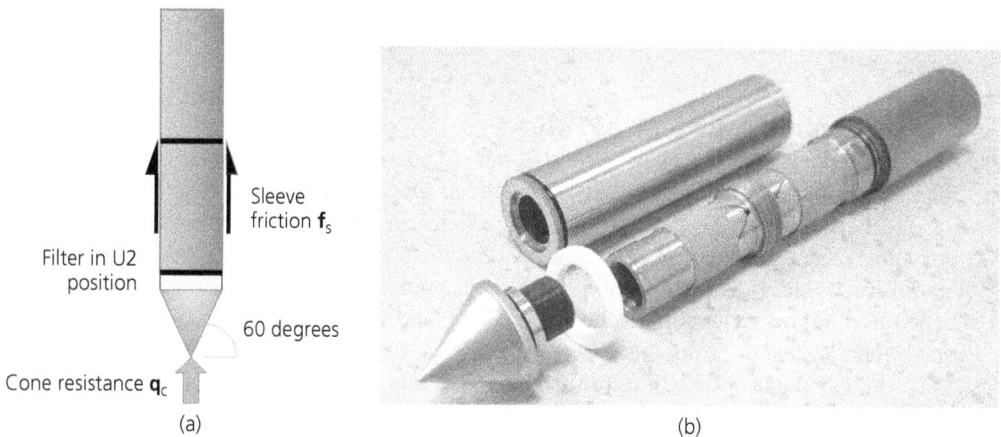

(a)

(b)

to provide settlement characteristics. The test duration will be at least 30 min and possibly longer depending on the soil type. While monitoring, the cone resistance should be recorded and if the soil is weak, it may be necessary to clamp the rods during the test to prevent the loading increasing.

It is essential that the pore pressure filter section is kept fully saturated. This is either achieved by soaking in de-aired water or more commonly by saturating the filter in a light oil under vacuum. A latex membrane is commonly used to maintain this saturation while the cone is assembled and lowered into the ground. The membrane is sufficiently weak that it tears away when the cone is inserted into the ground.

The cone is subject to an unequal area effect, this being related to the internal geometry of the cone and the sleeve. Using the piezo cone enables this effect to be measured and provides a means of calculating a corrected value of cone resistance q_t, where:

$$q_t = q_c + u_2 (1-a)$$

where q_t = corrected cone resistance

a = cone area ratio

u_2 = the pore pressure acting behind the cone.

The inclination of the cone is recorded by a biaxial inclinometer housed in the cone and is measured as it is pushed into the ground. This provides information for the operator regarding deviation of the cone from the vertical and allows a depth correction to be made, which is required by the standard. In addition, as cones are very expensive damaging or losing them is to be avoided. The operator will terminate the test if the cone deviates more than about 5 degrees from vertical per metre, as much more than this would lead to a risk that the cone will be broken from the rods.

Tests may be terminated before reaching the specified depth if any one of the following conditions is met.

- The capacity of any of the load cells is reached.
- The maximum push that can be developed by the reaction (weight of the truck) is exceeded.
- The cone is deflected from vertical by more than 5 degrees in any 1 m length of push, or any abrupt change.

In order to attain adequate depth of penetration, the test requires a sufficient reaction. Commonly, this is provided by mounting the rams in a truck that can provide up to 20 tonnes of reaction force although higher reactions, up to 30 tonnes, are also available. In suitable ground conditions, a truck mounted cone system can push to depths of about 20 m. However, if the soils are too stiff or contain gravel, the test may be terminated at a much shallower depth. Alternatively, it may be prudent to prebore through such soils using one of the borehole methods described in Chapter 1. This might be the case where there is a profile of made ground and terrace gravel overlying clay, for example.

Systems have been developed that use water flushed behind the cone to reduce the friction on the following rods. In this way and in favourable ground, tests have been conducted to depths of 45 m.

Others have used a combination of over drilling when refusal is met and then reinserting the cone to push beyond the first refusal depth.

The equipment may be adapted to enable testing to be conducted in a variety of circumstances. For example, using demountable rams that can be coupled to an excavator arm. In this way, testing can be conducted in situations where access using the conventional equipment is not possible, such as on embankments. In addition, the rams may be set up in confined spaces, in a basement for example, where the structural members of the building can be used to provide the reaction. If this is done then the members must be structurally assessed to ensure they are capable of withstanding the loads that will be imposed as jacking takes place.

Other methods incorporate the jacking system in track mounted trailers or as part of a dynamic sampling rig providing testing and sampling from the same rig (Figure 5.21(d)). These rigs are relatively light, at approximately 1–2 tonnes and have limited capacity to push the cone. This

Figure 5.21 Various means of deployment of CPT: (a) track mounted trailer rig (12–15 tonne reaction); (b) track mounted lightweight rig (5–8 tonnes reaction); (c) lorry mounted (20 tonne reaction); (d) Pagani tracked rig with dynamic sampling and CPT capability (1.5 tonnes and screw pickets to give 5–8 tonnes push capability) (images (a), (b) and (c) courtesy of SOCOTEC; image (d) authors' image, courtesy of Brunel University)

(a)

(b)

(c)

(d)

can be improved with the use of screw pickets, which can be deployed to provide much higher reaction loading. If this method is adopted then the area where the screw pickets are deployed must be shown to be free of utilities. Trailer mounted rigs generally develop loads of the order of 10–15 tonnes (Figures 5.21(a) and (b)). For specialist applications, there are trucks that incorporate a rail chassis capable of converting from road to rail driving. The flexibility of the pushing method has also allowed the development of seabed systems whereby the CPT test can be conducted using remote control underwater, which has revolutionised the design of offshore structures enabling high-quality data of the seabed to be obtained for both structures and pipelines.

A typical output from a cone test is shown in Figure 5.22 below. The basic measurements of cone and sleeve resistance, together with friction ratio, are used to provide an indication of other parameters. These relationships while they may be useful, do require proving to ensure the assumptions made in their derivation are true for any particular site. It is recommended that additional intrusive investigation is conducted and that CPT tests are conducted as part of a wider investigation, which includes the recovery of soil samples enabling comparison of the cone results and soil types (see Figure 5.23).

There are a number of methods of analysis of the results, these can be quite complex and are a subject in their own right. It is recommended that reference is made to Lunne *et al.* (1997) for furtherdetails. In any event, the following parameters may be obtained and are regularly reported.

From the tests, the following directly measured parameters are obtained and are generally presented as graphs of the value against depth:

Cone resistance q_c;

Sleeve resistance f_s;

Friction ratio f_r,

Depth and inclination.

Where a piezo cone CPTu is used the pore water pressure u and the hydrostatic profile is obtained and the pore water pressure ratio B can be derived.

From these measured parameters, numerous parameters may be derived. These include the following.

- Soil behaviour that alludes to soil type.
- Relative Density, D, in granular soils that may be related to SPT and Friction angle Ø.
- Over consolidation ratio OCR of cohesive soils.
- Sensitivity.
- Undrained shear strength s_u of cohesive soils.
- Coefficient of volume change m_v.
- Constrained modulus M.
- Hydraulic conductivity k.
- Preconsolidation stress σ'_p.
- Small strain shear modulus G_s.
- Small strain Young's modulus E_s.

Figure 5.22 Typical test output from a CPTu test (images courtesy of In Situ Site Investigation)

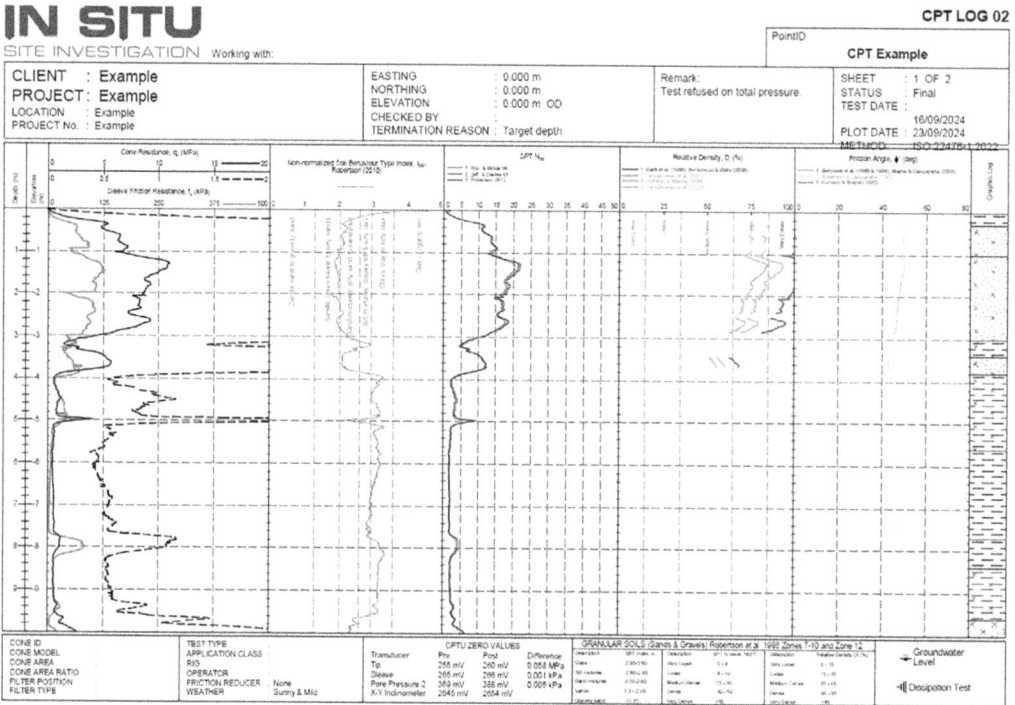

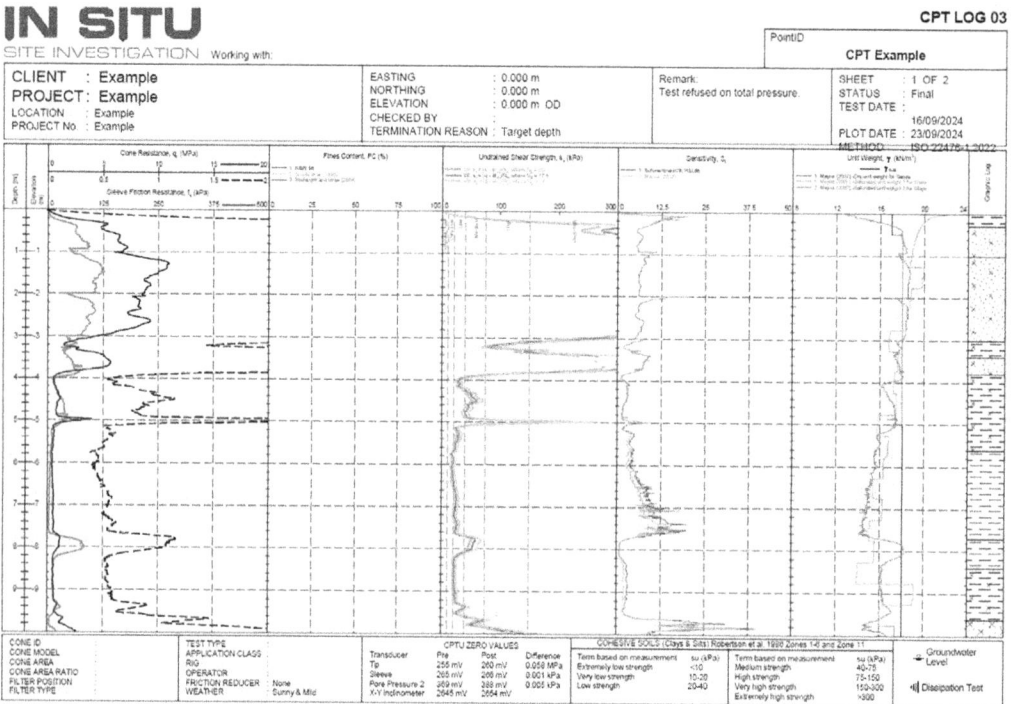

(Continued)

Figure 5.22 (*Continued*)

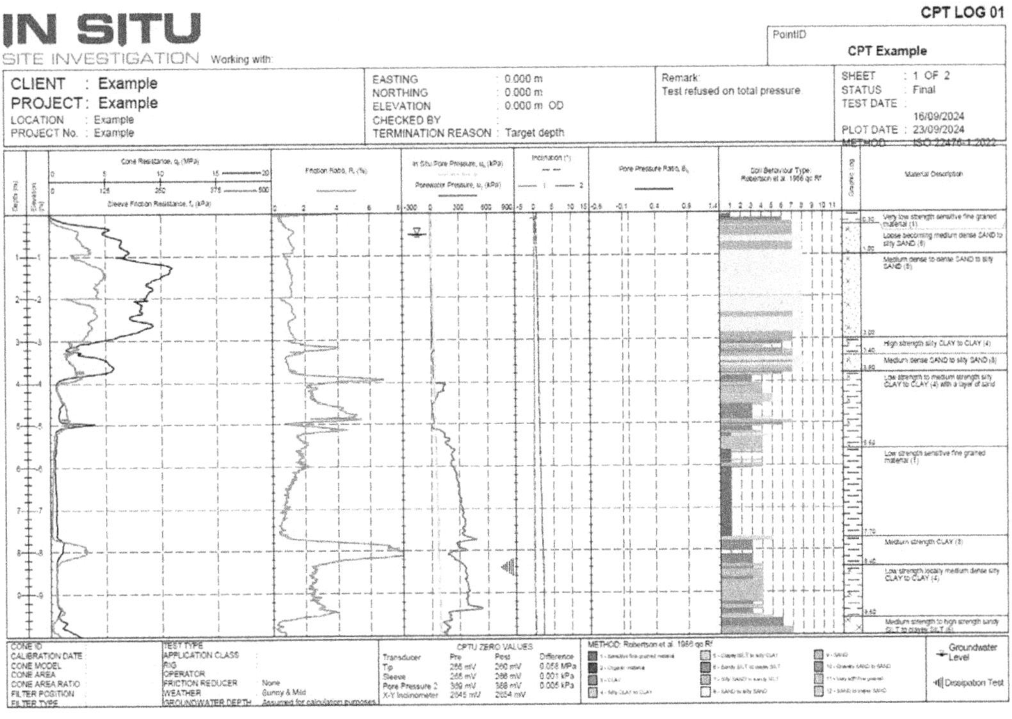

Figure 5.23 Showing graphs of friction ratio and cone resistance with the interpretation of soil type from this relationship. (images courtesy of In Situ Site Investigation)

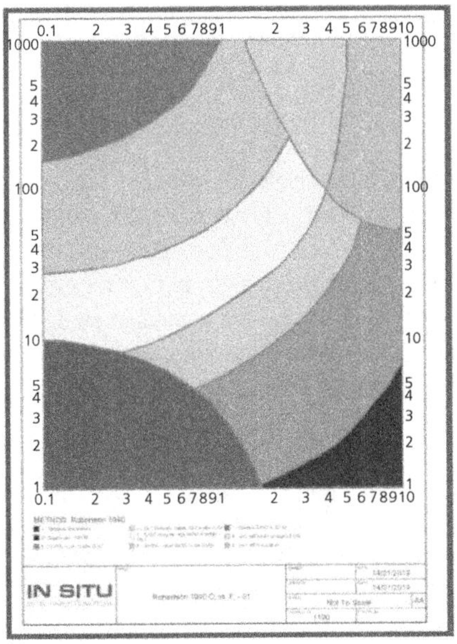

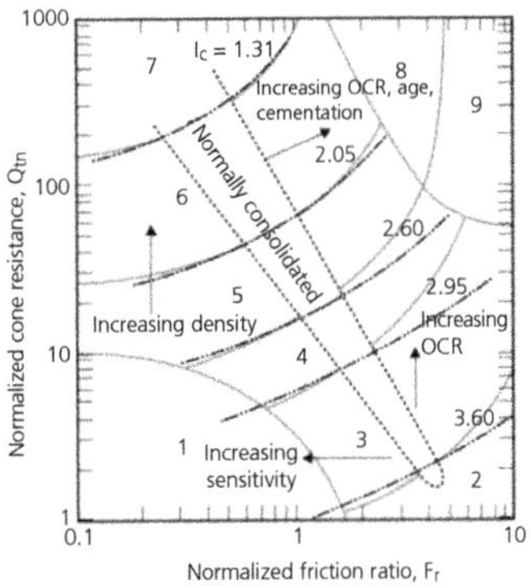

Zone	Soil Behavior Type	I_c
1	Sensitive, fine grained	N/A
2	Organic soils – clay	> 3.6
3	Clays – silty clay to clay	2.95 – 3.6
4	Silt mixtures – clayey silt to silty clay	2.60 – 2.95
5	Sand mixtures – silty sand to sandy silt	2.05 – 2.6
6	Sands – clean sand to silty sand	1.31 – 2.05
7	Gravelly sand to dense sand	< 1.31
8	Very stiff sand to clayey sand*	N/A
9	Very stiff, fine grained*	N/A

The modern electric CPT is often used to determine additional parameters by incorporating probes and other devices into the body behind the cone. These allow the parameters that are normally obtained from the cone as well as an array of other parameters and information. The most commonly used methods are discussed in sections 5.6.3.2 to 5.6.3.14 below.

5.6.3.2   Cone pressuremeter

The cone pressuremeter incorporates an expandable membrane into the section behind the cone; testing is conducted in a similar manner to that described in Section 5.9.4. Because the cone and pressuremeter are pushed into the test position, there is close contact with the soil to be tested. Some devices are a slightly smaller diameter than the cone to reduce the friction over the pressuremeter section. The membrane is made of rubber and protected by vulcanised rubber or thin metal strips. The membrane is inflated either by nitrogen gas or a fluid such as oil. The device has a length to diameter ratio of 10. Expansion is measured using either two or three centrally placed strain arms. The membrane is inflated until maximum strain is reached pausing at intervals to conduct unload/reload loops. As with standard pressuremeter testing, a correction needs to be applied for membrane stiffness.

5.6.3.3   Seismic cone

By incorporating geophones/transducers approximately 0.5 m apart into the body, a seismic test can be conducted. The shear wave is generated at surface using a hammer, which can be impacted manually, onto a plate or using a mechanical hammer integral to the CPT truck. From these results, stiffness parameters such as shear modulus (G_{max}) and Young's modulus (E) may be determined, similar to the method described in Section 5.2.3. Poisson's ratio (ν) can be determined if p-wave velocity is measured.

5.6.3.4   Magnetometer

This is used to allow scanning for metalliferous objects such as unexploded ordnances (UXOs). This is commonly used to probe ahead of boreholes and at pile locations where a potential risk has been identified at the site in general. The magnetometer is able to scan a distance of up to 2 m around the probe, although the detection capability is affected by the size of the object. In this way, the probe can be used to detect buried metalliferous objects that may be below or close to a proposed drilling position. The method is often carried out to ensure pile and borehole locations are free from UXOs and other metalliferous obstructions.

5.6.3.5   Electrical conductivity

Determination of electrical conductivity or resistivity (the inverse of conductivity) may be measured by the inclusion of a resistivity module behind the cone. Electrodes are positioned along the body at suitable spacings to allow either single or double electrode configurations using the Wenner spacing method, the electrodes are insulated from each other with two outer electrodes being used to pass a current and pairs of inner electrodes which measure the resistance. To determine the resistivity, a calibration factor is required that can be found by conducting calibration test in a calibration chamber. The resistivity results may also be used to determine porosity and density.

5.6.3.6   Gamma cone

This carries a sensor that can detect naturally occurring gamma radiation. This type of cone is particularly useful in soil or weak rock where there are marked differences in the natural gamma radiation, such as dissolution features in chalk. The chalk has very low gamma emission, while in clay it tends to be much higher, enabling the identification of clay filled dissolution features (see Figure 5.24).

Figure 5.24 Gamma cone ready to be deployed (image courtesy of In Situ Site Investigation)

Figure 5.24 Gamma cone ready to be deployed (image courtesy of In Situ Site Investigation)

5.6.3.7 Soil moisture cone

This instrument is similar to the conductivity cone and uses the dielectric properties of the soil to assess the volumetric percentage of water within the soils in conjunction with the soil resistivity. The method uses pairs of electrode rings set into the body behind the cone. The innermost of the two pairs is used to determine the soil moisture content and the outer two are used to determine the conductivity/resistivity. A constant amplitude AC voltage signal is applied across the rings and a measurement is made across the sampling resistor and is proportional to the current passing through the soil. This cone configuration is capable of detecting contaminated soils.

5.6.3.8 Fuel fluorescence detector cone

The presence of aromatic hydrocarbons in the soil can be measured using a fuel fluorescence detector. This comprises an ultraviolet light source from which light is selected using filters to provide a wavelength, Λ, which is used to characterise light hydrocarbons (280 nm $< \Lambda >$ 450 nm) and another used to excite heavy hydrocarbons with a wavelength greater than 450 nm. The selected light passes through a sapphire window. The nature of the reflected light can be used to assess the type of hydrocarbon and the degree of contamination. By deploying the cone in a line or on a grid, the extent and type of a fuel spill plume may be readily assessed using this method, greatly reducing the exposure to contamination to which other methods might be susceptible.

5.6.3.9 Video cone

The cone incorporates a mini colour video camera together with an LED light source and a micro optical lens. These components are protected by a sapphire window sealed into the body behind the cone. The images are displayed onto a screen in real time and are recorded along with the date, time and depth increment. Typical uses include assessment of contamination and grain size/material type recognition and is particularly useful to assess the location of strata boundaries when used in conjunction with the standard cone test results.

5.6.3.10 Hydraulic profiling tool (HPT)

This profiling tool incorporates a porous membrane in the section behind the standard cone, and as it moves through the soil water is injected through the screen at a flow rate of <300 ml/min. By recording the injection pressure, the hydraulic behaviour of the soil may be assessed. Typically, a low pressure would indicate a large grain size while a high pressure tends to indicate finer-grained soils such as clay.

5.6.3.11 Membrane interface probe

This probe is used to detect volatile organic compounds (VOCs) in both chlorinated and nonchlorinated forms. The device incorporates a permeable screen within a heated block. The block is heated to a temperature between 100–120°C. At this temperature, the volatiles are transferred to their gaseous phase. A clean inert gas flow of 35–45 ml/min is induced on the inside of the screen, which draws the gaseous volatiles to surface. The gas rises up through a tube, which itself is heated to ensure they maintain the gaseous phase throughout. The concentration of the volatiles is measured using standard detectors. These include a photoionisation detector (PID) used to measure BTEX and chlorinated ethylene compounds, a halogen detector for halogenated compounds and chlorinated solvents, and a flame ionisation detector FID, which is a general hydrocarbon detector. The versatility of the membrane interface tool can be increased by combining with the hydraulic profiling tool.

5.6.3.12 Flat plate dilatometer (Marchetti dilatometer)

See also Section 5.9.3. This device is normally deployed using a cone truck with the flat plate device being held on similar rods. The technique requires compressed air and an electrical connection to the diaphragm. As with the standard cone test, this device can also be deployed with a seismic section behind the flat plate. The test obtains a shear stress profile of the soil from which many other parameters may be derived.

5.6.3.13 MOSTAP sampling

See also Section 2.5.5. The Monster Steek Apparaat (MOSTAP) soil sampler provides a 1 m long, 65 mm dia. sample whereby the CPT pushing equipment is used to deploy the sampling tube. The sampler cutting shoe at the bottom end of the tube is sealed with a cone. Once the assembly has been pushed to required depth for the sample to be taken, the cone is locked into position and the sample tube is then released from the holding clamp and can be pushed beyond the cone into the soil. A nylon lining, stockinet, wraps the sample as it enters a plastic protective liner. Once pushed to depth, the whole assembly is brought to surface, where the sample can be recovered within its sample tube and stockinet.

If further samples are required, the cone may be reinserted and pushed beyond the depth of sampling to allow further cone testing or additional samples to be taken. Although this process can be quite slow, it does provide the ability to inspect or test samples to corroborate the cone results (see Figure 5.25). Alternatively, a delft continuous sampler uses a similar technique to obtain samples continuously to depths usually of up to 18 m (although depths of up to 30 m have been achieved in suitable soils). The technique provides samples of either 66 mm or 29 mm diameter. A stockinet is used to retain the sample in plastic tubes.

Figure 5.25 MOSTAP samples after the plastic lining has been cut away (image courtesy of SOCOTEC)

5.6.3.14 Ball and T-bar tests

Either a ball or a T-bar shaped tip can be used to replace the traditional cone when testing in very soft soils. The instrument is pushed into the ground and the soil flows around the penetrometer. Both types have a higher surface area up to 10 times that of the traditional cone; this allows for more accurate readings. The probes are only used in very soft material and are limited to giving only a tip resistance from which the shear strength can be derived. The use of a ball or T-bar allows for more precise modelling of the soil behaviour, leading to better understood derived parameters.

5.7.   Other handheld probes and penetrometers

5.7.1   General

Throughout the development of ground investigation methods engineers have devised lightweight probes and penetrometers with which the ground strength can be tested. In the main, these are handheld and generally rely on either hand pushing or dropping a relatively light weight and then measuring the amount or rate of penetration into the ground. They all rely on the individual to provide a force that cannot be determined, while ensuring penetration is vertical and at a steady rate, which is rarely the case. These tools, while they might be of some assistance in assessing soil stiffness, are not able to provide the accuracy required to provide parameters for design and should only ever be used to provide indicative assessments. In some situations, however, they may provide useful information – for example, to determine the depth of peat in a peat bog, something that is commonly required when choosing a location for wind turbines in the Scottish Highlands.

The most commonly used methods seen in the UK are briefly discussed in the following sections.

5.7.2   Mackintosh probe

The Mackintosh probe is a prospecting tool, which found its way into geotechnical investigation during the late 1970s. The equipment comprises a 4 kg drop hammer and rods of 12 mm diameter, which can be used to drive various attachments including a sample tube and a rounded cone. While this is of some use in awkward situations where the ground might be very soft or where access is difficult, its use is limited in a modern investigation.

Originally, the equipment was devised to provide prospecting information in the quarrying industry. The Mackintosh probe's main use was to determine the depth of overburden over a deposit – for example, the thickness of Brickearth overlying a viable Terrace Gravel deposit.

Its value in the geotechnical industry is similar – for example, determining the thickness of alluvial soils resting on more competent strata. The depth of penetration is limited to about 4 m. Driving the cone to greater depth introduces the problem of removal, which is carried out using rod clamps and pulling the rods out. In most cases, however, rods are removed by back hammering.

5.7.3 Soil assessment cone penetrometer (MEXE cone)

It is quite remarkable that this tool is still used by some as a method to obtain a cone index and CBR value. The tool was devised by the Royal Engineers during the Second World War to assist in the design of roads and operational facilities in difficult terrain, such as the tropical jungle of the Far East. In today's technologically driven world, tools such as this should be seen in a museum and not on site and certainly not for any design use.

The test is conducted by pushing a 1.3 cm^2 cone on small diameter rods into the ground. The rods are directly connected to a coil spring that is attached to a rolling tumbler. The resistance of the ground to the penetration of the cone causes the tumbler to rotate, which indicates the CBR value (see also Section 5.3.2). A cone index is obtained when using a cone of 3.2 cm^2. When using this equipment, it is prudent to do so with two people: one to push the cone and the other to take readings against penetration. The equipment includes depth markers on the rods at 75 mm intervals such that the depth of penetration may be readily recorded corresponding to a particular CBR or cone index value.

The cone index is used to make approximate assessments for the suitability for plant and vehicle access. The manufacturers suggest that values ranging from 40 for a four-wheel drive vehicle to 75 for a 2250 l tanker indicate a condition when they would 'just go'. It is also pointed out that the readings are only valid at the time of the test and that factors such as precipitation and temperature fluctuations will significantly influence the results (although this applies to most field tests conducted within the upper 2 m of soil, as noted in Table 5.1).

Because the depth of push and indeed the energy of pushing is dependent on the individual carrying out the test the depth of penetration is variable as well as the consistency of the force applied. This variability causes the tumbler to fluctuate, making both the rate of penetration variable and reading the actual divisions on the tumbler very difficult. Consistent or repeatable readings are rare.

5.7.4 Weight sounding

This test is rarely used in the UK; however, it has been included in the Eurocodes (BS EN ISO 22476-10 (BSI, 2017a)) and is further discussed in BS 5930 (BSI, 2015). The test was devised in Sweden and is often referred to as the Swedish weight sounding test. The test is used to assess soils that might be prone to liquefaction or those that are particularly soft. It is primarily used to assess the depth at which a competent founding soil is met. The test provides two parameters, namely the smallest standard load under which the penetrometer sinks without rotation and the number of half turns required to give a penetration of 200 mm when the penetrometer is under its maximum load and is rotated. Any further correlation to load carrying capacity is purely empirical.

The test equipment comprises a twisted screw cone carried on a series of 12 mm dia. rods. The basic test is conducted by adding weights to the rods to give load increments of 50 N, 150 N, 250 N, 500 N, 750 N and 1000 N. As each increment of load is applied the resulting penetration is measured. The minimum applied load to achieve a penetration of between 20–50 mm/s is recorded as the weight sounding required to achieve penetration. Once penetration has begun, the rods are rotated; this penetration for each half turn is recorded. The test is stopped when the screw cone stops penetrating, either owing to the soil being too stiff or the presence of gravel or the penetration has achieved 10 m depth. It can be difficult to determine if the cone stops penetrating owing to an obstruction such as gravel or because the soil is too stiff for further progress. This can be overcome to a certain extent by striking the rods with a hammer to assess if the stop is temporary.

An alternative method is to count the number of half turns to achieve a given depth, say 250 mm. Rotation of the screw cone can either be carried out manually or by the addition of a motor to turn the rods.

This test is relatively cheap to conduct and in weak soil does provide some useful information, however the interpretation is not standardised and does not provide direct determination of geotechnical properties, although relationships for bearing capacity and density can be derived for specific soil types.

5.8.   Shear strength tests

5.8.1   Hand vane test

The hand vane is either of the form where the vanes are mounted onto a circular plate directly coupled to a spring, or the vane is on a spindle that connects into a torque head that has an integral spring and a gauge by which to read off the shear strength (see Figure 5.26). Once the soil has

Figure 5.26 The two types of hand vane commonly used to assess soil strength: (a) the Pilcon hand vane (image courtesy of J. Ford Crush, SOCOTEC); (b) the Pocket Vane (authors' own image)

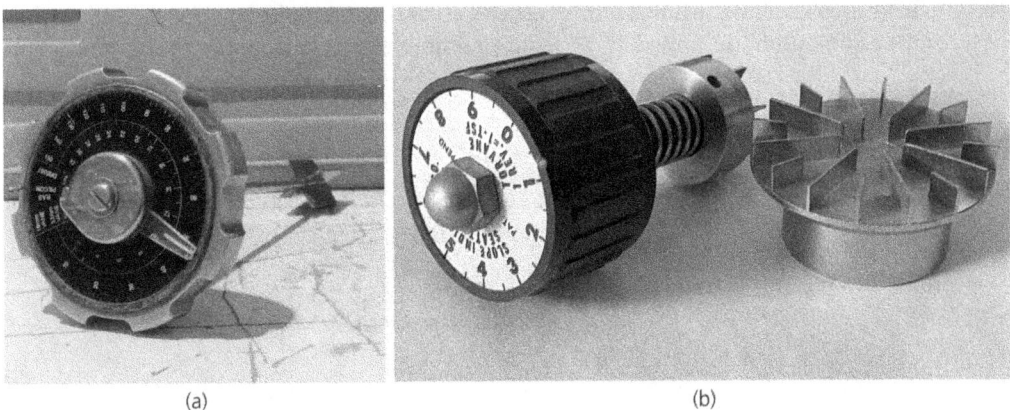

(a)   (b)

failed, an indication of the residual shear strength may be obtained by continuing to turn the head and recording the lower bound value once it becomes constant.

Both equipment types require a steady hand and the ability to apply the torque at a constant rate. The vane is pushed into the soil to be tested in a steady motion without twisting or wobbling the tool. Once at the required depth of penetration, the head is turned slowly (a rate similar to that of a second hand is generally appropriate) until failure occurs. Both instrument types are equipped with a maximum reading pointer that is set to zero before starting the test. Some hand vane tools provide a direct reading of undrained shear strength while others are provided with a calibration factor that is used to obtain a shear strength value.

Owing to the small volume of soil tested, results from hand vane tests may be unrepresentatively high, and should therefore be treated with caution. The test is difficult to perform consistently and with any accuracy, and in particular it is difficult to continue to rotate the vane at a constant rate by hand without affecting the orientation of the instrument. Results may also be affected by minor soil variations, which can have a significant influence.

5.8.2 Pocket penetrometer

The pocket (or hand) penetrometer is an aid to soil description and, much like the hand vane test, the results should be treated with caution owing to the very small scale of the test. These tests are only applicable to fine-grained soils such as clay and clayey silts which exhibit cohesion.

The penetrometer comprises a spring-loaded piston housed in an aluminium body and is approximately 150 mm in length, the piston having an area of 32 mm^2 (0.05 sq in), as shown in Figure 5.27. The penetrometer usually has a sliding ring or indicator that should be moved to zero before starting testing. The penetrometer is then inserted into the soil to penetrate a depth of 6.35 mm (1/4 in). The soil compressive strength can be read from a scale etched into the penetrometer body as either kg/cm^2 or tons/ft^2. Dividing the compressive strength values by two will provide an estimate of undrained shear strength, which can then be expressed in more appropriate units, for example multiplying by 100 to convert kg/cm^2 to kN/m^2.

For particularly soft soils, penetrometers can be obtained either with a weaker spring or an adaptor that increases the surface area of the penetrometer tip.

Many specifications will ask that this test is carried out on cored samples of cohesive soils. While such results can be useful in providing an indication of strength variations with depth, absolute

Figure 5.27 The pocket penetrometer (image courtesy of E. Bell, SOCOTEC)

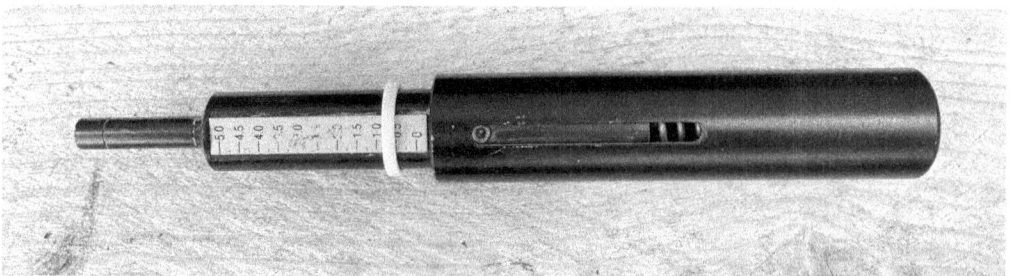

strength values are not considered to be sufficiently accurate for design purposes. As such, while pocket penetrometer results may be reported in field notes, it is recommended that they are only reproduced in the technical report, or shown on the borehole record, if accompanied by a warning that the results are not suitable for design.

5.8.3   Field vane test

The field vane test, also referred to as the borehole vane test, is similar in principle to a hand vane test, only larger in scale and conducted at the base of a borehole (see BS EN ISO 22476-9 (BSI, 2020a)). It is less commonly used today as part of a ground investigation, although there are merits in the test. The test is conducted in soils ranging in consistency from soft to firm, generally because soils with low strength are often difficult to sample and prepare for testing in the laboratory. Therefore, a field test that provides a realistic assessment of the undrained shear strength in these difficult soils, such as the borehole vane test, is of value.

Two types of vane system are commonly used. One consists of a simple cruciform shaped vane, mounted on a series of rods that is lowered to the base of a borehole and then pushed to the required test depth, which should be at least three borehole diameters below the base of the borehole. The other type, the penetration vane, incorporates an outer hollow rod and an inner solid rod. The outer rods are connected to a cone shaped body with the vane on the inner rods, protected inside the body until it is required. The whole assembly is pushed to the required test depth using the outer rods. Once the vane is at the required depth, the vane is then pushed from the body using the inner rods to an additional depth of at least 0.5 m; this method does not necessary require a borehole. The test procedure is similar for both types.

The test requires the base of the borehole to be clean and clear of debris. It is imperative the vane is pushed without wobbling. When in a borehole, centralisers should be positioned on the rods to ensure the rods remain straight and true. Care needs to be taken when coupling the vane rods to ensure the vane head is not wobbled, particularly once it has been pushed to the test depth. To prevent the rod joins moving, the couplings are designed as a square notch with a sleeve to lock them into position.

Once the vane is at the required test depth, the vane rods are attached to the torque head that incorporates a calibrated spring and a gauge that registers the torque. The torque is applied by rotating the worm screw on the head. This is usually done by hand. The vane is rotated until failure occurs. Torque readings are taken at intervals of rotation, allowing a graph of torque (rotational stress) to be plotted against time that is directly proportional to the angular strain. Rotation is applied at a rate of between 6–12 degrees per minute, The maximum torque is used to calculate the vane shear strength. The test may be continued beyond peak failure until the torque readings become constant, allowing the residual strength to be determined and hence the sensitivity of the soil.

Vanes can be of various dimensions but the height to diameter ratio should be a factor of 2 (see Figure 5.28). In the UK, the preferred dimensions for the vane head are given in Table 5.4. The thickness of the vane can also be critical and ideally as thin as possible; significant disturbance has been demonstrated in sensitive clays where the measured shear strength may be as much as 30% lower than the actual value (Rochelle *et al.*, 1973). Typically, the blade thickness should be < 3 mm and < 2 mm for use in soils with a sensitivity >30.

The maximum torque recorded is used to calculate the vane shear strength T_f as follows (some publications refer to T_f as T_v):

Figure 5.28 Failure model for the vane test (authors' image)

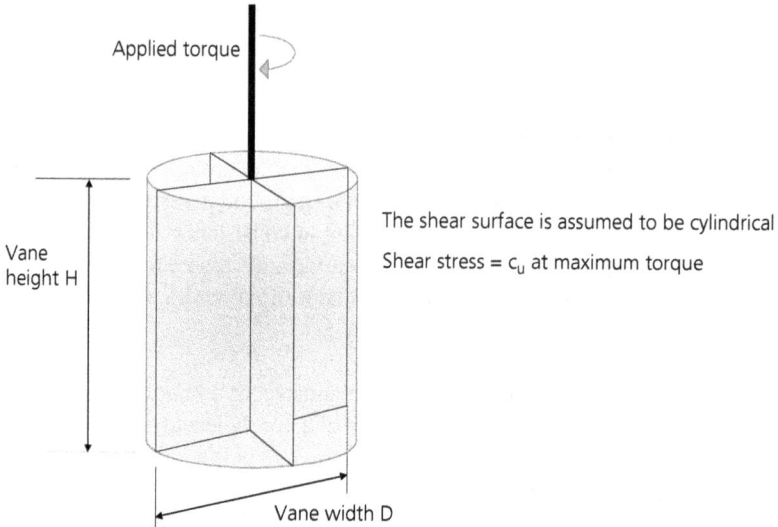

Applied torque

Vane height H

Vane width D

The shear surface is assumed to be cylindrical

Shear stress = c_u at maximum torque

Table 5.4 Suggested vane sizes for soils of different strength

Undrained shear strength (kPa)	Vane dimensions (mm)	
	Height	Width
<50	150	75
50 to 75	100	50
>75	Not suitable	

$T_f = M / K$ (kPa)

Where M = peak torque

K is a constant dependent on the vane dimensions

$K = \pi D^2 (H/2 + D/6)$ (mm^3)

Where D = diameter of vane (mm) H = height of vane (mm)

The results of the peak and residual stress are reported along with the angular torque at failure. On completion, it is good practice to obtain a sample of the material at the test position to enable water content and plasticity values to be determined in the laboratory.

5.8.4   Shear box test
Although many texts include the determination of shear strength in the field by in situ shear box testing, in the authors' experience this is a very challenging test to perform. The test will require significant

precision and care, as the slightest errors can cause serious disturbance of the specimen and poor test results. Points to consider are the ease at which the soil will be disrupted as the assembly is built and the influence of weather, in particular heavy rainfall, which can cause the whole exercise to be aborted.

Considering the above, while the field shear box test might seem a very useful test to undertake, the authors would suggest that taking a block sample for testing in the laboratory – which itself can be challenging enough – is likely to provide better results.

The theory is to perform a shear box on a cut section of the soil or rock in question that should be fashioned into a cube. To do this, the soil around the block is removed to allow access, which results in a pit needing to be hand dug to avoid disturbance of the soil/rock specimen. The whole pit needs to be stable and may need support. Steel sides are assembled around the cube and any space between the block and box can be infilled with Plaster of Paris or quick setting cement. Once set, the lid is located on the soil surface and the loading mechanism can then be fitted; this incorporates a jack that is operated horizontally with a load or pressure measuring device. The jacking assembly is placed against a jacking block situated in the end of the pit. A datum bar is also required so that the movement of the box, or strain, can be measured. A means of applying a vertical load is also required; this is placed on a steel plate that is situated on a series of roller bearings such that the loading remains normal as the box is moved horizontally.

The test is taken to failure and a graph of stress against strain plotted from which the peak shear strength can be determined and if run beyond the failure a residual strength value may also be determined.

5.9. Pressuremeter testing

5.9.1 General

The pressuremeter comprises a cylindrical cell with a rubber membrane carried on the drill rods. The membrane can be pressurised at a controlled rate, usually in stages. The membrane is expanded against the wall of the borehole, the resultant expansion of the membrane is a function of the soil stiffness requiring greater pressure the stiffer the soil. The expansion of the cell is determined by either measuring the fluid volume change in the cell or by direct measurement of the volumetric strain using in cell linear transducers or spring-loaded callipers to measure movement. A graph of applied pressure (stress) is plotted against strain usually volumetric. Borehole pressuremeters are discussed in Section 5.9.2 and other variants of the pressuremeter such as the flat plate dilatometer and cone pressuremeter are discussed in Sections 5.9.3 and 5.9.4, respectively.

There are several variants of the pressuremeter test, all of which measure an applied stress and resultant strain in situ. By testing within the soil, many of the issues related to sample disturbance can be eliminated giving greater confidence in the strength results obtained. Tests may be conducted in soil or rock. The test section is prepared either by drilling a test socket or by the device drilling itself to the required position. The advantage of the self-boring pressuremeter is that the membrane is in contact with the wall of the self-bored hole. Where the pressuremeter is placed into a preformed socket the membrane will first be expanded to bring it into contact with the drilled hole wall. Other methods use pushing or driving the pressuremeter to the required position for the test. It is important that whichever method is used, the soil is in contact with the body of the pressuremeter or the annulus between the pressuremeter body and the soil is small.

The Cambridge pressuremeter was originally referred to as the Camkometer (Cambridge K_0 -meter) because its lift off reading provides an indication of K_0. Lift off occurs at the start of the test. The pressure inside the pressuremeter cell is increased and initially the borehole wall does not deform. Lift off is the pressure that causes the soil to start deforming, which can be clearly seen in Figure 5.33 corresponding to 415 kPa.

As with all field-testing equipment, it is essential to carry out rigorous calibration of the system and the test cell. A correction for the membrane stiffness is obtained by expanding the membrane in air and within a stiff steel tube such as borehole casing.

5.9.2 Borehole pressuremeters

The simplest and earliest form of the test was developed by Ménard (1957). The 'Ménard pressuremeter test' is performed by the radial expansion of the cylindrical tricell probe within a test pocket. The test is carried out by injecting liquid into the central measuring cell and gas into the guard cells to expand the probe to induce displacement of the ground. It is designed for testing in superficial deposits, such as loose to medium dense sands and soft to very stiff clays.

Figure 5.29 shows a section through the test equipment. The volume change is measured at surface by measuring the increased volume of water delivered to the cell. The test socket is drilled such that the pressuremeter will just fit into the socket.

The Ménard test relies on the pipework having no leaks and expansion of the cell being uniform. At high pressures both can be problematic. The loss of the pressurising fluid will provide very misleading test results by overestimating the volumetric strain and reducing the actual applied pressure. The equipment is calibrated to allow for the expansion of the connecting lines, which are shown in Figure 5.29.

Figure 5.29 Cross section through the Ménard pressuremeter (image courtesy of C. Dalton, Cambridge Insitu Ltd)

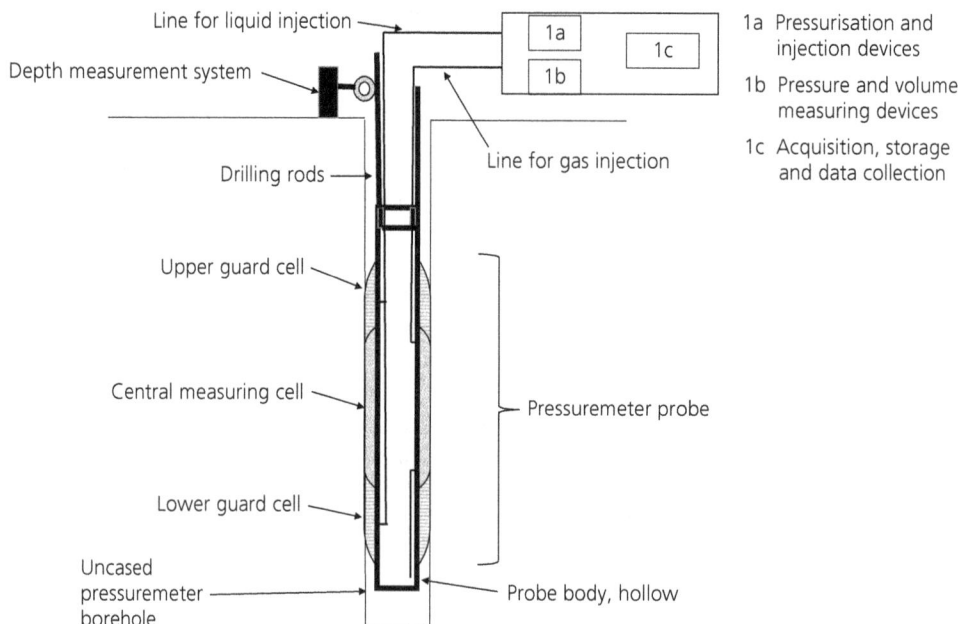

Figure 5.30 From top, the 95 mm and 73 mm dia. high pressure dilatometer (HPD), the self-boring pressuremeter and the cone pressuremeter (image courtesy of S. Pearce, Cambridge Insitu Ltd)

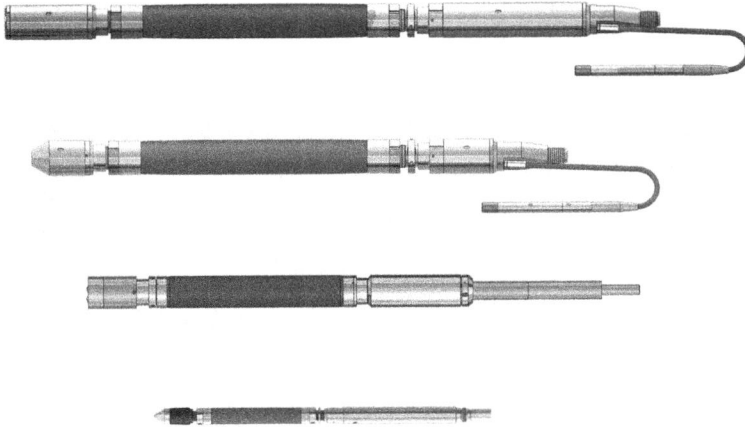

Even with such issues the test is relatively simple compared with other forms of the test, enabling stress–strain relationships to be conducted as the borehole is drilled. The pressuremeter has a central cell with rigid guard cells above and below (see Figure 5.30), thus defining a fixed section over which the test can be conducted. A plot of the applied stress against the volumetric strain enables the undrained shear strength to be determined.

Work was carried out at Cambridge University to reduce the effect of error in the measured displacements by designing an instrument that measures the expansion of a membrane from within the membrane, resulting in the Cambridge Pressuremeter that measures the stresses and strains directly. The device uses strain arms to measure the wall displacement and a pressure transducer to measure the pressure generated within the test cell. By measuring parameters directly within the test cell, the accuracy is greatly improved. The body of the pressuremeter is a flexible rubber tube sealed at each end and protected by a Chinese Lantern comprising flexible strips of thin metal.

Simple forms of the Cambridge Pressuremeter use a single strain arm that measured the diameter of the cell at its centre, while others have combinations of three or six strain arms. The greater number of strain arms means much greater accuracy for the determination of the cell expansion and also provides an indication of soft zones or voids that might be present and will cause the cell to expand in an irregular shape. The system will usually incorporate a pore water pressure transducer within the instrument body. The pressuremeter is installed into a preformed socket. The pocket needs to be slightly bigger than the size of the pressuremeter and must be stable and not prone to collapse as the pressuremeter is inserted. This latter fact can be problematic in some soils and particularly below the water table.

The Cambridge High Pressure Dilatometer (HPD) pressuremeter system is also deployed by inserting the pressuremeter into a preformed pocket drilled at the base of a borehole. The HPD is primarily used in stiff soils and rock and is a more robust device capable of providing high expansion pressures to the wall of the borehole. The 73 mm dia. HPD requires a socket of between 75–83 mm diameter while the 95 mm dia. HPD requires a test socket of between 97–110 mm diameter.

Figure 5.31 (a) The self-boring pressuremeter (image courtesy S. Pearce, Cambridge Insitu Ltd); (b) the HPD ready to insert into the borehole (image courtesy of SOCOTEC)

(a)   (b)

Figure 5.32 The self-boring pressuremeter with outer casing removed to show internal electronics (image courtesy of S. Pearce, Cambridge Insitu Ltd)

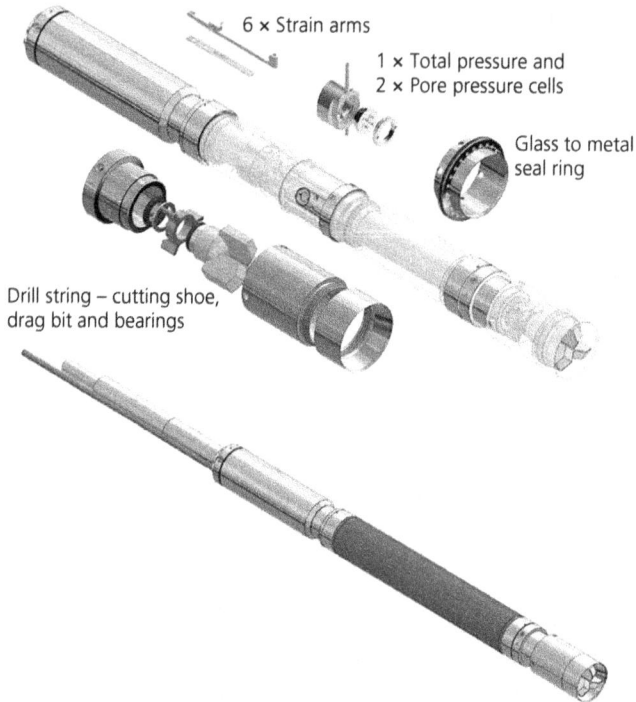

The test device was further refined to improve the test cavity by the development of a self-boring pressuremeter (SBP), which creates its own test pocket with a diameter of between 83–89 mm (see Figures 5.31 and 5.32). This ensures close contact between the test cell and the borehole wall, much improving the test quality and removing many of the issues associated with the standard socketed method. The pressuremeter should be drilled into the soil for a depth of at least 1 m. The device is capable of measuring the pore water pressure and can be configured with either three or six displacement sensors and may also include an inclinometer.

If hard layers or gravel are met, they may cause disturbance of the pocket wall, which may seriously affect the results. Ideally, the test should be carried out in a relatively homogeneous soil with little or no gravel.

The recorded test parameters are stress as total pressure at the borehole or socket wall and strain measured directly using the strain arms, and are plotted to provide a stress-strain curve. Interpretation of the curve provides the K_0 value. The test continues with increasing pressure and corresponding increasing shear strain. At intervals, unloading and reloading is conducted; the example shown in Figure 5.33(b) have three such loops. Analysis of these unload reload loops enables the determination of stiffness G_s. The test is continued until a peak stress is reached where the load is slowly released. The peak curve gives the undrained shear strength c_u. Although it is general practice to average the displacement from the six arms, it can be useful to plot each arm separately (Figure 5.33(a)) as this will show if there is a weakness in a particular direction.

The unload/reload cycles are an important aspect of all expansion pressuremeter tests. Such cycles are conducted at stages as the strain develops, both in the early stages of the tests and also later, where the procedure can be used to take the surrounding materials to a well-developed plastic condition in which the shear strength of the material is fully mobilised. The direction of loading is then reversed. From the reversal point onwards the ground responds elastically and will do so until it reaches the yield condition again.

This means that even if the act of placing the pressuremeter has disturbed the surrounding soil, the true elastic properties can still be recovered. Expanding the cavity puts the material into a new stress and strain condition and erases the previous stress history. It is the material at some distance from the pressuremeter, still in its elastic state, that drives the unload/reload event. As shown in Figure 5.34, the slope of the straight line from a new origin to the crown of the loop can be analysed to give G_s.

Stiffness and strain relationships are derived from the power law where $\Delta p = \eta \gamma^\beta$

p = total radial strain; γ = shear strain; η = intercept of log-log plot; β = slope of log-log plot

Using the power law, the pressuremeter results can be resolved using the partial differential for shear stress:

Shear stress is given by: $\tau = v. \gamma \, [dp/d\gamma]$ (Palmer, 1972)

where v = poissons ratio

The secant shear modulus is given by: $G_s = \eta \beta \gamma^{\beta-1}$

The pressuremeter accuracy is clearly shown when graphs are plotted of the secant shear modulus against the log shear strain (Figure 5.35). The shape of the graph is predicted by the power law, and pressuremeter results are able to define the small strain section of this curve, which would require very advanced laboratory testing and very high-quality samples to replicate in the laboratory.

Figure 5.33 Typical graphs obtained from a self-boring pressuremeter test: (a) all six arms displacement; (b) average displacement (images courtesy of S Pearce, Cambridge Insitu Ltd)

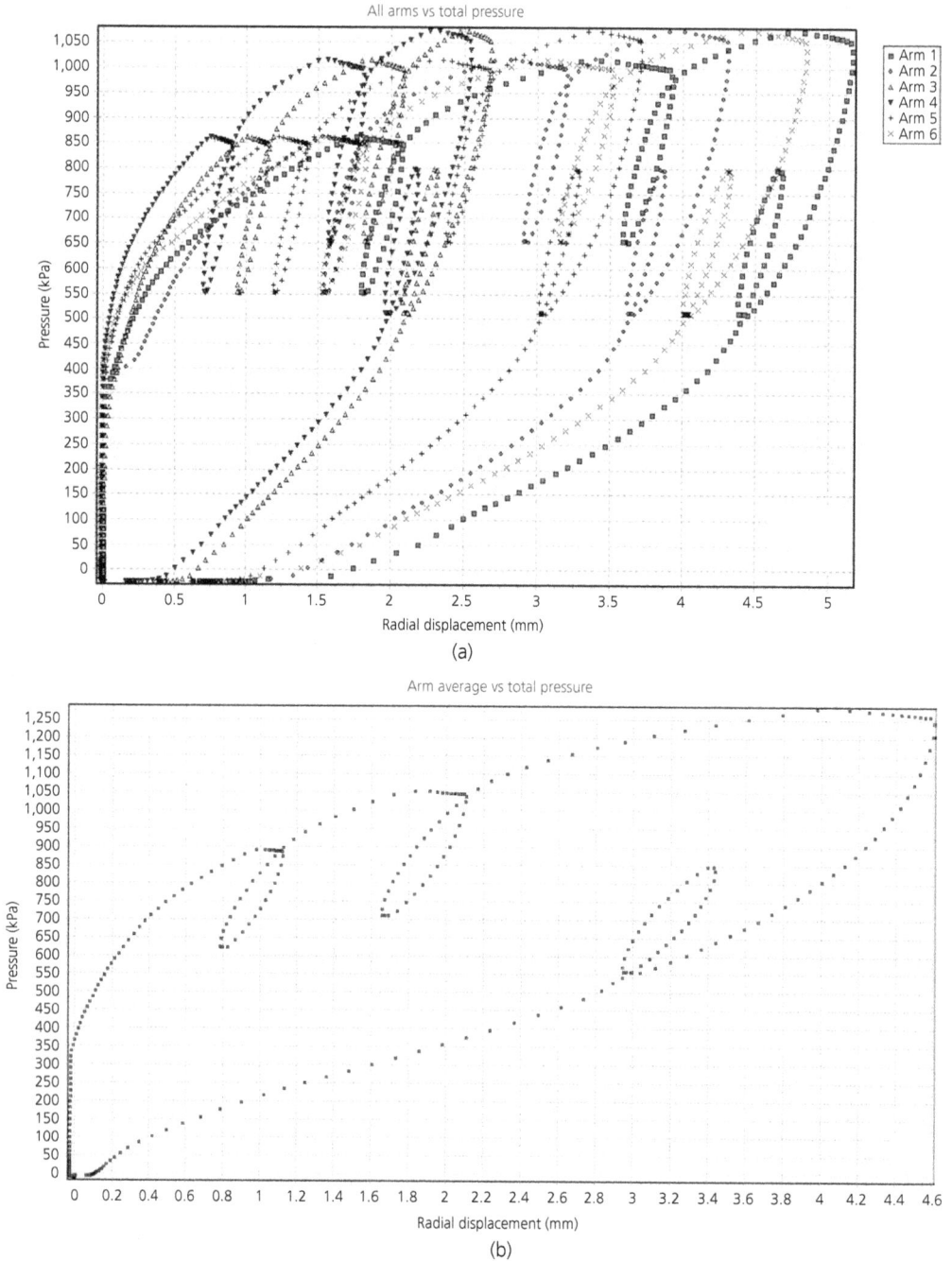

(a)

(b)

Figure 5.34 Graph showing analysis of an unload reload loop from a SBP test used to determine G_s (authors' image, after C. Dalton, Cambridge Insitu Ltd)

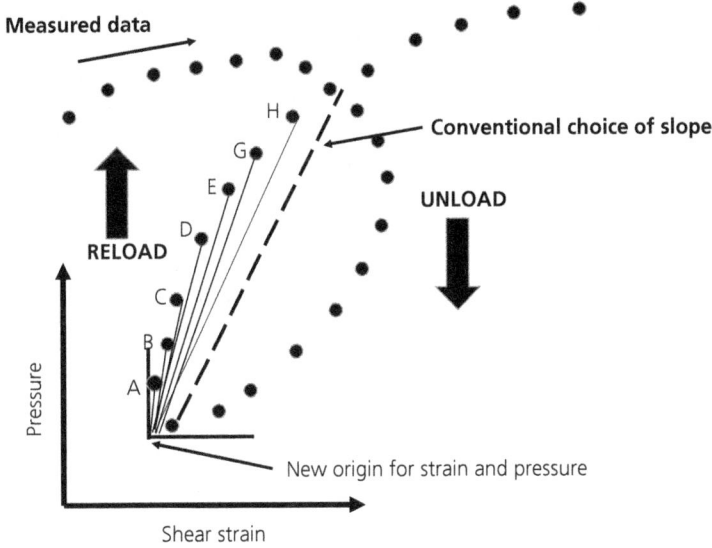

Figure 5.35 Graph of the secant modulus against log shear strain (authors' image, after C. Dalton, Cambridge Insitu Ltd)

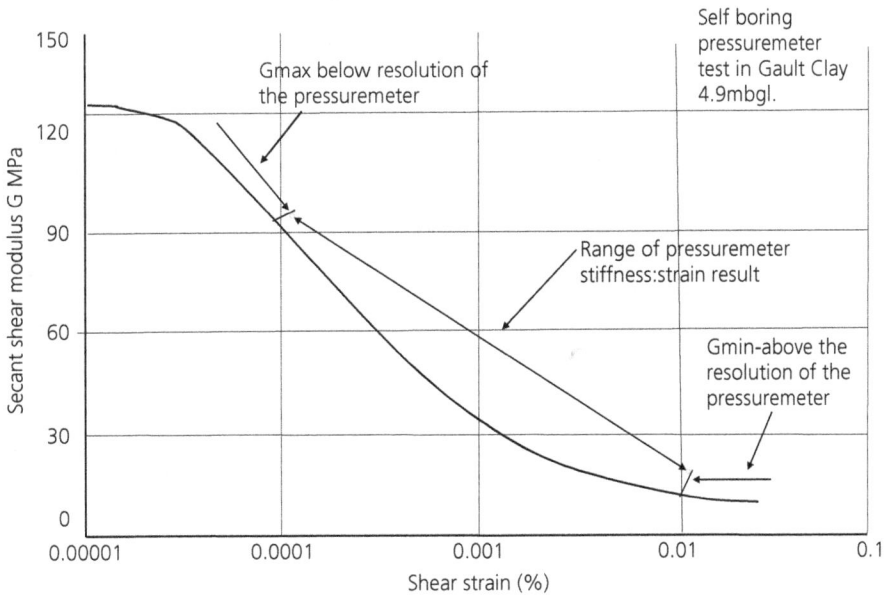

5.9.3   Flat plate dilatometer (DMT)

The flat plate dilatometer, introduced by Marchetti in 1980, consists of a stainless steel blade 250 mm long, 94 mm wide and 14 mm thick with a thin, expandable circular steel membrane mounted on one side (Figure 5.36(a)). The blade is connected to an electro-pneumatic tube running through the penetration rods up to a control unit at surface, equipped with pressure gauges, an audio-visual signal and valves for regulating gas pressure supplied by a gas tank. A USB cable may connect the control unit to a computer for automatic logging of DMT readings. The blade is advanced into the ground using common field machines, that is static penetrometers or drill rigs. The test procedure consists in advancing the blade into the ground and stopping penetration at each test depth. The membrane is initially flat against the surrounding plane behind it, owing to the horizontal pressure of the soil. The operator opens the flow valve on the control unit to inflate the membrane and, in about 30 sec, takes two readings: the P_0 pressure, required to start the expansion of the membrane (lift-off pressure); and the P_1 pressure, required to expand the membrane centre 1.1 mm against the soil. A third reading, P_2, may optionally be taken by deflating the membrane with the slow vent valve, just after the second reading P_1 is taken. The blade is then advanced to the next test depth, with a typical increment of 0.20 m (Mair and Wood, 1987).

The flat plate dilatometer may be used in clays, silts and sands and is limited only by the penetration force of the field machine. This field-testing method provides reliable estimations of fundamental soil properties, in particular:

Figure 5.36 (a) Blade with membrane, (b) DMT field test layout, (b) seismic dilatometer field test layout (images courtesy of Marchetti DMT)

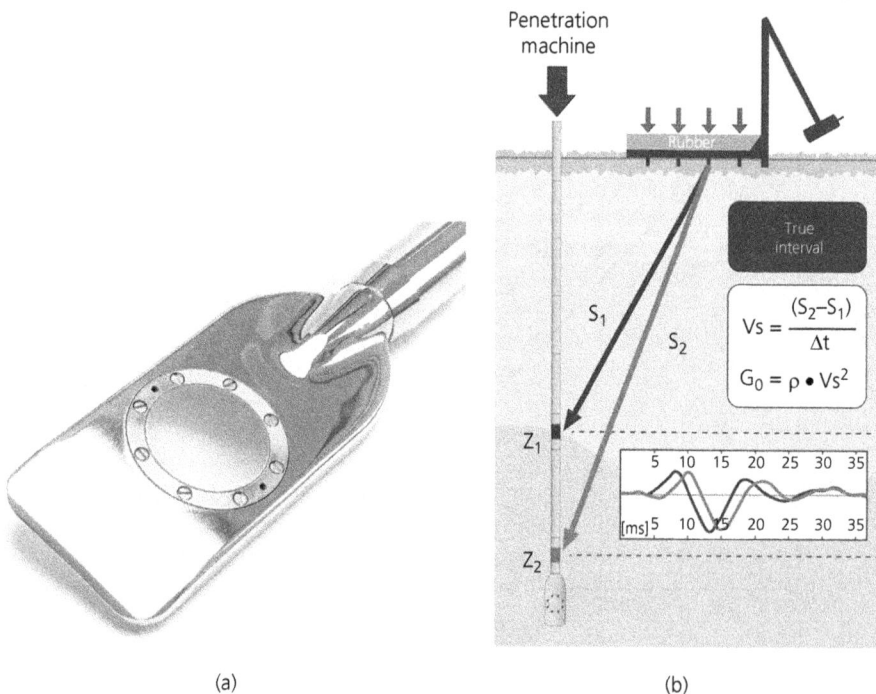

(a)

(b)

- information on soil type: material Index (I_D)
- compressibility: vertical drained confined tangent modulus (M)
- strength: undrained shear strength in clay (Su), friction angle in sands (φ)
- stress history parameters: horizontal stress index (K_D), earth pressure coefficient (K_0) and overconsolidation ratio (OCR) in clay
- pore water pressure: equilibrium pore water pressure in sands (μ)
- consolidation and permeability parameters (c_h and k_h).

In 2004, the Seismic Dilatometer was developed as the combination of the standard flat dilatometer with a true-interval seismic add-on module for measuring also shear wave velocity, Vs. The module consists of an instrumented rod with two sensors spaced 0.5 m and able to record seismic waves. The waves are generated with a hammer and shear beam placed at ground level. The shear wave is recorded at depth within the module and the data sent at surface for a fully automated and real time evaluation of the shear wave velocity, as shown in Figure 5.36(b).

In 2020, a fully automated version of the Flat Dilatometer was introduced (Medusa DMT – Figure 5.37). It is able to perform standard DMT tests by generating and measuring the pressure at depth using a motorised syringe integrated in a rod behind the blade, driven by an electronic board and powered by a rechargeable battery pack. The probe may operate as cableless or with an electric cable for accessing results real time. In 2020, the device was enhanced with seismic sensors to measure also shear wave velocity (Medusa SDMT).

Figure 5.37 Fully automated dilatometer (Medusa DMT) (images courtesy of Marchetti DMT)

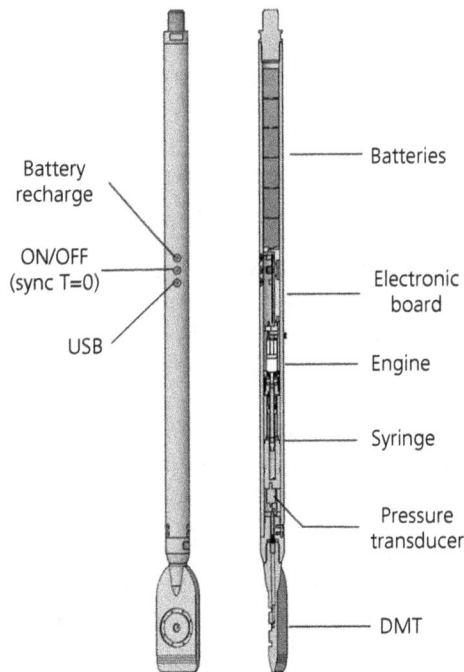

5.9.4   Cone pressuremeter

A successful method of deploying the pressuremeter is to incorporate the pressuremeter behind the cone of the CPT. This has the added advantage of obtaining both CPT test results as well as direct measurement of the shear strength characteristics. A disadvantage is that the overall rate of progress is slowed to enable the pressuremeter testing to be conducted. The test method and results are similar to the general arrangement discussed in Sections 5.9.1. and 5.6.3.

5.10.   Permeability and infiltration testing
5.10.1   General

Tests to determine the permeability of soil and rocks in situ are commonly conducted in boreholes and the test methods are covered in Eurocodes BS EN ISO 22282-1 to 6. Tests may consist of artificially raising or lowering the water in the borehole and monitoring the rate the water level returns to the equilibrium position, or by determining the rate at which water needs to be added or removed in order to maintain a head of water in the borehole (i.e. a constant head). Other tests induce a rapid change in the water level or pressure and then monitor the rate of response to bring the water level or pressure back to equilibrium.

While these methods sound relatively simple, there are many parts to the process that can influence the final results. Whichever method is adopted, it is necessary to plan these tests before starting and to record meticulously the necessary details required for the analysis. Above all else, adequate time needs to be allowed for the test to run its course. Soil permeability testing cannot be rushed. It is perhaps worth bearing in mind that if a soil has a permeability of 10^{-8}m/s the water will move at about the same rate that human hair grows.

Not all water permeation tests require a borehole, the commonest are infiltration tests and soakaway tests that can be conducted in a trial pit.

5.10.2   Infiltration rate testing (soakaway tests)

The determination of the infiltration rate can be used to design soakaways, and the tests are therefore often referred to as 'soakaway tests'. Commonly, infiltration testing uses a method originally devised by the Building Research Establishment and described in BRE Digest 151 (1973) and the subsequent BRE Digest 365 (1991 and revised 2003).

The BRE 365 test is conducted by filling a hole with water (this initial depth of water being referred to as the 'effective depth') and measuring the time taken for the water level to fall from 75% to 25% of the 'effective depth'. In the earlier version of the test (BRE Digest 151), a shallow borehole was used, although the later versions (in BRE Digest 365) revised this to use only a machine dug trial pit. This later method requires significant volumes of water, which should be potable. The time to complete the test can also be significant and may run into more than one day when testing in medium or low permeability soils. To do this the pit will need to be made safe so that livestock, wildlife and people cannot inadvertently fall into the pit should it be left unattended.

The trial pit should be dug with vertical sides; the BRE Digest specifies a width of between 0.3–1 m and a pit length of 1–3 m. Also, the depth of pit should be sufficient to determine the infiltration rate of the ground below the likely invert level of a pipe entering a proposed soakaway. As such, the BRE Digest states that trial pits typically need to be up to about 2.5 m deep (see Figure 5.38).

If the sides are likely to be unstable, or if the pit will need to be made safe for tests running over multiple days, it may be filled with gravel of a single size. If this is required, it will be necessary to insert a length slotted pipe in one corner of the pit and to the base of the pit, prior to backfilling, to allow the water level in the pit during the test to be measured using a dip meter or diver (see Chapter 4). Where filling the pit with single sized gravel is required, the porosity of the gravel should be determined to enable the volume of water to be calculated. This can be achieved simply by placing a sample of the gravel in a container of known volume, such as a graduated bucket, and assessing the volume of the void spaces by gradually adding water to the gravel from a 1 l measuring cylinder. The porosity, which is the ratio of the void space (measured by the water) to the total volume of the gravel and the void space (i.e. the volume of the container holding the gravel), typically ranges between about 0.2–0.4. The calculated porosity of the gravel can then be related to the pit volume to give the actual volume of water used.

After filling the trial pit with water (either completely full, to ground level, or otherwise as full as possible with the water available), a series of measurements of water level depth below ground level is taken as the water soaks onto the ground. A graph of the water level against time is plotted, enabling an accurate determination to be made of the time taken for the water to drain over the effective depth. Where it has not been possible to fill the pit completely, 100% storage depth is taken as the depth of water achieved at the start of the test, and the other relevant assessment levels are based on this initial depth (see Figure 5.38). The process of filling and draining the pit is repeated twice to give a total of three measurements of infiltration rate for each trial pit. The repetition creates a wetting of the soils around the test pit and therefore replicates more closely the conditions surrounding a real soakaway. Of the three results, the lowest infiltration rate is taken as the design value for the test.

The soil infiltration rate is given by the following equation:

$$\text{Infiltration rate } f = V_{p75-25} / a_{p50} \times t_{p75-25} \, m \, / \, s$$

Where V_{p75-25} = the effective storage volume of water in the trial pit between 75% and 25% effective storage depth (m^3)

a_{p50} = the internal surface area of the pit up to 50% effective storage depth and including the base area (m^2)

t_{p75-25} = the time taken for the water level to fall from 75% to 25% effective storage depth (seconds).

Figure 5.38 Trial pit dimensions for infiltration rate testing, and a typical plot of water levels against time (authors' image)

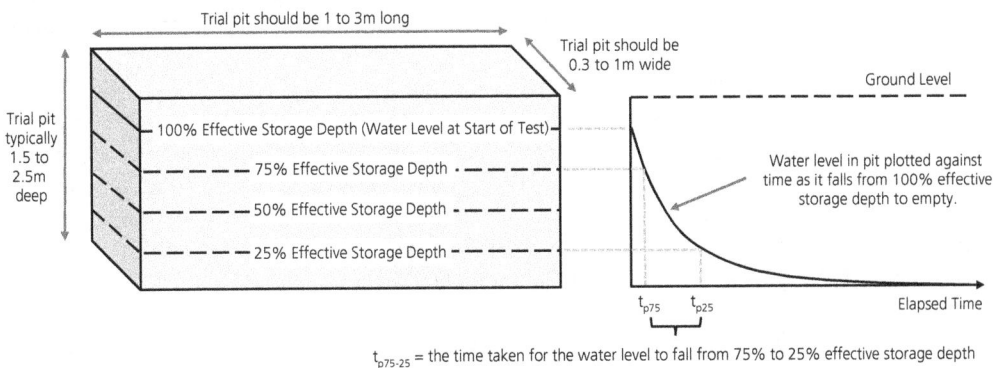

t_{p75-25} = the time taken for the water level to fall from 75% to 25% effective storage depth

It should be noted that even although the infiltration rate is reported as m/s, this should not be confused with permeability. Permeability generally requires the head above the water table, or that applied to a soil sample in the laboratory, to be known, which is not the case in these tests. In addition, the area of the pit sides available to soakage is continuously reducing as the test progresses.

Much can go wrong with the soakaway test and hence these tests require suitable planning. Sufficient water will be required to ensure the testing can be adequately completed. The water must be introduced rapidly to the pit while not causing the pit to collapse; the water may need to be pumped to deliver sufficient volume quickly. A piece of layflat hose tubing of sufficient length can be used to direct the inflowing water to the base of the pit, which will reduce the erosional effect of the water on the pit sides. The test time can run for several hours, which may make it difficult to complete three filling cycles in one day (this typically being feasible only in high permeability soils). It is generally a requirement to carry out testing at different locations around a site, which can lead to several tests running at the same time.

There is a tendency for the soakage pathways to silt up as each test is conducted, which will generally result in the infiltration taking longer and the infiltration values reducing with each successive test. The pit dimensions may also be changed by material spalling from the sides during filling and during the test. These errors may be reduced if a large pit size is adopted however the volume of water required for each test will be increased which in itself can provide logistical problems.

Good communication should be maintained between the site team and the designer. Soakaways are constructed in unsaturated materials above the water table and, as such, should shallow groundwater seepage be encountered during excavation of the trial pit, the designer should be informed immediately. At the very least, shallower pits would need to be dug so that the testing can be carried out above the level of groundwater, but it may otherwise be decided to cancel the testing as the presence of shallow groundwater conditions may preclude the use of soakaways altogether. Similarly, should the trial pits encounter only low permeability soils the designer may decide that the ground conditions are unlikely to be suitable for soakaways, and consider that long term testing, to achieve the standard three fills per pit, each of which may take a number of days to complete, is not cost effective.

BS EN ISO 22,282 Part 5 (BSI, 2012g) describes the infiltrometer test, which is rarely seen in the UK but is widely used in North America. The test is used to determine the infiltration of water into the ground surface. The infiltrometer comprises a cell that is partly pushed into the ground. The cell is filled with water, which can be pressurised. Testing may be conducted using either a single ring or double ring and using either falling head or constant head methods. The results can be used to calculate permeability related to surface flow rate. The test does not provide the infiltration rate f.

5.10.3 Percolation testing

Percolation tests share numerous aspects with infiltration rate tests in that they are carried out in trial pits situated at a shallow depth, within unsaturated soils, and their objective is to measure the rate of water soakage from the pit into the surrounding ground. In contrast with infiltration rate tests, however, percolation tests use small, hand excavated pits with a standard size of 300 mm cube, and the results are used specifically for the design of drainage systems that discharge effluent from small sewage treatment processes into the ground (see BS 6297-2007 + A1-2008) (BSI, 2007b).

Percolation test pits should be excavated just below the intended depth of the infiltration pipes, and this is often shallow enough to dig to the relevant depth and then dig the small test pit (see Figure 5.39).

Figure 5.39 Pit dimensions for percolation testing (authors' image)

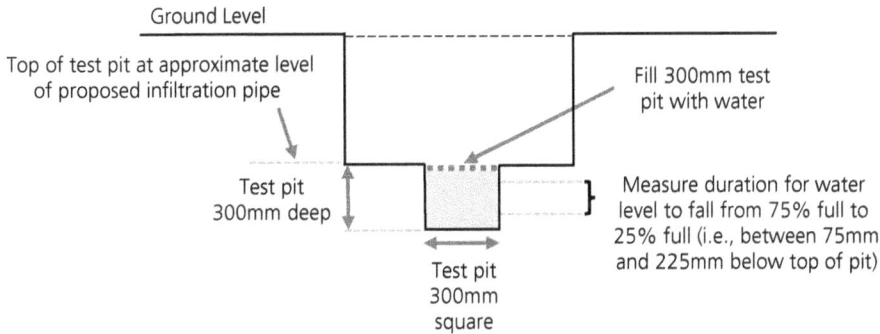

Ground Level

Top of test pit at approximate level of proposed infiltration pipe

Fill 300mm test pit with water

Test pit 300mm deep

Measure duration for water level to fall from 75% full to 25% full (i.e., between 75mm and 225mm below top of pit)

Test pit 300mm square

Once the pit is excavated, the ground immediately surrounding the 300 mm test pit needs to be prepared by presoaking with water. The pit should be filled with water and allowed to empty completely. If this happens very slowly, requiring more than 6 h, it is an indication that the ground may not be suitable for this kind of effluent drainage system. Similarly, if the pit empties very rapidly (in less than 10 min) this may also indicate unsuitable ground conditions as the effluent could migrate too quickly, creating a risk to groundwater. Should the pit empty rapidly, the process should be repeated up to 10 times to verify that such conditions remain consistent. Assuming the pit empties over a moderate duration, of between 10 min and 6 h, the percolation test can be undertaken.

The test procedure is to fill the pit again with water and record the time required for the water level to fall between 75% full and 25% full (i.e. the middle 150 mm of the total 300 mm pit depth). The 'Percolation value', V_p, which is expressed in seconds, is calculated by taking the time interval (in seconds), between 75% and 25% full, and dividing it by 150. The percolation value therefore represents the average number of seconds for the water level to fall by 1 mm. The test should then be repeated a further two times in the same pit and the average value of V_p adopted as the overall test result. Additional tests should be carried out if any individual values of V_p are greater or less than 50% of the average.

5.10.4 Borehole permeability testing

5.10.4.1 General

Permeability tests carried out in boreholes can have significant advantages over tests conducted in the laboratory as the latter may be compromised owing to disturbance caused by sampling, transportation and preparation of the test specimen, as well as the inability of the sample to replicate the drainage paths seen in the field.

Reasonable quality information on the rate at which water will pass through a particular stratum can be obtained by conducting permeability testing in a borehole. Tests can be carried out as the hole is drilled, the top of the required test section being defined by the lower edge of the casing, a temporary bentonite seal or the lower edge of a packer, and the bottom of the test section formed by the base of the borehole at the time of the test. Alternatively, tests can be undertaken once the hole has been completed, the test sections then needing to be formed between two packers, or between the upper and lower seals of a standpipe installation (see Figure 5.40).

A successful permeability test will rely on the test section remaining stable throughout the test period. Where there is a risk of collapse of the borehole walls during the test (for example, when

Figure 5.40 Typical methods of defining a permeability test section in a borehole (authors' image)

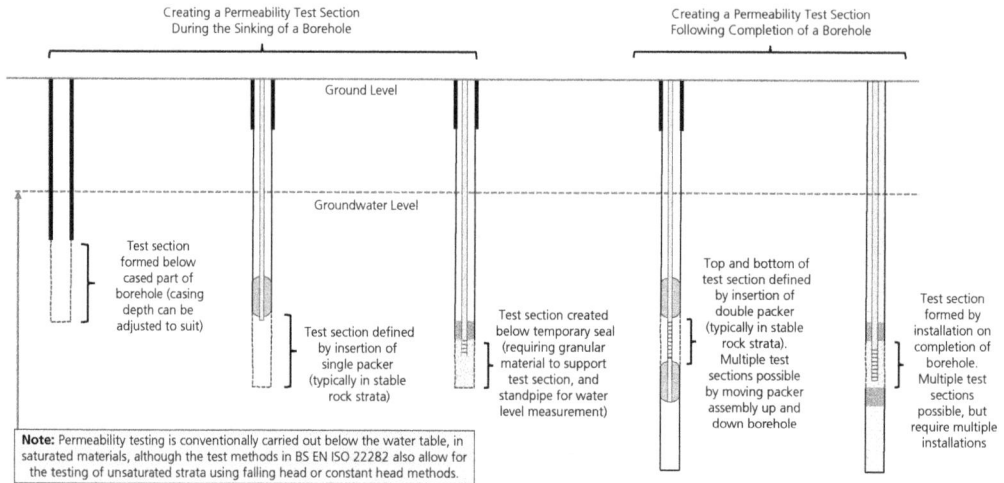

testing loose, coarse-grained soils, fissured clays or fractured rocks), consideration should be given to providing support of the test section. This can be achieved either by the temporary installation of a perforated pipe, with a diameter slightly less than that of the borehole, or by backfilling the test zone with a granular material. Where the latter option is adopted, care should be taken to ensure that the backfill material possesses higher permeability characteristics than the surrounding ground, so as not to adversely affect the test result. A further option in unstable materials is to case the borehole fully and conduct the permeability test through the base of the borehole only. One disadvantage of this approach, however, is that the surface area of the test zone (i.e. the cross-sectional area of the borehole) will be very small compared with the area formed by establishing a longer cylindrical test section. These various options are discussed and illustrated in BS EN ISO 22282-1 (BSI, 2012c).

Great skill and care, along with meticulous preparation and recording, are required to deliver meaningful results. The dimensions and nature of the test section must be clearly defined, and a datum point chosen against which all water level measurements are related.

With either constant head or variable head, the main issues are measuring the true water level within either the borehole or tube, requiring a 'device to measure the water level in the casing or borehole with an accuracy of 0.01 m'. Measurement can be made using a dip meter (see Chapter 4), although this can be challenging and prone to spurious readings particularly in the early stages of the test when the water level is being adjusted to the required test start level. It can be quite difficult to make a measurement while holding the dip meter tape to the datum point and determining when the dip meter sensor is just at the water level. This point is by its nature changing constantly. An alternative is to use a pressure transducer or a diver (see Chapter 4, Section 4.2.8.4). These measure the water pressure induced by the column of water above the point at which the pressure transducer is installed. It is easiest to lower the pressure transducer down the borehole taped to a small diameter (19 mm) pipe; in this way, the true depth of the transducer is readily determined. The readings from the vibrating wire transducer will require multiplying by a calibration factor to provide the head of water in metres, each device is individually calibrated.

The borehole should be as clean as possible and any water in the borehole should be clear of suspended particles. These can seriously affect the permeability results by causing a thin low permeability film over the surface of the borehole wall, which will then influence the measured permeability. This is almost impossible to eliminate without pumping for a suitable time prior to beginning a test (i.e. 'developing' the well where an installation is being used). Should this preparation for the test to be omitted – for example, owing to time or cost constraints – the results may be of little use.

Prerequisite to these tests is a knowledge of the ground water level at the time of testing, because the calculation of permeability is generally based on the relationship between the head of water in a test section at some time during the test (H) and the initial head of water at the start of the test (H_0). If the groundwater level is not known, it is possible to estimate it by assuming different depths to the water table using an iterative process. This is based on the fact that when using the true depth to groundwater, the graph of $\log_{10} (H/H_0)$ against time should in theory provide a straight line; all other values will give a curve. Therefore, if a test result is plotting as a curve, the likelihood is that the water table depth is incorrect. An alternative approach to adjusting a nonlinear result, using a calculated correction, is given in Annex B of BS EN ISO 22282-2 (BSI, 2012d). It should also be considered that the validity of a linear relationship is dependent on the assumption that there are no smears blocking the paths the water will flow through and the soil is fully saturated and does not swell or consolidate as the test is conducted.

Before starting any test, the following should be recorded: details of the ground water level, hole diameter and depth, casing sizes and depth. Similarly, if the test is being conducted within a water level observation pipe then the details of the borehole, the filter size and its particle size distribution, details of the pipe slotted section and filter sock should also be recorded. When considering using an installation, the overall permeability of the installed materials must be greater than the permeability of the material being tested.

5.10.4.2 Variable head permeability

The variable head test is the most commonly used method of determining permeability in the field. The test may be conducted with the test water level inside the borehole falling or rising. In general, the variable head test is conducted in soils with permeability between 10^{-6} and 10^{-9} m/s.

The falling head method is the simplest of these tests and is carried out by adding water to raise the water level either within the borehole or an installed standpipe, and then monitoring the rate at which it falls back down to its equilibrium level. A graph of the change in head within the borehole against time is useful to plot as the test proceeds (see Figure 5.41).

Ideally, the test should run for the length of time it takes the water level to drop to the equilibrium groundwater level, or at least until there are three successive readings taken at intervals of five minutes or more where the head does not drop by more than 1 cm. In some cases, it is sufficient to achieve 75% fall in the applied head. In any event, the quality of the data is improved by a longer test period, although to wait until the water level falls to the ground water level can take several hours and possibly days (for which the use of a pressure transducer, or diver, can be beneficial).

The rising head test (shown in Figure 5.42) is more difficult to conduct, although it is likely to be more accurate because water will flow into the borehole or pipe and will therefore reduce the influence of siltation and can overcome smearing of water pathways caused by the drilling process. The main difficulty is removing the water to a sufficiently low level to begin/start the test, particularly

Figure 5.41 Falling head permeability test in a borehole or installed standpipe (authors' image)

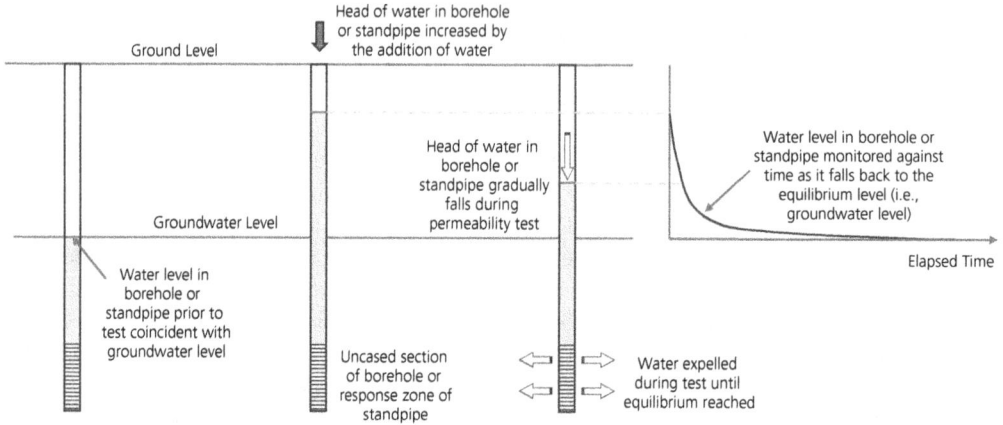

Figure 5.42 Rising head permeability test in a borehole or installed standpipe (authors' image)

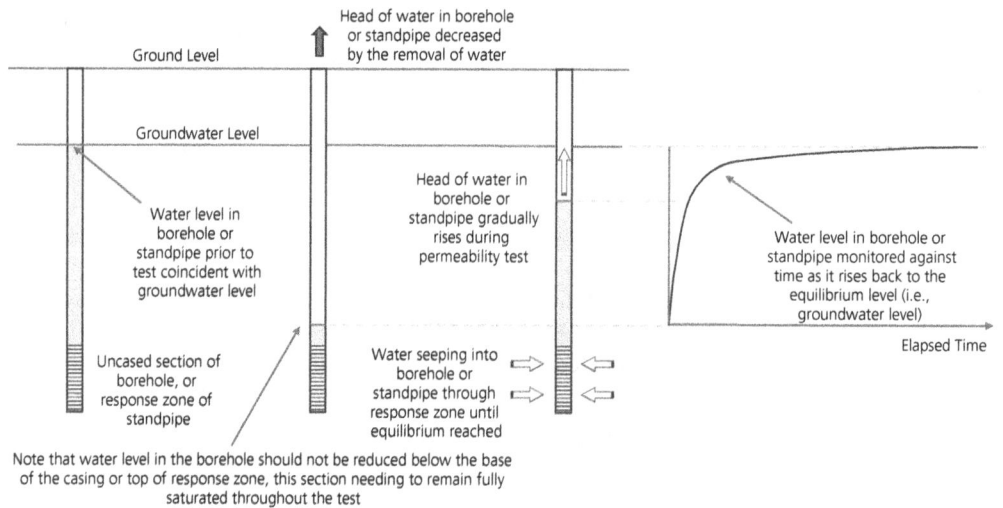

in higher permeability materials that allow a rapid inflow of groundwater. If the test is carried out during drilling, it is difficult to remove the water cleanly just by 'shelling out'. Ideally, water should be removed by pumping, which requires appropriate equipment and a suitable pump to draw the water level down, as well as suitable disposal arrangements for the water removed. Great care must be exercised when doing this, and consideration should be given to the potential risk of imposing additional loading onto the surrounding soil owing to a reduction in pore pressure and increase in effective stress; a process that may induce settlement as the water level is lowered. The lowering of the head can also induce heave in the bottom of the borehole, which may change the hole geometry and in granular soils can result in significant ingress of material that will loosen the surrounding soils, changing their permeability (see also Section 5.10.5 Pumping tests). Some of these issues are easier to control when performing the test in an installation rather than an open borehole.

Before beginning the test, it is important to record the ground water level. In the case of the borehole, it is essential to know the depth to the base of the borehole, its diameter and the depth and diameter of casing. The test may be conducted with either the casing flush with the bottom of the borehole, within an uncased section at the base, or with the test section formed by a standpipe installation or packer (see Figure 5.40). There are several variations of the borehole configuration, which will influence the calculation of the soil permeability in terms of the geometry of the soil and borehole. Methods described in Eurocodes amalgamate the borehole geometry to determine an intake factor, which is then used in the basic Darcy's law formula to calculate permeability for the test configuration.

The interpretation of results and methods of permeability calculation are presented in Annex B of BS EN ISO 22282-2 (BSI, 2012d). A commonly adopted method uses a standard formula derived by Hvorslev (1951). For a variable head test, the 'basic time lag' method requires the piezometric head, presented as the ratio of H/H_0, to be plotted to a log scale against time on a linear scale. The basic time lag is taken as the time where $H/H_0 = 0.37$ (i.e. the time taken for the water level to rise or fall to 37% of the initial change in head) and is taken to represent the section of the curve where steady state flow is achieved.

In addition to the basic time lag, the Hvorslev calculation is dependent on the dimensions of the test zone and aquifer conditions. An example calculation, that can be applied to tests carried out in piezometers, is given in BS EN ISO 22282-2 (BSI, 2012d), and for other circumstances reference is given to Hvorslev (1951). These variations in the calculation method are what were formally accommodated by the 'intake factor' in older versions of BS 5930 (BSI, 2015).

5.10.4.3 Slug test

A variation on the variable head test is the slug test. This method is commonly used in the USA and increasingly conducted in the UK. The test involves rapidly changing the level of water within the test well, which is achieved by inserting a weighted 'slug'. This is generally a segmented, plastic cylinder with a diameter sufficiently small to prevent a piston effect but with sufficient volume to displace the water in the hole, thus increasing the water level sufficiently for the variable head test to be conducted. The rate at which the water level returns to its equilibrium position is recorded in a similar manner to that described above. The test can then be reversed whereby the slug is removed from the hole, causing the water level to drop, and the rate of rise then being monitored, as in a standard rising head test. It may be appreciated that the process is effectively a variable head test, but with the displacement of water in the borehole rather than abstraction or injection. Such tests are often performed in pairs, one falling head test following insertion of the slug followed by a rising head test on withdrawal of the slug, the water level changes all being recorded by a pressure transducer (diver) installed below the test section. The permeability is calculated using the same formula discussed in Section 5.10.4.2.

5.10.4.4 Constant head (or flow) permeability tests

Constant head and constant flow testing are generally used for soils with a permeability in the range 10^{-4} to 10^{-7} m/s.

These tests require the geometry of the hole or instrument to be defined as discussed for the variable head test above. The difference between the constant flow/head tests and variable head tests is that water is either added or pumped out to maintain a constant head (positive or negative) in the borehole or standpipe (see Figure 5.43). Constant flow tests differ from constant head tests only in the way that head is adjusted in the former (to maintain a constant flow) and flow is adjusted in the

Figure 5.43 Constant head permeability test in a borehole or installed standpipe (authors' image)

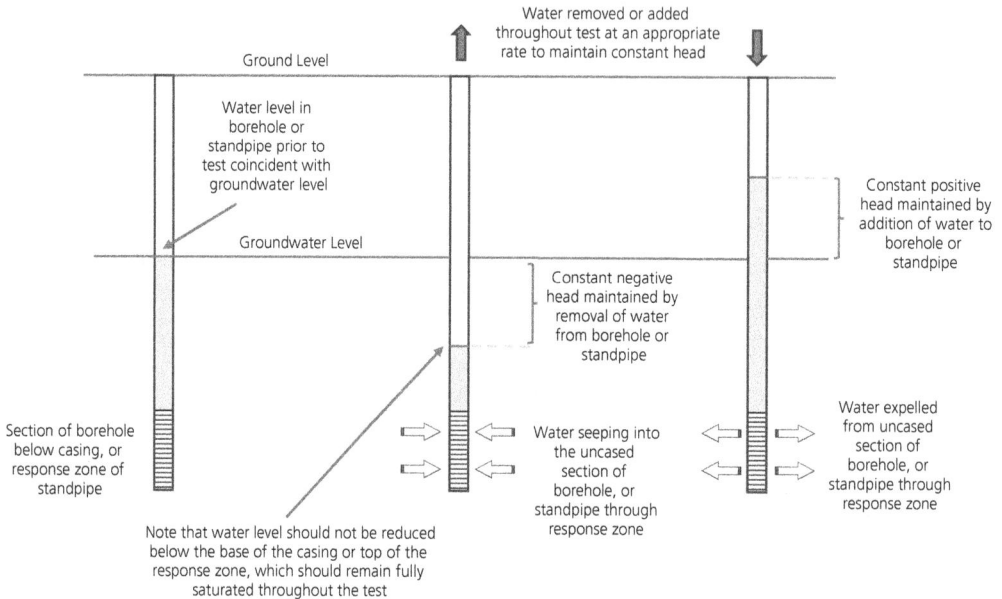

latter (to maintain a constant head). Their common objective, however, is to establish an equilibrium between the two parameters.

The pump-in test requires a sufficient reservoir of water and a means of pumping the water into the hole so that the water input can be controlled to provide a suitable flow rate and is capable of being adjusted or fine-tuned to maintain the required head. The pump-out test requires the same control of flow but by removing water at the same rate it is entering the borehole, thus maintaining a negative head in the borehole. Multistep tests may be conducted at different hydraulic heads.

The head used needs to be carefully chosen, because if too high a pressure is applied this can lead to hydraulic fracture that will fracture the rock or soil and hence dramatically change the permeability. For pump-out tests, it is beneficial to avoid reducing the water level below the top of the test section/response zone in order to simplify assessment, especially where a pump-in test and pump-out test is being performed in the same test section for comparison.

While constant head testing is usually undertaken in saturated materials below the water table, it is also possible to carry them out in unsaturated materials, although the permeability calculations still rely on an understanding of the position of the water table beneath the test zone. Annex B of BS EN ISO 22282-2 (BSI, 2012d) provides three different permeability calculations based on the groundwater level of an unconfined aquifer either being at significant depth beneath the test zone, close to the base of the test zone or above the base of the test zone (i.e. the lower part of the test zone being saturated), shown in Figure 5.44 as Cases A, B and C, respectively.

For any such testing in unsaturated ground, however, the material needs to be presaturated so that the flow of water through the soil is not influenced by air occupying the pore spaces. Based on

Figure 5.44 Constant head permeability test in a borehole or standpipe with unsaturated or partially saturated response zone (authors' image)

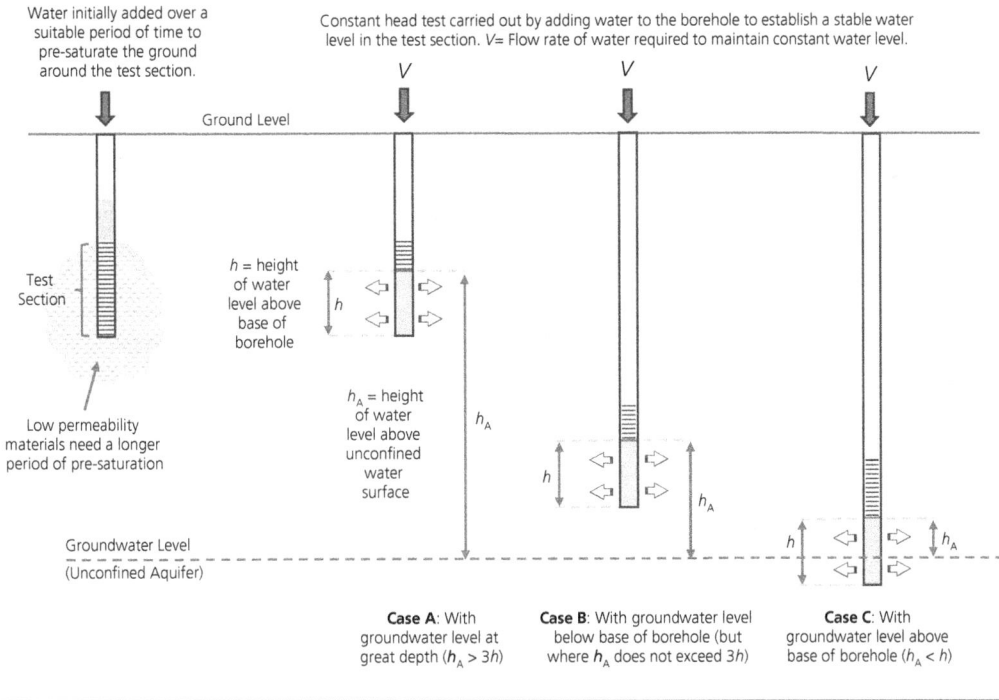

recommendations set out in BS EN ISO 22282-2 (BSI, 2012d), soils that are expected to have a permeability of 1E-06 m/s or greater should be saturated for 0.1 h, and this should rise to 1 h for a permeability of 1E-07 m/s, 10 h for a permeability of 1E-08 m/s and 100 h for a permeability of 1E-09 m/s. The need for presaturation means that the overall duration of such tests can be significant, and this needs to be taken into consideration whenever planning this type of test.

5.10.4.5 Water pressure testing (single and double packer)

The water pressure test, sometimes referred to as a packer test or the Lugeon test, provides a method of conducting a constant head test in a rotary drilled borehole and uses either a single inflatable packer to isolate the base of the borehole, or a double packer to define a test section within the borehole as shown in Figure 5.45. The water pressure within the test section is controlled by pumping from the ground surface, enabling higher pressures to be used than could be generated by simply raising the water level in the borehole. As the borehole in the test zone needs to be stable, and the packers need to provide a seal within reasonably competent materials, packer testing is predominantly carried out to assess the hydraulic conductivity of rock.

The packer consists of a tube-shaped thick rubber membrane that can be inflated using gas, usually nitrogen, to seal against the borehole wall. An access pipe runs through the centre of the packer and allows water to be pumped in or out of the test section.

An advantage of the packer test, and in particular the double packer, is that a specific section of the borehole can be chosen for testing, usually following inspection of core, thus enabling particular

Figure 5.45 Packer testing with single or double packers (authors' image)

Conventional packer testing, using single or double packer, with water flow into the test section, and water pressure in test section, measured at ground level. Manual dip of water in BH to assess potential leaks past top packer.

Alternative packer setup, with pressure transducers monitoring / logging water pressures within test section, as well as above and below packers.

Ground Level

Groundwater Level

Water level in borehole or standpipe prior to test coincident with groundwater level

Double packer test

Single packer test

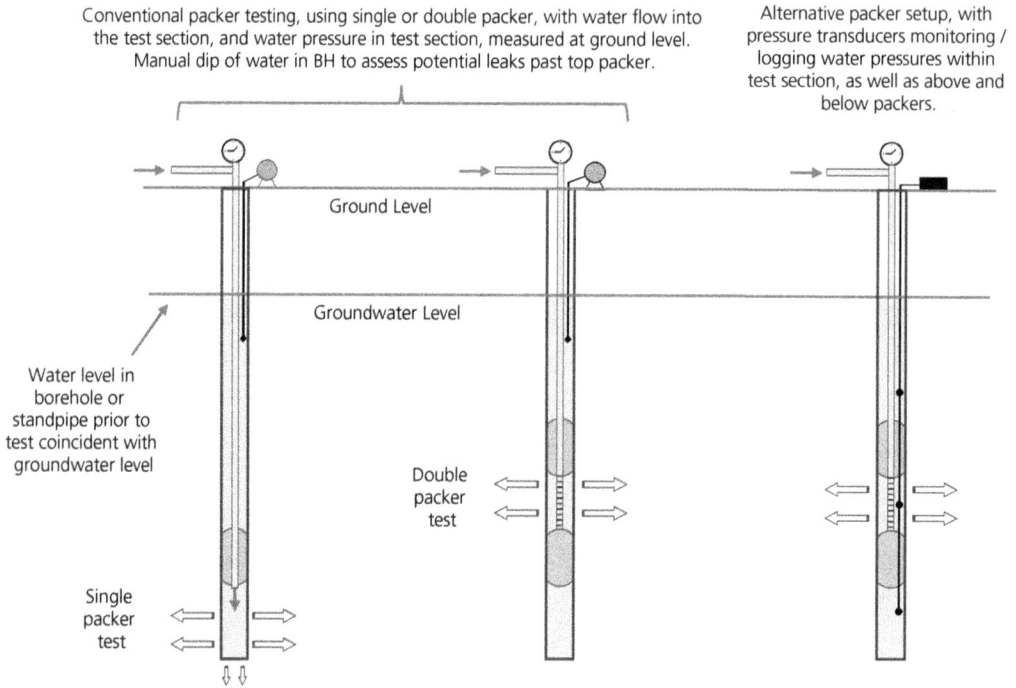

Figure 5.46 Fully instrumented packer system with water delivery by an unjointed polythene hose, and with integrated packer inflation gas line and pressure transducers for monitoring water pressures above, within and below the test section. The packer assembly, which will be inserted into the borehole, is resting on top of the reeler unit (image courtesy of SOCOTEC)

zones to be targeted, such as where there are close discontinuities or granular materials. Several zones might be tested to determine how permeability might change through a strata sequence.

As indicated on Figure 5.45, the double packer provides a measure of the horizontal permeability while the single packer provides overall permeability with drainage being possible in both the horizontal direction and, to a much lesser extent, the vertical direction (through the base of the borehole).

Permeability, or hydraulic conductivity, is calculated from the known dimensions of the test section, the water pressure created in the test section, and the steady state flow rate, Q. As with most geohydraulic testing techniques, water pressures used in the test are considered in terms of their exceedance of equilibrium pressures at the selected test depth. As such, groundwater monitoring also needs to be carried out to develop an understanding of equilibrium groundwater levels.

The packer set up can look complex because it requires flow measuring devices and controls connected to a pump and a pressure regulator. A regulator is required because some pumps produce a fluctuating or pulsing delivery, and the regulator is used to smooth out the flow rate. The traditional test setup uses a series of drilling rods to suspend the packers and deliver the water into the test section. This method can introduce errors that will need to be taken into consideration, such as the pressure head loss owing to turbulent flow within the drill rods, and for the possible loss of fluid through leakage at the joints between drill rods. Where such a packer system is used, it is important to measure flows in the system prior to inserting the equipment into the borehole to provide a measure of losses that can be used to calibrate the system. Water under pressure in the test section can also leak past the packer, or packers, if a good seal has not been achieved, meaning that the flow readings may overestimate the volume of water being taken up by the ground. This can be monitored, at least above the packer, by taking regular water level readings in the annular space between the drill rods and borehole wall during the test (as shown in the central diagram of Figure 5.45).

More modern systems use a flexible hose on a reel which removes the potential for leakage, ensuring that the flow measured at the surface reaches the test section (see Figure 5.46). Such systems can be fully instrumented to monitor water pressures within the actual test section, avoiding the need to assess pressure loss, and also monitor the water pressures above and below the packers, to assess potential leakage past the packers during the test.

Test methods should follow the recommendations of BS EN ISO 22282-3 (BSI, 2012e) and the general requirements of BS EN ISO 22282-1 (BSI, 2012c). The standards require measurement equipment to be calibrated.

The rotary drilling of the test sections should not be undertaken using bentonite or polymer mud as a flushing medium, as this could affect the hydraulic conductivity of the rock (which is largely dependent on the fractures). The test section should be cleaned prior to testing by flushing with clean water until the returns are clean. The packer assembly is then lowered to the correct depth and the packer, or packers, inflated. The inflation pressure should be sufficient to prevent leakage of water past the packer, the standard recommending an inflation pressure at least 30% higher than the anticipated test pressures.

The test traditionally consists of five stages during which the pressure of the water pumped into the test section is varied, the first three stages rising to a maximum pressure and the final two descending again. It is essential that the maximum pressure used is not so high as to cause hydraulic fracturing of the material forming the test zone, that is, the maximum test section water pressure should not exceed the total overburden pressure at the test depth.

The test begins by starting the pump and then adjusting the flow control valves until the required pressure for the first stage of the test is established. Flow and test pressure readings should then be taken at 1 min intervals for at least 10 min. After this time, if readings are stable (no more than 5% change per minute) then the flow can be adjusted for the next stage. Should the readings after 10 min not be sufficiently stable the stage should continue for up to 30 min. This procedure should be followed for all five stages, and the results plotted as flow and pressure (once equilibrium has been established) for each stage, as shown in Figure 5.47.

An idealised result would be a straight line relationship between flow and pressure, passing through the origin. There may be a number of reasons, however, for test results to vary from this simple model, and a discussion regarding the interpretation of such results is given by Preene (2019).

5.10.4.6 Calculation of permeability from field tests

The general formula to calculate permeability, which can be adapted for each scenario, is determined from the hole geometry and follows the form of Darcy's law:

Figure 5.47 Packer test results plotting flow against test pressure, together with monitoring of water pressures above and below the double packer (image courtesy of SOCOTEC)

Stage	Average flow q (l/s)	Average test section pressure Tp (kPa)	Total head H=Tp-Tz (m)	Stage permeability k (m/s)	Stage lugeon value Lu
1	0.206	498	27.42	2.0E-06	23
2	0.295	527	30.46	2.6E-06	29
3	0.523	555	33.27	4.1E-06	47
4	0.515	526	30.36	4.5E-06	51
5	0.433	498	27.48	4.2E-06	47

Test permeability by regression of all stages	8.9E-06 m/s

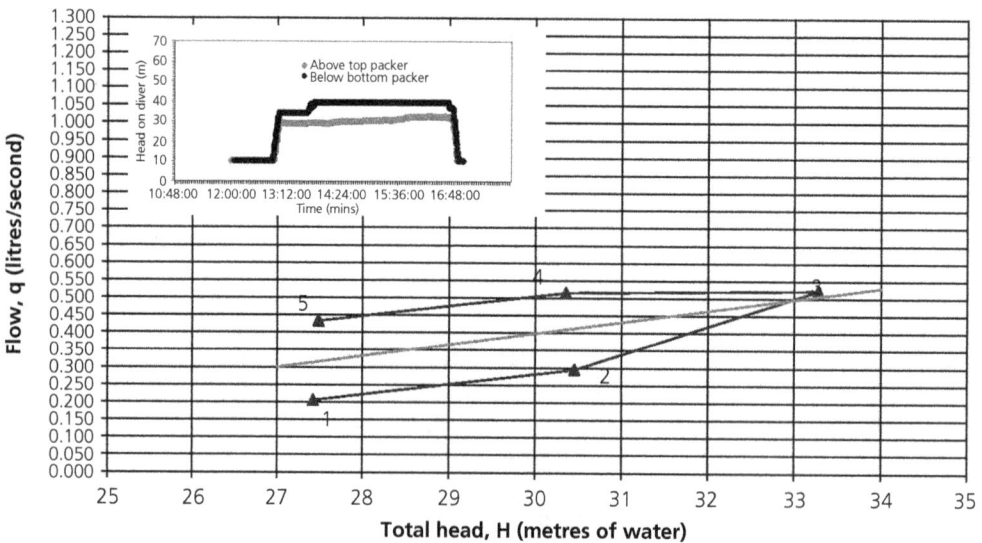

Q = aki which when transformed gives k = Q/ai

where Q = the steady state flow in m^3/s

  a = the surface area over which the flow is acting (generally the area of the well) in m^2,

  i = the hydraulic gradient

  k = permeability in m/s

(It should be noted that in the UK and Eurocode permeability is determined in metres per second, although in many texts cm/s is used)

When using the intake factor F the formulae for permeability are

  $k = Q/FH_c$ for constant head

where F = The shape factor (See ISO 2012: 22282.1)

  H_c = hydraulic head

  k = A/FT for variable head

where A = area of the test section of the well

  T = time in seconds between two points on the straight line portion of the test curve (equates to steady state condition).

  $k = \{A/F(t_2 - t_1)\} \log_e(h_1/h_2)$ variable head general approach

where t_1, h_1 and t_2, h_2, relate to two points on the straight line portion of the test curve where steady state flow is achieved.

5.10.5 Pumping tests

Pumping tests are carried out to investigate the response of the water table to pumping. They are normally conducted to provide information that enables the design of dewatering systems, as might be required, for example, to enable excavations to be made under dry conditions when there is a shallow water table. The most representative results are obtained when such tests are conducted in homogeneous coarse soils.

The process of causing a draw down should be carefully planned, and the recommendations in BS ISO 14686 (BSI, 2003) and BS EN ISO 22282-4 (BSI, 2012f) that are relevant to the required test should be followed. The planning stage also needs to involve close liaison with the appropriate environmental body for the part of the UK in which the site is situated, as the investigation of a groundwater resource, groundwater abstraction and discharge require suitable permitting.

Significant planning of the logistics for the test will be required. Substantial volumes of water will need to be handled, and ideally this might be discharged to a ditch or stream. However, before this can be done the water will need to be tested to ensure it is suitable for such methods of disposal and is not contaminated or too saline. It should be noted that many of the soils found in the UK are of marine origin and as such contain high salt contents that will prohibit disposal to local waterways. If this is the case, the pumped water will need to be collected and stored on site and arrangement for disposal to a suitable treatment site will need to be provided throughout the duration of the test.

The action of reducing the level of the water table will increase the effective stresses within the soil and can result in significant settlement in the area of influence, sufficient to cause considerable damage to infrastructure, roads, services and structures. It is essential to conduct a condition survey prior to beginning the test and to provide monitoring of the surrounding area during testing and through the recovery period following the test.

Pumping tests are conducted to determine the zone of influence caused by pumping groundwater from a point location and to assess the average permeability of the affected soil mass surrounding the location of pumping. The result of pumping is a drawdown or cone of depression on the water table over a large area (see Figure 5.48). As the objective of the test is to determine how the water abstraction affects groundwater levels in the wider test area, it is essential to develop a good understanding of the equilibrium water levels prior to the influence of the test. As such, all pumping tests should be preceded by groundwater monitoring of a sufficient duration to determine the natural groundwater levels and possible fluctuations.

Field preparation for a pump test starts by installing a pump well from which groundwater will be abstracted. The well needs to be of sufficient diameter to enable a suitable pump to be inserted at a suitable depth below the water table to ensure the pump does not pump the hole dry, which would cause it to stall. Typically, the well will be at least 300 mm in diameter and will need a well screen installed with a suitable filter medium (Figure 5.49). The filter should be designed to exclude fines from the well that could block the pump. The pump is a relatively expensive piece of equipment and can be rendered useless if it is allowed to draw in abrasive fine particles.

Once the pump well is installed, observation wells are drilled and instrumented to enable the water level to be determined at various distances from the pump well sufficient to enable the draw down at each location to be defined. It is normal to set out observation wells in two directions, at right angles to each other, radiating from the pump well.

Figure 5.48 Pump test arrangement in an unconfined or confined aquifer (authors' image)

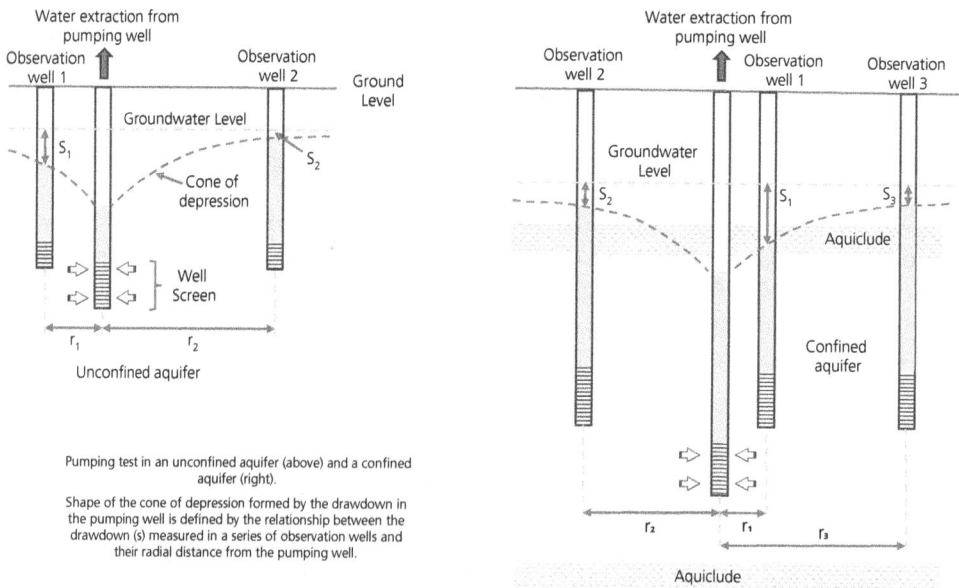

Pumping test in an unconfined aquifer (above) and a confined aquifer (right).

Shape of the cone of depression formed by the drawdown in the pumping well is defined by the relationship between the drawdown (s) measured in a series of observation wells and their radial distance from the pumping well.

Figure 5.49 Diagrams showing groundwater behaviour before and during a pumping test (authors' images)

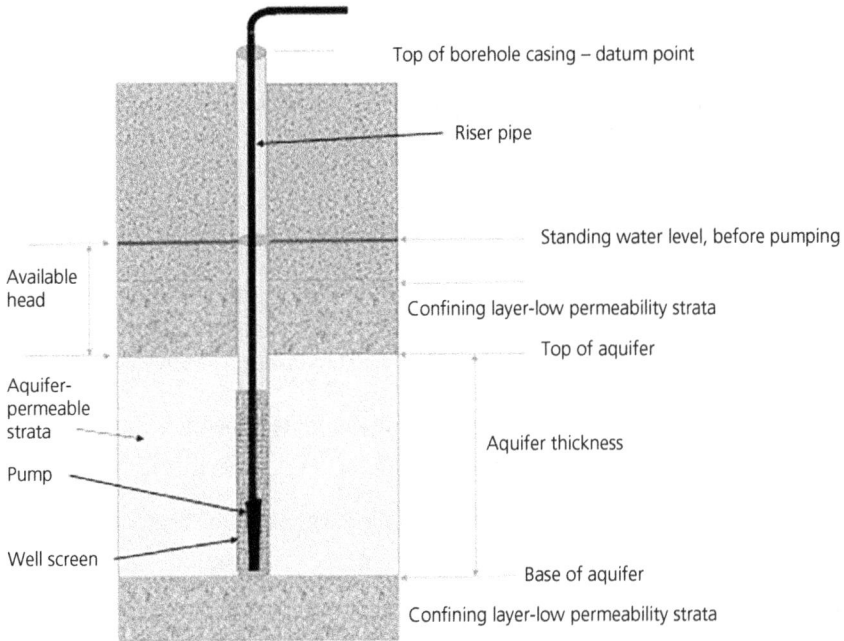

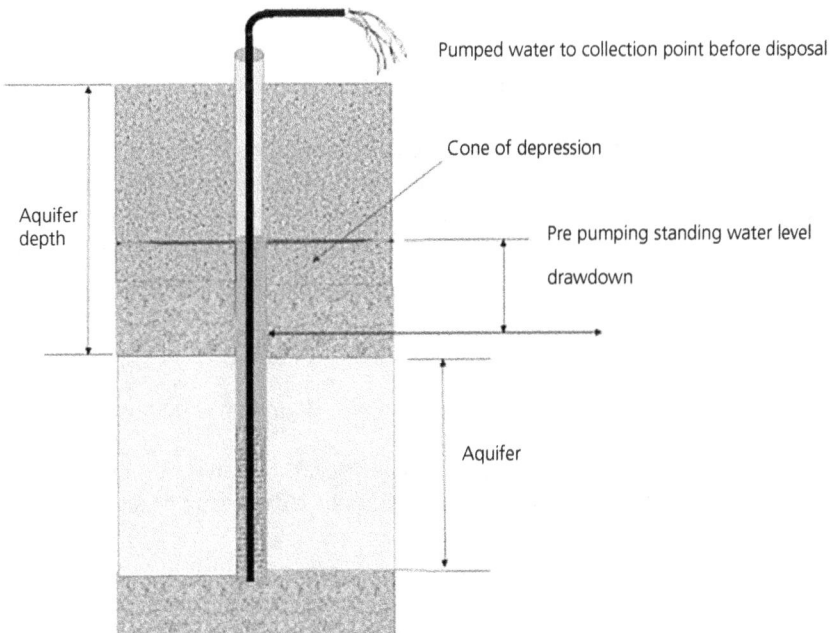

There are a number of different methods of testing, including constant head and constant flow tests, similar in principle to those discussed in Section 5.10.4.4, and a recovery test, similar to the rising head test discussed in Section 5.10.4.2. In addition, a step test may be performed, whereby the flow is increased over a number of stages and held for a duration while water levels and flow rates are measured before progressing to the next stage.

A constant flow pump test is started at a slow rate and gradually increased to provide an optimum flow rate. The rate is maintained whist regular readings are made of the water level in the observation wells. Pumping tests of this type are generally run for several days or even weeks.

When a test is running for a long duration consideration needs to be given to how the works will be supervised, and the facilities required to do this. Consideration should be given to maintaining pumping and the implications should a pump fail. This may mean a standby pump which can be quickly brought online, switching pipework to ensure the pumped flow is not interrupted. If a pump does fail and water level recovers it may mean that the test will need rerunning to achieve the objectives.

The results of the pumping can be analysed to provide the average permeability of the soil which is being pumped using the equations set out below.

The analysis assumes that the hydraulic gradient at a distance r from the centre of the well is equal to the slope of the water table, thus:

$Ir = dh/dr$

where I = hydraulic gradient

r = distance from pump well

Where h is the height of the water table at the distance r from the well. This is the Dupuit assumption.

There are two cases that can be considered: the unconfined case and the confined case.

The unconfined case assumes that the stratum extends a significant distance below the base of the well and the zone of interest. Assume there are two wells radiating from the pumped well at distances r_1 and r_2 with water levels of h_1 and h_2, respectively. The water level is measured from the base of the stratum, which as noted above is some distance below the well. The area through which seepage takes place is given by the expression $2\pi rh$ where both r and h are variables:

Applying Darcy's law $q = aki$, thus in this case $q = 2\pi rhk \, dh/dr$

This can be transformed to provide permeability where $k = 2.3q \, \log(r_2/r_1) / \pi(h_2^2 - h_1^2)$.

If the aquifer is confined, that is the base of the borehole is located on an almost impermeable boundary where H, the thickness of the confined aquifer, is constant and r is variable the equation becomes:

$$k = 2.3q \, \log\left(r_2 / r_1\right) / 2\pi H\left(h_2 - h_1\right)$$

Interpretation of the test can provide useful information regarding transmissivity and storage, which can be used to assist with aquifer management.

emerald PUBLISHING

ice Publishing

Peter Reading and Miles Martin
ISBN 978-1-83549-891-0
https://doi.org/10.1108/978-1-83549-890-320251006
Emerald Publishing Limited: All rights reserved

Chapter 6
Geotechnical laboratory testing

6.1.   Introduction

The reasons why we carry out geotechnical laboratory testing fall into three categories.

(a) To improve our general understanding of the materials forming the ground and their physical properties.
(b) To provide engineering parameters for use in design.
(c) To check material performance in relation to a specification.

The suite of classification tests generally improves our understanding of the soil and its behaviour, particularly when combined with good soil description. Strength and deformation testing helps to inform our design and provide data that can be used to predict potential failure limits and ground behaviour in response to civil engineering works. There are also tests that are used to verify conformance (i.e. validation testing).

In order to ensure the most appropriate tests are carried out, it is important to have an understanding of the result and how it might be used. Consideration of the following is essential before deciding what tests to conduct and which methods to use to obtain the information required (see also Chapter 1, Reading and Martin (2025)).

The choice of undertaking laboratory testing rather than conducting testing in the field is not simple and, in general, a combination of field and laboratory testing is advisable. Table 6.1 provides some of the advantages and disadvantages of testing in the field and laboratory (see also Chapter 5).

The overriding considerations when deciding to conduct laboratory tests are the quality of sample required for the proposed test and the quantity of sample needed to perform the test. The range of tests should be chosen to provide the necessary parameters for design (see also Reading and Martin, 2025). The samples also need to be representative of the in situ materials, this being particularly relevant for tests such as particle size distribution (PSD) testing where the sampling process may alter the particle size range.

In general, the more precise the measurements being made during the test the higher the quality of sample needed to perform the test. This is particularly important for the advanced tests (Section 6.9), which invariably will require samples with little or no disturbance.

6.2.   Standards for test

Geotechnical laboratory testing laboratories require appropriate samples to be supplied for the scheduled tests. The samples must be fit for purpose and provide sufficient material for the testing to be carried out. Details of laboratory test procedures are given in BS 1377 (BSI, 1990) and BS EN ISO 17892 Parts 1–12 (BSI, 2014a–2014l). For the most part, the Eurocode Technical

Table 6.1 Advantages and disadvantages between testing in the field and laboratory

Laboratory testing		Field testing	
Advantages	Disadvantages	Advantages	Disadvantages
Controlled test conditions	Sample quality can be compromised	Test conducted at field stress state	Test set up can be time consuming
High specification equipment held in secure, controlled environment	Sample quantity often insufficient	Local drainage paths influence results not replicable in the laboratory	External conditions may influence results that is weather
Numerous tests can be undertaken with repeatability of test conditions	Sample identification needs to be clear	Most tests can be repeated relatively easily	Equipment failures can cause long delays
Quality assurance in controlled conditions	Limited material available if test needs to be repeated	Materials are tested in a relatively undisturbed state, meaning that in a soil the full range of particle sizes will be present in their natural proportions	Some tests cannot be carried out in some conditions – for example, frozen ground
Economy of scale enables numerous tests to be conducted more efficiently and economically	Sample can be disturbed during transportation	Structural elements such as fissures will be present and in their natural orientation and aperture	Transportation of sensitive equipment can be problematic
Calibration and control of equipment very good			
Highly experienced staff can supervise several test activities and be available to assist with advice			

Specification (TS) follows the procedures defined in BS 1377 with the exception of Part 6, which covers the fall cone test. Not all tests in BS 1377 are covered by the Eurocodes. For this reason, BS 1377-2 (BSI, 2022a) has been released, providing guidance for tests not covered by BS EN ISO 17892, such as compaction and other earthworks tests, and some addition methods for tests that are not fully developed within it. Where this is the case, both the 1990 and 2022 references are shown in Table 6.2.

Some more recent developments noted here as 'advanced tests' are covered in the American ASTM standards. In the UK, it is generally accepted that any test method given in BS 1377 is described as general and tests that lie outside this group are 'advanced', which includes tests such as small strain and bender element testing. Table 6.2 gives guidance on where particular test methods are

Table 6.2 Reference standards for specific test methods (continued on next page)

Test	BS 1377(1990) /BS 1377(2022)	Eurocode	Other
Water content	Part 2 test 3 /2022.4	BS EN ISO 17892-1	
Bulk density, right cylinder (weight density)	Part 2 test 7.2 /2022.8	BS EN ISO 17892-2	
Bulk density, irregular sample	Part 2 test 7.3 and 7.4	BS EN ISO 17892-2	
Particle density Specific gravity	Part 2 test 8 /2022.9	BS EN ISO 17892-3	
Particle density fine-grained soil	Part 2 test 8.3		
Particle density coarse-grained soil	Part 2 test 8.3 / 8.2		
Intact dry density /intact dry density of chalk	Part 2 test 7.3		Lamont-Black and Mortimore (1996)
Saturation water content	Part 2 test 3.3		
Liquid limit	Part 2 test 4 /2022.5	BS EN ISO 17892-12	
Plastic limit	Part 2 test 5 /2022.6	BS EN ISO 17892-12	
Particle size distribution dry sieve	Part 2 test 9.3 /2022.10	BS EN ISO 17892- 4	
PSD wet sieve	Part 2 test 9.2	BS EN ISO 17892-4	
PSD sedimentation by hydrometer	Part 2 test 9.5	BS EN ISO 17892-4	
PSD sedimentation by pipette	Part 2 test 9.4	BS EN ISO 17892-4	
Linear shrinkage	Part 2 test 6.5 /2022.7		
Volumetric shrinkage	Part 2 test 6.3/6.4		
Compaction test 2.5 kg rammer	Part 4 test 3.3 & 3.4 /2022.11		
Compaction test 4.5 kg rammer	Part 4 test 3.5 &3.6 /2022.11		
Compaction test vibrating hammer method	Part 4 test 3.7 /2022.11		
CBR test	Part 4 test 7 /2022.15		
Maximum and minimum density	Part 4 test 4 /2022.12		
Chalk crushing value	Part 4 test 6.4 /2022.14.4		
Moisture condition value	Part 4 test 5.4 & 5.5 /2022.13		

Table 6.2 Continued

Test	BS 1377(1990) /BS 1377(2022)	Eurocode	Other
Strength tests			
Laboratory vane test	Part 7 test 7/ 2022.24		
Unconfined compressive strength test	Part 7 test 3.3 / 2022.27		
Unconsolidated undrained test single stage	Part 7 test 8 / 2022.28	BS EN ISO 17892-8	
Unconsolidated Undrained test multistage	Part 7 test 9	BS EN ISO 17892-8	
Consolidated undrained with PWP	Part 8 test 7 / 2022.29	BS EN ISO 17892-9	
Consolidated drained with volume change measurement	Part 8 test 8 / 2022.30	BS EN ISO 17892-9	
Fall cone test		BS EN ISO 17892-6	
Shear box 60 mm × 60 mm rapid single shear	Part 7 test 4/ 2022.25.2	BS EN ISO 17892-10	
Shear box 60 mm × 60 mm multiple reversals to residual values	Part 7 test 5 / 2022.25	BS EN ISO 17892-10	
Ring shear	Part 7 test 6 / 2022.26	BS EN ISO 17892-10	
Consolidation tests			
Oedometer test	Part 5 test 3 /2022.16 & 17	BS EN ISO 17892-5	
Constant rate of strain			
Rowe cell	Part 6 test 3, 4, 5/ 2022.20		
Isotropic consolidation in hydraulic consolidation cell	2022.22		
Permeability tests			
permeameter	Part 4 test 5	BS EN ISO 17892-11	
Falling head test	Part 4 test 3.3	BS EN ISO 17892-11	
Cell test	2022.21	BS EN ISO 17892-11	

Table 6.2 Continued

Test	BS 1377(1990) /BS 1377(2022)	Eurocode	Other
Advanced testing			
Anisotropic triaxial			Head and Epps (2014), Volume 3
Small strain triaxial testing			
Direct simple shear			ASTM D6528
Resonant column			ASTM D4015-07
Cyclic triaxial test			ASTM D5311 / ASTM 3999
Rock tests			
Shear box			ISRM (2007)
Unconfined compression testing			ISRM (2007)
Point load			ISRM (2007)
Ultra sound velocity pulse test			ISRM (2007)
Schmidt rebound hammer			ISRM (2007)
Slake durability Schmidt rebound hammer			ISRM (2007)
Los Angeles abrasion test			
Slake durability			
Pinhole dispersivity	Part 5 test 6.2 /2022.18		
Crumb test	Part 5 test 6.3 /2022.18.2		
Frost heave	Part 5 test 7 Part 5 test 6.3 /2022.19		
Double hydrometer	Part 5 test 6.4		
Frost heave	Part 5 test 7		

documented. The ASTM standards are only provided here where there is no equivalent British Standard or Eurocode Technical Specification.

The documents referenced in Table 6.2 provide definitive methods of testing that laboratories will, in the main, follow. Other publications also detail testing procedures, notably Laboratory Testing Parts 1, 2 and 3 (Head, 1992, 2nd edn; Head and Epps, 2011, 3rd edn; 2014, 3rd edn). To provide structure to the testing discussed here, we have adopted the general order given in BS1377 (BSI, 1990). This is still considered by many as the definitive standard for laboratory testing, albeit partially replaced by the Eurocode. In the following sections, brief test methods are discussed to provide sufficient information for the engineer to understand the requirements of the tests and the form in which results will be reported. However, the primary function of the following is to high-light what might influence the results obtained and to act as a guide to scheduling testing.

Scheduling the right test is the final part of a process that starts with planning the investigation. The process begins by combining the knowledge of the proposed works and the design requirements with an understanding of the site, which will usually be provided in a desk study (see Reading and Martin, 2025). Combining this knowledge with an understanding of the parameters required for design will enable the engineer to determine the type of investigation required to provide them. This process should include consideration of the type and class of sample required to be able to conduct the appropriate laboratory testing (see Chapter 2).

For the purposes of the following discussion, a sample refers to the original sample as obtained in the field. A specimen is a sub sample of the field sample, generally prepared in some way for a particular test. Sample handling and sub sampling are discussed in Chapter 2.

6.3.   Classification tests

6.3.1   General

Classification tests provide fundamental physical properties of the soil such as water content and density. Some of these tests require an undisturbed sample to give meaningful results.

Soil comprises three phases: solid particles, water and gas (typically air). The proportions of either mass or volume of these phases are used to obtain various parameters that are useful in defining the soil type, its condition and behaviour.

The volume of voids in relation to the total volume of the solids is referred to as the voids ratio. Air and water occupy the void space between the soil particles. The water in these void spaces is free water and is influential in the behaviour of the soil. Water content is the relationship between the mass of water and the mass of soil particles. The model given in Table 6.3 is used to demonstrate the relationship between the phases and the parameters that can be determined by the relationship of volume and masses for the component parts.

The component parts of the ideal soil model can be used to determine the following parameters:

Water content, W = mass of water/mass of solids = M_w / M_s %
Bulk density, ρ = total mass/total volume = $M / V = (M_w + M_s) / V$ Mg/m³
Dry density, ρ_d = mass of solids/total volume = M_s / V Mg/m³
Particle density, ρ_s = mass of solids/volume of solids = M_s / V_s Mg/m³
Specific gravity, G_s = mass of solids / equivalent volume of water (dimensionless)

Volume of voids, $V_v = V_a + V_w$ m^3

Degree of saturation, S_r % = $(V_w / V_v) \times 100$ thus for a fully saturated soil $S_r = 100\%$

Air voids = volume of air / total volume = V_a / V usually expressed as a percentage and is
   inversely proportional to the degree of saturation, thus when $S_r = 100\%$ the air voids = 0%
   and vice versa

Porosity, n = volume of voids / total volume = V_v / V (dimensionless)

Voids ratio e = volume of voids / volume of solids = V_v / V_s (dimensionless)

The relationship of density terminology and unit weight terminology is shown in Table 6.4 (where
g = acceleration due to gravity)

The saturated unit weight and saturated density refer to the condition when all the voids are full of
water – that is, the degree of saturation is 100%. This condition is independent of the position of
the water table, which is free water.

$$\gamma_{sat} = [(G_s + e)/(1 + e)]\, \gamma_w$$

The submerged unit weight and submerged density refer to the condition when the soil is below the
water table and, therefore, the buoyancy effect gives an *'effective'* unit weight or *'effective'* density
to the soil. The water provides an uplift such that the submerged density is

$$\gamma' = (\gamma - \gamma_w)$$

Similar equations can be developed for density using ρ and the mass of the soil.

Table 6.3 Relationship between mass and volumes in the ideal soil model

Volume			Soil column	Mass M (Mg)	
Total volume V	Voids V_v	Air V_a	Air	Air M_a	$M_a = 0$
		Water V_w	Water	Water M_w	Total mass M
	Solids V_s		Solids	Solids M_s	

Table 6.4 Relationship between unit weigh and density terms (after Shukla, 2014)

Density terms	Units	Unit weight terms	Units	Relationship
Total/bulk/wet/moist density (ρ)	Mg/m^3	Total/bulk/wet/moist unit weight (γ)	kN/m^3	$\gamma = \rho\, g$
Dry density (ρ_d)	Mg/m^3	Dry unit weight (γ_d)	kN/m^3	$\gamma_d = \rho_d\, g$
Density of solids (ρ_s)	Mg/m^3	Unit weight of solids (γ_s)	kN/m^3	$\gamma_s = \rho_s\, g$
Saturated density (ρ_{sat})	Mg/m^3	Saturated unit weight (γ_{sat})	kN/m^3	$\gamma_{sat} = \rho_{sat}\, g$
Submerged density (ρ')	Mg/m^3	Submerged unit weight (γ')	kN/m^3	$\gamma' = \rho'\, g$
Density of water (ρ_w)	Mg/m^3	Unit weight of water (γ_w)	kN/m^3	$\gamma_w = \rho_w\, g$

6.3.2   Water content

Water content, which may be referred to as moisture content in some earlier texts, is the mass of free water in a soil expressed as a percentage of dry weight of the soil or mass of solid particles.

Sampling requirements: any class of sample can be tested for water content. However, when assessing the natural water content of a soil it is essential that a representative sample be taken as soon as it is extracted from the ground and is placed in an airtight container. The container should be as full as possible with little or no air included. The sample should be representative of the in situ material such that, if the soil contains gravel, cobbles or boulders, it will require a large sample. The cobbles and boulders maybe removed prior to testing and the proportion of the soil tested is determined as a percentage of the total soil mass including the cobbles and boulders.

For purely granular, free draining soils, the water content is almost meaningless. It will have limited effect on the soil properties, and it is very difficult to make an accurate determination of the in situ water content by taking a sample. Invariably, samples of granular soils are disturbed, and any free water will readily escape as the sample is retrieved. Generally, when granular soils are tested the results give very low values, often less than 10%, recording only the water that is held around the surface of each particle.

It should be noted that samples contained in plastic bags cannot be fully sealed and as such are not suitable as containers for water content testing (see Chapter 2: Geotechnical sampling). If bulk samples are being taken, it is a useful practice to take a tub sample along with the bulk sample. The sealed tub may be placed in the bulk bag for transportation to the laboratory as a sub sample of the bulk sample.

The water content of a soil is often not a value that remains constant over time, particularly within the upper 3–5 m in fine-grained, cohesive soils. It is important to understand the influence water content will have on the properties and parameters of a particular soil type. Changes in water content in fine-grained soils will induce volume change that will materialise as shrinkage and swelling. In fine soils, it is useful to carry out water content tests on sufficient samples to provide a water content depth profile, which can be an invaluable and cost-effective tool when defining soil type and other parameters. For example, Standing (2018) has shown that by measuring the water content at close intervals through the London Clay, the divisions suggested by King (1981) can be determined, thus enabling the identification of zones that may be problematic for certain engineering activities such as piling.

It is commonly found that the water content will decrease with depth. This is owing to the increasing vertical stress in the soil column, caused by the weight of the soil above, pushing the soil particles closer together and resulting in a reduction in the void space available for the water to occupy.

It should be borne in mind that weather and vegetation will influence water content at various times of the year. These natural cycles can produce significant water content variations (shown in Figure 6.1) and hence volume changes such that, for instance, in summer shrinkage will occur while in winter or other wetter periods swelling will occur. These changes to the volume of voids can have a significant detrimental effect on paved areas and foundations. A knowledge of the water content, the soil density, volume of voids and volume change potential can enable predictions of the significance of changes in water content. The water content may be used to investigate the potential effect of seasonal changes by obtaining samples for water content testing from intrusive investigation

Figure 6.1 Comparison graphs showing typical water content variation for London clay taken in winter (left) and summer (right) at the same site (authors' image)

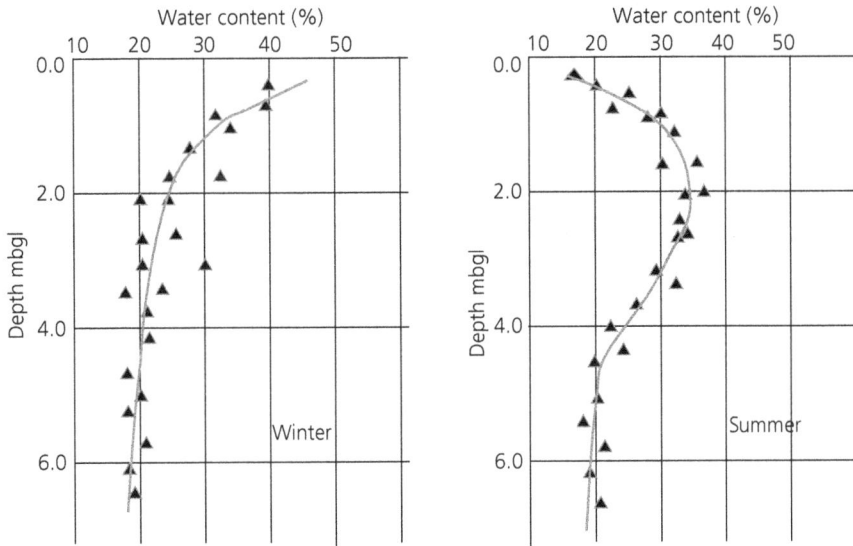

conducted during both the summer and winter periods. Such investigations can show the depth to which these changes affect the soil and hence, for example, enable the assessment of a suitable foundation depth to avoid this zone.

To determine the water content, samples of fine soils should have a mass of at least 30 g while granular soils may require up to 3 kg depending on the particle size range. The test procedure is similar for all sample types with the exception of organic soils.

Most water content test are carried out using the oven drying method, which requires an oven of suitable capacity capable of maintaining a constant temperature of between 105–110°C in all areas of the oven. Sample containers are usually of metal trays or tins with lids. It is normal practice to conduct the test using two specimens from the same sample and the results averaged.

If the soil contains salt – for example, if the sample is taken in soils that are subjected to saline intrusion – the results of the water content determination can be unreliable.

To determine the water content of a soil sample the following procedure is carried out.

- Weigh the container within which the soil will be placed before placing in the oven $= M_t$.
- Place a representative and sufficient sample in the container and weigh the container and soil $= M_{t+s}$.
- Place the container and soil in the oven and leave for least 4 h (often the drying process will be carried out overnight).
- Weigh the container and dry soil $= M_{t+sd}$.
- Return the container and soil to the oven for further 4 h and reweigh. If the change in the weight is less than 0.1% of the sample weight the water content may be calculated.

Water content% = (mass of water/mass of dry soil)×100

$$W = \left[\left(M_{t+s} - M_{t+sd} \right) / \left(M_{t+sd} - M_t \right) \right] \times 100\ \%$$

It should be noted that the water content can exceed 100%, which is commonly the case with organic soils and peat.

6.3.3 Density determination methods

6.3.3.1 Introduction

Eurocode 7 has introduced the term weight density, which refers to the unit weight. The terms are synonymous but can cause confusion when consulting older texts.

It is useful to be clear about the density of the soil, which is usually referred to as the bulk density (soil and water), or dry density (soil only), which means the density of the soil mass. This 'mass density' is given the symbol ρ (rho) and the units Mg/m^3.

The term 'weight density' is used to define the force exerted by a mass of soil, or its unit weight, and is given the symbol γ (gamma) kN/m^3 (see Table 6.4).

The **bulk density** ρ is the density of the soil in its natural undisturbed state, or the condition the sample is in when received at the laboratory, and includes the water naturally held in the pore spaces.

The **dry density** ρ_d is the calculated density of the soil when the degree of saturation is 0%. The **saturated density** ρ_{sat} is the density when the degree of saturation is 100%.

The **submerged density** ρ_{sub}, or effective density, is denoted by ρ_{sub} or more commonly ρ', and is obtained when the soil is below the water table and the soil is subjected to a buoyancy effect.

Where $\rho' = (\rho - \rho_w)$ and ρ_w is the density of water.

The bulk density is determined from intact soil. The samples should be undisturbed, although in practice this is often difficult to achieve, and ideally a Class 1 or 2 sample. Density determinations can be conducted on tube samples as well as intact irregular lumps of soil.

6.3.3.2 Density by linear measurement

Using a tube sample, the test specimen can be prepared as a right cylinder of soil by extruding a suitable length and preparing the ends to be flat and parallel. The dimensions of the cylinder are measured and the sample weighed to enable density to be calculated.

Bulk density ρ = total mass of soil/volume = $M_s / \pi r^2 h$

Where r is the radius of the sample and h is the height, both in metres.

It is highly recommended that the water content, W, of the test sample is also obtained (see Section 6.3.2). Knowing the water content will enable the dry density to be calculated.

$\rho_d = \rho / (1 + W)$ where W is expressed as a decimal.

Where the sample is granular or friable, such that if it is extruded from the tube it would collapse, the test can be conducted by measuring the internal diameter of the tube and the length of sample in the tube and then weighing the sample and tube. The weight of the sample can then be determined by subtracting the tube weight once the sample has been extruded to give the sample weight from which the density can be calculated.

6.3.3.3 Density by fluid displacement

This method describes the density determination of an irregular intact lump of soil or rock. The method adopts Archimedes' principle to determine the volume of the irregular lump using an overflow pot, sometimes known as 'Eureka pot', water displacement method. It is rare to find such a container in a present-day laboratory and the test is usually carried out using an underbalance hanger suspended in a reservoir of water and is therefore known as the water immersion method. In both methods the sample is coated in Paraffin wax of a known density G_{wax}.

To avoid trapping air, the lump is trimmed to remove any rough parts and then the following measurements are made.

Weigh the sample in air, Ma_s.

The sample is then immersed into a wax bath with complete immersion to give a thin single coating of wax. Allow the wax to set and then weigh the waxed sample in air, to give $Ma_{(s+wax)}$. From this, the weight of wax, Ma_{Wax}, is determined and hence the volume of wax, V_{wax}, can be established using the density of wax G_{wax}. Weigh the wax coated sample in water to give the total mass of wax and soil, $Mw_{(S+Wax)}$.

$$V_{wax} = G_{wax} / Ma_{wax}$$
$$Ma_{wax} = M_{(s+wax)} - Ma_s$$

Weighing the wax coated soil in water gives the volume of the soil and wax

$$V_{s+wax} = Mw_{(s+wax)}$$

The volume of soil is given by the volume of wax coated soil minus the volume of wax

$$V_s = Mw_{(s+wax)} - V_{wax}$$

$$\text{density of soil } \rho = \left(Mw_{(s+wax)} - V_{wax}\right) / \left[V_{sw} - \left(G_{wax} / M_w\right)\right] = M_s / V_s \quad Mg/m^3$$

6.3.3.4 Intact dry density and saturation water content of a chalk lump

These tests are used in bearing capacity assessments for chalk and also in conjunction with the chalk crushing value (CCV) to determine its degradability and indicate the type of plant that can be used to excavate the material. TRRL LR806 (Ingoldby and Parsons, 1977) describes the test procedure and provides graphs showing the relationship between the CCV and saturation water content in relation to excavation of chalk. Importantly, the relationship enables the engineer to determine if the chalk is suitable for winter working.

Lamont-Black and Mortimore (1996) suggest that the above method (Section 6.3.3.3) gives the best results when testing chalk lumps to determine the dry density by applying the wax in a single coat by immersion in a wax bath.

This density test is primarily used to determine the saturation water content (i.e. water content when fully saturated) of intact pieces of porous rocks such as chalk. When carrying out this test, it is important to record the frequency of lumps in the chalk mass. For example, in weathered, structureless chalk the intact lumps may represent less than 35% of the mass, each lump being isolated in a weak mélange of comminuted chalk.

The test follows a similar procedure to that described in Section 6.3.3.3.

The following values are recorded and the soil density calculated using the formula. For clarity, the individual weights in grams have been assigned simplified symbols for their masses M_1, M_2, and so forth.

Weight of wax $= M_{wax} = M_1$

Density of wax $= \rho_w \, Mg/m^3$

Weight of sample in air $= M_s = M_2$

Weight of sample with wax in air $= Ms + w = M_3$

Weight of sample and wax in water $= Mw_{s+w} = M_4$

Dry weight of sample $= Md = M_5$

The following calculations are then made:

Volume of soil $V = \left[(M_1 - M_4) - (M_3 - M_2) \right] / \left[(M_1 - M_2) / \rho_w \right]$

and hence density $\rho = M_2 / V \, Mg/m^3$

and dry density of chalk $= \rho_d = 100 \, \rho / 100 + w$ where w is the water content of the chalk lump

Saturation water content $= Ws = 100 \left[(1/\rho_d) - 1/Gs \right]$ or $100 \left[(1/\rho d) - 0.37 \right]$

Where particle density $Gs = 2.7 Mg/m^3$

The particle density (see Section 6.3.3.6) of chalk ranges between 2.69 and 2.71 Mg/m³ (Clayton, 1983).

6.3.3.5   Maximum and minimum density

The maximum and minimum density test provides a test standard for the determination of the dry density of free draining granular soils in the laboratory under different degrees of compaction. This is particularly useful because the determination of the density of granular soils in the field is difficult to achieve and results are often influenced by constraints outside the control of the test method.

It should be borne in mind that it is almost impossible to obtain an undisturbed sample of granular soils from the field and hence it is similarly difficult to determine their natural density in the laboratory. The maximum and minimum density test provides a method of obtaining these values of dry density and gives a density range under specific test conditions.

The maximum and minimum density determinations are performed on granular dry soil containing no more than 10% of the material dry weight to be passing the 63 μm sieve and excluding particles greater than 37.5 mm.

For coarse soils that contain gravel, the sample is poured gently into a California Bearing Ratio (CBR) mould (see Section 6.4.5 on CBR testing). The weight of material that is required to completely fill the mould is recorded. A total of 10 determinations are performed on the test sample. The lowest mass recorded is used to calculate the minimum density.

To carry out the test on sand, 1 kg of the soil is poured into a 1 litre measuring cylinder. The measuring cylinder is inverted to allow the grains to fall with limited force. The volume occupied by the sand is then measured. Again, the test is repeated 10 times, and the highest volume measure is used to calculate the minimum density.

To obtain the maximum density of a gravelly soil, the sample is compacted into a CBR mould (nominal volume 2.3 litres). Compaction is carried out using a vibrating hammer. Approximately 8 kg of soil is required for each determination.

The material to be tested is usually left overnight in a bucket of warm water to expel as much air as possible. The test is performed by placing the mould into a water bath with at least 50 mm depth of water. The test specimen is placed gently into the mould, which has an extension, using a scoop to avoid air entrapment. The soil is placed in three layers, which when compacted fill one-third of the mould volume per layer. The water level in the mould and container should be maintained above the top of the soil at all times. The vibrating hammer is applied for 3 min with a total downward force of 300–400 N to each of three layers. Once compacted the extension is removed and the sample is levelled with the top of the mould.

The sample is carefully removed from the mould into a tray and is then dried in an oven at 105°C. The density is determined using the dry weight and the calculated volume of the mould. The test is repeated at least twice.

A similar procedure is adopted for finer-grained soils but using 3 kg of soil and a 1 litre mould with an extension. Compaction is again carried out under water using a vibrating hammer with a downward force of 300–400 N.

6.3.3.6 Particle density

From the soil model and Section 9.3.1, Particle density ρ_s = Mass of solids / Volume of solids = M_s / V_s.

The particle density is expressed in Mg/m^3.

The particle density of soil is often assumed to be approximately 2.65 or 2.70 Mg/m^3. However, using an assumed value rather than a measured value can introduce a significant error to calculations such as the air voids lines on compaction curves or the void ratio in the oedometer test. It is recommended that the particle density is determined, particularly where the parameter is to be used in a calculation considered to be critical for design. Table 6.5 gives typical particle density ranges for various clays and other minerals commonly found in soil.

When carried out on fine-grained soils (clay and silt), the method requires a very accurate and precise balance typically capable of measuring to 0.001 grams. The test for fine-grained soils uses

Table 6.5 Particle density for some common minerals (after Shulkla, 2014)

Mineral	Particle density (Mg/m³)
Kaolinite	2.62–2.66
Illite	2.66–2.72
Montmorillonite	2.75–2.78
Calcite	2.72
Quartz	2.65
Biotite	2.8–3.2
Gypsum	2.32
Magnetite	5.2

finely manufactured density bottles, known as Pycnometers (previously spelt 'pyknometer' in BS 1377), see Figure 6.2. The stopper for the bottle has a small bore through the middle and each stopper is individually ground to provide a perfect fit in the neck of the bottle.

Determination of particle density for coarser soils is made using gas jars and, as such, do not demand such accurate measurement. Typically, balances capable of measuring to 0.01 g may be used. The test method is the same for either piece of equipment. The test specimen should be about 30 g of fine soil for the bottles and 200 g of granular soil for the gas jars.

The following measurements are required:

- Weigh the bottle or jar dry, clean and empty with the stopper $= M_1$.
- Weigh the bottle with the stopper and dry soil $= M_2$.
- Weigh the bottle with stopper with soil and filled with de-aired water $= M_3$.
- Weigh the bottle with stopper full of de-aired water only $= M_4$.

When testing fine-grained soils, this final step needs to be carried out by adding sufficient water to cover the soil. The bottle and soil should then be de-aired in a desiccator until no air is seen to escape from between the soil particles. Care should be taken not to overfill the bottle because the sample could be lost owing to 'boiling' as the air escapes. Once all air has been extracted, the bottle is topped up with de-aired water and the stopper inserted expelling a small quantity of water through the bore in the stopper. This ensures the bottle is completely full.

The particle density or unit weight is determined by the following formula using the determined masses

$$\text{Particle density, } \rho_s = \gamma_s / \gamma_w = (M_2 - M_1) / (M_4 - M_1) - (M_3 - M_2)$$

$$\rho_s = M_s / V_s \rho_s = \gamma_s / \gamma_w \text{ and is numerically equivalent to } \rho_s / \rho_w$$

It should be noted that particle density has replaced specific gravity to be compliant with Eurocodes and is the mass of soil divided by the mass of an equivalent volume of water and is dimensionless. Numerically, they are both the same.

Figure 6.2 A particle density test conducted using a density bottle (image courtesy of M. Colman, SOCOTEC)

6.3.4 Plasticity index (Atterberg limits)

The influence of water on soil behaviour is defined by the plasticity indices. The following definitions and terminology are used.

Liquid limit (W_L) – the water content at which the soil stops behaving as a solid and takes on liquid characteristics.

Plastic limit (W_p) – the water content at which the soil starts to lose cohesive properties, between a plastic and a brittle state.

Plastic index (I_p) – the difference in water content between the liquid limit and the plastic limit.

Plasticity index = liquid limit – plastic limit.

The tests are carried out on fine-grained soils that have a particle size less than 425 μm and exhibit cohesion. The sample should be representative of the finer fractions and can be determined using a representative sample of at least 250 g.

Albert Atterberg recognised that clay soils are influenced by water content, which he termed its plasticity. He worked on tests to determine an upper and lower bound value, which defined the plasticity range – referred to as the Atterberg limits tests in some publications. Arthur Casagrande, who was working as an assistant to Karl Terzaghi at MIT, developed the test further, resulting in the test equipment now known as the Casagrande apparatus, originally used to determine the liquid limit of soils. The tests provide a good indication of soil condition and, when coupled with natural water content, the soil state can be assessed. Typically, all results from a site are presented on a Casagrande plasticity chart (see Figure 6.6). In this way, soils of similar behaviour can be readily identified, enabling classification and assisting with segregation when carrying out earthworks, or to assign soil properties to the soils exhibiting similar behaviour. In recent times, the fall cone apparatus has superseded the subjectivity of the Casagrande apparatus and is now the preferred test method to obtain the liquid limit.

Samples containing particles greater than 425 μm should be washed through a 425 μm sieve to enable removal of these coarser particles. The percentage of material finer than 425 μm should be recorded. It is generally considered that if the percentage passing the 425 m sieve is less than 20% the test is not representative, and its properties will be dominated by the coarser factions. Notwithstanding this, there are some good engineering reasons for performing the test where this criterion is not met – namely, for permeability assessments and where slope stability is being investigated (see also modified plasticity index below).

The soil can be tested without any pretreatment, provided it is made up entirely of clay and silt-sized particles. If this is not the case, the sample should be washed to remove any soil greater than 425 μm. To do this the sample is dried and then lightly broken, taking care not to break the soil particles. The soil is then washed through the 425 μm sieve ensuring all the washed water and soil is retained in a tray. Once settled, the excess water can be drawn off and the sample then left to dry. It is not necessary for the soil to completely dry out. The soil is mixed with distilled water to a smooth paste. Once evenly mixed, the paste should be cured for at least 12 h (usually overnight). This is achieved by wrapping the sample in a thin plastic film to ensure it will not dry out and placing it into a plastic bag, which is then sealed.

The sample is again thoroughly mixed and is then divided into two subsamples, one of approximately 50 g and the other approximately 200 g, the smaller being used to obtain the plastic limit and the remainder to determine the liquid limit.

The plastic limit is obtained by progressively rolling thin threads of the soil to a diameter of 3 mm. The rolling process gradually dries the soil out owing to the heat from the test technician's hands. The plastic limit is reached when the threads begin to crack as they thin to 3 mm. Rolling of the threads should be undertaken using the middle and index finger, exerting a constant light pressure to obtain regular cross-sectioned threads. The threads are collected into two water content tins and the water content determined as described above. The average of the two water content determinations is taken as the plastic limit.

There are two methods that can be used to obtain the liquid limit, and both can be carried out as single and multiple point methods.

The preferred method of BS EN ISO 17892-12 (BSI, 2014l) uses the fall cone, while the alternative uses the Casagrande Liquid Limit equipment. The latter test method is now rarely used. It should be noted that the two methods provide slightly different values of liquid limit.

For both tests, the soil is mixed with water to obtain a smooth paste. For both four-point methods, the soil is wetted in four stages increasing the water content with each point. Ideally, the liquid limit should lie midway between the upper and lower water contents, such that two points lie above and two points below the liquid limit value.

The fall cone test uses a 30-degree cone that has a hollow stem capable of holding lead shot or small rods such that the total weight of the cone and its shaft assembly can be adjusted to be 80 g.

Figure 6.3 Liquid limit fall cone apparatus and test operation (image courtesy of M. Colman, SOCOTEC)

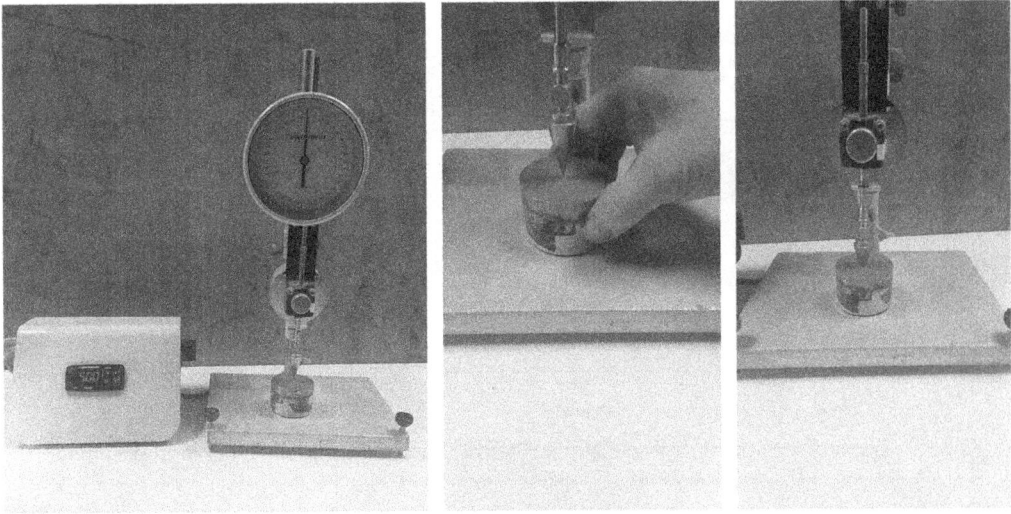

The test is conducted by placing the soil into the cup ensuring all air is expelled. This should be achieved using a spatula and not by tapping the mould with soil on the work bench. The cup should be completely filled and the soil struck level with the top of the cup. The typical test equipment is shown in Figure 6.3.

The cup with soil should be placed under the fall cone and the cone brought down onto the soil so that it is just touching the soil surface. The cone is then allowed to fall under its own weight by releasing the clamp for 5 s. Some types of equipment will time this automatically, while others are manually operated. The penetration into the sample is measured to an accuracy of 0.1 mm, and should be between 15–25 mm. The four points should all lie within this range and should increase as the water content is increased. For each water content point, the test is conducted twice and if the resulting penetrations differs by more than 0.5 mm the specimen is removed and remixed and the point repeated.

The Casagrande method uses a saucer-shaped cup within which the soil is placed in a smooth arc (see Figure 6.4). The centre of the specimen is grooved using a grooving tool that forms a trough through the soil with a fixed side wall angle and width of 2 mm at the base. This process divides the soil in half. The cup is attached to a cam that, when the handle is rotated, will make the cup rise and then fall to impact or 'bump' on the base. This ensures that each drop is from the same height. The cam is rotated at a rate of two turns per second. The number of 'bumps' is counted until a 13 mm length of the groove has closed. The number of 'bumps' for this to occur should lie between 10 and 50 for all four points with the number of blows decreasing with increasing water content.

The liquid limit for the fall cone method is obtained by plotting the cone penetration against water content, which should produce a straight line, and then taking the water content corresponding to 20 mm penetration, as shown in Figure 6.5.

For the Casagrande method, the liquid limit is taken as the water content corresponding to 25 bumps from the straight line graph of water content against the number of bumps for the closure

Figure 6.4 (a) Casagrande apparatus for liquid limit determination with grooving tool and drop height adjustment block and (b) soil with groove ready for test (images courtesy of M. Colman, SOCOTEC)

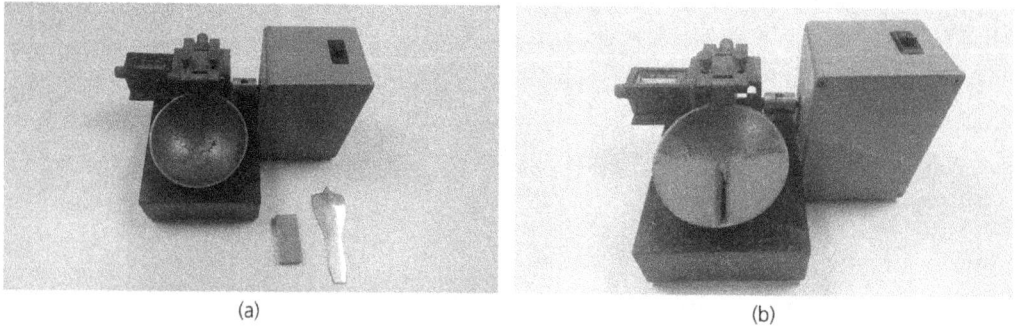

(a)

(b)

Figure 6.5 Graph of four-point cone penetration testing to determine liquid limit (authors' image)

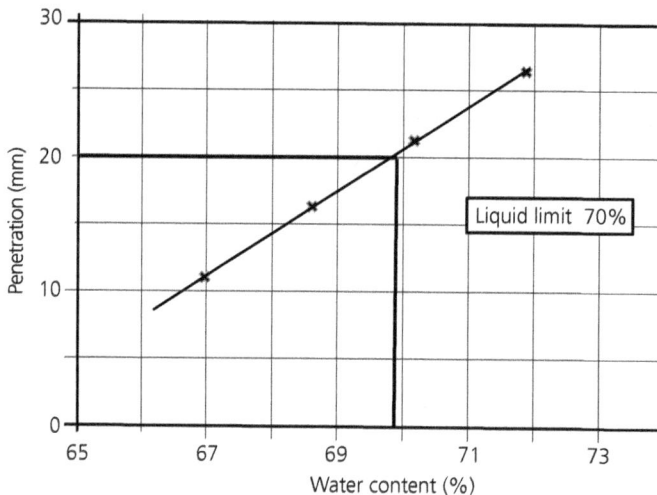

of a 13 mm length of the groove. Note the gradient of this graph is in the opposite direction to that of the cone test.

As noted earlier, both tests may be conducted using a single point test method. The sample is mixed with distilled water to a water content judged to be close to the liquid limit. As before, the single point test is carried out twice at the same water content and should give similar results of either penetration or bumps depending on the method adopted. The water content obtained for the tested soil is then multiplied by a factor to give the corrected value. It should be noted that this method assumes the gradient of the straight line will be the same for all materials, which is not the case in practice, although in many cases the variation is generally small. The author's view is that, wherever possible, the preferred four-point method should be carried out, if only to reduce uncertainty.

In some soils, the mixing process may break down the soil particles, which can produce erratic test results. This can be seen in some zones of the Mercia Mudstone – for example, where the material is seen to comprise aggregations of clay particles, forming weakly cemented silt-sized particles. In these soils, it has been found that consistent results are obtained with a minimum mixing time of 10 min per test (Dumbleton, 1966). Caution should be observed, however, because the soil may in practice behave as a silt while the plasticity test, having broken down this flocculated structure to a degree, might indicate clay-like behaviour.

The Casagrande Plasticity Chart provides a graphical plot of the plasticity index against liquid limit, as shown in Figure 6.6. Where the results plot above the 'A-line' (i.e. the diagonal line running roughly from the bottom left to the top right of the chart), the soil will behave predominantly as a clay, while soils with results that fall below the A-line will behave predominantly as a silt. In reality, tests carried out on soils with a high silt content produce poor results because the material is effectively granular, albeit fine grained. As the plastic limit is approached, the soil will lose the apparent cohesion afforded by the surface tension from the water between particles, which does not enable the soil to be rolled to a diameter of 3 mm, and therefore the test criteria cannot be met. Soils that are unable to be tested, either for the liquid limit or plastic limit, will be noted to be 'nonplastic' by the laboratory, meaning the plastic limit cannot be obtained.

Using the results of the plasticity index test in combination with the water content, some useful parameters can be derived.

Figure 6.6 Casagrande graph used for Atterberg limits tests (earlier versions use the symbol C for clay and M for silt) (authors' image)

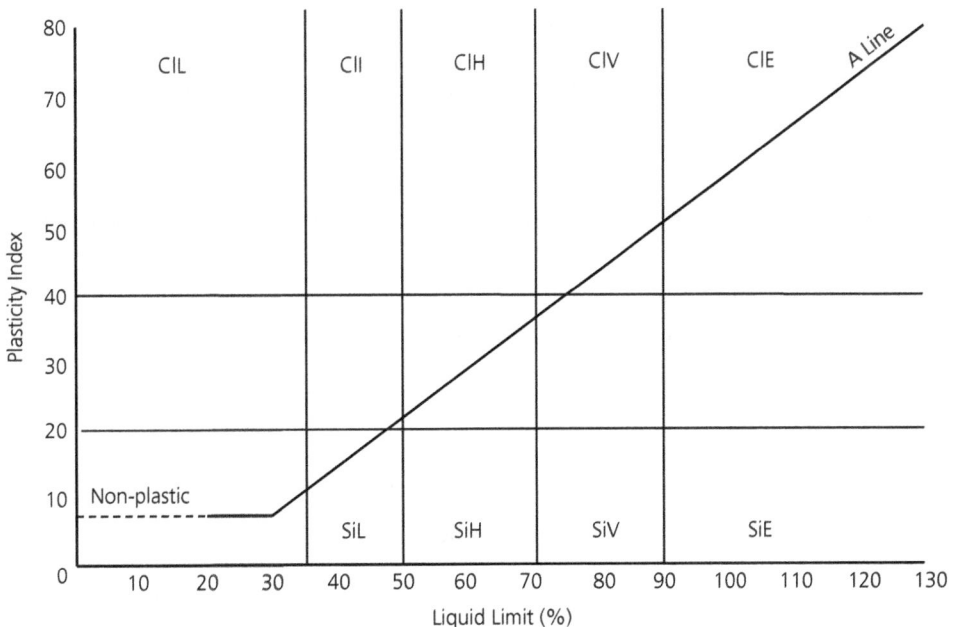

Plasticity index $\quad Ip = (W_L - W_p)$

Consistency index $I_c = (W_L - W) / (W_L - W_p)$ $\qquad$ where W_L = water content at liquid limit

$\qquad\qquad\qquad\qquad\qquad\qquad\qquad\qquad\qquad\qquad$ W_p = water content at plastic limit

$\qquad\qquad\qquad\qquad\qquad\qquad\qquad\qquad\qquad\qquad$ W = water content

Liquidity index $\quad Il = (W - W_p) / (W_L - W_p)$

Consistency index, I_c, can be expressed using a scale of soil consistency terms. However, it should be noted that these are not the same definitions as those used for the qualitative soil consistency terms assigned on the basis of hand test during soil description. Table 6.6 gives the current interpretation for consistency index.

Where the sample contains a high proportion of coarse soil particles, it is more accurate to adjust the measured plasticity index using the percentage of the sample that passes the 0.425 mm (425 micron) sieve. This is termed the 'modified plasticity index' (see below).

With a knowledge of the clay fraction (% of particles less than 0.002 mm) the activity of a clay soil can be determined. This will require a sedimentation test to be conducted (see Section 6.3.6.2).

Activity $\qquad A = Ip / CF \qquad$ (Dimensionless)

Where CF = clay fraction of the soil (particles less than 0.002 mm) expressed as a percentage of the dry mass.

Typically, the activity for UK soils is close to 1, although there are a few exceptions such as kaolin rich clays and lacustrine clays, where A varies between 0.4 and 0.75, and for organic soils where A is generally greater than 1.25.

The plastic limit is used as a criterion when assessing the suitability of cohesive soils as a potential fill material (see Section 6.4). Here, the water content is compared to the plastic limit minus 4% (NH DMRB), materials being described as 'dry cohesive' where the water content is less than the Wp-4% and 'wet cohesive' when above this value.

National House Building Council (NHBC) suggests the plasticity index is used to assess the volume change potential when related to foundations placed in proximity to trees. The relationship

Table 6.6 Consistency index with consistency terminology as defined by BS EN 14688-2 (BSI, 2018b)

Consistency	Consistency index
Very soft	0 to 0.25
Soft	0.25 to 0.50
Firm	0.50 to 0.75
Stiff	0.75 to 1.00
Very stiff	>1.00

is adapted for materials that contain a proportion of particles larger than 425 µm, to provide the 'modified plasticity index'. Typical values are given in Table 6.7.

Modified plasticity index I'p $= $ Ip $\times$ (% material less than 425 µm)$/100\%$

There have been many correlations with other parameters from the results of plasticity tests. In most cases, these are specific to a particular material – for example, the relationships developed by Soresen and Okkels (2013) that relate the effective angle of friction to the log of the plasticity index. For all correlations of this type, care must be exercised. In most cases, the researcher has used a single material type, and it cannot be assumed that the relationship will hold for other material types.

The plasticity index test results may be used to obtain an indication of the CBR value which can be obtained with reference to Powell *et al*. (1984), Table C1 of TRRL LR1132. While this gives an approximate indication of the CBR value, it can be very unreliable because it does not take account of the actual water content of the soils. Water content clearly can be very variable near the ground surface and is a key factor in determining the observed soil strength and thus the CBR value.

Table 6.7 Relationship between modified plasticity index and volume change potential (after NHBC, 2024)

Modified plasticity index	Volume change potential
40% and greater	High
20–39%	Medium
10–19%	Low

6.3.5   Linear shrinkage limit and volumetric shrinkage

Volumetric shrinkage provides a definitive measure of the shrinkage potential of a soil, either in an undisturbed state (definitive method) or disturbed (subsidiary method). However, these methods use mercury to assess the volumetric change as the soil dries out and are therefore almost obsolete owing to the acute risks of working with mercury in a geotechnical laboratory environment. The only test in common use is the Linear Shrinkage Test, which is described here.

This test is used to determine the shrinkage limit: SL. The SL is the water content below which no further volume change will take place (see Figure 6.7). The SL is generally found to be slightly lower than the plastic limit. This indicates that the soil will continue to reduce in volume as the water content moves from the liquid limit reducing until the SL is reached, and this fact may be significant if the plasticity index is high and if the water content is prone to change, such as when vegetation and/or trees are present.

The test is carried out using a trough with a semicircular section, as shown in Figure 6.8. This assumes that there is a direct relationship between the shortening of the sample in the mould to the SL. The soil sample is thoroughly mixed, adding water to bring the water content close to the liquid limit. It is then placed into the trough, excluding all air from the soil as it is placed. The sample is struck smooth with the top of the trough.

Figure 6.7 Graph showing relationship between plasticity index, liquid limit, plastic limit and shrinkage limit (authors' image)

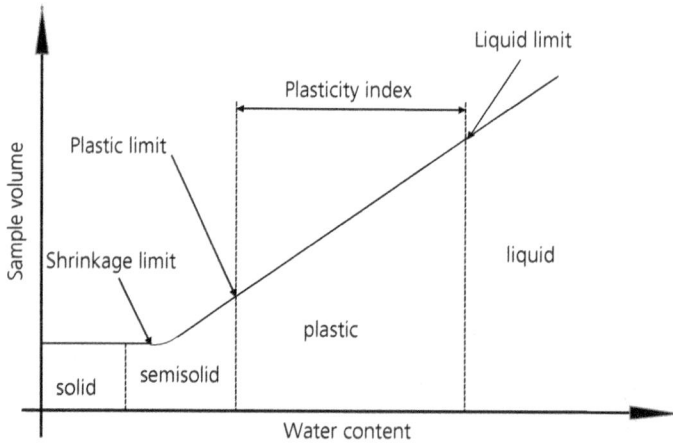

Figure 6.8 Trough with tested specimen to determine linear shrinkage (image courtesy of K. Winiarska, K4 Soils Laboratory)

The sample and trough are placed in a warm dry place, free from dust, and allowed to air dry such that the soil shrinks from the sides of the mould. The sample should then be dried in an oven at a temperature of not more than 65°C. When shrinking is all but complete, drying may be continued at a temperature of between 105° to 110°C. The SL is calculated by measuring the shortening of the soil in the trough.

Linear shrinkage $\% = (1 - L_d/L_o) \times 100$

Where L_d = length of oven dried specimen

L_o = original length of specimen

6.3.6 Particle size distribution test (PSD)

6.3.6.1 Coarse-grained soils

The PSD test is relatively cheap and can be carried out on almost any sample type. In many cases, the test is performed without any real intention to use the results in a design. It is the authors' opinion that the use of this test, to determine the general particle size in the sand and gravel ranges, should be considered and justified. If there is no real justification it should not be carried out as the money is probably better used elsewhere. In the authors' experience, a good visual description by a well-trained logger is more accurate and of better use. The case for the finer fractions is more easily justified, although caution needs to be exercised because particles such as fine sand and silt are easily washed out of the sample during the sampling process, and PSD results can therefore be very misleading.

The test specimen should be representative of the soil to be tested, and such samples may be undisturbed or disturbed. Generally, samples from quality classes 1 to 3 are probably suitable.

The larger the largest particle in the soil is, the bigger the sample needed to provide a representative test. Table 1 of BS EN ISO 17892-4 suggests, for example, that samples for coarse soils samples containing particles in the size range 20mm to 63mm, should be at least 17 kg.

Usually cobbles and boulders are excluded from this test because, by virtue of their size, the sample size required to comply with the representativeness criteria would necessitate an extremely large volume of material (up to about 1000 kg). It is normal practice to record the number and individual sizes for cobbles and boulders. A good description and photograph of the material in situ would provide a more meaningful record of the particle size and material condition. The photograph needs to incorporate a scale.

PSD tests are usually conducted either to verify natural variations, to classify the soil for earthworks, to provide an assessment of permeability or to assess the potential for the reuse of soils as drainage materials.

There are several variants of these tests, but essentially, they all consist of a representative soil sample being dried and then passed through a series of sieves (Figure 6.9) with the mesh size becoming progressively smaller. The mass retained on each sieve size is measured and these are expressed as percentages of the initial dry sample mass. The results are presented as a graph of the percentage of the initial mass that passes each sieve size, and this is often termed a grading curve, although particle size distribution (PSD) curve is to be preferred.

Sieves are made from accurately constructed plate or mesh such that the size of each individual aperture on the sieve is identical and precise. The aperture of the sieve is square, and the size of the sieve is recorded as the length of the sides to the square. However, the dimension of the particle that can just pass through the aperture is also dependent on its shape, the controlling dimension being the midsize of the length, width and depth of the soil particle, as shown in Figure 6.10. For example, if a particle is roughly rectangular and measures 60 mm × 50 mm × 30 mm, the particle will just pass the 50 mm sieve but will be retained on a sieve with a mesh size of 49.5 mm. However, a thin flaky particle may pass the same sieve with a middimension of up to 70 mm, where the particle can pass through the diagonal of the aperture. The sieve method theory is based on equivalent spheres of equal masses demonstrated in Figure 6.11.

Figure 6.9 (a) Soil being introduced to a nest of sieves (image courtesy of K4 Soils Laboratory) and (b) larger particles offered up by hand (image courtesy of M. Colman, SOCOTEC)

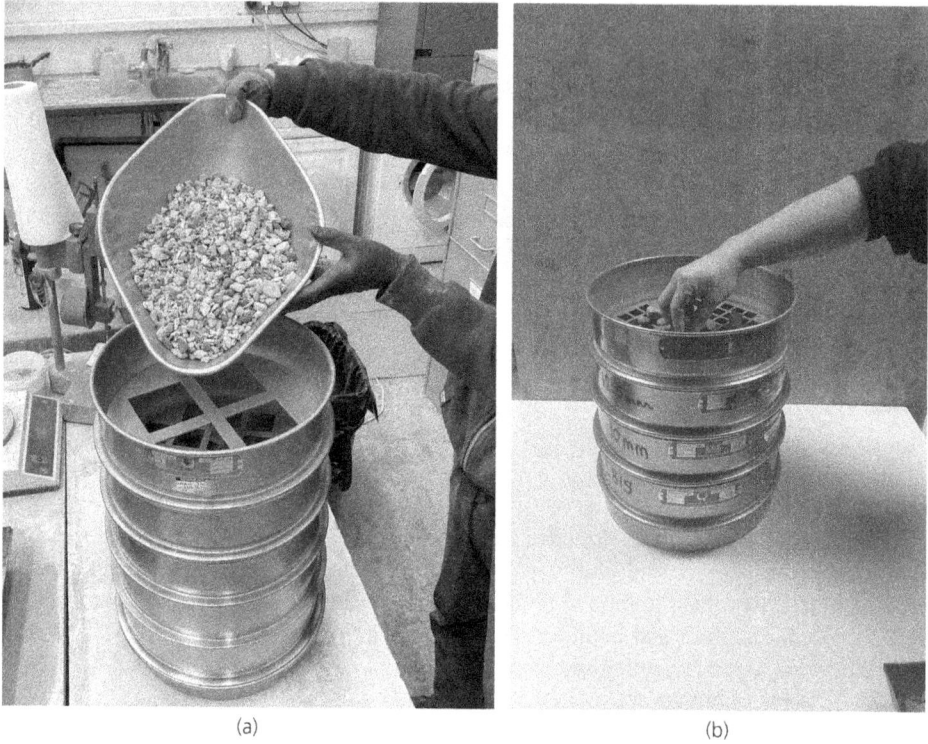

(a)

(b)

Figure 6.10 The shape of the aperture in the sieve does not necessarily mean the particle that passes through is of that size (authors' image)

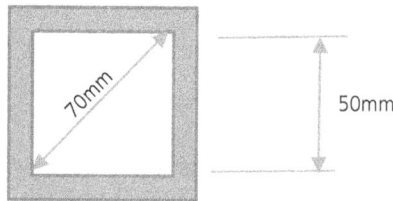

By knowing the initial mass of soil, the percentage of material passing each sieve size can be determined and hence a PSD curve showing the percentage of the original material passing each sieve size can be drawn. For convenience, the particle size axis (x-axis) is plotted on a logarithmic scale. The results of the test can be used to classify the soil and to provide an aid to the soils description, although it should be noted that the particle description from a test sample may differ from the actual material in situ and hence may be different to the engineering description of the stratum of its origin.

Figure 6.11 Illustrating the concept of equivalent spheres of different diameters to represent different particle properties. The equivalent sphere of diameter passing the same sieve aperture size as the soil particle is highlighted (image courtesy of Dr P. Hepton)

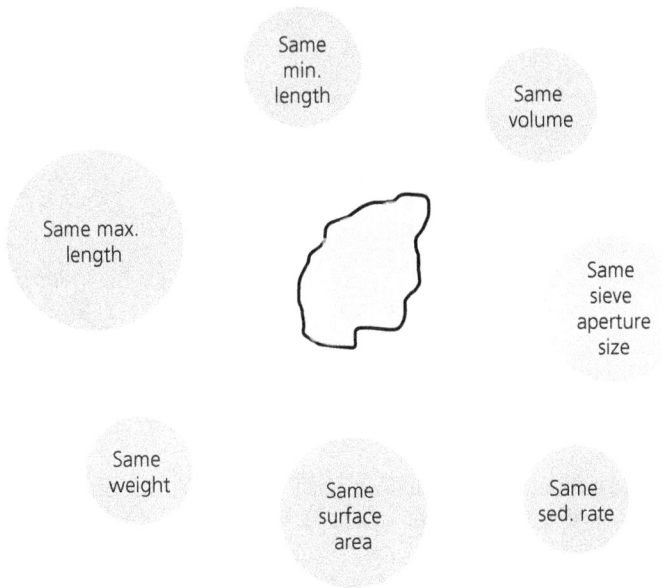

It is normal practice to use a larger diameter sieve through the gravel range (say 450 mm in diameter) and a smaller sieve diameter (200 mm) for the sand fraction. As shown in Figure 6.9(b), for larger particles it is permissible for the technician to offer individual particles to the aperture, particularly if particles are elongated.

The particle size is a key factor when describing granular soils. Table 6.8 gives the current accepted terminology for ranges of particle sizes. Some parts of the world use different size ranges.

It should be noted that the particle sizes 63 mm, 6.3 mm, and so forth, are obtained from the direct conversion from the original imperial size – that is, 6.3 mm = ¼ inch. Rationalisation to less cumbersome metric alternatives was attempted in BS 1377 (BSI, 1990) but this was removed because the actual sieve sizes were never changed.

The uniformity coefficient, C_u is determined from the PSD results. This parameter is one criterion used to segregate well graded granular soils from uniformly graded granular soils by the National Highways Specification for Highway Works, for soils to be used in earthworks, where $C_u = d_{60} / d_{10}$, d_{60} is the particle size of the 60 percentile from the grading curve, and d_{10} is the particle size of the 10 percentile.

Following on from this

$C_u < 10$ is termed well graded
$C_u > 10$ is termed uniformly graded

(**Caution:** in some areas of the world this terminology is reversed.)

Table 6.8 Classification of particle size ranges

Very coarse	BOULDERS	Large boulders	> 630 mm
		Boulders	630–200 mm
	COBBLES		200–63 mm
Coarse	GRAVEL	Coarse	63–20 mm
		Medium	20–6.3 mm
		Fine	6.3–2 mm
	SAND	Coarse	2–0.63 mm
		Medium	0.63–0.2 mm
		Fine	0.2–0.063 mm
Fine	SILT	Coarse	0.063–0.02 mm
		Medium	0.02–0.0063 mm
		Fine	0.0063–0.002 mm
	CLAY		< 0.002 mm

In addition, well graded normally means an even distribution of particle sizes, whereas the geological term 'well sorted' indicates a single or limited particle size range, which would be poorly graded.

The shape of the grading curve can be defined by its coefficient of curvature C_c, which is given by

$$C_c = D_{30}{}^2 / (d_{10} \times d_{60})$$

In any event, great care must be applied when classifying soils using uniformity coefficients because the tested sample may be very different from the overall stratum grading. This may be a natural variation or, as is often the case, it can be the result of the drilling process. In particular, samples of granular soils taken from cable percussive boreholes may be unrepresentative of the materials in situ owing to the drilling process as previously mentioned. Larger particles may be broken down by the percussive action, which produces a greater proportion of particles in the smaller size ranges. In addition, finer soils may be washed out of the sample, either because water is used to assist drilling, or samples are obtained below the water table. In both cases, as the sample is recovered water is lost from the shell and with this some finer particles will also be lost (see Chapter 1). The result is a highly distorted PSD curve, with reduced fractions at both the course and fine end of the curve, and a relative increase in particle fractions in the midrange.

A simplified approximation of the permeability (k) of the soil can be found by using Hazen's formula.

Where $k = d_{10}^2 / 100$

and d_{10} = particle size of the smallest 10 percentile

Table 6.9 shows the values of C_u, C_c and permeability k calculated for the three curves shown in Figure 6.14.

Table 6.9 Values of C_u, C_c and k determined from the grading curves shown in Figure 6.12

Curve	d_{10}mm	d_{30}mm	d_{60}mm	C_u	C_c	k (m/sec)
1	0.15	0.2	0.26	1.73	10.25	2.25×10^{-4}
2	0.30	0.4	20	66.7	0.042	9×10^{-4}
3	0.35	4.0	15	42.8	3.04	1.2×10^{-3}

Furthermore, it follows that for clay, where the largest particle size is 0.002 mm, the permeability using Hazen's formula would be 4×10^{-8} m/sec, which is misleading because the void spaces between such small particles is dependent on the electrochemical attraction between particles of clay minerals that significantly affect the permeability.

Hazen's formula provides a useful indication of permeability, although it should be borne in mind that Hazen (1892) developed the formula in the laboratory by using glass ballotini (glass spheres), determining the permeability using a single uniform size for each determination. Clearly, soil does not behave in the same way as single-sized uniform ballotini.

The shape of the PSD can indicate layered systems with concentrations of particles in two or more steps in the curve, as shown by the grading analysis for Curve 2 in Figure 6.12.

The sieve analysis can be carried out on any soil type, although where there is a significant proportion of fine material such as silt and clay the test method is adapted to remove these finer soils from around the coarser soils. To do this, the sample is washed through a 63 μm sieve. In practice, the 63 μm sieve cannot withstand the wear and tear of the washing process and the sieve is generally protected by a larger sieve such as a 150 μm sieve. The washing process can be very time consuming. Prior to beginning washing, the original dry mass of the soil is measured, and all calculations are determined from this initial sample mass. Where samples have a wide range of particle sizes, both the large sieves and smaller diameter finer particle size sieves are used. In order to ensure that the mesh aperture is not distorted by excessive sample mass, the sample may be divided by riffling. The reduced amount can then be passed through the sieve, and an allowance can be made either by sieving all the riffled factions individually and adding the masses together, or by sieving one of the riffled mases and making a mathematical correction for the total mass requiring to be sieved.

6.3.6.2 Particle size determination for fine-grained soils

The use of sieves to grade soils and determine the particle size becomes impractical once the particle size is less than 63 μm (0.063 mm), the boundary between sand and silt. This is because electromagnetic forces between the particle and the sieve mesh attract particles and prevent them falling through the sieve under gravity. Alternative methods of determining the grading envelope below this fraction use the fact that larger particles will settle in a column of fluid before smaller particles. This is described by Stokes's Law, which defines the rate at which particles of a certain size will settle in a fluid. The theory assumes that particles are spherical, which is not true for soil particles and in particular clay particles that tend to be plate-like in shape. However, it is considered the discrepancies from the variation in particle shape will be small.

Figure 6.12 Typical grading curves for three different materials (authors' image)

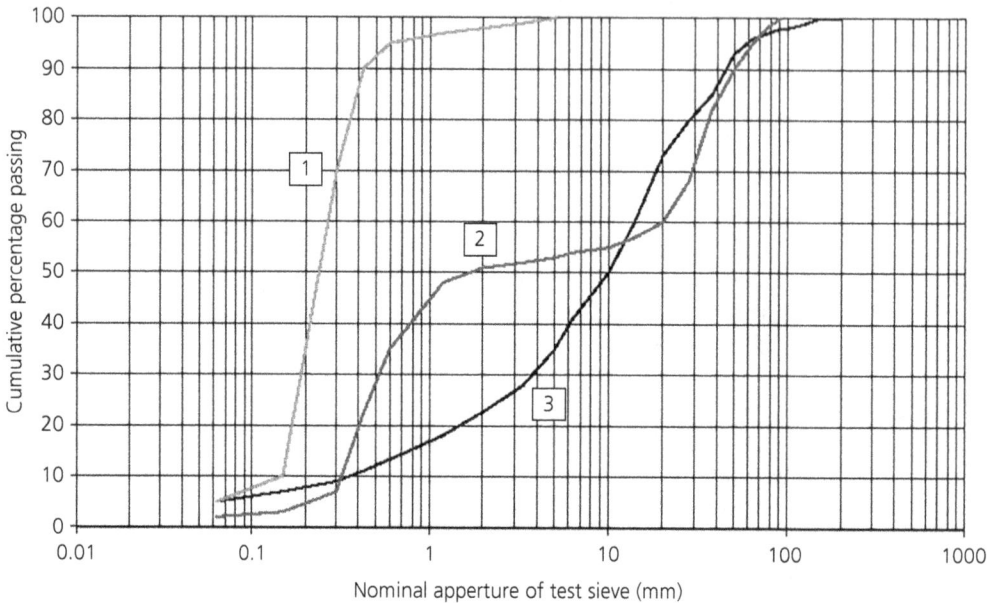

Terminal velocity of soil particle $= v = \{D^2 g(\rho_p - \rho_f)/18\mu\}$

ρ_p = density of spherical particle; ρ_f = density of fluid; g = acceleration due to gravity; μ = viscosity of the fluid.

The formula assumes that the soil particles have the same density and are spherical; they are evenly dispersed within the soil water column and the water column remains at the same temperature throughout the test with no turbulence, hence the condition of viscous flow is constant in still water. These conditions are almost impossible to meet in practice. However, with care, they can be approached with sufficient accuracy to provide reasonably reliable test results.

There are two versions of the test to determine the finer particle sizes, using either a hydrometer or a pipette. In both cases, measurements are made at set time durations that are based on the particle size and time defined by Stokes's Law. To do this, a sample of the soil that passes the 63 μm sieve is put into suspension, usually in water although in some circumstances other fluids may be used which have a lower specific gravity and are inert. The soil is placed into a 1 litre measuring cylinder that is then rotated end over end to put all the particles into suspension. The cylinder is placed into a water bath, which is maintained at a constant temperature of 20°C, and measurements are made at the selected time intervals.

The hydrometer method measures the density of the water column at the test point while the pipette takes a small sample of the water column at a depth of 100 mm from the surface of the liquid. The pipette is the preferred test method mainly because it causes less turbulence in the soil/water column as the measurement is taken.

Some soils require pretreatment to prevent flocculation (aggregation of finer particles to form larger 'particles') during the test. In addition, soils that contain organic matter may require pretreatment using hydrogen peroxide and, in some circumstances, hydrochloric acid.

The percentage of soil in suspension at the time of reading is then calculated, either from the density of the water column as measured by a calibrated hydrometer, or by the mass of soil measured in the volume of the solution sampled in the pipette, which produces a soil-fluid sample of a fixed volume. The time of reading is used to determine the particle size following Stokes's Law.

The results obtained may be presented as a continuation of the grading curve determined from the sieve analysis. It some cases, it may be found that the two curves do not provide a smooth transition at the point at which they meet. This is mainly because the assumptions made by Stokes's Law are not adhered to by a soil, particularly for larger particles which are assumed to settle in the early part of the test. A correction may sometimes be required to smooth this transition from one curve to the other. An explanation and corrections are described by Head and Epps (1992), should these be required.

6.3.6.3 Double hydrometer

The dispersive properties of clay soils were first determined by Volk (1937) using the double hydrometer method. Two specimens from the same sample of similar composition are used to conduct a hydrometer sedimentation test, as previously described in Section 6.3.6.2. One sample is conducted using a sodium hydroxide solution and vigorous agitation of the specimen while the other is conducted using only distilled water and light agitation. The degree of dispersion, D, is taken as the ratio D = A/B, where A is % of particles smaller than 0.005 mm, without mechanical agitation and chemical dispersant, and B is % of particles smaller than 0.005 mm with mechanical agitation and chemical dispersant. The soil can be classified as nondispersive, ND, when D < D50% (Kinney, 1979).

6.4.   Earthworks related testing

6.4.1   General

Several tests were derived in the USA during the 1930s to provide valuable information on the behaviour of soils when used to construct earthworks such as embankments for rail and road infrastructure. The tests fall into two main groups: compaction related tests and tests performed to provide checks on the construction works.

The process of compaction is the densification of a material by the expulsion of air, either by static or dynamic forces. Early compaction tests were devised by R. R. Proctor, after whom the test was originally named. These initial tests were intended to assist in the construction of dams, and later to provide information to assist in the formation of earthworks. The original test is similar to the current 2.5 kg rammer method, often termed 'light compaction'. The soil density achieved from this test was originally thought to be close to the maximum achievable using compaction plant. However, as machinery became heavier and more efficient, a heavier test was required and the 4.5 kg method was conceived, often termed the 'heavy compaction' method.

In the 1967 British Standard, the vibro-compaction method was introduced, primarily to deal with the compaction of granular soils following the development of vibratory site compaction plant. This larger grain size required a larger mould size to the other two tests, and hence the CBR mould was used.

All compaction tests are carried out in a similar fashion, compacting a specimen of the soil into a mould at a measured water content. The process is repeated with specimens at different water contents, and a graph of dry density against water content obtained. The test water contents are chosen to provide a graph, as shown in Figure 6.14, and from this the values of maximum dry density and optimum water content (i.e. the water content corresponding to the maximum dry density) are determined.

Alongside the compaction testing, the moisture condition value (MCV) was derived to provide a quick method of determining the soil water content in the field. This can then be used to decide if the soil is in the right condition (water content) to ensure good compaction. Often included as an earthworks parameter, the CBR test is also specified to determine the soil performance in relation to pavement design. This has been used extensively in the past to confirm the performance of earthworks compaction, although its use has declined in recent times (see Chapter 5).

The compaction related tests are discussed in the following sections. These tests are rarely done in isolation and would normally be conducted along with PSD tests for coarse- and fine-grained soils and with plasticity limit tests for cohesive soils.

6.4.2   Compaction tests

The compaction test can be used to design earthworks, to predict the amount of soil required and how it will behave when placed. The key values obtained are presented as a graph showing the relationship between water content and the dry density achieved by a given compactive effort. From this relationship, the maximum dry density and the optimum water content may be determined. Soil that is too dry will risk the inclusion of high air content and hence there is a risk of collapse under inundation. On the other hand, soils that are too wet are difficult to compact and become unstable, the incompressible water preventing the reduction of the volume of pore spaces. For inorganic soils, a water content typically in the range of about 10–25% will be suitable to provide the best conditions for compaction. At high water contents, the soil strength is significantly reduced, and manoeuvrability of plant can be significantly hampered (see Section 6.4.6).

There are three main variants of this test related to the mass of the rammer required to perform the test. These are the 2.5 kg rammer, falling 300 mm, referred to as the standard or light rammer; the 4.5 kg rammer, falling 450 mm, referred to as the heavy or modified test; and the vibrating hammer method. Figure 6.13 shows the equipment used for the light and heavy tests, while Figure 6.15 shows a vibration hammer.

The test procedure is similar for all three methods and comprises mixing a sufficient mass of sample of the soil to fill the chosen mould. An allowance needs to be made for water content test samples at each compaction point, unless the material is of a type prone to fragmentation (such as weak mudstone) in which case a fresh sample will be required for each test point. Typically, this will need at least 20 kg of soil that passes the 20 mm sieve (except where a larger CBR mould is used, where coarser particles may be retained).

The compaction process drives air out of the sample being tested, hence compacting the particles closer together. By knowing the weight of the soil compacted into the known fixed volume of the mould the density is obtained and is recorded for the measured water content. The test is then repeated for a range of water contents. A graph of dry density against corresponding water content is plotted similar to those shown in Figure 6.14, at the highest dry density on the curve the optimum water content is obtained.

Figure 6.13 (Left) 1 litre mould and a 2.5 kg rammer; (right) CBR mould and a 4.5 kg rammer (image courtesy of K4 Soils Laboratory)

Figure 6.14 Graph of increasing maximum dry density with increased compactive effort on the same soil type (image courtesy of SOCOTEC)

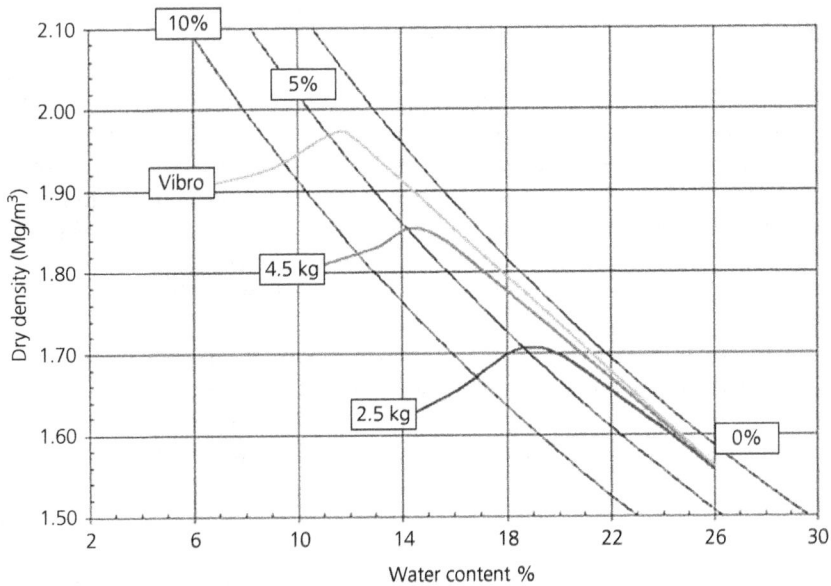

Air voids lines represent 10%, 5% and 0%; from left to right with assumed specific gravity of 2.65.

Figure 6.15 The vibrating compaction hammer and stand (image courtesy of K4 Soils Laboratory)

Preparation of the sample includes sieving to exclude all material greater than 20 mm, unless the CBR mould is used. Once this initial preparation has been completed, the sample is dried. The dry sample is then mixed to a prescribed water content. The mixing needs to be thorough to ensure even water distribution throughout the sample. The sample should then be wrapped in a sealed plastic bag and left to cure for at least 12 h. A minimum of five samples should be prepared, with the water content increasing in the five samples. Typically, for a sandy clay, the starting water content might be 8% and the water content increased by 3% in each subsequent sample, such that for the second sample the water content is 11% and so on, with the fifth sample being mixed to 20%. After the samples have cured, the material should then be removed from the bag and mixed to ensure even consistency throughout. It should be noted that deciding the water content for the start of the test is often down to the skill of the technician, rather than a prescribed series of water content increments.

The test is conducted using a steel mould that has a removable base and collar as shown in Figure 6.13. The mass of the mould should be recorded. This is normally taken to be the central mould section with the base. It is simpler to weigh these together when to soil has been compacted in the mould and deduct the mould and base mass to give the compacted soil mass.

The mould should be assembled with the mould, collar and base clamped securely. A volume of the prepared soil is then placed in the mould such that, after compaction, the soil takes up one third of the mould volume. This requires a degree of competency by the technician to be able to judge the correct volume. Two further layers are then compacted such that the final level of soil is just above the top of the mould.

Table 6.10 Hammer dimensions, number of blows and layers for the variants of the compaction test

Type	ASTM (4 inch)		1 litre		CBR	
	Number of layers	Blows per layer	Number of layers	Blows per layer	Number of layers	Blows per layer
2.5 kg falling 300 mm	3	25	3	27	3	62
4.5 kg falling 450 mm	5	25	5	27	5	62
Vibro (32–41 kg)	N/A	N/A	N/A	N/A	3	1 min duration

The number of layers required is dependent on the type of rammer adopted (see Table 6.10).

The sample is compacted using an even distribution of blows from the rammer, the aim being to ensure that each layer is of uniform thickness and even compaction. This is usually carried out using a mechanical compaction machine, which delivers the blows in an even pattern around the sample. The hand-held method is still in use in some laboratories using the rammers shown in Figure 6.13. The rammer is designed with a weight and handle confined in a sleeve such that, when the sleeve is held vertically on the soil in the mould and the weight is dropped from the full height, it will fall the precise distance dictated by the sleeve.

With the final layer almost level with the top of the mould, and the soil slightly higher than the top of the mould, the collar is removed and the soil is parred level with the top of the mould. The mould and sample should then be weighed to enable the mass of soil to be determined.

Two water content samples should be taken from the middle of the compacted sample to enable the water content to be determined as described above. The next test point can then be performed in exactly the same manner, but using a sample prepared to a higher water content, and this process is continued until all five points have been completed. A preliminary plot of water content and dry density can be drawn to provide an indication of the test result, using assumed water content values to enable an approximate dry density to be calculated. The ideal curve should increase to a peak value and then fall. Ideally, this should comprise two points below the peak one point at or close to the peak value and two points at higher water content lower than the peak. If most points are found to lie on either the high or low water content side of the peak value, further points should be conducted to ensure that the peak of the curve lies around the midpoint with an equal number of points on either side. It is useful to carry out one point at the natural water content measured at the time the sample was taken.

It is beneficial to make this assessment while the sample is in test rather than waiting for the results of the water content tests to provide a more accurate plot of the test results. Typical compaction curves are shown in Figure 6.14. The water content at the maximum dry density is termed the optimum water content.

The soil type will affect the shape of the curve. For example, sandy soils produce flatter curves, generally flattening in proportion to an increase in sand content. Pure sands produce a relatively flat curve, reaching a limit beyond which it is not possible to increase the water content, this limit occurs at the saturation point.

When the test is carried out at a higher compactive effort – for example, by increasing the hammer weight or the number of layers – there will be an increase in the maximum dry density. This will occur at lower water content, as demonstrated in Figure 6.14, where the curves represent the same soil each being compacted under an increasingly higher compactive effort.

It is normal practice to plot the air voids lines on this graph, which can be calculated using the formula:

$$P_d = 1 - (V_a / 100) / (1/G_s) + (w/100/\rho_w)$$

Where P_d = Dry density Mg/m³, V_a = air voids % total volume of soil, G_s = Specific gravity of soil,

w = soil water content %, ρ_w = density of water Mg/m³ (assumed to be 1 Mg/m³).

In order to plot the voids lines accurately, it is necessary to know the Particle Density G_s. Often this is estimated and, if so, the accuracy of the air voids lines may be compromised. Typically, lines are plotted for zero, 5% and 10% air voids. Some authors refer to these as the saturation curves and, in this case, they will have the values of 100%, 95% and 90% saturation, respectively.

The maximum dry densities achieved by the 2.5 kg and 4.5 kg rammer tests are considered to represent, respectively, the average achievable density and maximum achievable density using conventional compaction plant.

When conducting compaction tests it is often useful to understand the relationship between water content and CBR value or shear strength. To achieve this the sample can be compacted into a larger CBR mould, which will enable a CBR test, or a shear vane test, to be conducted in each sample after compaction. Owing to the larger area of the mould the number of blows per layer is increased to 62. This is done to provide comparable total compactive effort per unit surface area.

When scheduling compaction testing it should be remembered that as much as 50 kg of material may be needed to provide sufficient material for the test. In some cases, material can be reused for different stages at different moisture contents, but if the compaction effort is likely to break down the soil particles then fresh material must be adopted for each test point.

A further consideration is that there may be slight variations in the same material across a particular site. It is therefore useful to mix samples from the various locations to provide a more general material, which is representative of a particular stratum. This strategy will also go some way to allow for an overall larger sample for testing, particularly where single bulk samples are not of sufficient size.

The vibrating hammer apparatus, shown in Figure 6.15, is mainly used to provide a compaction density for coarse granular material, and the test is rarely applicable to cohesive soils. However, if used on granular soils there is a limit at which no further water may be retained in the material and water will flow out of the sample if the water content is increased. The test is considered to provide the maximum possible dry density for the soil.

6.4.3 Moisture condition value (MCV) test

This test was devised by Parsons (1979) at the Transport and Roads Research Laboratory (TRRL) in an attempt to reduce the many issues cropping up on road construction projects, where the water content of compacted material lay outside the water content range considered to be suitable to achieve an adequate compaction density. The water content of a material that has been removed from a location for use elsewhere is difficult to ascertain without taking numerous samples and performing water content tests. This is a time-consuming process and may result in material being laid before the results are known and hence having the potential of not fulfilling the specification. The use of the MCV test was derived to provide a quicker, more reliable method of determining water content in the field, which removes the delay in waiting for laboratory test results.

The water content with respect to plastic limit PL- 4% is often found to be close to the actual water content of many fine-grained soils during dry weather. The water content measured in the field may well be less than this value, while during wet periods it may be significantly higher for the same soil. This often results in disputes over the resultant density achieved with the prescribed compaction plant. The MCV test was devised to reduce the occurrence of these problems, allowing the contractor to test the material to be laid on site and obtain an assessment of the water content of the soil before starting the compaction works on a particular day.

The test is carried out in two parts: the first being a multipoint calibration test determined in the laboratory, and the other a simpler single point test carried out in the field. The apparatus is shown in Figure 6.16.

The calibration test is performed on a series of separate specimens, mixed to a range of water contents. After mixing, each specimen is compacted in a cylindrical mould. It is preferred, and useful, if one point is conducted at the natural water content. The compaction is carried out using a fixed weight rammer of 7 kg dropping a fixed distance of 250 mm. After each blow, or a fixed number

Figure 6.16 (a) The MCV apparatus (image courtesy of SOCOTEC) and (b) test in operation (image courtesy of K4 Soils Laboratory)

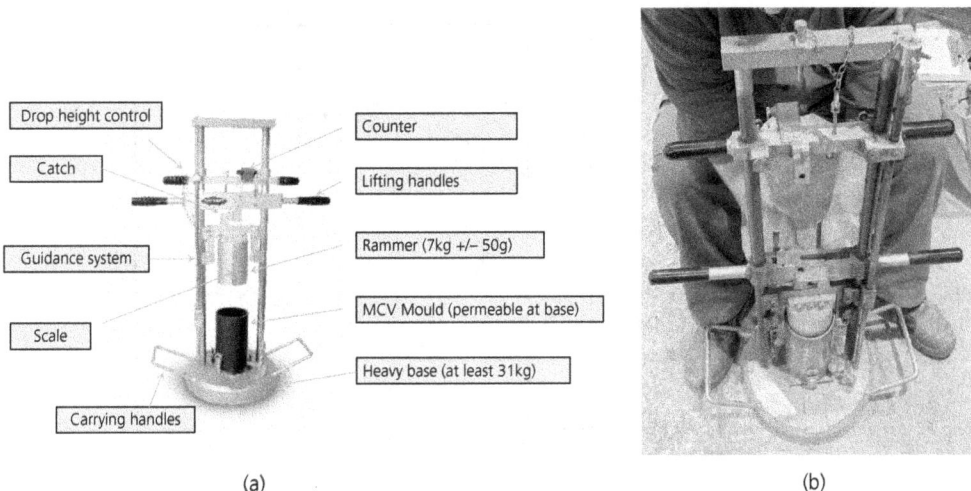

(a)   (b)

of blows, the change in height of the specimen is measured and the drop height is then reset to the standard height for the next blow or series of blows to be applied. The process is repeated. As the number of blows increases, and the specimen becomes more compacted, the change in height at each measurement stage reduces. The test is carried out until the specimen height in the mould is reduced by less than 5 mm between two consecutive sequences of blows. A curve is produced of specimen change in height on a linear scale against the number of blows on a log scale. The best fit straight line is drawn on the portion of the curve that passes through the point giving 5 mm penetration and the number of blows at this point is recorded. Using the formula below, the number of blows (B) to give a specimen change in height of 5 mm is converted to the MCV.

$$MCV = 10 \, Log \, B$$

The test data sheet usually provides this conversion on the graph as shown in the data sheet in Figure 6.17.

The test is repeated for each test specimen which has been premixed to provide a range of water contents. The results are normally presented on a single graph. Each derived MCV is then plotted against the specimen water content to give a calibration curve (see Figure 6.18).

The calibration curve can be used on site to determine the water content of a soil of similar characteristics that is being used in an earthworks operation. The success of the test is dependent on the soil that is tested in the laboratory being the same as the soil being compacted on site. To ensure that there is compatibility between the two, it is essential to carry out other tests such as plasticity index and particle size testing both on the original laboratory sample and on selected samples of the site compacted material.

Figure 6.17 MCV test graph of change in sample height against blows for single specimen at natural water content (authors' image)

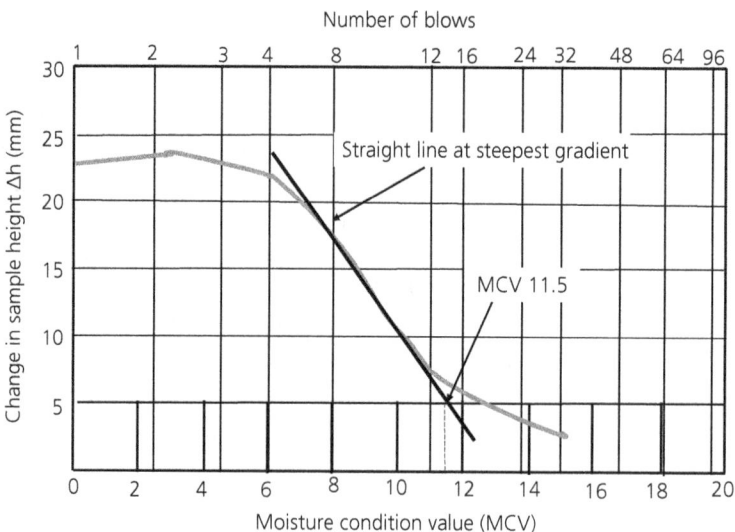

Figure 6.18 Typical MCV calibration graphs: (a) multi point test with each curve representing a different water content; (b) same MCV results plotted against water content for each sample (authors' images)

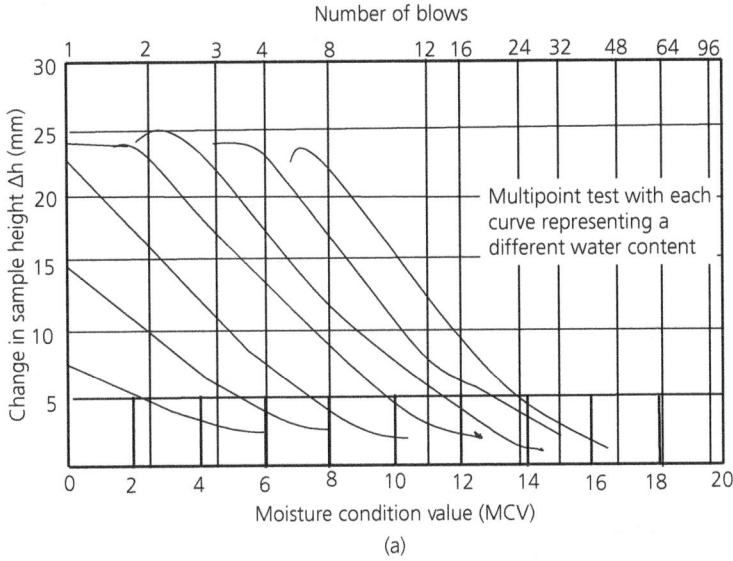

(a)

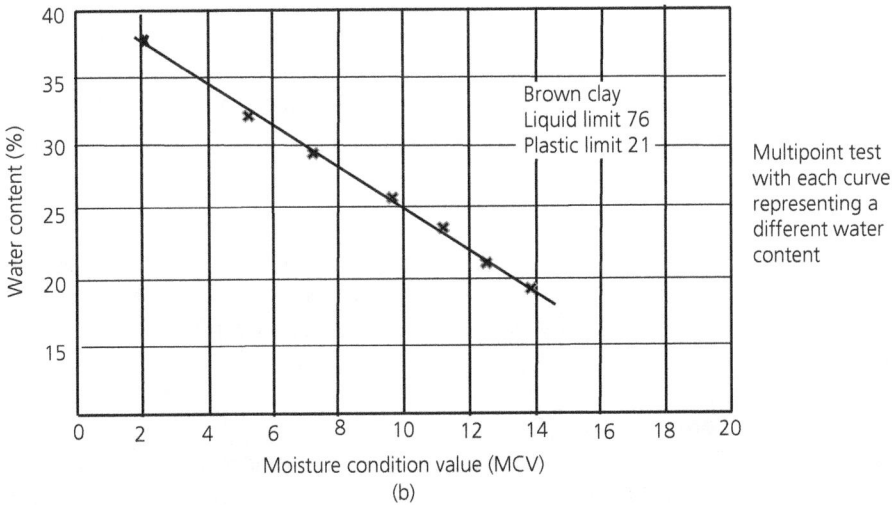

(b)

Note the plasticity index value is used to assist classification.

The results of the MCV test are often presented as part of a suite of tests including the compaction test, undrained shear strength and CBR, all performed over the same range of water contents (see Section 6.4.6).

When the MCV test is conducted on gravelly soils, it will be necessary to exclude particles greater than 20 mm from the test specimen. Where this fraction makes up a significant proportion of the material the results can be misleading. It is therefore recommended that the test should be performed along with other tests such as PSD, plasticity index and undrained shear strength. Similar misleading results may also be found if a compaction test is conducted on such materials.

Some soils give results that are difficult to interpret as they plot as a curve rather than a straight line, and this can result in an uncertainty in the MCV value in relation to the measured height change of 5 mm (Barnes, 1987; Dennehy, 1988).

6.4.4   Chalk crushing value (CCV) test

The MCV apparatus can also be used on samples of chalk to determine the CCV. This test uses 1 kg of chalk that passes the 20 mm sieve, the material being broken down if lumps are too large to provide a suitable sample size.

The test specimen is placed in the mould with a separating disc on top of the material. Blows are then applied from the rammer, set to fall 250 mm for each blow. The procedure is continued until either no further penetration occurs between blows, water oozes from the base of the sample, or the total number of blows reaches 50.

A graph of penetration P (linear scale) against blows (log scale) should be plotted, which should provide a straight line, the slope of which gives the rate of crushing for the chalk sample. The crushing value is given as one tenth of the gradient of the line.

$$CCV = Pa - Pb/10 \left(Log\ a - Log\ b \right)$$

Points a and b represent two points that lie on the straight line. If these are integer values (e.g. 2 and 20, respectively), the formula simplifies to

$$CCV = Pa - Pb/10$$

When conducting the CCV test it is good practice also to determine the saturation water content of the chalk lumps from the sample. The combination of CCV and saturation water content provides useful correlation to the state of chalk, its ease of excavation and susceptibility to breaking down by weathering processes.

6.4.5   California bearing ratio (CBR) test

The CBR test is a relationship test whereby the stiffness of a soil under test is compared with the result that would be obtained if a sample of crushed Californian Limestone were to be subjected to the same test method. The results are normalised to the Californian Limestone results, which are taken as the reference condition to give a 100% CBR value, all other results being expressed as a percentage of this value. The CBR is therefore an arbitrary value and not a true soil parameter. The use of the laboratory based CBR test is gradually being replaced by field tests such as lightweight falling weight deflectometer testing, which determines the soil stiffness using geophysical methods, or by the CBR plate test (see Chapter 5). The CBR value is used solely for the design of road pavements and runways.

This test has the advantage of being able to be performed both in situ and in the laboratory and can produce consistent results between the two. Tests in the laboratory can be carried out on samples taken on site using a CBR mould equipped with a cutting shoe. These can be pushed into the ground, usually with an excavator (see Chapter 2). Alternatively, samples of disturbed material may be compacted into a CBR mould using the methods described for compaction tests. In practice, the value obtained from the recompacted sample is probably more representative of the in situ material that will probably be disturbed by the construction process and plant movements. BS 1377 (BSI, 1990) describes six methods that may be used to prepare disturbed samples for test.

Once prepared the sample is placed onto a loading frame. A surcharge load is applied, using annular steel rings to provide constraint to the sample with the objective of limiting bulking during the test. These are generally chosen to be equal to the weight of the pavement construction. Each ring weighs 2 kg, which is equivalent to approximately 70 mm of road construction. For a typical road construction of between 250–300 mm four rings would be required.

The test method comprises applying load at a constant rate of strain to the sample and measuring the force that is required to maintain the constant strain. In practice, this is achieved using a load frame to raise the sample and mould upward at a constant rate of 2 mm per minute. The soil resistance measured by a proving ring or load cell is recorded at regular intervals of say 15 s, equivalent to 0.5 mm penetration, until a penetration of at least 7.5 mm has been achieved. The values of load applied for a penetration of 2.5 mm and 5 mm are then expressed as a percentage of the value which is obtained for the sample of California Limestone under similar test conditions. The CBR value is taken as the highest of the two values. The test is normally repeated by turning the mould over and testing the base of the sample. The specimen CBR value is taken as the average of the values for the top and base tests.

The Standards Values obtained when testing California Limestone are

2.5 mm penetration standard Force = 13.2 kN
5.0 mm penetration standard Force = 20.0 kN.

In order to perform the test on soils that contain coarse granular particles, it is necessary to remove all particles that would be retained on the 20 mm sieve. Therefore, for some soils the tested sample used in the laboratory may not represent the in situ material in terms of the CBR value achieved, and potentially the water content of the tested specimen. Where such coarse materials are present, an in situ plate loading test, which applies a load to a much larger area and therefore tests a more representative volume of material, may be selected as an alternative method of estimating the CBR value (see Chapter 5).

The drainage condition and pathways present on site will not be replicated in the laboratory. For this reason, and to represent the potential saturation of the subgrade materials on site, soaked tests are often conducted whereby the compacted sample is immersed in a tank of water, as shown in Figure 6.19. A surcharge is usually applied to simulate pavement construction. The amount of swelling is measured using a linear dial gauge positioned on the centre of the sample. It should be noted that in many cases soaking produces conservative CBR values; this is because of the influence of the mould and contact with the sample providing a preferential pathway for water penetration that would not be replicated in the field.

In recent times, the CBR test has been replaced with other forms of stiffness determinations, although in the UK the test is still undertaken routinely (for alternatives, see falling weight deflectometer testing and in situ testing for CBR value in Chapter 5).

Figure 6.19 Samples for swelling testing in a water bath (image courtesy of K4 Soils Laboratory)

Several researchers – for example, Black and Lister (1979) – have also tried to establish a correlation between undrained shear strength and the CBR value. However, this has proven very unreliable. At best, it might give a rough indication if calibrated for a particular site, but a universal conversion value does not exist. It should be remembered that the CBR value is a normalised value, which is determined from a randomly chosen material, the California Limestone.

There are several in situ test methods that are able to measure either the CBR value directly or may be used to derive the CBR value through correlation, and these are discussed in Chapter 5.

6.4.6   Earthworks specification testing and soil stabilisation

When conducting testing for suitability of soils for earthworks it is usual practice to conduct a series of tests that generally include compaction tests, MCV, CBR, triaxial strength tests and, where the soils are cohesive, plasticity index tests. Testing is conducted over a range of water contents that are normally the same as the water content points for the compaction test (Nowak and Gilbert, 2015).

A typical series of results is presented in Figure 6.20. This type of relationship graph can be a useful tool to understand the properties that can be achieved at a given water content and as such may influence the condition at which soil will be placed. The relationship graph serves to convert water content values – for instance, to expected shear strengths or compaction performance values.

It should be noted that, while graphs such as those in Figure 6.20 would suggest that there is a direct relationship between any of the values plotted, the relationship should not be transposed to similar materials at other sites or to other materials. The relationship should be viewed as site specific and material specific only.

The plastic limit is used as a criterion when assessing the suitability of cohesive soils as a potential fill material. In National Highways (2017), the water content is compared with the plastic limit minus 4% materials being described as 'dry cohesive' where the water content is less than the Wp-4% and 'wet cohesive' when above this value.

Figure 6.20 Typical relationship graphs for earthworks parameters note all graphs plotted to the same water content scale, lines show values of parameters at measured natural water content of 16.5% (authors' image)

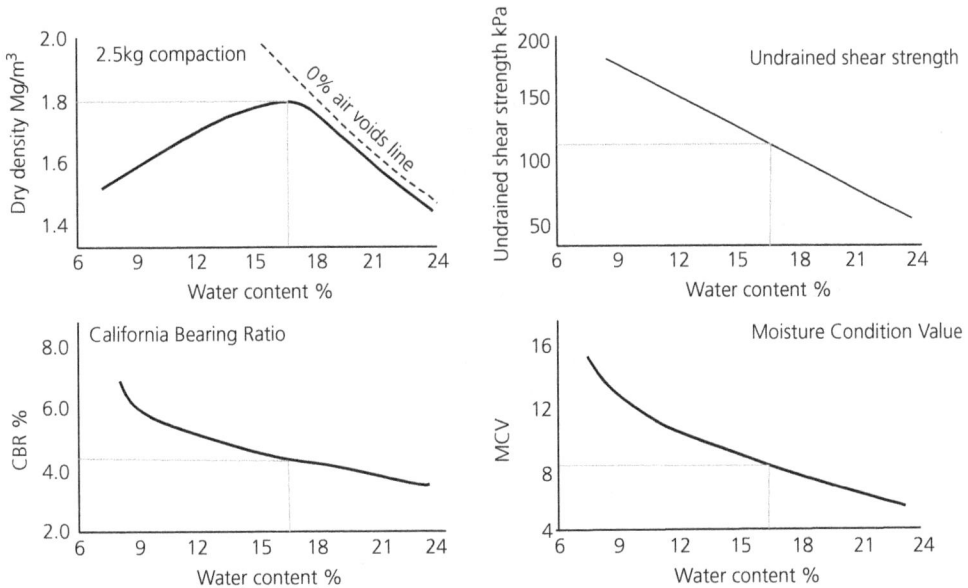

The PSD is primarily used to determine the material characteristics when classifying granular fills, as defined in Table 6.1 of National Highways (2017). The Uniformity Coefficient is used to distinguish between uniformly graded and well graded soils.

It is often found that the soils being proposed for use in earthworks in the UK are wetter than considered suitable. This can often result in poor performance of compaction plant and hence fill being placed outside the design water content values and producing compacted densities that are too low. In acute cases plant can become bogged down. Such situations are undesirable and where possible need to be avoided.

Several methods of reducing the water content might be adopted. The simplest would be to lay the soil in shallow lines in wind rows allowing the wind to dry the soil out. This is obviously weather dependant and can take some time to achieve the desired effect. An alternative might be to mix the wet soil with a soil of lower water content to provide a combined water content within the required range, although again this does not always deliver the necessary effect.

Because of the uncertainties of the above methods, it is common to uses additives such as lime, cement or pulverised fuel ash (PFA) or combinations of these.

Quick lime (CaO) is generally adopted where the fill is very wet while hydrated (slaked) lime might be used for dryer soils.

In any event, it is essential to conduct a series of trials using the relationship tests described above. These should be conducted at increasing amounts of additive to determine the optimum mix proportions.

Where treatment with lime or cement is considered, it is essential to determine the presence and concentration of sulfates and sulfides on the soils being treated. In certain soils, the concentrations may be high. The presence of lime or cement can precipitate a strong chemical reaction forming the highly expansive sulfate hydrates of Ettringnite and Thaumasite resulting in a significant volume change. Typically, overconsolidated clay such as Kimmeridge Clay and Lower Lias Clay are susceptible to these reactions.

A rigorous regime of testing of the natural soils together with trials of the mixed soils is recommended to assess the potential swelling reaction. This can be achieved using samples compacted into CBR moulds and placing them in a water bath, as shown in Figure 6.19, for period of least 28 days. It should be noted that although this method is suggested in the design specification it should only be considered alongside the laboratory determinations for concentrations of sulfate and sulfide.

Relationship testing for stabilised soil will require significant amounts of representative samples of the soils to be treated. In addition to the earthworks tests describe above, it will also be necessary to determine the Initial consumption of lime where this is proposed as the additive. The test involves increasing the lime content in the soil until the pH is seen to exceed 12. The percentage of lime to achieve this is termed the ICL (initial consumption of lime).

6.5. Consolidation testing

6.5.1 Introduction

Consolidation tests are carried out to provide parameters that can be used in settlement calculations. These primarily include the determination of the change in voids ratio with loading pressure, values of the coefficient of volume compressibility, m_v, and the coefficient of consolidation, c_v. These parameters are used to calculate the magnitude of settlement and the time over which settlement will take place. Any test to determine these parameters must be carried out on high quality, undisturbed samples. In general, the tests are conducted on fine-grained soils. Granular (coarse-grained) soils are very difficult to sample such that undisturbed samples, or samples with very low disturbance, are rarely obtained from the field. This is discussed in Chapter 2 and field tests are discussed in Chapter 5. In any event, most granular soils will consolidate very quickly, making measurement in the laboratory impractical.

There are several methods of deriving settlement characteristics in the laboratory, and each has their advantages and disadvantages.

6.5.2 The oedometer test

The test commonly conducted uses an oedometer cell and incremental loading. The oedometer enables the soil to be confined laterally, which forces any drainage from the sample to be in the vertical direction only. Because the cross section of the specimen is constant, any change in height is directly proportional to a change in volume, and thus to the change in voids ratio, provided the specimen is fully saturated. The initial voids ratio of the specimen is calculated and any subsequent change in height can be directly related to a change in the voids ratio of the specimen.

A plot is usually obtained of voids ratio against the log pressure, generally referred to as the e - log p curve. The gradient of this curve, between any two loadings, provides the value of m_v. The m_v is the coefficient of volume change, known as the 'coefficient of volume compressibility', and is primarily used to determine the amount of settlement that might be experienced under a specific load, such as beneath a foundation.

The test specimen is usually cut from an undisturbed sample into a confining ring. The ring has a cutting edge on one end. The specimen is generally about 20 mm in thickness, although variants do use slightly smaller thicknesses. Commonly the diameter of the ring varies from 40 mm to 75 mm, depending on the range of characteristics required and the diameter of the sample received for test. In any event, smaller samples tend to provide fewer representative results owing to the limitations on micro discontinuities being included in the test specimen, and the effects of sample disturbance.

The specimen within its confining ring is placed in the cell, which has a porous plate at the base (see Figure 6.21). A top cap is then positioned on the top surface of the specimen and the whole assembly is placed onto the loading frame. The loading frame consists of a yolk, which carries a lever arm pivoted on a fulcrum (see Figure 6.22). Weight placed on the hanger at the end of the lever arm will increase the loading to the specimen by a factor of usually about 10. For example, when a weight of 1 kg is placed on the hanger this will deliver an applied pressure onto a specimen in a 75 mm dia. confining ring of 22.63 kPa. A similar load applied to a 40 mm dia. specimen would give an applied pressure of 79.56 kPa.

As the first loading weight is applied, the cell is filled with water to saturate the specimen fully. Readings are taken of the compression of the specimen with time. The loading is carried out in stages, with generally each stage of loading taking 24 h. It is normal practice for the laboratory to conduct several tests at the same time, the results from each test being recorded on a multichannel data logger.

It is useful to start the first loading stage at a test pressure below the effective overburden pressure for the depth at which the sample was obtained. This will ensure that the test will cover the whole stress range under which the soil might be affected. For example, if a sample is obtained from a depth of 5 m and above the water table, the effective overburden pressure would be of the order of 100 kPa. However, a suggested starting pressure of 80 kPa might be preferable. The test will need close observation to detect any swelling of the specimen. It is essential that the specimen under test

Figure 6.21 (a) The oedometer test set up showing the sample cell and yolk with a dial gauge, which is now often replaced by a linear transducer (image courtesy of K4 Soils Laboratory); (b) a cutaway section through the test cell (authors' image)

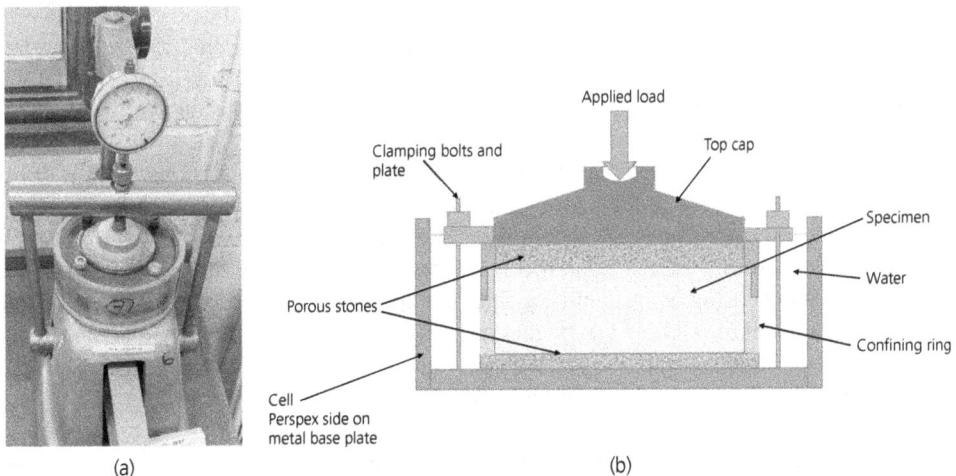

(a)

(b)

Figure 6.22 A bank of front loading oedometer frames, with testing underway (image courtesy of K4 Soils Laboratory)

is not allowed to swell above the top of the confining ring because this will invalidate the volumetric calculations that are required to obtain the consolidation properties. When swelling is detected, it is normal practice to add small increments of additional weight to prevent the specimen from swelling. In this way, the swelling pressure may be determined – that is, the pressure under which the specimen size remains unchanged for this first loading stage. This method can be discarded if the swelling pressure is not of interest, and if this is the case the starting pressure is usually the original effective overburden pressure at the depth at which the sample was obtained.

If swelling is of interest – for example, if loading from the proposed foundations is light or deep excavations are to be formed, or the soil is prone to shrinkage and swelling – either the method described above can be used to assess the swelling pressure or alternatively the swelling volume change can be determined. To do this, the specimen can be prepared such that the top of the specimen is 1–2 mm below the top of the confining ring. The specimen can then swell within the confining ring and the amount of swelling can be measured. This may be directly related to the situation on site – for example, to assess the size of the void required to accommodate swelling and therefore avoid uplift forces on the underside of a floor slab.

Once the first stage of loading has been completed, further increments of loading are applied, usually by doubling the load from the previous stage. Through all stages readings are taken of change in height at timed intervals, the spacing of which increases the longer the test is conducted, with final readings being taken several hours apart. The critical readings are those in the first few minutes where the time intervals are closer.

As mentioned above, it can be seen that, because the cross-sectional area of the sample remains constant, the change in height induced by the applied load is directly proportional to the change in volume. It is assumed that the specimen is fully saturated, therefore any change in volume must be attributed to pore water being squeezed out of the sample. The pore water occupies the pore space, thus the change in volume is directly related to a change in the voids ratio. Moreover, the movement of water, which is related to permeability, affects the rate of volume change, and hence the rate of any settlement.

It is usual to conduct at least five increments of load and, where it is of interest, unload stages may be introduced. These are useful to understand the elastic rebound, which will occur during soil unloading. Unloading stages are useful when, during the construction process, loads might be removed – for example, during demolition or excavation for basements. When the specimen is reloaded, the resultant e - log p curve will rejoin the previous virgin compression curve. Where heavily overconsolidated clay is being tested, the unload-reload loops provide a more accurate assessment of the m_v value.

From the graph of change in height against time, the coefficient of consolidation, c_v (m^2/year), is calculated. The c_v is a parameter that relates to the change in excess pore water pressure induced by the applied load, and is related to time, thus the rate at which settlement will take place can be determined. Alternatively, the time taken for a given amount of settlement to occur can be assessed. The rate of settlement is related to a time factor, T_v, which is dependent on the particular drainage conditions in the soils at the site.

Typically, the results of each loading stage of the test will be presented as a graph of change in height plotted against time on either a log scale, where the Casagrande analysis is adopted (Figure 6.24), or as the square root of time, where the Taylor analysis will be used (Figure 6.23). It should be noted that the consolidation plot of time against change in height for any one stage will comprise an initial compression stage, associated with the settlement which occurs under the immediate application of loading. This happens rapidly and is often completed within the first 15 s of the stage loading (the time the first reading is taken).

The majority of the time compression curve shows the primary consolidation and is seen as a steeply inclined curve. Towards the end of the stage, the primary compression is completed but the sample will continue to compress, with time, under the applied load. This is best seen on the Casagrande compression curve and is represented by a change in height that remains the same for each log time cycle. This is termed the secondary compression. For most soils this part of the curve is relatively small, although for organic soils such as peat the secondary compression can form a significant proportion of the overall consolidation. A knowledge of the amount of settlement per log time cycle can be directly used to predict the amount of settlement owing to secondary compression over the lifespan of a structure.

It should be noted that because the change in height, Δh, for each stage is directly proportional to ΔV and to Δe, any of these parameters could be plotted against log pressure to give the same graph

Figure 6.23 The Taylor method of analysis showing the calculation of $\sqrt{t_{90}}$ (authors' image)

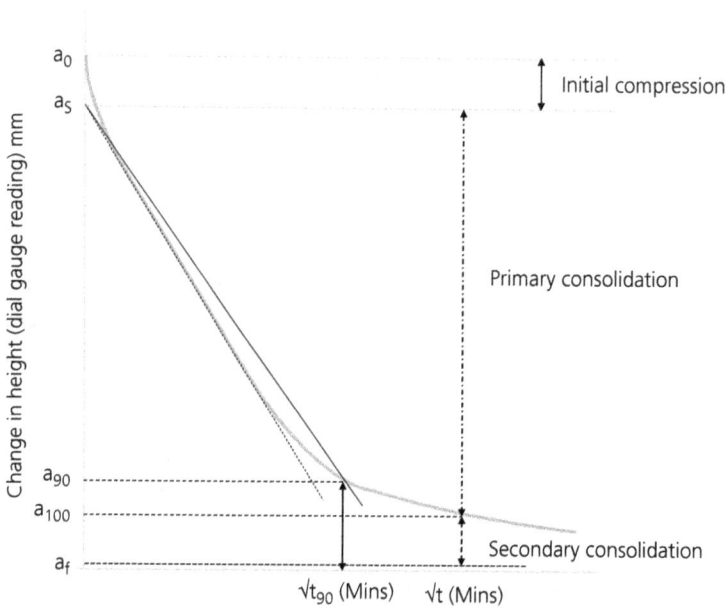

and be used to calculate the m_v values. Typically, along with the m_v and c_v values, the initial voids ratio and the preconsolidation pressure might also be determined and reported by the laboratory.

The laboratory will generally include the e - log p graph in the reported results, along with the determination of the calculated parameters for each load stage, although the time compression curves are rarely presented. In the authors' view, these are an integral part of the test and should be provided because they show which of the two methods of analysis of c_v is most appropriate.

Care should be taken when deciding which method of analysis to use to determine c_v values. The Taylor method (see Figure 6.23) calculates a value of the time when 90% of consolidation has taken place, t_{90}, while the Casagrande method (see Figure 6.24) calculates the time for 50% consolidation to take place, t_{50}. These values are used in the appropriate formula to obtain c_v for each loading stage. In general, the Casagrande approach will always provide reliable values of t_{50}, while the Taylor approach can give very misleading values of t_{90}, particularly if the compression curves deviate considerably from the theoretical curve, and do not exhibit a significant straight line section at the start of the curve. If this method is adopted, it should be checked that the position of t_{90} is located close to the lower 10% of the curve. If this is not the case, resorting to the Casagrande method is recommended. In any event, the results from the two methods should not be mixed for the same test.

The plot of voids ratio, e, against the logarithm of applied pressure (the e - log p curve, Figure 6.25), is a useful graph providing the settlement characteristics for the soil and the critical state condition for the particular soil under test. The graph should produce a smooth curve, and if this is not the case it is likely that the test has not been conducted correctly, or the sample contains gravel or another large inclusion that has affected the test result. This graph presents a critical state line, known as the

Figure 6.24 The Casagrande method plotted on a log time scale and showing the calculation of t_{50} (authors' image)

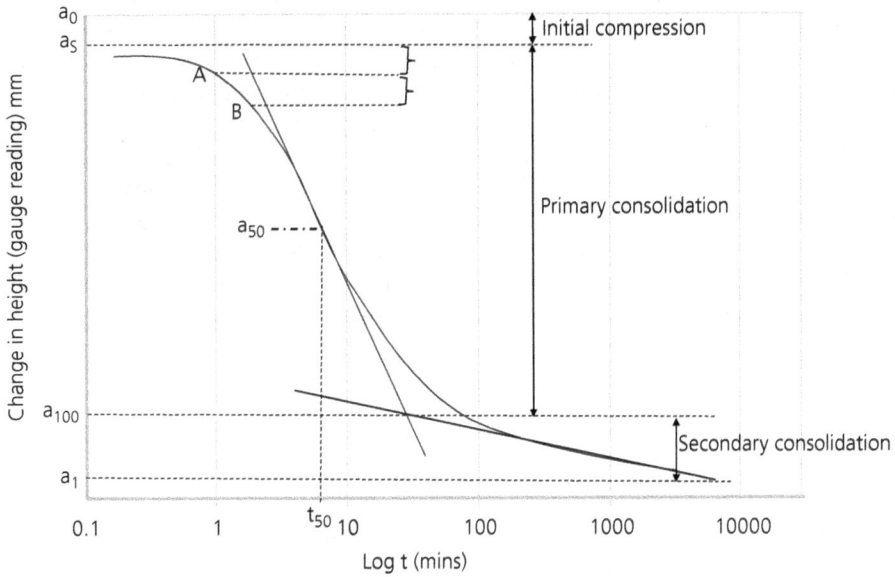

Figure 6.25 Graph of voids ratio against applied effective stress (e - log p curve), showing an unload reload loop, and the difference between the oedometer and in situ virgin compression curves (authors' image)

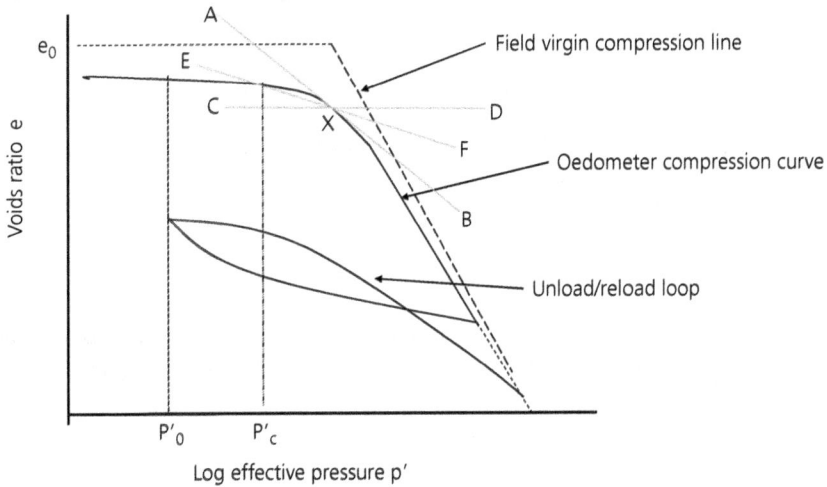

virgin compression curve, which means that the soil can only exist below the line, and it is not possible to provide conditions for the soil in its natural state to exist above the line.

The preconsolidation pressure or yield stress may also be determined, which is the maximum pressure under which the sample has been subjected during its geological past, after deposition. Typically, this can be a high value in overconsolidated soils, which have either had a significant thickness of material above them removed by weathering, or have been subjected to the weight of ice formation during the Ice Age, the load having been removed once the ice melted. The effect of this loading is to consolidate the soil, thus reducing the voids ratio. The potential swelling behaviour of the soil will govern how much it might recover (rebound) once the load has been removed. Because the soil has already undergone compression, further settlement will only take place once the loading has reached the previous maximum pressure, and elastic recovery has taken place. This point can be determined, to give the value of the preconsolidation pressure, p_c. The ratio of the preconsolidation pressure to the existing effective pressure, p_o, is referred to as the over consolidation ratio (OCR).

$$OCR = p_c / p_o$$

The preconsolidation pressure is determined using the following construction on the e - log p curve, as shown in Figure 6.25.

(a) Draw a tangent AB to the curve at the point of greatest curvature X.
(b) Draw a line horizontally through X to give CD.
(c) Bisect the angle between CD and AB to give line AD.
(d) Project the virgin compression line back to give line EF.
(e) Read the apparent preconsolidation pressure p_c where oedometer compression curve intersects line EF.

The following formulas are used to calculate the various parameters often required from the test.

The relationship between the change in void ratio and the change in applied pressure is the coefficient of compressibility $a_v = \delta e / \delta p$.

The resultant volume change from this relationship is given by

Coefficient of volume compressibility $m_v = a_v / (1+e) = -\left[1/(1+e) \right] \delta e / \delta p \ m^2 / MN$

Where e = void ratio at the load stage such that $e = e_0 - \Delta e$

Where e_0 = initial voids ratio and Δe = change in voids ratio

From earlier, it will be noted that Δe is proportional to Δv and ΔH, the change in volume and change in height at each stage, thus the same result can be obtained for mv by substituting H into this formula.

Thus $\Delta H / H_0 = \Delta e / 1 + e_0$ and $\Delta e = (1 + e_0 / H_0) \Delta H$ therefore $\Delta e = \Delta H / H_s$

Where $H_s = H_0 / 1 + e_0$

H_s is equivalent to the height of solid particles, which is a constant for the specimen under test. In addition, H_0 is the original height of the specimen.

The value of c_v can be determined from the test results where

Coefficient of consolidation $= c_v = (T_v / t) h^2$

Where $T_v = c_v t / h^2$

T_v is called the time factor, and is dimensionless, and h is the length of the drainage path.

The value of c_v is usually expressed in m^2/year. When t is t_{50} as determined using the Casagrande method and of curve fitting this becomes

$c_v = 0.01036 h^2 / t_{50}$

The length of the drainage path is, h, in the oedometer apparatus where drainage can take place both to the top and base of the specimen. If the height of the specimen is H, the length of the drainage path is $H/2 = h$, and the equation becomes

$c_v = 0.026 h^2 / t_{50} m^2 /$ year

Similarly, where the Taylor method of curve fitting is adopted

$c_v = 0.112 h^2 / t_{90} m^2 /$ year

The c_v is related to the soil permeability, k, and can be expressed as $c_v = k / (\rho_w g m_v)$ which, when rewritten, becomes $c_v . m_v = k / \rho_w g$. This approximates to $k = 10 (c_v . m_v)$ m/s.

Where the specimen is subjected to successively increasing pressure during the test then the above equation can be used to ensure the results obtained are correct. As the specimen is compressed the void space must decrease and hence permeability will decrease. It can therefore be inferred that the sum of c_v multiplied by m_v must also decrease.

The field consolidation curve will be different to the laboratory test; this is owing to sample disturbance that cannot be avoided. There are several suggested methods of obtaining the field e - log p curve, such as that suggested by Schmertmann (1953). The laboratory curve approximates towards the field curve as the pressure increases, the field result plotting as a straight line as seen on the e - log p curve, shown in Figure 6.25.

The slope of the field consolidation curve is termed the compression index, C_c, and is derived from the straight line section of the curve. It is calculated over one log cycle, where:

$e = e_0 - C_c \log \{(p_0 + \delta p)/p_0\}$

This equation can also be used to determine the swelling index, C_s, by using the slope of the unload curve.

C_c is the change in voids ratio over a log cycle and is dimensionless. It can be related to the liquid limit, w_l, Skempton (1944) proposing the following relationship

$C_c = 0.009 (w_l - 10\%)$

Along with the issues of the drainage direction, another limitation of consolidation tests is that they generally use small test specimens. This is done to reduce the test time, particularly with clay soils where the permeability can be as low as 1×10^{-9} to 10^{-13} m/s. A low permeability will result in test times being extended if the specimen size is large. It is possible to undertake consolidation testing within the triaxial cell, using cylindrical specimens. This has the advantages of being able to saturate the specimen fully, to allow a more natural horizontal drainage direction, avoid the scale effects of a small test specimen and, in common with the oedometer test, apply a pressure equivalent to the effective overburden pressure. However, most commercial laboratories do not have cells or pressure systems that would allow increasing the pressure in the cell to those likely to be experienced under foundation loads, or where the overburden pressure requires high starting loads and hence higher subsequent loads as the test progresses.

Consolidation testing in the laboratory to obtain parameters for settlement analysis is not perfect, and there are many issues that should be considered when using the results in design. These factors include the type of sample tested and the material properties. In general, the test methods are best suited to fine-grained soils. In any event, it is essential to obtain a high-quality sample (Class 1) for such testing. Preparation of the test specimen requires great care to ensure that the sample is not disturbed, and in particular that natural drainage paths are not blocked by the preparation process.

One of the main drawbacks for the test is the effect of the confining ring that forces drainage to be only in the vertical direction. In the field, permeability is likely to be anisotropic and is typically higher in the horizontal direction. It is therefore often found that the results obtained from consolidation testing predict settlement to be slower than will be experienced post-construction. The confining ring also introduces friction between the ring and the soil that will potentially reduce the total settlement at each loading stage.

Even with these issues, this test is useful, being one of the few ways that settlement characteristics for cohesive soils can be readily estimated. Table 6.11 gives approximate ranges of m_v for typical soils. It should be noted that settlement of granular soils tends to occur very quickly, owing to the high permeability of such soils, and where this is the case the c_v becomes indeterminable, the results often being reported as 'Rapid'.

Table 6.11 Typical ranges of coefficient of volume compressibility for various soil types

Soil type	Compressibility	Coefficient of volume compressibility m_v (m²/MN)
Organic clay, peat	Very high	>1.5
Normally consolidated clay	High	0.30–1.5
Firm to stiff clays	Medium	0.10–0.30
Very stiff clay	Low	0.05–0.10
Glacial till and other heavily overconsolidated clays	Very low	<0.05

6.5.3 Constant rate of strain (CRS)

A relatively recent advance in consolidation testing is the constant rate of strain test. In this form of the test, the load applied is gradually increased at a constant rate, the change in voids ratio being plotted as the test progresses. This provides a much quicker way of obtaining a graph of voids ratio vs log pressure. The curve can then be analysed at any point to provide the m_v value for a particular loading range. The m_v is the gradient of a tangential line to the curve at the required effective stress. This test is becoming more popular because it is quicker and will generally cost less. The loading is applied by way of a computer-controlled pressure system, which is much more compact and avoids health and safety issues that accompany handling weights (see Figure 6.26).

Figure 6.26 A constant rate of strain consolidation (CRS) set up using a standard load frame (image courtesy of Geolabs Ltd)

The system is able to monitor the settlement under any applied load and can increase the load smoothly and at a rate that will enable the compression of the sample to continue at a constant rate of volumetric strain.

Because the test is continuous, time settlement plots are not obtained, and hence the c_v cannot be determined. The void ratio against log applied pressure is generated as the test proceeds and the soil stiffness parameter, m_v, can be determined at any point. The c_v could be determined indirectly using the formula $c_v = k / (\rho_w\, g\, m_v)$, as shown in Section 6.5.2.

6.5.4   Rowe cell consolidation test (hydraulic cell)

The Rowe cell test, developed by Professor Rowe at Manchester University, was conceived to address some of the issues presented by the standard oedometer test. The key factors addressed by the test are to enable drainage in either the vertical or horizontal direction, or both – a condition that is more representative of in situ conditions. In addition, the ability to measure the pore water pressure provides the ability to determine that the specimen is fully saturated.

The Rowe cell has several advantages over the conventional consolidation apparatus. The hydraulic loading system is less susceptible to disturbance and vibration, and high loads can be applied, up to 1000 kPa, removing the need to handle heavy weights. The system also allows better control of the test condition with the ability to saturate. Moreover, control of the drainage enables the undrained condition to be achieved. In most cases, the specimen is much larger than used in the conventional test method, giving a more reliable and representative test result.

Testing can be conducted on samples at diameters of 75 mm, 151 mm and 252 mm, with corresponding sample heights of 30, 50 and 90 mm, respectively. Before the test specimen is prepared, it is necessary to determine the test conditions required; in particular, how drainage is to occur and if pore pressure is to be measured. These variables should be specified when scheduling the test. The specimen is confined in a cell fitted with a porous outer membrane to permit horizontal drainage. Alternatively, a drain may be formed at the centre of the specimen; this is achieved by removing a small diameter core from the centre of the specimen and filling the hole formed with sand. The various configurations enable the following test conditions to be achieved

- vertical flow upwards
- vertical flow downwards
- vertical flow in both upward and downward directions
- radial drainage outwards
- radial drainage inwards.

In addition to each of these drainage conditions, the upper soil surface can either be in contact with the flexible membrane, for a free strain loading condition, or a rigid porous plate that gives an equal strain loading condition. A section through a rolling diaphragm test set up is shown in Figure 6.27.

Because the test is conducted in a hydraulic cell, the specimen can be saturated before beginning the test. This alone will improve the test result because in the standard test, although the specimen is tested in a water bath, it cannot be guaranteed that all the air within the specimen has been removed. Saturation can be tested using the pore pressure measurements and applying increments of cell and back pressure (see also saturation in triaxial testing, Section 6.8.7.3).

The test proceeds in a similar incremental fashion as used in the standard oedometer test, with readings of change in height being taken at increasing time intervals. The results are then plotted either using the Taylor or Casagrande method, and the parameters m_v and c_v obtained. These are often determined using the pore water pressure readings rather than the change in height or voids ratio values. In addition, owing to the controlled drainage, the permeability may also be determined.

The main drawback of the test is that is difficult to set up without causing disturbance to the test specimen. The test specimen is extruded into a steel ring in the same way as the standard oedometer test. Once the ends have been prepared level and flat, the specimen may be pushed from the ring into the cell, and great care must be taken when carrying out this part of the preparation to avoid considerable disturbance of the test specimen. This action is particularly difficult for the larger diameter specimens which generally require specifically designed extrusion equipment. However, with training and great care these difficulties can be overcome. An alternative is to recompact the

Figure 6.27 (a) Cross section of Rowe cell; (b) the assembled unit (images courtesy of VJ Tech Limited)

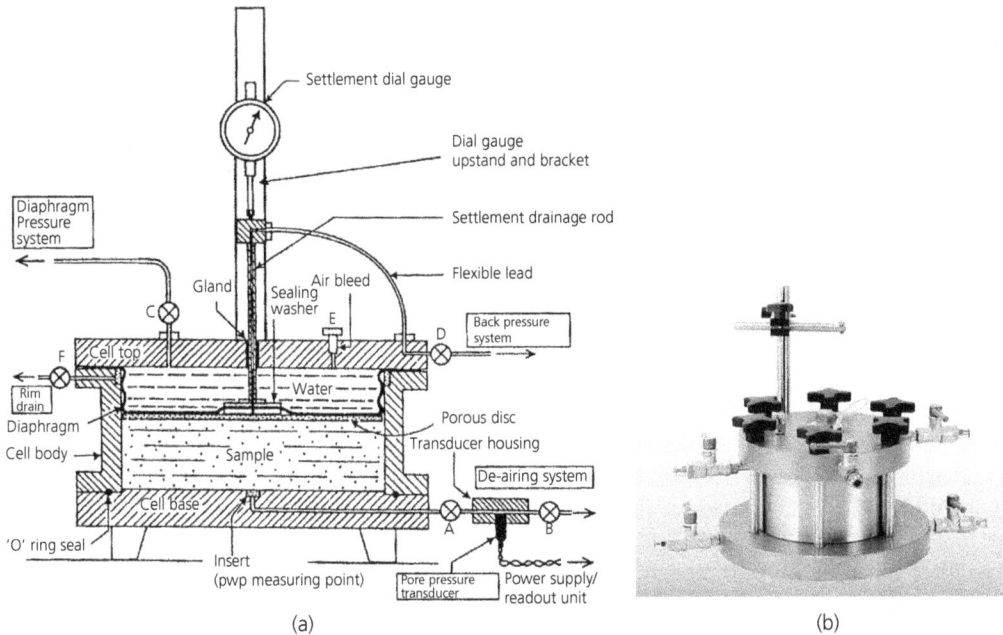

(a)   (b)

specimen into the cell, achieving a density similar to that obtained from the compaction test. This method is often adopted to assess the settlement of earthworks.

6.6.   Suction tests

6.6.1   General

Suction within a soil occurs when the pore water pressure reduces, often to a negative pressure, with accompanying increase in effective stress in the soil skeleton. In practice, soil suctions are developed with fluctuations of the groundwater level, resulting in a change in effective stress that in plastic soils can be a precursor to slope failure. Suctions are also developed owing to desiccation, where evaporation can significantly influence the near surface soils. Similar effects are caused by vegetation which is particularly significant in the vicinity of trees. While plasticity index tests related to the natural water content will give an indication of the potential for suction forces to be present, they do not quantify the degree of suction. This is given by the potential soil suction, pF, which may be determined in the laboratory.

Soil suction is significant in the design of underground structures, and in the slope stability of earthworks. Soil suction also plays an important role in the behaviour of unsaturated soils. Suction forces influence the effective stresses in the soil and may produce higher strength characteristics, which may dissipate over time as the soil moves towards a drained condition. Soil suction also provides a key indicator of the degree of disturbance in samples, which is particularly useful when conducting advanced tests such as small strain and soil stiffness determination. In the laboratory, tests can be conducted to assess humidity associated with a soil sample. This is the effect of the water within the air around the sample increasing as water evaporates naturally from the specimen and from this, the potential suction may be calculated.

It should be borne in mind that sampling, particularly by percussive methods, will significantly affect the soil suction and water content distribution through the sample. It is therefore unlikely that soil suction will remain unchanged by the time the test specimen arrives at the test laboratory.

The improved understanding of the role of soil suction has led to the development of test methods to determine the suction properties of soils. The first of these methods was the filter paper suction test. However, more recent developments in pore water pressure devices, capable of direct measurement of the suction forces in the soil, have improved accuracy and shortened testing times. These are generally deployed in situ and are covered in the instrumentation and monitoring section, Chapter 4. Pore pressure devices that can measure suction in the triaxial cell are also now available.

6.6.2   Filter paper suction test

The laboratory test measures the humidity released by a sample held in a sealed container, and from these results derives the potential soil suction, pF. The test measures the matrix suction by sandwiching a filter paper between two halves of the soil sample and measuring the water content of the filter paper after a period of time.

The filter paper used for the test is a particular type made to a precise specification such that they have consistent prescribed properties.

To conduct the test the equipment needs to be clean, dry and free of oil/grease. Handling of the sample and filter papers is carried out with extreme precision, wearing gloves and using tweezers to hold the filter papers. The test is conducted on cylindrical samples of cohesive soils. A length of the soil is cut from the sample approximately 100 mm in length and this is cut in half to provide two identical cylinders of the soil. The filter paper is positioned between the two halves and placed into a sealed container and left usually for at least 7 days. The test requires the filter paper wetting curve to be determined. The calibration curve can then be used to determine the total suction from the filter paper water content. Commonly the specimen will be cut into a number of equal discs and filter papers placed between each. The average value obtained for the weight of water in the filter papers is used to calculate the average soil suction.

An alternative, more rigorous test procedure is sometimes performed where the filter paper is placed between two protecting filter papers. The soil sections are taped together around the periphery to seal the filter papers in place. An additional filter paper is positioned on a small plastic ring to hold the paper above the sample. The whole set up is placed into a sealed container and left for a minimum of four weeks. The test determines the matrix suction and the osmotic suction. The osmotic suction arises from dissolved salts contained in the soil water.

Where the paper is positioned between the soil discs, the value obtained is representative of the matrix suction. Where the paper is positioned above the test specimen, the result relates to the osmotic suction. The two values combine to give the total suction for the specimen.

In both methods, at the end of the test period the filter papers are recovered and weighed to an accuracy of 0.001 g. The weight of water in the paper is related directly to the soil suction using the filter paper wetting calibration curve for the type of filter paper used for the test.

6.7. Laboratory permeability testing

6.7.1 Introduction

Permeability testing can take several forms in the laboratory. Tests are typically based on a constant head test method, where the applied head is achieved either by gravity or by inducing the required pressure head by way of a pressurising system. The latter is achieved using a triaxial cell but is run without the application of any external loading from a load frame.

The basic calculation of permeability uses Darcy's formula Q = aki.

Where Q = rate of flow (m³/s),

a = the area of the soil through which the water is flowing (m²),

k = permeability (m/s),

i = the hydraulic gradient (dimensionless), that is the difference in head between the two reference points of distance y apart.

Permeability of soils in the UK are measured as metres per second (m/s), and typical permeability values for various soil types are given in Table 6.12. It should be noted that various other units are used is different parts of the world, such as mm/s or cm/s. To be compliant with the SI system, units should be m/s.

To put these values into context, the hair on an average human head grows at a rate of 10^{-8} m/s. Puddled clay, often used to seal landfills, may have a permeability of the order of 10^{-11} to 10^{-12} m/s.

Table 6.12 Typical values of permeability for various soil types (various sources)

Soil type	Coefficient of permeability k m/s	Degree of permeability
Gravel	10^{-2}	High
Coarse sand	10^{-2}–10^{-4}	Medium
Fine sand	10^{-4}–10^{-5}	Low
Silty sand	10^{-5}–10^{-6}	
Silt	10^{-6}–10^{-7}	Very low
Sandy clay	10^{-8}	
Clay	$\leq 10^{-9}$	Extremely low
Chalk	10^{-4}–10^{-5}	Low

6.7.2 Constant head test using a permeameter

Permeability testing under gravity uses a permeameter. This type of test is best suited to soils that exhibit relatively high permeability characteristics, such as sand and/or gravel. The sample should be placed into the permeameter at the required density. The system is then saturated. The permeameter has manometer ports that connect to manometer tubes. Water is allowed to flow into the system by way of a header tank until a constant rate of flow is achieved. A typical section through the apparatus is shown in Figure 6.28.

The rate of flow and the difference in head between the two manometer ports are recorded. The distance between each port is also measured. The following formula is then used to calculate the permeability of the test specimen, using Darcy's Formula. Transposing the formula gives

$$k = (q / i)(Rt / A)$$

Where k = permeability m/s; q = average flow ml/sec; $i = h / y$

h = the difference in water level in the two manometers

y = the distance between the two manometer outlet points

Rt = temperature correction factor for viscosity of water = 1 at 20°C

Figure 6.28 Diagrammatic section of a permeameter cell (authors' image)

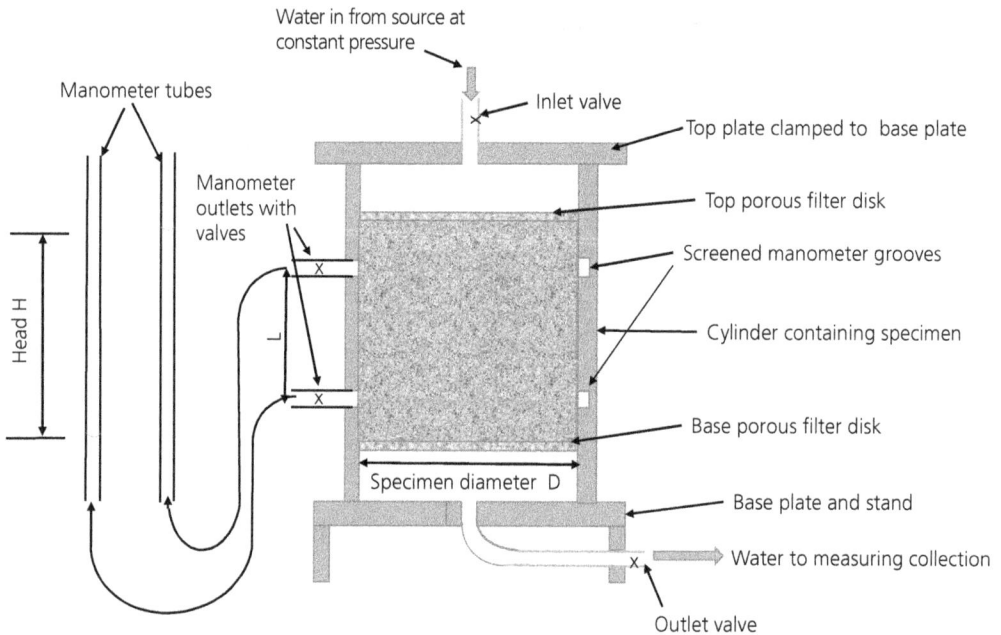

6.7.3   Permeability in hydraulic cells
Most laboratories will carry out permeability testing on fine-grained soils, and particularly clay, using a hydraulic cell. The test specimen is prepared in a similar way to that for the triaxial test. For permeability testing, the specimen height is usually the same as the diameter. Details of the specimen volume and weight are measured to enable the soil density to be calculated. The water content of the sample is also measured.

The specimen is prepared as a right cylinder, sealed in a latex membrane, and placed in the cell. A filter paper is placed on the pedestal and the sample is sealed into place using rubber bands that provide a watertight seal around the specimen. The pedestal has a port connecting to the base of the specimen.

To conduct the test, the cell is filled with water, which is then pressurised to the equivalent of the overburden pressure relating to the sample being tested. The sample may be saturated using a back pressure system, ensuring the exclusion of air, in a similar way to the effective stress triaxial test (see Section 6.8.7). A pressure differential is then applied between the top and base. The top cap has a port that enables water to flow through the sample, emerging through the base pedestal into a volume change device. Measurements of the volume of water passing through the sample are taken at intervals to enable a graph of the volume of water passing through the sample against time to be plotted. The graph can be analysed to provide the soil permeability. Once complete, the test specimen is removed and weighed, and an after-test water content determined.

It should be noted that, when tests are conducted on clay with very low permeability, the test can take several weeks to complete.

6.8.   Strength testing

6.8.1   Introduction

There are several types of strength test that can be undertaken in the laboratory, and these fall into two groups: total stress and effective stress determinations. Total stress testing is carried out at a fast rate of strain while effective stress testing is conducted at a slower rate with measurement of the pore water pressure response or volume change during loading (for tests conducted under undrained or drained conditions respectively). The form of test selected depends on why the test is being conducted. This will influence the choice of parameters required and the test method used. This section deals with routine methods of strength testing of soils, including both total stress and effective stress techniques. More advanced tests, such as the simple shear test, are discussed in Section 6.9, while Section 6.10 covers strength testing of rock specimens.

Testing to determine the shear strength and angle of internal friction (angle of shearing resistance) in soil is made in order to assess the risk of failure under loading. All buildings or structures impose changes in stress in the soil which will cause deformation and, should such stresses exceed the strength of the soil, potentially failure. The relationship between the shear stress and the properties of cohesion and the angle of friction has been known since Coulomb (1776), who developed the total stress relationship, albeit for a totally different reason.

$$T_f = c + \sigma_n \tan \phi$$

Where T_f = shear force; c = cohesion; σ_n = normal force; ϕ = angle of internal friction.

In general, the total stress situation is not the condition that will be the most critical for foundations. It is more accurate to consider the soil behaviour to be controlled by effective stresses. To determine the effective stress state of a soil, we need to know the pore water pressure under given conditions. Thus, Coulomb's equation is modified to the effective stress state where the normal stress becomes the effective normal stress, as defined by Terzaghi's equation

$$\sigma'_n = \sigma_n - u \text{ where } u = \text{pore water pressure}$$

Hence, Coulomb's equation becomes

$$T_f = c' + \sigma'_n \tan \phi'$$

The Mohr-Coulomb envelop (shown in Figure 6.41) defines the zone within which the soil may exist. On reaching the boundary defined by the critical state line, failure will occur beyond which there cannot be any further change in shear stress or volume (and thus the soil cannot exist beyond the envelope). The theories of critical state soil mechanics have resulted in the development of many of the advanced testing methods in common use today and has increased our understanding of soil behaviour.

It should be remembered that shear strength is not a unique property of the soil but is dependent on the soil density and water content, both of which can vary over time and in some case can change very rapidly. In addition, the pore water pressure will change depending on changes in stress.

Strength tests are usually conducted with a shear box or by using a triaxial cell. Two less common types of total stress testing are the unconfined compressive strength test and the vane test.

The shear box test can be used to determine peak and residual shear strength, and if tests are conducted at various normal stresses will provide an angle of shearing resistance. The main limitation of this test is that the test apparatus forces the plane of the shear at a particular position and, while this might be representative of a homogeneous soil, it is unlikely to occur in practice, where differences in lithology and varying strength throughout a deposit will almost certainly have a significant influence. Furthermore, it is difficult to ensure that the test specimen is undisturbed because physically placing the test specimen into the shearbox can disturb the soil. However, if tests are to be conducted on granular soils these can be compacted into the shearbox at a prescribed density and water content.

The triaxial test lends itself to the testing of undisturbed tube samples of fine-grained soils. The test can be carried out with a range of test conditions, with the measurement of stress, strain, pore pressure and other parameters, and these are discussed in detail in Section 6.8.7.

The triaxial test is more difficult to carry out on very soft soils, owing to the lack of lateral restraint as the specimen is prepared. Where this is the case, the laboratory vane test can be performed on the sample while still in the sample tube, removing the need to extrude the sample, a process that will often cause significant disturbance.

It should be noted that the use of the hand vane or pocket penetrometer are not considered suitable for the determination of the strength of the soil, particularly when the results are required for design (see also Chapter 5).

Most strength tests rely on the test specimen being of undisturbed material, and BS EN ISO 22475 (BSI, 2021) requires all strength tests to be conducted using a Class 1 sample (see Chapter 2).

6.8.2   Laboratory vane
The laboratory vane test is similar to the borehole vane test used in the field (see Chapter 5) although the vane is much smaller to enable the testing to be conducted in soil contained in a tube sample.

As mentioned above, this test is particularly suited to soils with low shear strength and provides the total stress at failure. For soft soils, the sample preparation process for tests such as the triaxial test can cause significant disturbance to the sample. These problems are overcome by performing the vane test on the sample while it is retained in the sampling tube or liner. To do this, the vane apparatus requires a suitable clamping system to hold the equipment and the sample tube while the test is conducted.

The vane is clamped in place and is attached to the torque head that incorporates a calibrated spring. The vane is inserted at least two vane lengths into the sample. A typical sample set up is shown in Figures 6.29(a) and (b).

The angular displacement is read from a calibrated scale around the top of the torque head. The shear strength is obtained using a precision calibrated spring, which is clamped to the head and vane seating. Shear is achieved by rotating the vane either manually or by a motor. The head is rotated at a radial strain of between 6–12 degrees per minute. Readings of the angular torque are taken until failure occurs.

The following calculations are made

Vane shear strength, $T_v = (M/K) \times 1000$ (kPa)

Torque, M = angular rotation at failure × spring calibration factor.

Vane constant, $K = \pi D^2 (H/2 + D/6)$. Where H = height of vane, and D = overall diameter of vane, as shown in Figure 6.30.

Figure 6.29 (a) Vane shear test showing forces and key dimensions (authors' image); (b) vane test equipment with vane fitted and selection of test springs, torque is applied by way of an electric motor (image courtesy of K4 Soils Laboratory)

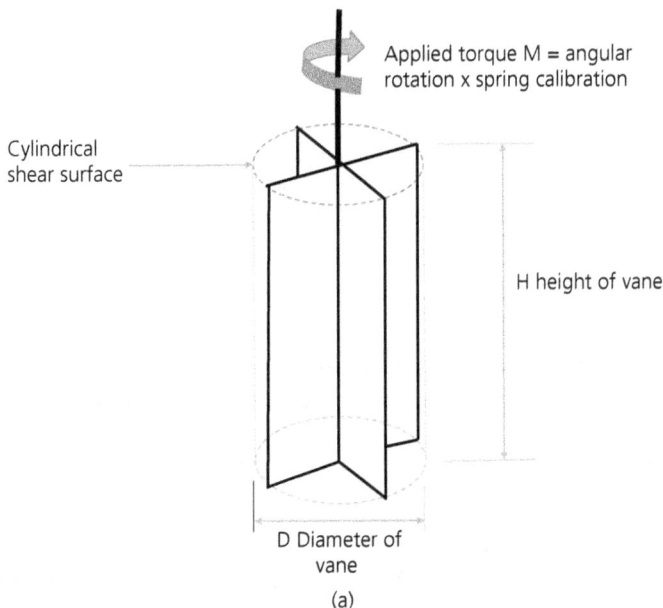

Figure 6.30 Component parts of the shear box (images courtesy of VJ Tech Limited)

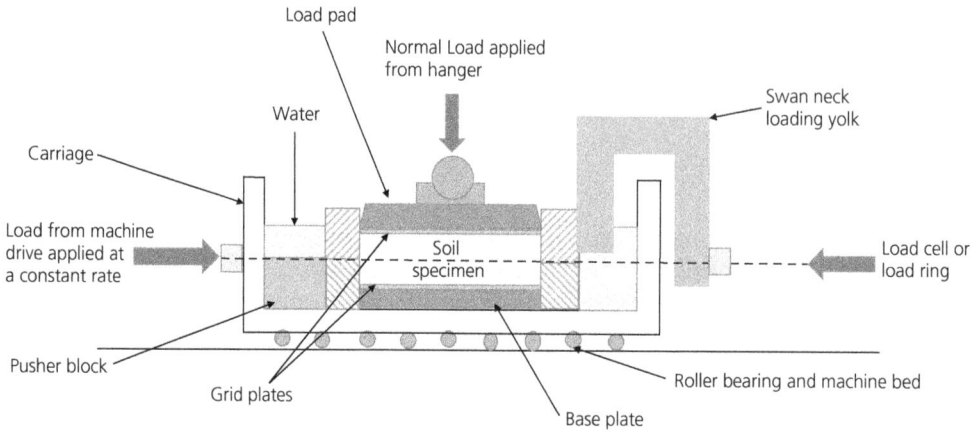

(a)

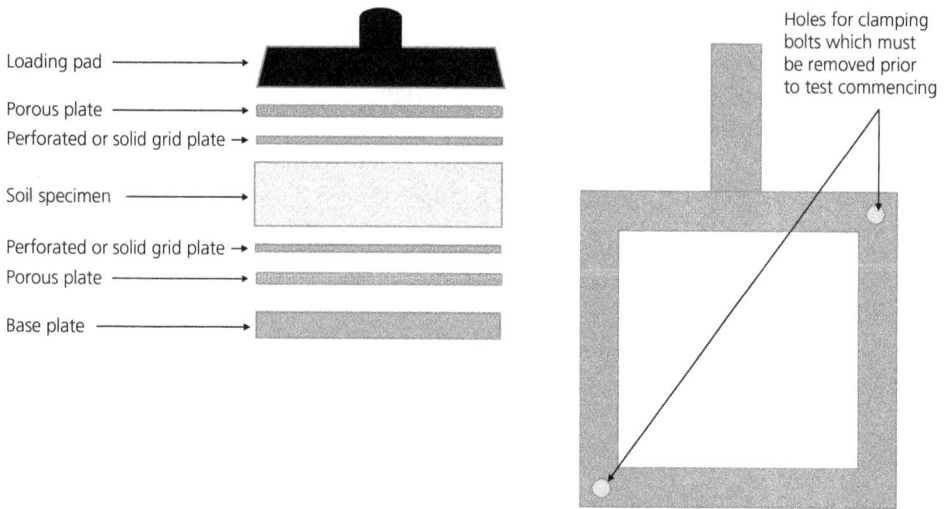

(b)

The test provides the peak total stress parameters. If a residual strength τ_{vr} is also required, the test is run beyond peak failure until a lower bound or residual value of total shear stress is obtained.

On completion of the test, it is conventional practice to determine the sample water content and soil density.

6.8.3 Shear box and related tests

The traditional shear box provides shear stress parameters from a square specimen, or more recently using a cylindrical specimen, both geometries being held in a steel box that has two halves. The upper half of the box is held in place, while the lower section is attached to a horizontal drive that causes the two halves to move relative to each. The displacement of the lower box at a constant rate of strain develops a horizontal shear surface in the sample. The test is conducted by measuring the load resistance or shear force in relation to the horizontal displacement. A normal stress is usually applied by a lever arm system and yolk arranged to provide a vertical force onto a top plate that is in contact with the upper surface area of the specimen under test. The vertical displacement is also measured during the test. Diagrams of the shear box setup and cross section are shown in Figure 6.30.

When the test is conducted at a relatively fast rate of strain, on cohesive soils, total stress parameters are obtained. Effective stress parameters may be obtained for granular soils, or on cohesive soils by testing at a sufficiently slow rate. It should be noted that the triaxial test, which enables the measurement of the pore water pressure, is a preferred method of obtaining effective stress parameters for cohesive soils.

Shear box tests using a standard 60×60 mm box can be conducted on soils such as clay, silt and sand. Soils that contain coarser particle sizes can be tested but will require larger boxes. Typically, these would be 100 mm, 300 mm or 600 mm square, although much larger boxes have also been constructed. Large boxes introduce manual handling issues, in that the weight of the box may require lifting equipment to set the apparatus up. Such arrangements are generally used for research or specific contract requirements.

To set the test up, a specimen is prepared generally by using a steel mould with a cutting edge, which is carefully pushed into an undisturbed sample while removing surplus soil with a paring knife. The specimen once prepared is pushed from the mould into the box. This requires some skill to achieve without disrupting the specimen.

An alternative is to compact the soil into the box such that the soil achieves a prescribed density at the requisite water content. The compaction is normally carried out using a square tamping rod and applying an even distribution of force over the entire sample area.

A normal stress is applied by way of the leaver arm and yolk. This is important because without the vertical force the specimen will tend to dilate resulting in soil particles being forced upwards as failure progresses and the stress along the horizontal failure plane will be reduced.

Measurements of the mass of soil and the dimensions of the compacted sample are recorded to enable the test sample density to be determined. Knowing the density, particle density and the water content will enable the degree of saturation to be determined.

The sample is set into the split box, which is then placed in the carrier box, and the securing screws removed.

For a total stress test, a normal stress is applied, usually starting at 0.5 times overburden pressure (p_o). The test is conducted until the sample is seen to fail or the maximum travel of the apparatus has been achieved, and this provides a peak stress value. The test is generally performed on three specimens using increasing values of normal stress, providing at least three determinations of the peak stress. The normal stress is doubled for each test, typically giving normal stress values of $0.5 \times p_o$, p_o and $2 \times p_o$.

Shear stress = applied load at failure / cross-sectional area

There is some debate regarding the area at failure to be adopted in this equation. The standard uses the original plan area of the sample. However, some practitioners suggest that the area should be corrected for the change in soil area in contact between the top and bottom parts of the box. The standard ignores this change of area, assuming that the soil will still contribute to the total resistance to failure, even if part of the specimen is in contact with the box rather than soil, and thus the area remains unchanged.

Further tests can be conducted to provide additional data, with each test being carried out at an increasingly higher normal stress value. From the results, a graph of shear stress against normal stress can be made (see Figure 6.31) and by drawing the line through the points the angle of shearing resistance can be measured. Care should be exercised when using particularly low normal stress levels because the behaviour is likely to be nonlinear.

Figure 6.31 Results from the shear box tests on four specimens each of (a) sand and (b) clay soils (note the intercept on the y-axis) (authors' image)

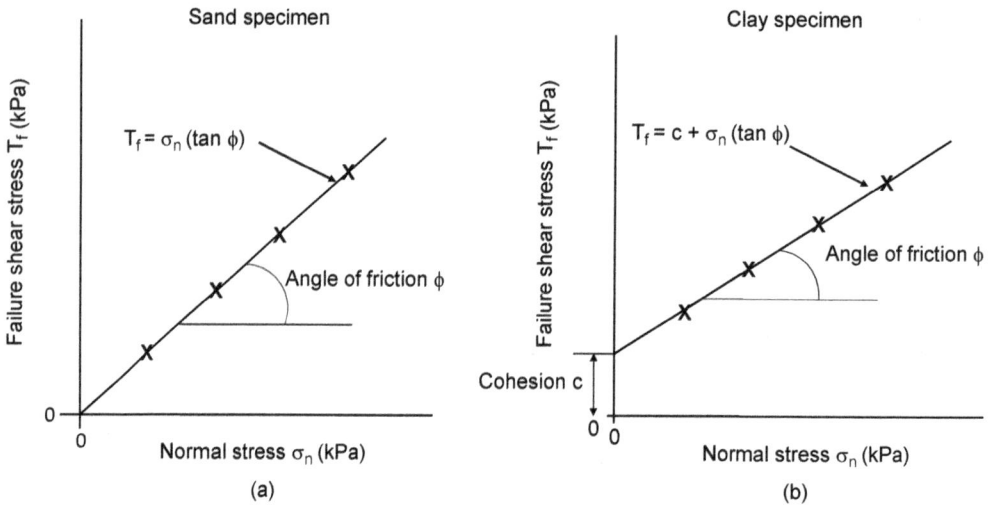

The shear box can be performed as a quick test to provide the total stress condition. The specimen is saturated by filling the box with water. A normal stress is applied, and the shear process is started immediately (i.e. without allowing the sample to consolidate). Values of the peak stress are plotted against the normal stress to provide cohesion and the angle of shearing resistance, ϕ, if applicable.

Alternatively, an effective stress test can be conducted by consolidating the specimen and running the test at a slow rate. The sample is initially saturated and then a normal load is applied. The vertical change in height is monitored until consolidation is assessed as complete. The shear test is then started. The applied rate of strain is determined from the consolidation stage, using the time settlement curve to ensure the pore water pressure does not increase during the shear stage. Typically, for the effective stress test the normal stress applied would start using p_o', and each subsequent sample being tested at twice the normal stress used for the previous test.

From the results of the shear stage, graphs of stress against strain are plotted for each normal stress stage. Plotting the peak stress against normal stress, values of effective cohesion (c') and effective angle of friction (ϕ') are obtained. The peak stress values should plot as a straight line, as shown in Figure 6.31, the resultant providing the parameters of cohesion (the intercept on the y-axis) and angle of friction (the angle of the tangent related to the horizontal). Note: the scale on both axes must be the same.

Determination of the residual strength can be made by running the shear box backwards and forwards with several reversals until a lower bound value is achieved. This process can be speeded up by manually reversing the sample by hand using a handle to operate the worm gear that drives the apparatus. This process is repeated until the residual stress for two consecutive shear runs remains the same. This can be a lengthy process.

An alternative method to obtain a lower bound residual value of shear stress can be determined by removing the box separating the two halves and polishing the interface to give a smooth surface. The box is then reassembled and a shear stage conducted allowing the shearing resistance on the two polished surfaces to be measured. This lower bound value, which will approximate to the residual shear strength, is likely to be more representative of the shearing resistance along a failure plane such as that which might develop in a slope failure. An alternative would be to conduct a ring shear test: see Section 6.8.4.

A similar technique might also be used to determine the friction along a discontinuity in weak rocks, such as chalk. A failure plane can be introduced by forming a cut surface at the junction between the upper and lower box, which will enable the strength along weaker planes (such as natural fractures) to be assessed.

A useful application of the shear box is to determine the angle of friction between soil and an engineering material such as concrete. This is achieved by using a concrete block to take the place of the lower part of the shearbox and placing soil into the upper part of the box. The test procedure is carried out as described above. This is particularly useful when designing concrete retaining walls and other similar structures where the friction between the soil and the structure may play an important role.

6.8.4   Ring shear

As described in Section 6.8.3, conventional shear box tests may be run to determine the residual shear strength, which can be used to model the lower bound shear strength in cohesive soils along a failure plane. However, when this lower bound value is used to back analyse a slope failure, it is evident that this value is not sufficiently low to enable failure to occur, thus a lower value must be in operation. The ring shear was developed by Professor E. Bromhead to determine the lower bound shear strength value that would be in operation as a slope fails. When the slip surface develops pore suctions increase along the slip surface, which draws water from the soil on either side of the slip plane. This increases the water content through the plane and thus reduces the shear strength of the soil along the surface of failure.

The ring shear apparatus is configured with the drive motor positioned in the centre of the specimen housing such that the specimen can be rotated through 360 degrees or more. The resisting force is measured using two proving rings. A normal force can be applied in a similar way to the standard shear box test.

The test specimen, typically of fine-grained soil, is mixed to a water content close to its liquid limit, which is considered to be the limiting state and probably close to the soil state along the slip plane. The specimen is moulded into the housing to form an annular ring of soil, as shown in Figure 6.32. The top section of the sample is held in place by the proving rings that measure the resisting force, while the lower section rotates such that a shear plane develops between the two sections. Owing to the configuration of the specimen housing, the test is able to induce extremely high strains, similar to those experienced along a slip plane in a slope failure.

Figure 6.32 The ring shear apparatus and a diagram of the specimen with the forces applied shown ((a) courtesy of K4 Soils Laboratory; (b) authors' image)

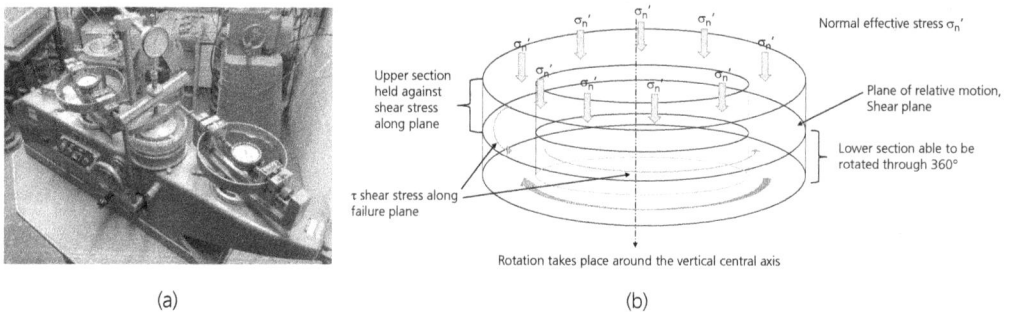

(a)   (b)

The test results are plotted as stress–strain curves, the stress being calculated from the average force recorded on the two proving rings over the cross-sectional area of the specimen. Tests may be conducted with a range of normal stresses similar to the standard shear box test.

6.8.5   Fall cone test

The fall cone test is described in BS EN ISO 17892-6. The test was initially developed and used in the USA and has been widely adopted in offshore geotechnics. The test is used to provide an indication of undrained shear strength from undisturbed samples of soil. The soil tested should be intact, fine-grained and free from gravel or other inclusions such as shells. The test is particularly useful to provide rapid determinations in the field, when testing in a laboratory is not practical. The test may be conducted on a sample retained in a sampling tube or on a recompacted sample held in a container, the dimensions of which should be at least 55 mm in diameter and 30 mm deep. By using a recompacted sample the remoulded strength can be estimated.

The fall cone apparatus is similar in form to the liquid limit fall cone apparatus. However, the cone mass and tip angle are varied according to the strength of the soil to be tested, as shown in Table 6.13.

Table 6.13 Cone sizes used in the fall cone test

Mass (g)	10	60	80	100	400
Tip angle (degrees)	60	60	30	30	30

To conduct the test, the cone tip is brought into contact with the top of the test specimen. The cone is released for 5 s and then locked. The penetration of the cone into the soil is recorded to 0.1 mm. If the cone penetrates more than 20 mm a lighter or higher angle cone should be used. Should the penetration be less than 5 mm then a heavier or sharper cone should be used.

The undrained shear strength can be calculated using the following formula

$$c_u = c\, g\, m\, /\, i^2$$

Where $c = 0.8$ for the 30° cone and 0.27 for the 60° cone

$$g\ \text{gravity} = 9.81 \text{m/s} \qquad m = \text{mass of cone grams} \qquad i = \text{cone penetration mm}$$

The calculated value is normally corrected in relation to the liquid limit, w_l, using the formula

$$c_{u(corr)} = c_u \times \mu \text{ where } \mu = \left[0.43\,/\,w_l\right]^{0.45} \text{ and } 1.2 \geq \mu \leq 0.5$$

While this test is included in the BS EN ISO 17892 suite, its use in the UK is limited and the results obtained are not considered suitable for geotechnical design.

6.8.6 Unconfined compression test

The original unconfined compression test was devised as a simple total stress test, which could be conducted either in a laboratory or in the field, using a portable device that applied a load against a calibrated spring. The sample, normally 38 mm in diameter, is obtained by driving a 38 mm diameter sample tube using a handheld jarring link. The sample is extruded from the tube and trimmed to provide a right cylinder of soil, 76 mm in height. The test specimen is placed into the apparatus and the shear stress applied by turning a handle by hand which raises the platen against a spring mechanism attached to a lever arm. This traces the increase in deviator stress against strain (change in sample height).

Because the sample is not confined, the result gives a lower bound value of shear strength which is not considered appropriate for design of structures. It is generally considered that this test only provides an indication of the soil strength and should not replace the triaxial test, as described in Section 6.8.7.

Although rarely requested, the unconfined compression test can also be conducted in the triaxial apparatus (see Section 6.8.7) without the application of a cell pressure. The test procedure is described in BS EN ISO 17892-7.

6.8.7   Triaxial testing

6.8.7.1   Introduction

As the name implies, the triaxial test applied stress to the test specimen in three directions, similar to the stresses experienced by an element of soil at a depth z below the surface, as shown in Figure 6.33. There are three principal stress directions: x, y and z.

In soil mechanics, these stresses are termed σ_1, σ_2 and σ_3 and these translate into the normal stress applied in compression, σ_1, and the cell pressure. Because the test specimen is cylindrical, the two orthogonal principal stresses σ_2 and σ_3 are coincident and equal and are represented by the lateral confining pressure or cell pressure, which is referred to as σ_3.

This test is used to obtain the failure strength mainly of fine-grained soils, and is normally carried out on undisturbed samples, requiring a sample of Class 1 (as defined by BSI 2008a) to provide a reliable result. For these tests to be meaningful, it is important to ensure that minimal disturbance is caused to the sample from the point it is recovered from the ground through handling, transportation, storage and preparation. Otherwise, a considerable amount of time and money will be expended for results that are unreliable.

Triaxial tests are conducted by sealing a specimen within a latex membrane inside a cell which enables the specimen to be subjected to a confining pressure, applied using pressurised water. The cell is positioned on a load frame (Figure 6.34(b)) where the speed of uplift of the base platen is controlled by a series of gears or by a stepper motor on a worm gear. This is effectively a precision jack. The rate at which the platen is raised can be controlled to give a wide range of accurate speeds which are consistent and precise. As the platen is raised, the sample is held rigidly between the platen and the upper cross beam of the triaxial apparatus. A means of measuring the force applied is situated above the specimen. This is by way of a piston with a sealing ring in the top of the cell. The sample is pushed upwards against the load measuring device, which is fixed to a rigid cross beam mounted on high tensile steel securing rods that are bolted to a heavy metal plate in the base of the machine. The sample is normally protected by a latex membrane and housed in a Perspex cell capable of withstanding the high water pressures that may be applied. A confining pressure is applied to the specimen by filling the cell with water and pressurising it to the required value. The confining pressure simulates the stress that the sample would have experienced in the ground. Figure 6.34(a) and (b) show a typical triaxial cell set up.

More sophisticated systems allow a back pressure (i.e. an increase in pressure within the pore spaces of the specimen) to be applied by introducing water under the required pressure through the

Figure 6.33 The three principal stresses acting on an element of soil (authors' image)

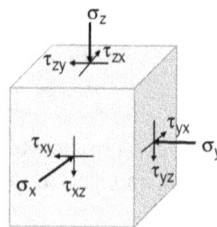

sample top cap and into the specimen. This is used to remove air from the specimen by balancing the pore pressure in the sample against the applied cell pressure. These systems also allow water to flow through the specimen by inducing a differential between the pressures at its top and base.

The latex membrane used to protect the specimen is secured to the base platen and top cap using rubber O-rings to provide a watertight seal.

Once the specimen has been secured, the cell is placed over it, ensuring the piston locates onto the button in the top cap. Again, care must be exercised while carrying out this operation to ensure the specimen is not loaded or knocked. If specialist tests are required further instrumentation can be attached to the specimen at this stage, and these are detailed in Sections 6.9.2 and 6.9.3.

The cell is clamped to the pedestal and can then be filled with water at atmospheric pressure. Once full the test procedure can begin.

It is normal practice to conduct the test on a single specimen testing at a single cell pressure. However, ideally the test should be conducted on three similar specimens with the cell pressure increasing from overburden pressure (OBP) and then OBP × 2 and OBP × 4. Earlier practice would involve testing three specimens cut from a single 100 mm diameter sample to give specimens of 38 mm diameter, enabling three Mohr circles to be drawn. It is now recognised that, although that approach is ideal for the failure envelope to be defined, the accuracy of the test at smaller diameters is compromised, particularly if the tested specimens exhibit a structure such as lamination or fissuring. There is much research that demonstrates that the larger the specimen the more reliable the test result. The primary reason for this is that it is more likely that natural structures, such as fissuring, will be present in larger specimens.

Considering the above, it is preferable to use three specimens of 100 mm diameter, taken in the same material and approximately the same depth. This will provide sufficient data to obtain the desired failure envelope.

Figure 6.34 (a) The component parts of the triaxial cell; (b) cell with sample ready for testing (authors' images)

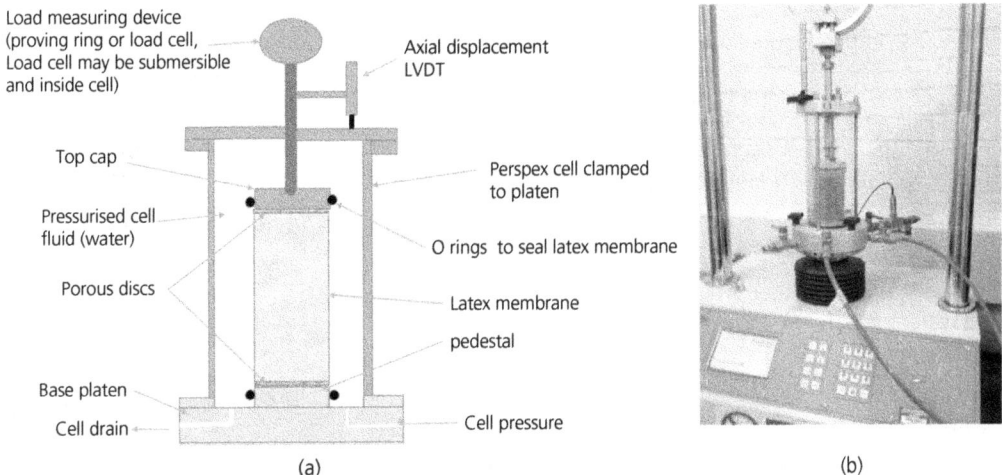

The above procedures are carried out in a similar manner for all triaxial tests.

The triaxial test can be conducted as either a total stress or effective stress test. For a total stress test, the load is applied rapidly during the shearing stage and pore pressure is not measured. The sample is undrained, hence the volume remains the same throughout the test.

The effective stress test is usually conducted in an undrained condition on cohesive (fine-grained) soils. The specimen needs to be saturated to remove air and then consolidated to a predetermined stress by increasing the cell pressure. The test is run to enable pore water pressure to respond to the change in stress. Pore water pressure is measured during the test, which enables the effective stress to be determined.

The drained effective stress test is normally conducted on free draining soils such as sand and sandy silt. The specimen is also saturated and consolidated before beginning the test. The pore water pressure is zero, and volume change is measured as the water content changes. The rate of shearing needs to be sufficiently slow to enable the water to drain and will be dependent on the soil permeability. During shearing under drained conditions volume changes of the sample are recorded, rather than the measurement of pore water pressure.

A summary of the test conditions discussed above is provided in Table 6.14.

6.8.7.2   Total stress triaxial testing

The unconsolidated undrained triaxial test is conducted to provide total stress parameters. Because the rate of strain is rapid, this test is often termed 'quick' triaxial testing. The effect of the pore water pressure is not considered. This test is commonly conducted to provide the undrained shear strength, S_u, used to calculate the ultimate bearing pressure which might be applied to the ground. This parameter is also useful when determining the load carrying capacity of piles, particularly where these are to be friction piles placed within cohesive soils.

The specimen, which should ideally be from a Class 1 sample, is prepared as a right cylinder with a height equal to twice its diameter (as shown in Figure 6.35). Care must be taken not to disturb the specimen as the membrane is put in place. This takes some considerable skill and can be quite

Table 6.14 Test configuration for triaxial tests

Test type	Total stress	Effective stress	
	Unconsolidated (Quick) Undrained Triaxial (UUT)	Consolidated Undrained Triaxial (CUT)	Consolidated Drained Triaxial (CDT)
Saturation	X	✓	✓
Consolidation	X	✓	✓
Pore water pressure measurement	X	✓	X
Volume change measurement	X	X	✓
Rate of strain	Fast	Depends on consolidation	Slow

a difficult operation if the specimen is soft. The ends of the cylinder of the test specimen must be flat and parallel, and this is achieved by using a metal mould of the required size and very careful trimming. Once trimmed the specimen is removed from the mould, its dimensions measured, and weighed so that the density can be determined. After the completion of the test, a small amount of material will also be taken for water content to be determined.

The prepared specimen is placed on the triaxial pedestal and the membrane folded down and secured in place with rubber bands. The top cap is also secured in a similar fashion. The cell is carefully lowered onto the test specimen, ensuring the cell piston locates onto the button in the end cap, again care being taken to ensure that the sample is not disturbed or loaded by this process.

The cell is clamped to the pedestal and then filled with de-aired water. Once full, a specified confining pressure can be applied. This is usually by way of an air pressure system and an air-water interface although compression pumps are now available as an alternative to the use of the interface system. Once the cell is filled and pressurised, the test can start.

Tests conducted as 'quick tests' apply a fast rate of strain, approximately 2% per minute, to provide total stress parameters; only the total stresses and related strain are measured. At this speed, the pore water pressure increases, but the sample remains undrained, particularly when testing cohesive soils. This situation is comparable to shallow foundation construction where the application of load is relatively rapid. The test results are plotted as stress against strain, as shown in Figure 6.36.

Figure 6.35 Equipment for preparation of the triaxial specimen, including callipers, preparation knife, end caps, membrane and membrane stretcher (image courtesy of M. Wright, K4 Soils Laboratory)

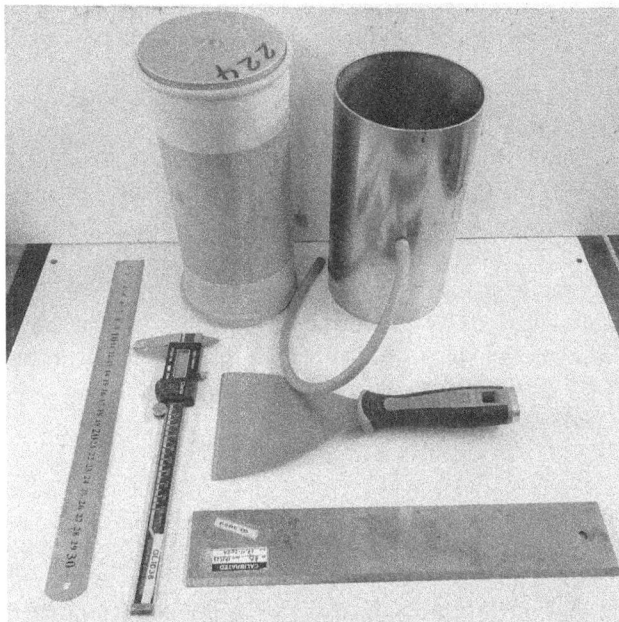

Figure 6.36 Typical stress–strain curve from a triaxial test (authors' image)

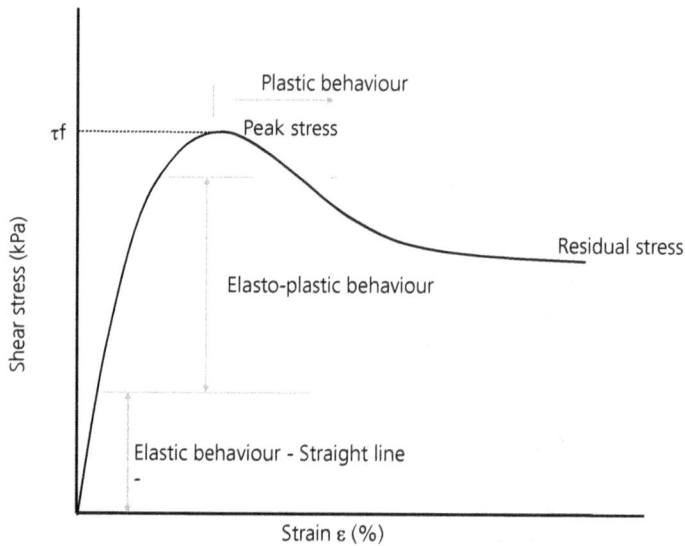

Tests should be conducted as single stage tests, at one confining pressure. Methods that use multistage testing on a single sample have been conducted where the shear stages at three different confining pressures are applied, the test being advanced to the next confining pressure just before failure is reached. Although this method has been used, it is not recommended.

The test results are reported as the strain at failure and the maximum deviator stress. The deviator stress is the difference between the axial stress, σ_1, which increases during the shearing process, and the cell pressure, σ_3, which remains constant during the shearing process. The undrained cohesion, c_u, is determined as:

$$c_u = \text{deviator stress}\,/\,2 = (\sigma_1 - \sigma_3)\,/\,2$$

Typically, a graph showing the shear stress against strain is plotted. This is useful by graphically showing the strain at failure (see Figure 6.36).

In general, the normal stresses and shear stresses are presented graphically as Mohr circles, which define the individual specimen's test behaviour in terms of the principal stresses. The full circle could be drawn to represent the full stress distribution for the sample, the area above the x-axis defining the compressive zone while that below defining extension.

The diameter of the circle $=$ deviator stress $= (\sigma_1 - \sigma_3)$

From the Mohr circle, the cohesion intercept, c_u, and angle of friction, ϕ_u, is measured. The results obtained are dependent on the material being tested.

For the case obtained when the test is conducted in the undrained condition, the angle of friction $\phi_u = 0$. In this case, c_u is generally referred to as the undrained shear strength S_u. This is shown

Figure 6.37 Results of a clay sample in the quick undrained test (unconsolidated undrained) plotted as Mohr circles and demonstrating $c = (\sigma_1 - \sigma_3)/2$ and $\phi = 0$ degrees. The cohesion value is generally referred to as the undrained shear strength: Su (authors' image)

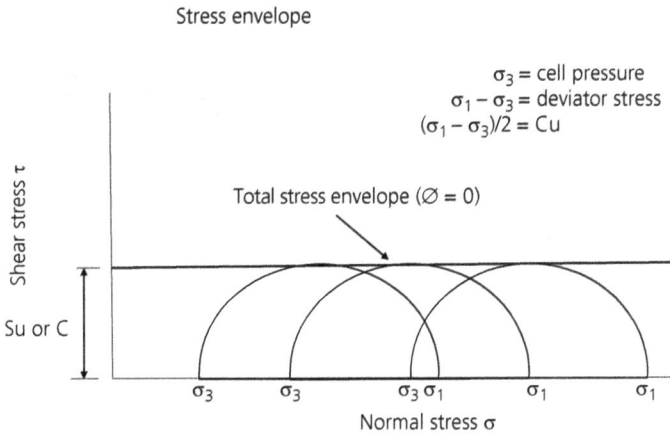

in Figure 6.37. However, it is possible to obtain values of c and ϕ if the soil is granular or contains a high proportion of coarse particles. There is a risk that the lack of full saturation may develop frictional properties, as air in some pore spaces would compress under different cell pressures, allowing the sample to become denser under higher pressures, even although no water is being allowed to drain from the specimen. However, this apparent friction should be ignored for design purposes.

The result shown in Figure 6.38 is typical of a sand where the cohesion intercept is zero and the material exhibits purely frictional properties.

The following are usually reported with the graphs of deviator stress and Mohr circles

- cohesion, c_u
- angle of friction φ_u
- confining pressure for each stage
- failure stress
- specimen density
- water content
- a visual description of the tested specimen
- the mode of failure (shear/plastic/compound).

In practice, the loading from a foundation or other source will cause the pore water pressure to increase and then slowly dissipate over time. This causes a reduction of the void space and hence particles realign and pack closer together, resulting in an increase in strength. This process is modelled by the effective stress test where the sample is allowed to consolidate in response to the increase in applied stress.

Figure 6.38 Results of a quick undrained test from a sand, plotted as Mohr circles and demonstrating $c = 0$ and ϕ degrees (authors' image)

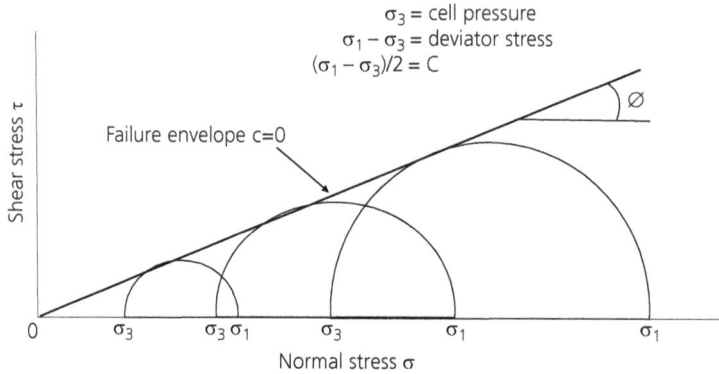

6.8.7.3 Effective stress triaxial testing

INTRODUCTION

The test described above, carried out quickly, provides a determination of the undrained shear strength in terms of total stress. However, many situations require consideration of the influence of the applied pressure on the effective stresses in the soil. Such tests are conducted at a slower rate, either to allow the pore water pressure to respond or to allow drainage and to provide effective stress parameters. These parameters are needed for many applications in civil engineering and enable the long-term behaviour of the ground to be predicted.

Effective stress testing using isotropic consolidation, achieved by applying a cell pressure that acts in all directions onto the specimen, has become a standard laboratory test. However, the procedure usually adopted does make some assumptions that, by using more advanced methods, can be minimised. In addition, parameters commonly required for numerical solutions can be obtained by adaptions to the standard test procedures and methods. The commonly used effective stress test methods are described here, and the more advanced methods, using bender elements, small strain measurements and cyclic loading, are described in Section 6.9. Typically, the test will be logged using a data logger, and modern equipment may be programmed to run the tests, applying prescribed stages as the test proceeds.

The test equipment will include several ports in the base platen, which enable pore water pressures to be measured, the cell to be filled with water, which is then pressurised, and through the top cap, allowing the ability to pressurise the specimen internally.

As with total stress testing, effective stress tests can be carried out either on a single specimen at a single confining pressure, or as a set of multiple specimens, ideally a set of three, each tested at different confining pressures, or as a multistage test on a single specimen. While the undrained shear strength of a soil can be assessed using a single stage of testing in terms of total stress, effective stress tests typically require three test stages in order to produce a failure envelope that will define the required parameters of effective cohesion and the effective angle of friction. Figure 6.39 shows the stresses on the specimen during testing.

SAMPLE PREPARATION

All effective triaxial testing requires the specimen being tested to be as follows.

- From a Class 1, high-quality sample, with little or no disturbance.
- Saturated, to reduce or eliminate the air present in the voids' spaces.
- Consolidated to an effective pressure at least equal to the original effective overburden pressure in situ.

Class 1 samples are discussed in Chapter 2. Sample preparation is carried out as described in the previous section. Before fitting the membrane, a filter drain can be fitted to the outside of the specimen. This will not adversely affect the results, but it does speed up the consolidation process by reducing the distance of the drainage path, thus enabling water to escape the sample relatively quickly. Corrections are applied for the effect of both the membrane and the side drains.

The sample is placed on the pedestal, which is fitted with a pore water pressure transducer. This may be fitted with valves either side of the pore pressure transducer, allowing it to be bypassed by opening both valves to permit flow through and into a volume change device.

Once the specimen has been set up in the triaxial cell, it is conditioned to bring the test parameters close to the conditions the soil was experiencing in the ground. This is a two-stage process of saturation (see 'Saturation' section) and consolidation (see 'Isotropic consolidation' section). This is followed by shearing, which can be undertaken either as drained (see 'Drained test with volume change measurement' section) or undrained (see 'Undrained test with pore water pressure measurement' section), the choice of which is based on the soil type and the conditions expected on site when the soil is under loading.

Figure 6.39 Diagram showing the stresses on the specimen under effective stress conditions (authors' image)

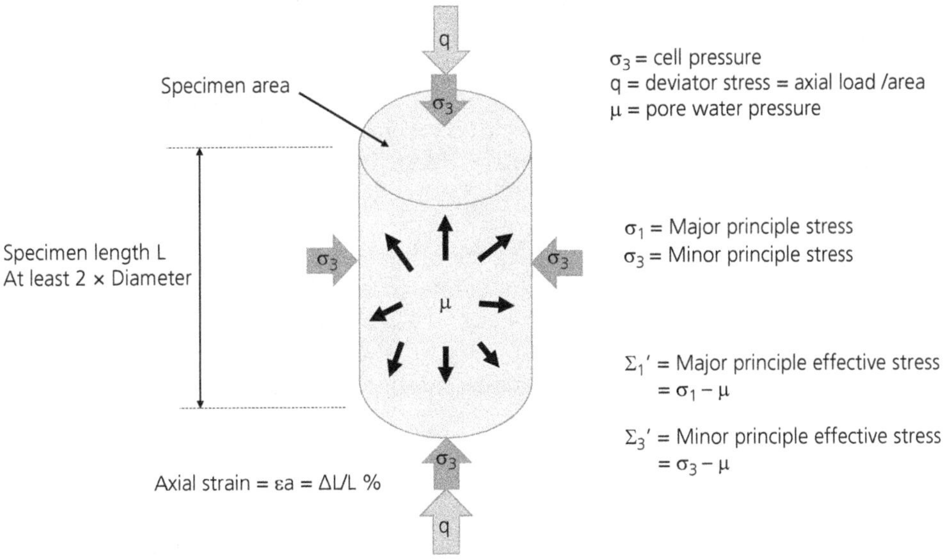

σ_3 = cell pressure
q = deviator stress = axial load /area
μ = pore water pressure

σ_1 = Major principle stress
σ_3 = Minor principle stress

Σ_1' = Major principle effective stress
= $\sigma_1 - \mu$

Σ_3' = Minor principle effective stress
= $\sigma_3 - \mu$

Specimen area

Specimen length L
At least 2 × Diameter

Axial strain = εa = ΔL/L %

SATURATION

Following sampling, it is almost certain that the sample will take in air, this being a consequence of the stress relief experienced by the sample as it is removed from the ground and transported to the laboratory. The stress relief the sample undergoes will develop negative pore water pressures within the pore spaces of the soil. Because the sample is now in an unsaturated environment, the suction developed will draw air into the sample. If the test is conducted without carrying out saturation, the air will compress as the specimen is loaded. This will induce volume changes that are not measurable and will introduce inaccuracies in the calculation of the sample volume and voids ratio. Ultimately, this effect can give erroneous measurements of the shear stress and will not be representative of the in situ soil strength.

The saturation process is carried out to remove any air present from the specimen without affecting the stress state. This is achieved by applying a pressure, known as a back pressure, to the pore water inside the test specimen. However, in order that the back pressure does not disturb the sample, it needs to be balanced by the water pressure applied to the water that surrounds the sample (the cell pressure). In practice, the back pressure is set just below the cell pressure, the procedure starting with a low cell pressure of say 50 kPa and the back pressure set at 45 kPa. The pore water pressure is measured by way of a transducer attached to a port that opens in the base plate and beneath the sample. The pore pressure parameter, B, which is the ratio of the pore-water pressure response to the increase of the cell pressure, is calculated. This effectively represents a measurement of saturation. The cell pressure and back pressure are again increased, and this process is continued until almost full saturation is measured – that is, the value of B is greater than 95%. Other less commonly used means of saturation include increasing the cell pressure only or by rapidly raising both cell pressure and the back pressure simultaneously to arrive quickly at the point where all air will go into solution.

ISOTROPIC CONSOLIDATION

Once full saturation has been achieved, the sample can be consolidated. The consolidation process can be the same in all directions – that is, isotropic, or alternatively anisotropic, where the vertical pressure is different to the horizontal pressure on the specimen (as discussed in the 'Anisotropic consolidation' section below). During consolidation, volume change is allowed and is measured by opening the valve on the back of the pore pressure transducer and allowing the volume to change that is allowing the pore water to be expelled through the base of the sample. The volume change is measured using a volume change device, which is generally a chamber with a rolling diaphragm that raises an arm attached to the diaphragm. The change in height is directly proportional to the change of water volume in the chamber.

The consolidation pressure is set by adjusting the cell pressure above the pressure achieved at saturation by the amount required to represent the stress at the level the sample was obtained in the field.

The time to failure and thus the rate of test for the shearing stage can be estimated from the consolidation characteristics using the formula

$$T_f = \left(\text{drainage factor} \times t_{100} \right) / \left(1 - U_f \right)$$

Where t_{100} = time for consolidation to be 100% completed; U_f = drainage factor; T_f = time to failure.

Typically, the drainage factor is taken as 0.721 for drainage at one end and radially (Bishop and Henkel, 1962).

UNDRAINED TEST WITH PORE WATER PRESSURE MEASUREMENT

Once consolidation has been completed, the sample should be in a condition that reflects the at rest state in situ, such that the sample is fully saturated and is confined at the effective overburden pressure equivalent to the depth from which the sample was taken. On completion of the consolidation, the shear stage may be started. This is carried out by increasing the axial pressure on the specimen (i.e. the deviator stress) until failure is achieved. It is conducted at a rate of strain (machine speed) calculated from the time taken to consolidate the specimen. This will ensure that the pore water pressure is able to respond to the changes in shear stress as the sample is compressed, measuring the pore water pressure throughout the shear stage. These tests can run for a long time, often taking a day for saturation and then consolidation and several days for the shear stage to complete. Test durations may range between four days to in excess of 20 days in some soils. As total test durations are difficult to predict accurately, testing laboratories typically charge a rate per test (or stage), which cover the first three or four days of testing, plus an additional rate for each additional day required to complete the test. It should be remembered that the duration and cost of such tests could be significantly greater than the basic values that might initially be assumed. These considerations should be taken into account when planning and budgeting for such tests.

DRAINED TEST WITH VOLUME CHANGE MEASUREMENT

Essentially, the shear stage for drained tests is carried out in the same way as for undrained tests, the main difference being that the sample is sheared at a sufficiently slow rate to allow drainage to occur throughout the shearing process. Volume change is measured, rather than the change in pore water pressure.

As the shearing stage is progressed, measurements of both shear stress and strain are recorded and it is normal practice to obtain plots of the stress and pore water pressure or volume change against strain as the test proceeds. Typically, these will displayed on a computer system in real time during the test.

Other plots often provided include stress path, Mohr's circles, total stress and effective stress.

STRESS PATH TESTING

While plotting of the failure condition during the effective stress test it is often more useful to study the soil behaviour within the failure envelope as stress and strain changes. To provide clarity it is simpler to plot specific points on the Mohr circle rather than the whole circle. Thus, it is possible to provide a more complete overview of the specimen behaviour during loading and the effect of the pore water pressure response. In this way, a single point can be used to represent the soil stress and strain state at any point during the test. A stress path plot presents the strain and pore pressure response as stress is either applied or reduced, thus enabling the soil behaviour under variable loading conditions inside the failure envelope to be studied before failure is reached.

The Mohr circles can be viewed as a series of circles that increase in diameter in response to an increase in loading until failure is reached, where the circle has its greatest diameter.

There are two methods used to follow the stress path: the Cambridge Method (after Cambridge University, UK) where p and q are plotted (see Figure 6.40) and the MIT method (after the Massachusetts Institute of Technology, USA) where s and t are plotted. Each give slightly different plots for the same data.

Figure 6.40 (a) Mohr circle plot with (b) corresponding Cambridge p-q plot (authors' images)

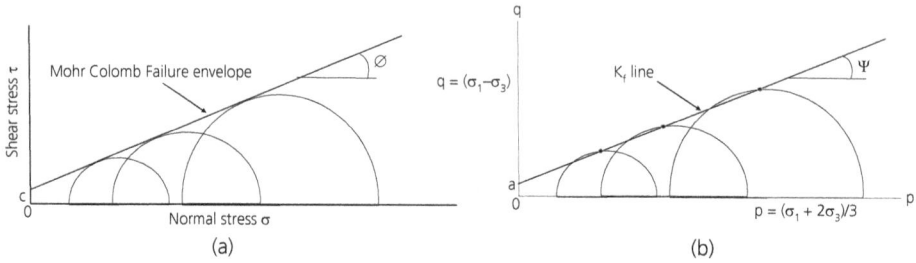

Note the K_f line is drawn at the peak of the circles.

Cambridge stress path method using p and q

Where $p = (\sigma_1 + 2\sigma_3) / 3$ or $p' = (\sigma_1' + 2\sigma_3') / 3$ and $q = (\sigma_1 - \sigma_3)$ or $q' = (\sigma_1' - \sigma_3')$

Thus $\sigma_1' = p' + 2/3q$ and $\tan \eta = q/p'$

$\sin \phi = 3 \tan \eta / (6 + \tan \eta)$

MIT method using s and t

Where $t = (\sigma_1 - \sigma_3) / 2$ or $t' = (\sigma_1' - \sigma_3')/2$

and $s = (\sigma_1 + \sigma_3) / 2$ or $s' = (\sigma_1' + \sigma_3')/2$

Thus $\sigma_1' = s' + t$ and $\tan \theta = t / s' = \sin \phi$

Both methods enable the stress path to be plotted as either total or effective stress fields.

ANISOTROPIC CONSOLIDATION

As shown in the Mohr circle diagrams, the compression tests plot the Mohr circle as semicircles, with both axes in the positive segment. However, as the name implies, the full failure zone is described by a full circle. Thus, the Mohr circles also extend below the x axis into the negative segment to become full circles. The zone below the x axis describes the area where the specimen is in extension. Using stress path testing it is possible to subject the specimen to extension within this segment.

This is particularly useful to model the soil stress state under a reduction of loading for instance, where swelling may occur such as in deep excavations or tunnels.

It is usual first to saturate and then isotropically consolidate the specimen back to its original stress state. Anisotropic consolidation is then conducted by increasing the cell pressure σ_3. Failure under triaxial extension can be achieved by decreasing the vertical stress p while increasing the cell pressure σ_3 until failure occurs. It is conventional to plot the stress path rather than the Mohr circle, as shown in Figure 6.41. The stress path plots the tangential point on the Mohr circle for each specimen as loading increases and until the failure envelope is met.

Figure 6.41 p q plot showing both triaxial compression and triaxial extension fields (authors' image)

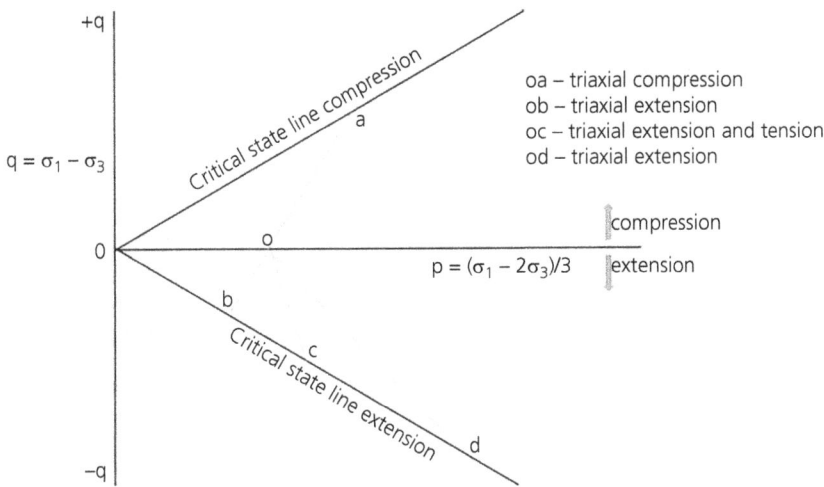

To perform extension tests the specimen top cap needs to be fixed to the cell piston.

Figure 6.41 includes the failure envelope, which is defined by the critical state lines (CSL) for extension, q (−ve) and CSL compression q (+ve). These define the range within which the soil state can exist. Stress path testing enables the conditions for the soil to exist at any point within the boundaries. On meeting the boundary condition, failure will occur.

6.9. Advanced testing

6.9.1 Introduction

The tests here are described as advanced testing because they are less likely to be used during a routine ground investigation and are less commonly offered by the standard geotechnical laboratories. These tests all lie outside those described by BS1377 or the Eurocodes. Some of the triaxial variants are covered by Head and Epps (2014: Volume 3). With the exception of these few triaxial variants the other tests described here are not covered in any British Standards and the laboratories that are able to conduct these tests will either have their own in-house methods or will, where available, adopt the American ASTM procedures. Table 6.1 identifies which tests are covered by a standard.

With advanced computer modelling and numerical methods of assessing the behaviour of soil to changes in stress caused by foundations and excavations, the ability to understand the soil responses at small strain has become more achievable. The main advances include being able to measure the change in the cross-sectional area around the middle of the specimen as it is tested. This is achieved using linear variable differential transformers (LVDTs) positioned on a radial belt affixed around the centre of the specimen and operating inside the cell in addition to the change in height is also measured more accurately on the sample and within the cell. These measurements enable a greater accuracy in the measurement of the cross-sectional area and strain closer to the point where failure will develop. This is further discussed in Section 6.9.2.

Many numerical modelling methods, such as finite element analysis, have been developed around the basic soil stiffness parameters rather than the more conventional undrained shear strength and angle of friction. This advance has been in response to many observations, suggesting that this numerical modelling does provide a better understanding of the soil response and enables the defining of critical zones more accurately. This is particularly relevant to finite element analysis that looks at the total response in a body of soil rather than individual points within the soil mass. To enable the analysis, parameters are required to be assigned to small zones of the soil across an element grid. Frequently, however, these parameters have not been assessed and hence the analysis is conducted with a certain amount of supposition. For example, while some researchers have used tests such as the SPT to derive soil stiffness, this method must be treated with caution owing to the unreliable nature of the test (see Chapter 5).

The development of cyclic loading testing has been driven by the expansion of offshore engineering projects where structures are transmitting loads to the foundations that are repeatedly changed by the force of wind and tides. These influences demand an understanding of the soil's response to such variable loadings.

Dynamic cyclic loading is experienced by structures subjected to external forces such as wind, waves and earthquakes. Other dynamic loads may also be imposed by railways or vibrating machinery. The soil response to such dynamic forces may be very different to its response to static loading. In order to investigate the soil behaviour dynamic loading can be simulated in the laboratory tests such as triaxial and simple shear.

One of the key changes has been the advancement of both software and measuring equipment driven by microelectronic developments and the almost exponential increase in the speed and amount of data we are now able to collect and analyse.

Modern laboratories are now able to undertake automated testing, often with remote monitoring and control. Some of the recent developments include the following.

- Automatic control of the saturation stage whereby the pore water pressure response to an increase in cell pressure is monitored, with the back pressure automatically adjusted to balance the cell pressure. The process is performed automatically until a required degree of saturation is achieved.
- Automatically controlled k_0 consolidation, whereby the cell pressure is increased at a specified rate while the sample is maintained with no change in the radial strain, thus maintaining 'at rest' specimen stress conditions. To do this manually is extremely difficult.
- Automatic stress path control to provide a user specified stress path. This might include control of the specimen height and radial strain, the deviator stress and mean effective stress.
- Stress controlled cyclic loading is really only possible with automatic control of the testing process, with repeated load and unload cycles.

Conducting any of the following tests requires a clear understanding of what is required, and what can be achieved. The test methods require a high degree of skill and understanding along with precision equipment. Even with this input it is not certain that high-quality results will be obtained. In particular, the condition of the test sample when it arrives at the laboratory is paramount, requiring meticulous care throughout the whole process of sampling and transportation (see also Chapter 5).

6.9.2   Midheight pore water pressure and local strain measurements

Many of the assumptions made during the standard triaxial test can be reduced or eliminated by taking accurate measurement of the midheight pore water pressure and the local strain around the midpoint of the specimen.

Most cohesive soils exhibit elastic properties at small strain. As strain increases the soil begins to act in a semiplastic manner, meaning some deformation will be permanent. In the elastic range, the soil will behave elastically, returning to its original shape if the applied stress is completely removed. Once beyond the elastic range the soil's behaviour is termed elasto-plastic, with deformation partly elastic and partly plastic. Once failure is reached, deformation becomes purely plastic and permanent.

In the standard consolidated undrained triaxial test, the pore pressure is measured at the base of the specimen and through the base platen. However, the pore water pressure will vary throughout the sample and, in particular, in the zone close to the failure plane. It is therefore more accurate to measure the pore water pressure at the specimen's midheight position. This is particularly important when using bender elements and small strain tests used to determine the elastic parameters.

In addition, for the standard test, axial strain is measured outside the cell and radial strain is derived by assuming the specimen deformation is as a right cylinder. The specimen cross-sectional area is calculated using the axial strain and the sample volume corrected for the change in volume measured during the consolidation stage of the test. This assumption is obviously in error and at small strain greater accuracy is required. To overcome these errors, the strain measurements are conducted by securing radial and axials strain gauges directly onto the sample.

The devices used to provide these additional measurements are quite difficult to set up, being attached to the specimen inside the cell with cables fed through ports in the pedestal to enable measurement. Both require the protective membrane to be punctured, which introduces issues in maintaining a watertight seal on the specimen. Each puncture through the membrane must be resealed with latex rubber glue to form a watertight seal around any penetrations through the membrane (see Figure 6.42).

At small strains, the soil will behave in an elastic manner, therefore if radial and axial strains are measured then Poisson's ratio can be calculated.

$$\text{Poisson's ratio} = \nu = \text{radial strain} / \text{axial strain} = -(\Delta R / R) / (\Delta L / L)$$

The relationship between the three stiffness constants for a soil is given by

$$G = E / 2(1 + \nu)$$

Where G = shear modulus   E = Young's modulus = stress / strain

These parameters can be determined at small strains with the soil within its elastic range. These parameters are of particular use when studying the soil's response to dynamic loading, because G reaches its maximum value (G_{max}) at very small strains, as shown in Figure 6.43.

Because the stresses and strains being measured are very small, the devices used to measure them are very sensitive, and they are particularly affected by external influences such as extraneous

Figure 6.42 A sample set up in a triaxial cell ready for test (image courtesy of Geolabs Ltd)

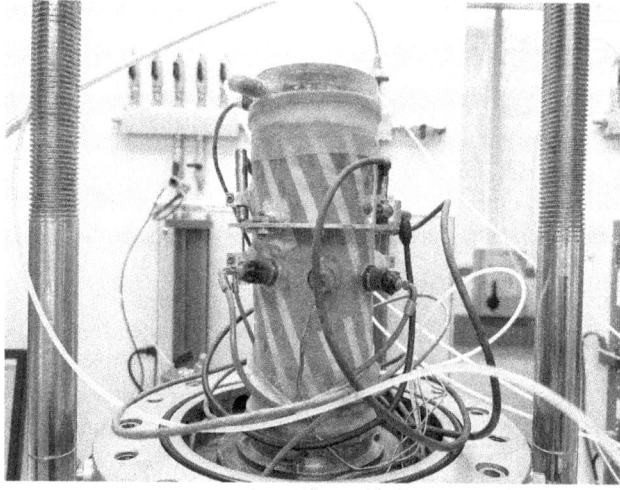

The radial belt and vertical strain gauges are clearly visible with the midheight pore pressure device and the horizontal bender element also positioned.

Figure 6.43 Graph showing the relationship of shear modulus to shear strain, plotted here on a log scale (authors' image)

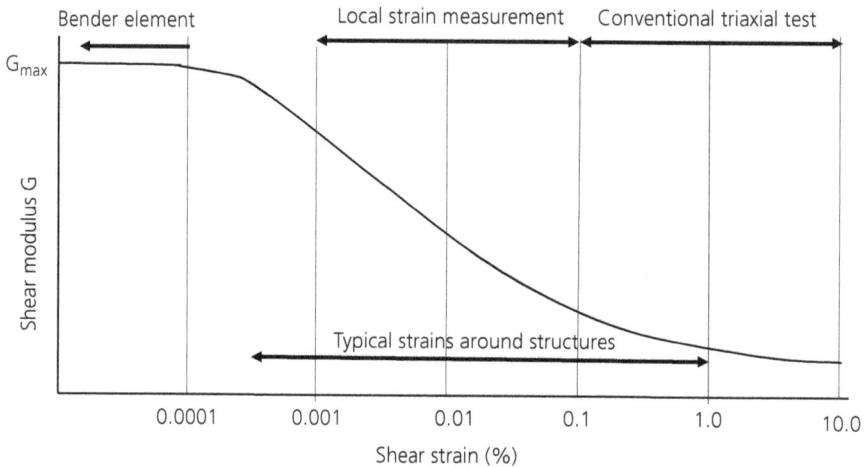

electrical currents, temperature changes and drafts. It is important that a consistent laboratory environment is maintained to avoid such influences.

6.9.3 Shear wave velocity measurement using bender elements

Using bender elements provides a nondestructive way to determine the shear modulus, G_{max}, within the elastic range. Typically, bender element testing enables the value of G_{max} to be determined at shear strains of less than 10^{-3}%. Bender elements can be used while conducting other tests on the same specimen such as triaxial or consolidation tests. This has the added advantage of allowing determinations of G at various stages of these other tests.

Bender elements are piezoelectric transducers, which are able to convert electrical energy into mechanical energy. This is achieved by passing a low DC electrical voltage through two ceramic plates separated by a thin metallic sheet. The voltage is applied alternately to one then the other plate causing the production of a pulse of energy, which takes the form of a sine wave. The generating element is attached to one side of the specimen and the receiving element is positioned on the opposite side. Alternatively, the generating element and the receiving element are set in the base pedestal and top cap, respectively. By knowing the distanced D, between the two bender element tips, and measuring the travel time for the energy pulse to reach the receiving element, the velocity of the pulse can be determined. To make these measurements, a function generator is used to generate the pulse, and a digital oscilloscope receives the signal. The oscilloscope enables the arrival time, t_{arr}, of the first pulse to be determined, by comparing the generated pulse and the received signal. Figure 6.44 shows a typical test result.

The value of G may be calculated using the following formula

$$G = \rho V_s^2$$

Figure 6.44 Example of the captured bender element signal from Clisp Studio software (redrawn from an image courtesy of VJ Tech Limited)

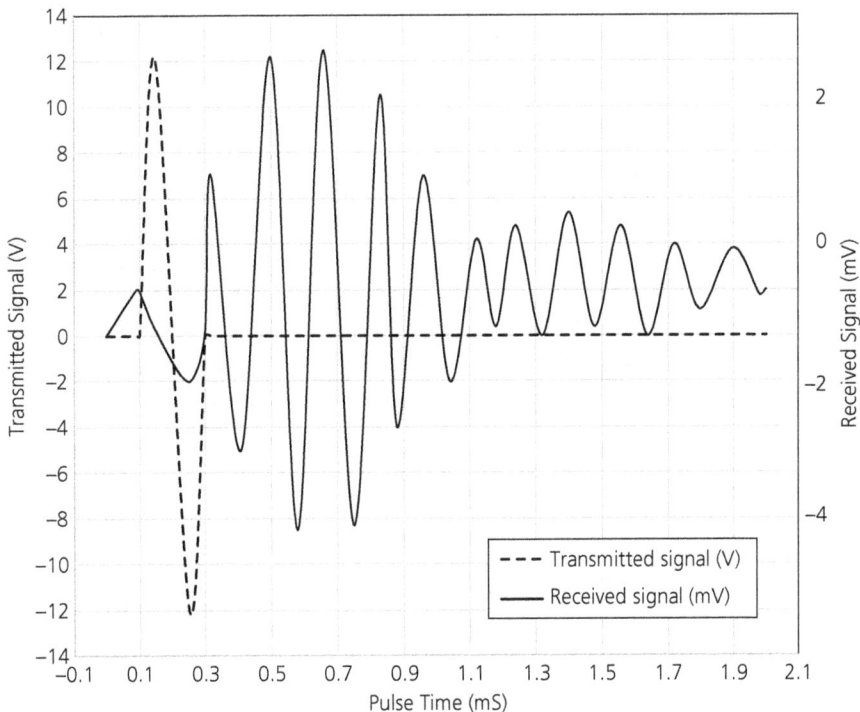

Where ρ = density

V_s = shear wave velocity = D/t_{arr}

D = distance between bender element output and input tips

t_{arr} = the arrival time of the first pulse

G is determined at various strain values to assess G_{max}.

The shear modulus anisotropy of the sample is the ratio of the horizontal to vertical shear modulus. This can be determined using two sets of bender elements, one pair in a vertical orientation with the elements set in the top cap and base plate, and the other pair set horizontally on opposite sides of the specimen, around its midpoint. Figure 6.45 shows the typical arrangement for the bender elements.

Many computer analysis programmes, such as finite element analysis, are based on elastic and small strain characteristics of the soil, and use parameters such as the Young's modulus, E, Poisson's ratio, v, and stiffness, G. These parameters are measured at small strains. As strain increases, the soil will behave in an increasing plastic fashion and a proportion of the deformation becomes permanent. This partially elastic and partially plastic behaviour is termed elasto-plastic.

Figure 6.45 Diagram illustrating the key elements of both horizontal and vertical bender elements (authors' image)

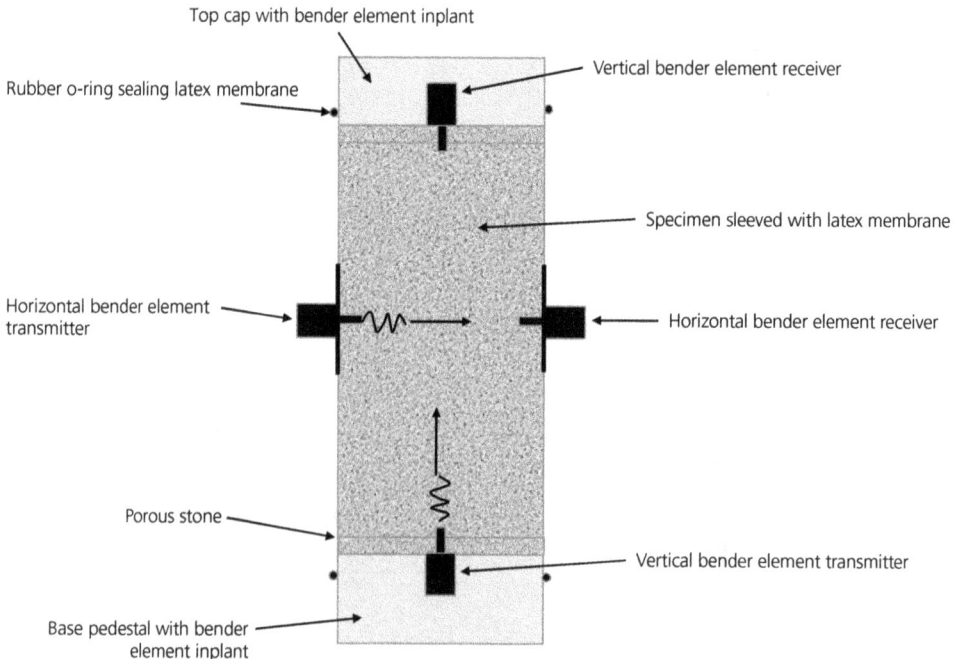

Some considerable skill is required when setting up tests for bender elements and small strain testing. The bender elements are inserted into the sides of the sample through the membrane. Sealing is achieved using a latex glue. The device can be incorporated with a pore pressure measuring element, termed a 'piezobender'.

With the bender elements set in position, and the cell filled with de-aired water, the test can begin. To conduct the test the specimen is fully saturated and consolidated, as described in the effective stress sections above. The diagrams in Figure 6.46 show the configurations of the bender elements and the direction of wave propagation. The shear wave velocity in any one orientation can be a function of the soil's structure and consolidation history. In Figure 6.46, the shear wave velocity is denoted as S (although in other texts it may be referred to as V_s). The suffixes relate to the direction the shear waves are polarised. One suffix is the direction of travel of the whole wave (propagation), while the other is the direction of particle displacement. Top to bottom is only S_{vh}, but across the sample it can be either S_{hh} or S_{hv}: both are S_h travelling/propagating horizontally, but the displacements can be either up/down or side to side.

Figure 6.46 Showing the differing orientations of the bender elements to provide shear wave velocities (authors' images)

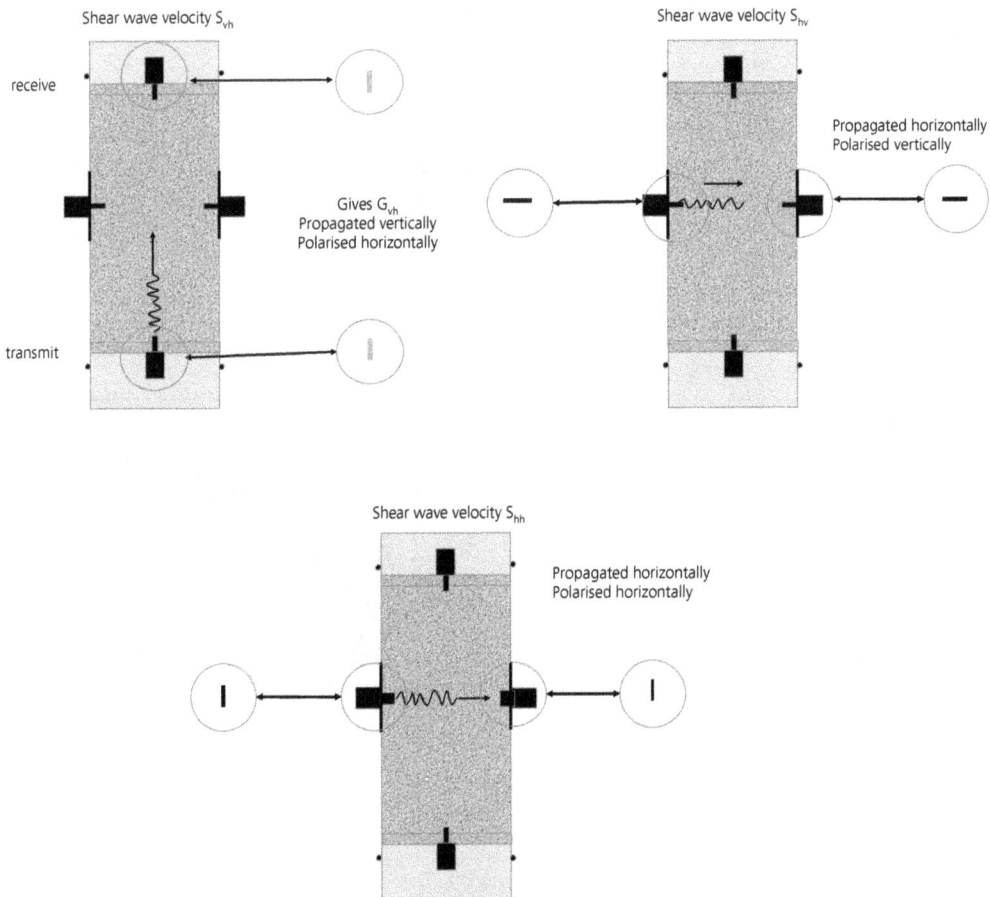

Because the received signal is significantly weaker than the transmitted signal, the interpretation of the travel time for the first arrival can be difficult, and very small inaccuracies in this determination can make significant differences to the resulting values of G (Gasparre, 2005). Typically, the determination will be automatically carried out, determined by the test computer programme.

6.9.4   Cyclic loading in the triaxial test

With the development of offshore structures, it has become important to understand the soil behaviour when subjected to repeated loading from waves and wind. Similar behaviour might also be seen in the context of a railway, with loading increasing and decreasing as the wheels of the train pass any particular point, and it is also seen to be generated by power station turbines. It has been found that under such loadings, failure may occur owing to strain hardening, which causes an increase in the yield stress, and applies to under consolidated and normally consolidated soils that compress. Over consolidated dilatory soils show strain softening and a reduction in yield stress. The failure mechanism, as shown in Figure 6.47(a), occurs through an increase in pore water pressure with load cycles and a consequent reduction in effective stress until failure occurs.

Test procedures have been devised to improve our understanding of the soil behaviour under these conditions, which enable design of foundations to withstand the forces involved.

In terms of the laboratory testing of soil samples, cyclic or dynamic loading refers to the process of applying an oscillating load by increasing and decreasing the deviator stress on the test specimen, as shown in Figures 6.47(a) and (b).

This type of test requires apparatus that are capable of withstanding the forces generated and is generally conducted using larger load frames designed on the same principles as the triaxial load frame. The tests require an actuator to generate the stress reversals situated on top of the cell

Figure 6.47 Typical results from a cyclic triaxial test showing (a) pore pressure response and (b) the p' - q' plot as the test progresses (images courtesy of GDS Instruments)

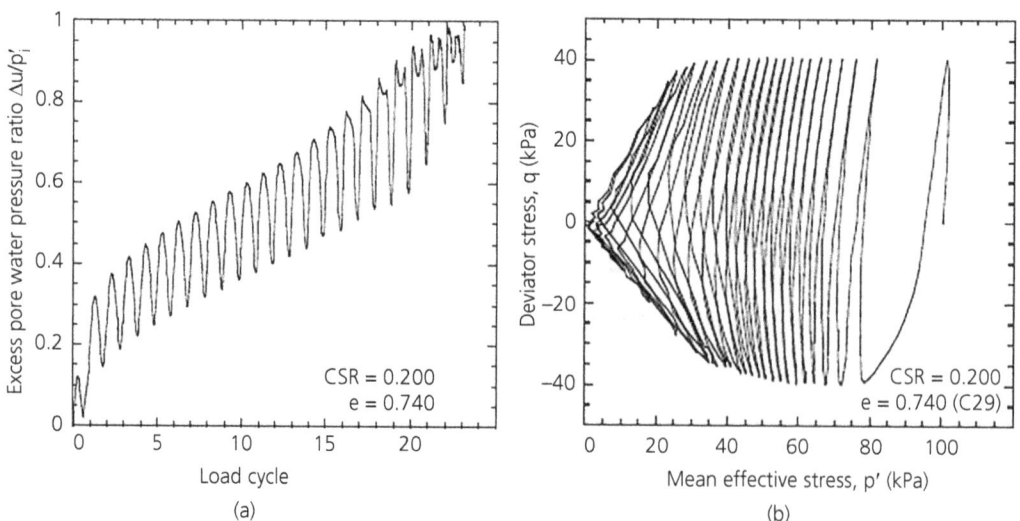

(see Figure 6.48). The load piston is fixed to the specimen top cap in a similar manner to the anisotropic test configuration. The loading cycles are faster than traditional static loading and the test therefore requires a suitable data acquisition system. The control system and actuator apply a uniform sinusoidal waveform and captures readings of stress and strain sufficient for the specimen response to loading to be recorded, typically at a rate of 40 readings per cycle or approximately 80 Hz. The cyclic loading can take two forms: one where the stress lies fully in the positive compression range and the other where the applied stress is fluctuated between compression and extension.

Figure 6.48 A cyclic triaxial test frame with specimen ready for testing (image courtesy of GDS Instruments and Geoloabs Ltd)

6.9.5   Direct simple shear testing

The simple shear test is similar in principle to the shear box test. However, rather than keeping the specimen confined in a rigid box and shearing across a defined plane, the simple shear test applies a horizontal force to the base of the specimen while holding the top stationary. The specimen cross section is allowed to deform from a rectangle into a trapezoid. This deformation is similar to that experienced by soil towards the lowest part of a slip surface on a failing slope, as shown in Figure 6.49.

Although the test specimen is not constrained laterally to the degree applied in a shear box test, some lateral support is provided, either by a stack of rings or a membrane with a reinforced coil within it. The restraint is provided to prevent the sample being displaced unevenly.

Figure 6.49 Changes of the orientation of the failure surface along the slip surface of a failing slope (authors' image)

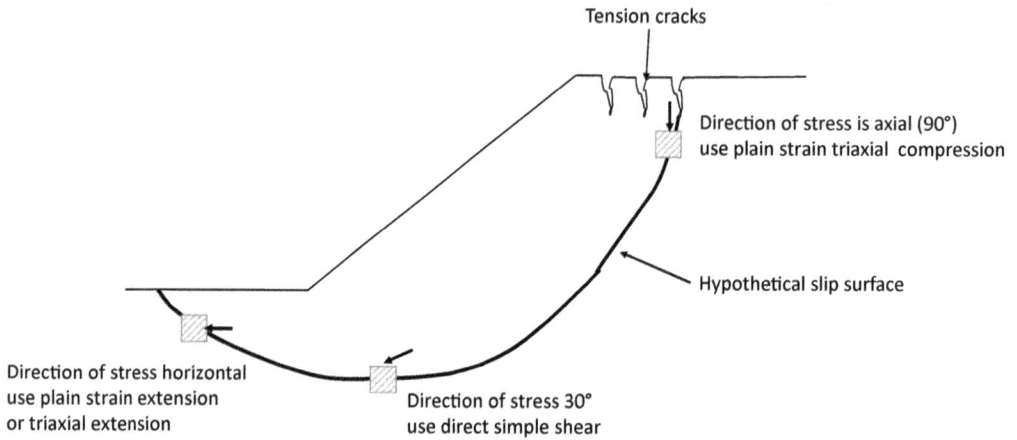

Tension cracks

Direction of stress is axial (90°) use plain strain triaxial compression

Hypothetical slip surface

Direction of stress horizontal use plain strain extension or triaxial extension

Direction of stress 30° use direct simple shear

Typically, the specimen is consolidated prior to starting the test by applying a vertical normal stress, as with the shear box test, and then sheared through the application of a horizontal load or displacement. Unlike the shear box, the simple shear apparatus can be configured to change the direction of shear by applying the shearing force in any direction, thus rotating the applied stress (see Figure 6.50).

The ability to rotate the direction of the shear stress enables the simple shear specimen to be cyclically deformed using one-way or two-way loadings. The data obtained from these tests can assist in designing foundations that are to be subjected to cyclic loads.

Figure 6.50 Showing the general arrangement for the simple shear test (authors' image)

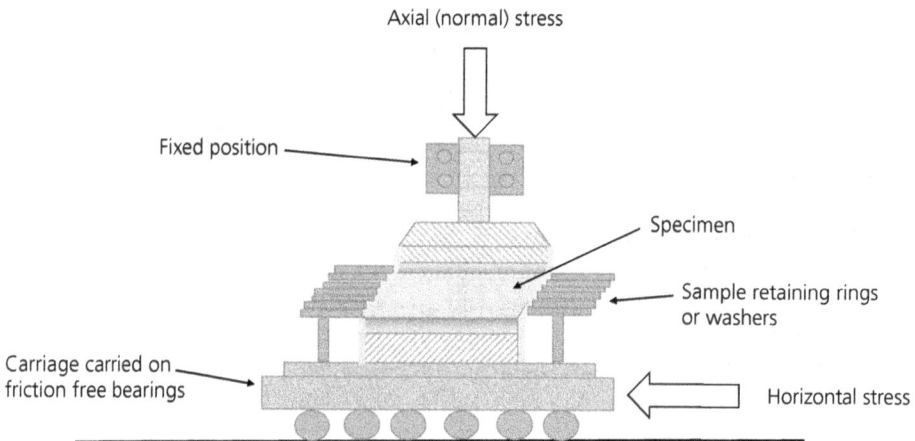

Axial (normal) stress

Fixed position

Specimen

Sample retaining rings or washers

Carriage carried on friction free bearings

Horizontal stress

Figure 6.51 (a) Simple shear test machines and (b) close up of specimen ready for testing (images courtesy of Geolabs Ltd)

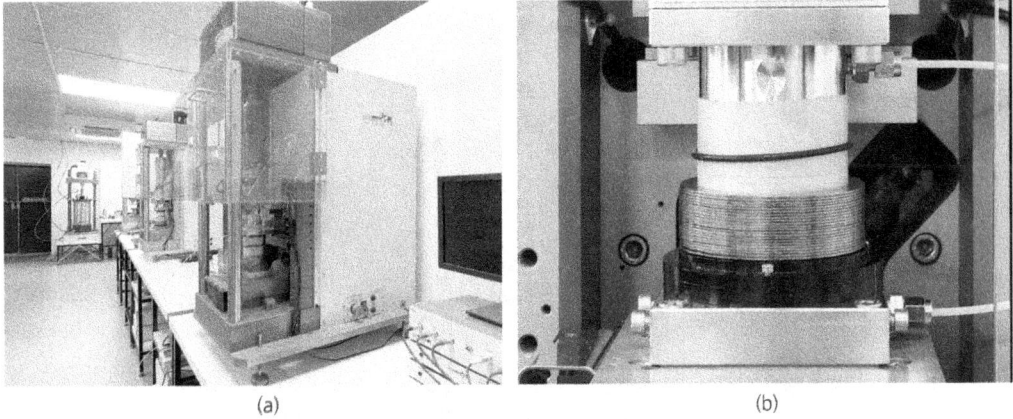

(a)

(b)

6.9.6 Resonant column

The resonant column test provides a useful tool for evaluating the strain-dependent modulus and damping properties of soils at small strains.

Resonant column testing enables the determination of the shear elastic modulus and damping properties of soils by inducing torsional or longitudinal vibration in the specimen. The base of the specimen is fixed, and the top end is free. The vibration is applied using an electromagnetic drive system with variable frequency set around the top of the specimen. The electromagnet develops a shear wave whereby the frequency can be adjusted to enable the fundamental frequency to be determined. The specimen is enclosed in a cell similar to that used for the triaxial test; this enables a cell pressure to be applied while the sample is subjected to the vibration. Because the electro-magnets are exposed, this pressure is applied by compressed air rather than using water under pressure, and therefore it is essential that a banded strengthened cell is used. Figures 6.52 and 6.53 show a typical resonant column test set up, with the cell raised ready to lower onto the specimen. Note a pulley system is used to raise and lower the cell because the cell is larger than the conventional triaxial cell and the electrical equipment occupy a significant space, making it difficult to assemble the cell over the specimen without disturbing the specimen or equipment.

The frequency of vibration is gradually increased until reaching the first-mode fundamental frequency of the sample. At this frequency, measurements are made of the resonance frequency and amplitude of vibration. Knowing the geometry and the end constraints of the sample, the measured resonance frequency is then used to calculate the wave propagation velocity, adopting the wave propagation equation and the theory of elasticity. The shear modulus is obtained directly from the derived velocity and the density of the sample:

$$G = \rho . V_s^2$$

Where G = stiffness, ρ = bulk density, Vs = small strain shear wave velocity.

Figure 6.54 shows typical test results obtained for sand samples, where D [%] = damping ratio; γ [%] = torsional shear strain; G/G_o = normalised shear modulus; G = shear modulus.

Figure 6.52 The resonant column test set up showing hoist for lifting the cell (image courtesy of GDS Instruments)

Figure 6.53 Resonant column set up showing detail of electromagnets and sample position (image courtesy of Geolabs Ltd)

Figure 6.54 Typical resonant column test results for various sands

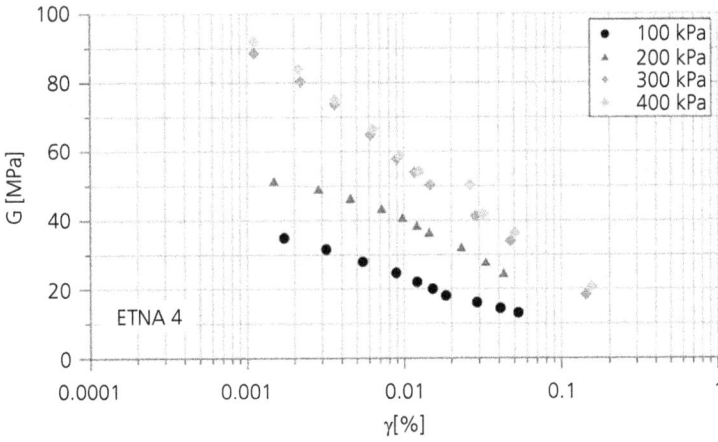

(a)

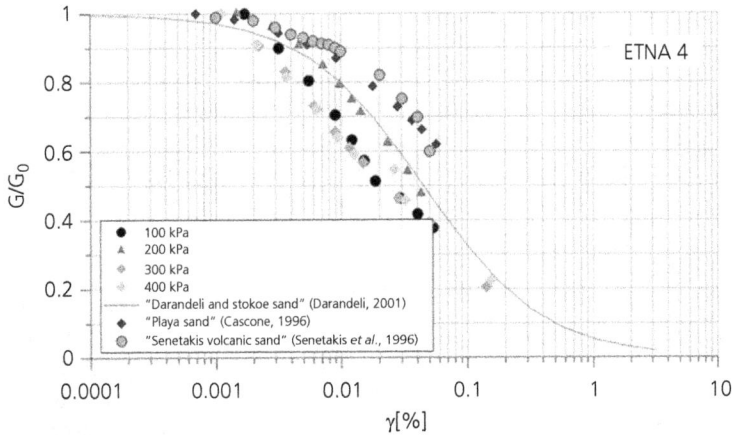

(b)

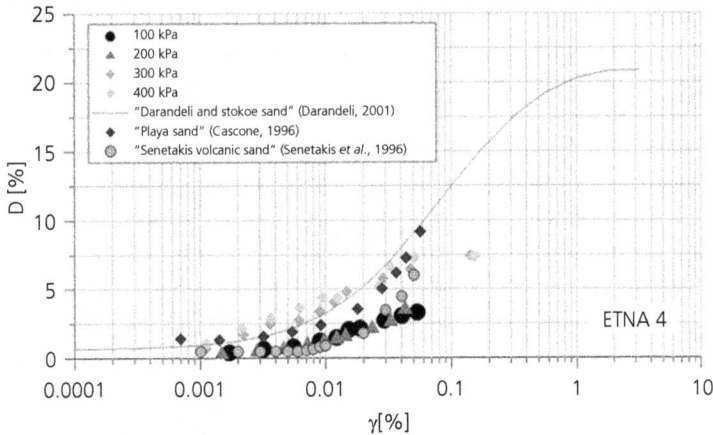

(c)

6.10. Methods for testing rock

6.10.1 Introduction

Tests on rock are generally derivations of similar tests on soil. However, the significantly higher strength of the material generally requires more robust equipment.

Tests such as density and water content use the same procedures as those used for soils.

Strength testing of rock can be misleading. Most tests provide the rock material strength. However, the overall mass strength is typically controlled by discontinuities and is very often found to be significantly lower than the material strength of intact specimens. When assessing the properties and behaviour of rock, it is essential to consider the influence of discontinuities as well as the intact strength, and therefore it is unlikely that any one test method will be able to provide the whole solution. It is recommended that when deciding a strategy for design of foundations or civil engineering projects that the first step should be to obtain a high-quality log of cores or exposures, which provides not only the rock type and composition, but also the rock mechanical properties and the identification of discontinuities, including their spacing, orientation and type.

In general, the specimens for testing are selected as part of the descriptive process. In reality, the best quality specimens are usually chosen for testing. For example, in a core box the sticks of core of greatest length and full diameter would be used to conduct uniaxial compressive strength tests because the test requires full diameter cores of length equal to at least three times the diameter. However, core of this quality may only represent a small percentage of the total core recovered. The mechanical log is a useful indicator of the degree of intactness of the core and should be used to assess how representative the test results are of the strata.

6.10.2 Point load strength index

The point load test equipment shown in Figure 6.55(a), the background to which is documented by Broch and Franklin (1972)), is lightweight and portable, which enables its use both in the laboratory and on site. The point load test procedure was initially set out as a suggested test method by the International Society for Rock Mechanics (ISRM, 1972), this initial document being revised a few years later (ISRM, 1985). The test, which is effectively an indirect tensile strength test, provides a quick and easy way to determine the approximate strength of rock and is used as an aid to rock description in the same way as the hand vane or pocket penetrometer test might be used in soil. It is not recommended that the point load test be used to obtain parameters for design of structures, although it is a helpful tool in the assessment of rock behaviour and classification.

The test can be conducted on pieces of core, cut blocks or irregular pieces of rock (see Figure 6.55(b)). The apparatus comprises an oil filled piston that is pumped by hand to increase the force (typically indicated in kN) delivered to the two steel 60 degree platens. Different specimen shapes and orientation will determine the test type, which may be summarised as follows.

- **Diametral test**: In some respects, this is the simplest form of the test. It requires a cylindrical specimen (i.e. a section of rock core) and the platen points are situated at the top and bottom of the core, fracturing the specimen normal to the core axis. The area of the fractured surface is defined by the core diameter, which is the platen distance (D) at the start of the test. The length of the specimen should exceed its diameter (i.e. the platen points should act through the smaller dimension), and the distance between the platen points and either end of the core should be greater than 0.5 D. If the rock contains an inclined, planar fabric, the specimen should be oriented such that a line drawn between the platens is parallel to the strike of the fabric.

Figure 6.55 (a) Point load test apparatus showing specimen between points prior to test (image courtesy of M. Igbode, K4 Soils Laboratory); (b) rock specimen types, dimensions and orientation (authors' image, based on ISRM, 1972).

(a)

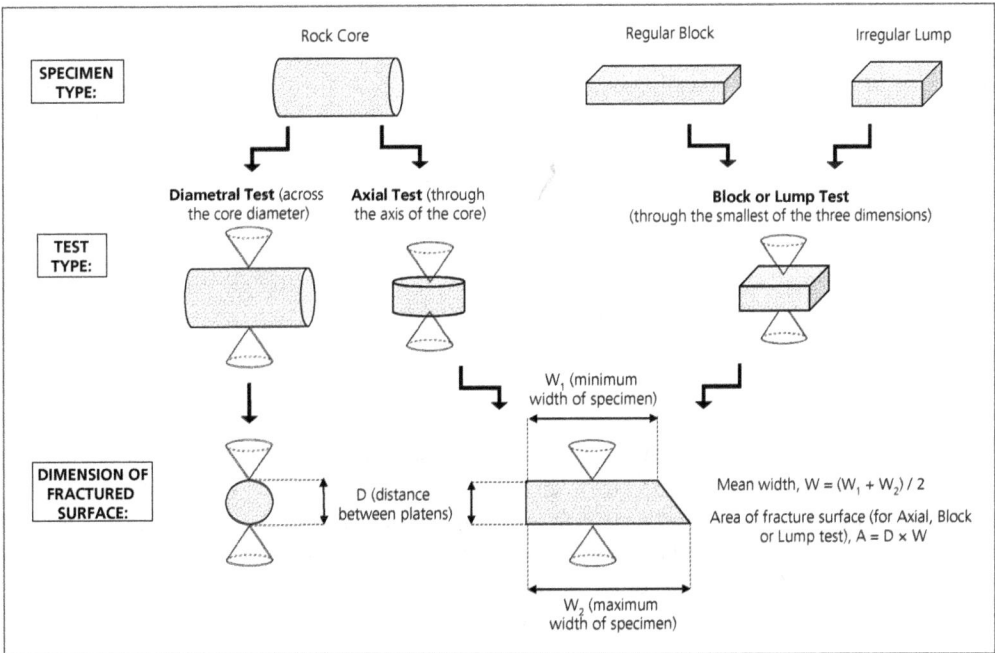

(b)

▧ **Axial test**: As with diametral tests, axial tests are carried out on cylindrical sections of rock core, but in this case with the specimen oriented such that the platen points engage with the core axis. Alternatively, if there is an inclined fabric, the specimen should be oriented such that a line drawn between the platen points is normal to the plane of the fabric. The area of the fractured surface is defined by the core diameter (which forms the specimen width, W) and the platen distance at the start of the test (which defines the specimen height, D). The specimen height should not exceed its width (i.e. the platen points should act through the smaller dimension), but also should not be less than 0.3 times the width.

▧ **Regular block and irregular lump tests**: Block and lump tests are similar apart from the fact that a block has regular dimensions (typically having been cut from a larger fragment), whereas the dimensions of an irregular lump are more complex. In both cases, the platens should act through the smallest of the three dimensions, and the rules noted above for diametral tests (i.e. the distance between the platen points and either end of the specimen should be greater than 0.5 D), and for axial tests (i.e. D should not be less than 0.3 times the width) also apply here. The area of the fractured surface is defined by the specimen width (the average of which may need to be determined if irregular) and the specimen height (i.e. the platen distance at the start of the test, D).

The test specimen is placed between the platens and the force is then increased with the hand pump until the specimen fails by a tensile fracture propagating through the specimen. Should the specimen not fail between the platens, from one side of the specimen to another, the test should not be considered valid, and the result should be rejected.

The rate of loading is not controlled as it is applied manually, although it is recommended that the steady application of load should generally lead to a failure of the specimen between 10 and 60 s. The force applied through the platen points at the point of failure, P, is recorded, typically in kN. Eye protection must always be worn as point load test machines are generally unguarded, and the failure of some stronger rocks can cause fragments to be ejected at great speed.

The test results, which are considered to provide a guide to the material strength of the rock, are expressed as the 'uncorrected point load strength index', I_s

$$I_s = P / D_e^2$$

The specimen size is expressed as D_e^2, i.e. the square of the 'equivalent diameter' of the fracture surface. For diametral tests $D_e^2 = D^2$, and for axial, block or lump tests $D_e^2 = 4 / \pi \times A$, where A (area) = D (platen separation) × W (average specimen width through the platen points) – see Figure 6.55(b). P is typically reported in kN, with D and W in mm. In ISRM (1985) it is proposed that, when a degree of penetration of the platen points occurs in weaker rocks, D' (measured at the point of failure rather than prior to testing) be used in place of D, with D_e^2 from diametral tests defined by D × D'.

A series of tests, at least 10 where possible, will usually be conducted on the same material or over a small range of core to provide an average value for a particular zone or material type. From a set of 10 results, the highest two and lowest two values can be ignored and an average taken of the remaining six (ISRM, 1985). From a smaller data set it may be more appropriate to remove the single highest and the single lowest values before determining an average of the remaining results.

The uncorrected point load strength index, I_s, is influenced by the size of the specimen, and it is therefore conventional to normalise the test results to 'I_{s50}' – that is, the equivalent point load strength for a 'standard' 50 mm diameter test specimen. This is best achieved by plotting P against D_e^2, for a particular rock type and location, and determining an appropriate value of P relating to $D_e^2 = 2500\,mm^2$ (i.e. $D_e = 50$ mm). Alternatively, I_{s50} can be estimated using a formula or calibration charts. In all cases, it is assisted by testing samples close to 50 mm in size where available.

Estimating I_{s50} by calculation is generally based on the following relationship:

$$I_{s50} = I_s \times F_s \qquad \text{Where the size correction factor } F_s = (D_e / 50)^{0.45}$$

The I_{s50} value may then be used to determine the approximate uniaxial compressive strength (UCS) by multiplying by a conversion factor, F. This factor is often taken as about 20 to 24, although it can vary significantly, depending on the rock type. Where both point load tests and UCS tests can be carried out on the same materials, it is always preferable to establish a site-specific value of F for each applicable rock type. See Norbury (1986) for further discussion on using the point load index to estimate UCS values.

Rocks often exhibit anisotropic strength such that the strength in the horizontal direction (or parallel to an inclined planar fabric) can be different to the vertical direction (or normal to an inclined planer fabric), this being particularly prevalent in some sedimentary rocks. Point load testing may be conducted – for example, in both the vertical (axial) direction and the horizontal (diametral) direction on core specimens (assuming vertically drilled core). This is particularly useful because there is little preparation required, and the sample size can be variable. These tests can provide a useful understanding of the ratio of the diametral and axial stress. See ISRM (1985) for further details regarding the testing of anisotropic rock.

6.10.3   Schmidt rebound 'hardness'

Both the Schmidt hammer and the shore scleroscope are described by ASTM as hardness indicators, although this is incongruous with the hardness terms that are related to the Mohs hardness scale. The device provides an estimation of hardness reflected in the rebound characteristics of the impact delivered by the device and is not a method of determining the Mohs hardness.

The Schmidt rebound hammer is a portable hand-held device enabling the test to be carried out on site where there are rock exposures, or in the laboratory when it can be used on cored or larger rock specimens. It has been included in this chapter because most sites do not have suitable, accessible exposures, and testing is therefore typically conducted in the laboratory on recovered samples. This test is considered to provide an empirical indication of rock strength to aid description, rather than providing design data. The test equipment is also commonly used to assess the condition of concrete.

It is a rapid test and therefore does enable many tests to be conducted, giving a more representative understanding of the rock strength and variations, but it should always be conducted in conjunction with other more substantive tests.

The Schmidt rebound hammer uses a spring-loaded flat-bottomed plunger and measures the rebound of the plunger when the spring releases. The rebound, recorded as a 'hardness' value, is used as a measure of the rock strength that is proportional to the amount of rebound. The hammer is pushed onto the specimen, or test area and the spring automatically releases and impacts the mass against the plunger, the rebound being recorded. The hammer is available in different impact sizes, although the Type L hammer, which has an impact energy of 0.74Nm, is most commonly used. Readings are calibrated against an anvil to provide a calibration factor that can then be applied to the recorded values for the rock under test. The ISRM suggested method (1986) suggests that at least 20 tests are conducted on any one rock sample.

Table 6.15 shows the approximate comparison of the hardness values from the Schmidt test and the UCS values. It should be noted, however, that these values are derived and do not take account of discontinuities that will generally be the controlling factor in the strength of the rock mass. Generally, hardness is not directly related to strength but is an important consideration when drilling or advancing a tunnel.

Table 6.15 Comparison of Schmidt 'hardness' values with UCS values

Schmidt 'hardness' value	20	30	40	50	60
UCS MPa	12	25	50	100	200

6.10.4   Shore scleroscope hardness

Similar in concept to the Schmidt rebound test, the shore scleroscope is primarily used to determine the hardness of metals or plastics, but can also be used for rocks. This test provides an indication of hardness and should only be used as a descriptive aid.

The shore scleroscope (see Figure 6.56) drops a plunger, which has a diamond set into it, onto the specimen. The diamond is set in the centre of the base and protrudes from it. The plunger weighs

Figure 6.56 The shore scleroscope (image courtesy of Geolabs Ltd)

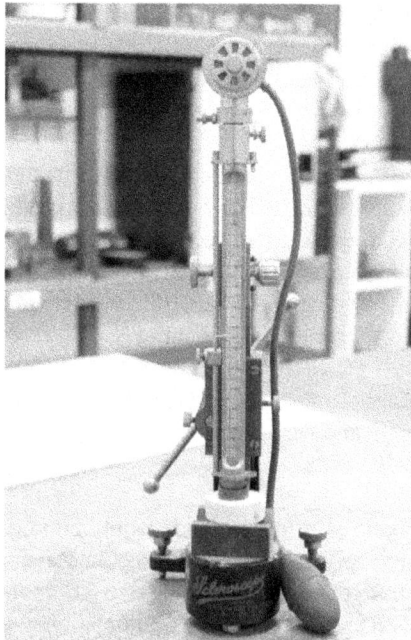

2.3 kg and is dropped under gravity from a height of 251.2 mm. The test measures the height that the plunger rebounds. The test is repeated several times to provide a range of values which are then compared to the calibrated rebounds for the particular instrument that have been measured against materials of known hardness. The rebound is proportional to the hardness; the harder the material the greater the rebound.

Each instrument is calibrated using reference bars of known hardness, and this calibration enables a comparison with the test specimen and hardness to be assigned.

6.10.5 Ultrasonic pulse velocity
This test uses a hand-held device delivering a pulse of ultrasound that passes through the rock under test to a receiver. The time taken for the pulse to travel the distance between the generator and receiver can be used to determine the intactness of the rock under test and has been used to estimate other parameters such as porosity and UCS value. The test is useful as a comparison index for the same rock type, particularly if there are discontinuities present. Because the test is nondestructive it enables the quality of cores to be assessed, and comparisons of other parameters (UCS for instance) to be made. This requires fewer specimens to be tested, which is particularly useful if the core is fractured and unsuitable for UCS testing.

6.10.6 Uniaxial compressive strength (UCS) determination
The uniaxial (or unconfined) compressive strength (UCS) is the most common way of expressing the strength of a rock, and is a parameter that other strength tests, such as point load tests, are calibrated against. The UCS test requires good quality cores of the material where the length of a test specimen is at least two times the diameter. The ends of the selected length of core are cut flat and parallel to form a right cylinder as a test specimen. Ideally, the core should be at least 54 mm in diameter, and the diameter should be at least 10 times the diameter of the largest particle of the rock being tested. This is not always practical – for example, if the material is a conglomerate with particles of gravel size the core would ideally need to be more than 200 mm in diameter. It is recommended that at least five tests are conducted for each of the strata under consideration. The UCS is determined in an unconfined condition and at a rapid rate of loading.

In many cases, cores are tested that are not compliant with the recommended dimensions because core of the sufficient length may represent only a small part of the samples obtained. Under such circumstances, consideration may be given to creating specimens of a smaller diameter by recoring in the laboratory. For example, a 150 mm long section of a 100 mm dia. core would not meet the recommended aspect ratio requirements, but the requirements may be met if a 50 mm dia. core were to be obtained from the original core section.

The test is conducted on a compression machine, which applies an axial load relatively rapidly. The rate of strain is not defined, although the test is typically conducted such that failure occurs within 5–10 min. The maximum load is recorded and the time to reach the maximum load. It is also normal practice to report the specimen's water content and wet density.

The uniaxial compressive strength (UCS) is given by

$$UCS = p/A \text{ (MPa) where } p = \text{the load at failure (MN), } A = \text{cross-sectional area (m}^2)$$

No allowance is made for the change in cross-sectional area as the test takes place.

As previously stated, caution needs to be taken when considering the results of the test because it is likely that it is not representative of the rock mass, generally representing only the better quality, intact sections of core.

6.10.7 Deformation characteristics: Young's modulus, E, and Poisson's ratio, v

Where the elastic deformation characteristics are required the UCS test procedure can be modified to determine these.

The test specimen is fitted with micro strain transducers around its midpoint. Transducers are fixed to the specimen normally using an epoxy resin glue. Two transducers are fitted in the vertical direction on opposite sides of the specimen to measure the local axial deformation, and a further two are fitted to measure the diametric strain. Figure 6.57 shows the transducer arrangement. The transducers are single use.

The UCS test is carried out as described above and simultaneously the strain responses to loading are recorded until failure occurs. The results are interrogated using the initial part of the stress–strain curves where the rock will be exhibiting elastic properties. Some finer-grained rocks such as mudstone and shale have very low elastic deformation moduli being predominantly elasto-plastic in their stress–strain behaviour.

The stress–strain curves are usually drawn as the average of the values for the pairs of transducers, plotting compression in the positive strain field and diametric strain in the negative field – as shown in Figure 6.58.

Figure 6.57 Sample ready for testing for Young's modulus and Poisson's ratio showing positioning of strain gauges (image courtesy of SOCOTEC)

Figure 6.58 Typical axial and radial (diametric) microstrain (με) measurements plotted against axial stress (τkPa), vertical axis (authors' image)

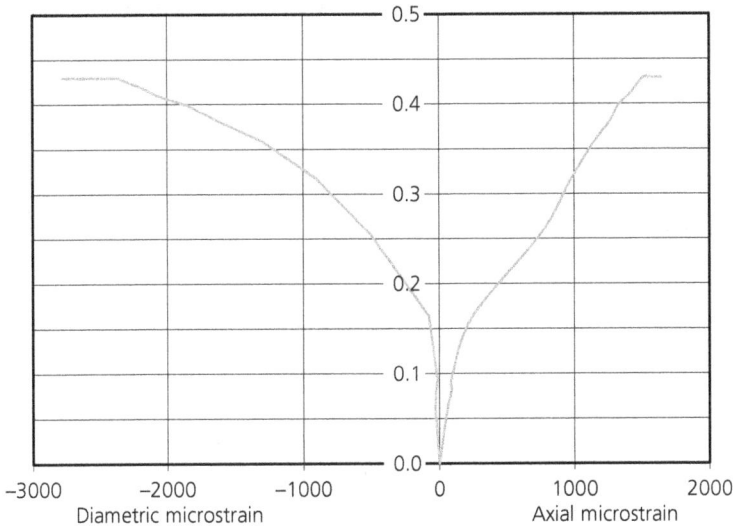

The following parameters are determined.

Axial strain $\varepsilon_a = \Delta l / L$ and Diametric strain $\varepsilon_d = \Delta d / D$

Where Δl = change in measured axial length and L = initial length of the specimen

and Δd = change in measured diameter and D = initial diameter of the specimen.

Poisson's ratio = v = slope of axial stress–strain curve / slope of diametric stress–strain curve.

The axial Young's modulus may be determined by one of the following three methods.

(a) Tangent Young's modulus, determined using the tangent to the stress–strain curve at a fixed percentage of the UCS value. This is generally taken at the point where the stress is 50% of the UCS value.
(b) An average determination of the slope of the stress–strain curve where the curve demonstrates linearity. This is generally seen in the curve at low strain levels.
(c) The secant Young's modulus, calculated as the slope of the line drawn through zero to the point, which coincides with 50% of the UCS value.

It should be borne in mind that Young's modulus should represent the part of the stress–strain curve that exhibits elastic deformation, and thus it should be a straight line that would normally be seen at the early stages of the test. Of the three methods above, method (b) is to be recommended. However, this does require a manual decision on the location of the tangent, while method (c) is often adopted because it is easier to be assessed automatically by computer software. Whichever method is used, the tangential line should be checked to ensure it is providing a realistic determination.

6.10.8 Indirect determination of tensile strength by the Brazilian test

This is a variation on both the point load and UCS tests and is carried out on cores of rock. The test specimen is put onto the platen of an unconfined compression testing machine and jaws are brought to bear on the sample such that failure occurs across the diameter of the specimen, as shown in Figure 6.59. The action is similar to that of a diametral point load test, only using broader jaws rather than conical points.

The tensile strength is determined using the following formula

Tensile strength $= \sigma_t = 0.636 \, P/Dt$ (MPa)

where P is the load at failure (N); D = diameter of specimen (mm); and t = thickness of specimen (i.e. the length of core being tested), measured at the centre (mm).

Figure 6.59 Clamps used to perform the Brazil tensile test (image courtesy of Geolabs Ltd)

6.10.9 Shear testing

Both triaxial and shear box tests can be conducted on rock specimens at the lower end of the strength range. The equipment and analysis are similar to those used for soils as discussed in earlier sections of this chapter. However, owing to the greater strength of rock specimens, the test equipment is generally larger and stronger in order to withstand the greater forces required for the test.

Great care needs to be taken when conducting these tests as significant loads are often required to cause failure. Many rock tests tend to result in sudden failure of the specimen and thus need to be adequately shielded or confined for health and safety reasons.

Figures 6.60 and 6.61 show examples of the Hoek shear box and a rock test triaxial apparatus, respectively.

Figure 6.60 Diagrammatic section showing the method of securing rock sample in the Hoek shear box (authors' image)

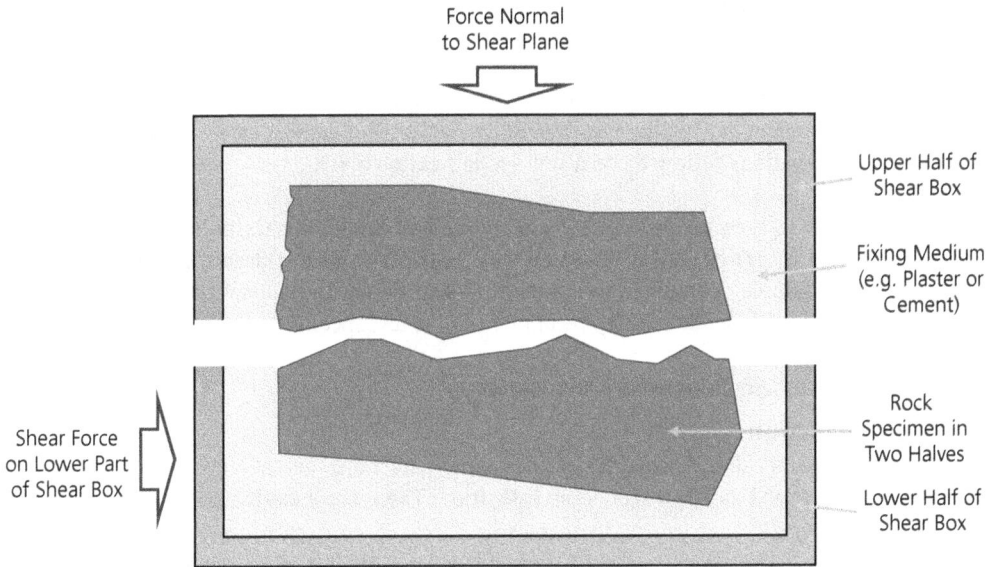

Force Normal
to Shear Plane

Upper Half of
Shear Box

Fixing Medium
(e.g. Plaster or
Cement)

Shear Force
on Lower Part
of Shear Box

Rock
Specimen in
Two Halves

Lower Half of
Shear Box

Figure 6.61 Rock triaxial test cell (image courtesy of Geolabs Ltd)

Owing to the high stresses required to fail a piece of intact rock, the shear box is more often used to test the strength along existing discontinuities. These may be artificial saw cuts, which would provide 'basic' strength parameters relating to the rock material, but with no allowance for undulations or asperities that may be present in a natural fracture. Alternatively, real fractures can be assessed, this approach requiring a sample of suitable size, typically from a rock core, which contains a natural discontinuity, such as a joint. The specimen is configured such that the discontinuity aligns along the plane between the two halves of the shear box, this generally being achieved by setting each half of the specimen in a medium such as Plaster of Paris or cement (see Figure 6.60). The test procedure and calculations follow those discussed in Section 6.8.3.

Triaxial testing also follows the procedures and methods of calculation described in Sections 6.8.6 and 6.8.7. The test is carried out in the same way as the triaxial test described earlier, with the exception that the sample is confined in a metal cell and the testing apparatus is larger and more robust to enable much higher stresses to be applied to the test specimen.

6.11. Tests for aggregate suitability

6.11.1 Introduction

In addition to those tests already covered by this chapter, there are several other tests that may from time to time be requested during a ground investigation. These are included here for completeness.

There are several tests that are carried out to determine the suitability of particular materials for use as aggregates such as those used as road stone or within concrete. These tests might be conducted because they are to be excavated from the site as part of the development but could be sold on as a mineral resource, or they may be utilised within the development, thus reducing costs and environmental impact and carbon footprint.

6.11.2 Aggregate impact value (AIV)

The AIV test indicates a relative measure of the resistance of the aggregate to a sudden impact. A sample of the material is chosen that passes the 12.5 mm sieve and is retained on the 10 mm sieve. This uniformly graded specimen is lightly tamped in three layers with 25 blows from a tamping rod, 10 mm diameter and 230 mm long, into a cylindrical 'cup' 102 mm in diameter and 50 mm deep. Tamping is carried out avoiding damaging the particles. The weight of the sample and the cup is recorded. The weight of the sample is calculated from the original weight of the cup empty and when full of aggregate.

The cup and sample are placed into the AIV machine, which consists of a metal hammer of 13.5–15 kg that falls 38 mm onto the sample. The hammer is 100 mm in diameter, to fit the 102 mm dia. cup. After 15 blows the sample is retrieved and sieved on a 2.36 mm sieve. The sample passing the 2.36 mm sieve is calculated as a percentage of the original sample mass. This percentage is taken as the AIV.

6.11.3 Los Angeles abrasion test

This test investigates the degree of breakdown of a sample of aggregate when subjected to abrasion by a series of abrasive 'charges' (i.e. steel balls). This is done by loading the soil and charges into a steel drum and then rotating the drum through a fixed number of revolutions.

The test is conducted on a sample of known grading. Depending on the specimen grading, the required mass of the test specimen may be determined based on fixed tabulated grade and soil masses. The grading curve enables a grade to be determined, which ranges from A to G. Depending on this grade, the number of metal charges is determined. Each charge is 48 mm in diameter and has a mass of between 390–445 g. The test specimen and the charges are placed in the Los Angeles abrasion testing machine, which comprises a steel drum with an inside length of 500 mm and inside diameter of 700 mm. The drum is rotated at a speed of 20–33 rev/min. For samples of grade A, B, C and D, the machine shall be rotated for 500 revolutions; for grades E, F and G, it is rotated for 1000 revolutions (see Figure 6.62).

The apparatus is usually housed in a soundproof box or room owing to the noise generated by the tumbling sample and charges during the test.

The percentage of sample that passes a 1.7 mm sieve, expressed as a percentage of the original sample mass, is termed the abrasion loss value.

Figure 6.62 (a) The Los Angeles abrasion test drum; (b) the steel charges (images courtesy of M. Colman, SOCOTEC)

(a)   (b)

6.11.4 Cerchar abrasion test

There are two types of this test, each giving the same result. In the original test design, a 90-degree conical tipped steel pin, or 'stylus' (Figure 6.63), is dragged across the surface of a rock specimen a distance of 10 mm, with a static load of 70 N applied to the pin during the test. In the more recent test method, the pin is held still while the rock is passed under the tip. In both cases, the abrasiveness of the rock is related to the degree of wear of the stylus tip, which is determined by measuring the diameter of the flatted end of the pin. The Cerchar Abrasivity Index CAI = 10d, where d is the diameter (mm) of the flat surface, measured to an accuracy of 0.01 mm (Figure 6.63). Should the worn area not be circular, the diameter should be taken as the arithmetic mean of that measured in two orthogonal directions across the surface (Alber *et al.*, 2014).

Figure 6.63 (a) The Cerchar abrasion test apparatus with sample clamped in place for testing; (b) the sharp point of the Cerchar abrasion test stylus prior to testing (a) and the worn, flattened tip following completion of the test (b) (authors' images).

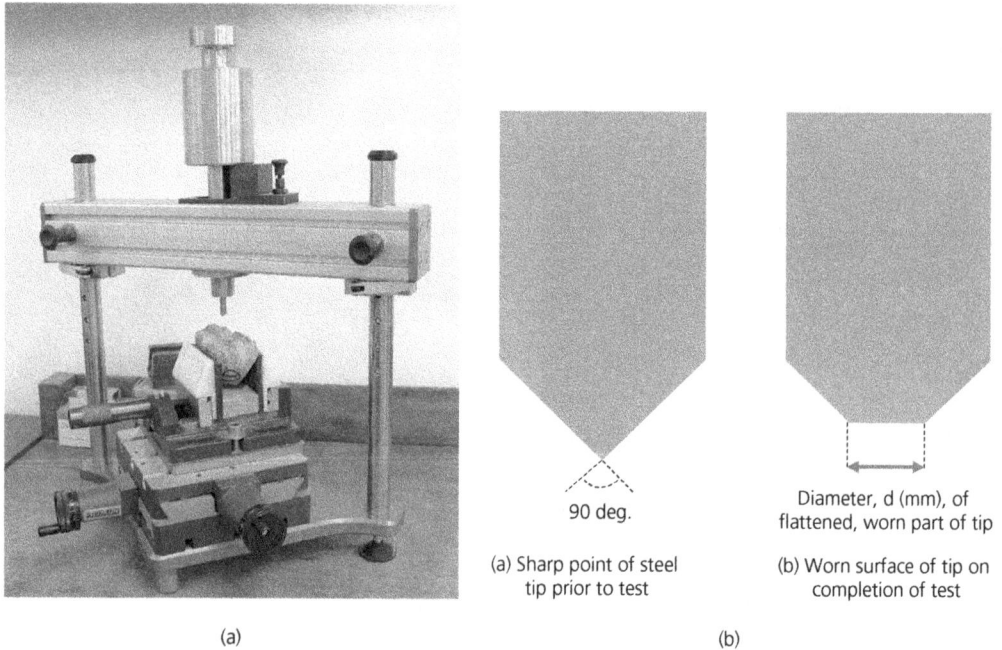

90 deg.

(a) Sharp point of steel tip prior to test

Diameter, d (mm), of flattened, worn part of tip

(b) Worn surface of tip on completion of test

(a)

(b)

6.11.5   Slake durability test

The slake durability test is used to determine the resistance to weathering of a rock; typically, tests are conducted on weak, fine-grained rocks such as shales, mudstones and siltstones.

The test specimen consists of 10 representative fragments of the rock that are intact, roughly equidimensional, and weighing 40–60 g each. These fragments may be naturally occurring or may be produced by breaking with a hammer and removing any sharp corners, all dust should be removed from the test specimen prior to weighing. The specimen should have a total mass between 450–550 g. It is placed into a wire drum consisting of a 2 mm standard square wire mesh cylinder, 100 mm wide and 140 mm in diameter. The ends of the drum are rigid plates, one of which is removable. The drum is partially immersed in a water trough such that the water level is 20 mm below the drums axis and there is 40 mm clearance with the base of the trough. The drum is connected to a motor, which rotates at a speed of 20 rpm for 10 min. When completed, the remaining sample pieces in the drum are dried and reweighed. The slake durability index, SDI or Id, is the percentage ratio of final to initial dry weights of the rock specimen.

6.12. Tests to determine soil erodibility

6.12.1 Introduction

This series of tests are conducted to determine a soil's dispersivity under imposed conditions and are conducted to ascertain the soil response to erosion. In most cases, the results obtained are somewhat subjective and a soil that is found to be nondispersive may still be susceptible to erosion. There are two tests described below that might be considered to demonstrate this property. A further test, the double hydrometer, is discussed in Section 6.3.6.3.

6.12.2 Pinhole dispersivity

The pinhole test is used to measure the erodability of fine-grained soils by causing water to flow through a 1 mm diameter hole formed in a compacted soil specimen that has a water content close to its plastic limit. The test apparatus consists of a 33 mm dia. tube that has a sealing plate at each end. The specimen is compacted into the tube, which is held vertically, to form a plug with a length of 38 mm. A 1 mm dia. hole is made through the test specimen with a punch. Gravel is then added to confine the specimen at both ends, as shown in Figure 6.64. With the tube held horizontally, distilled water is passed through the hole by way of a nipple, which has a 1.5 mm dia. hole.

Different hydraulic heads of 50, 180, 380 and 1020 mm (hydraulic gradients of approximately 2, 7, 15 and 41) are applied to cause the water to flow through the pinhole, which is designed to simulate the flow of water through the pore space. During the test flow rate, effluent turbidity and pore size in the sample are measured. These parameters are used to determination the dispersivity class of the specimen. The class is based partially on observation of turbidity in the water, the measurement of the water flow rate and the change in the pinhole diameter owing to erosion. BS 1377-5 (BSI, 1990) includes a flow chart that sets out the various criteria used to classify materials ranging

Figure 6.64 The pinhole erodability test apparatus (authors' image)

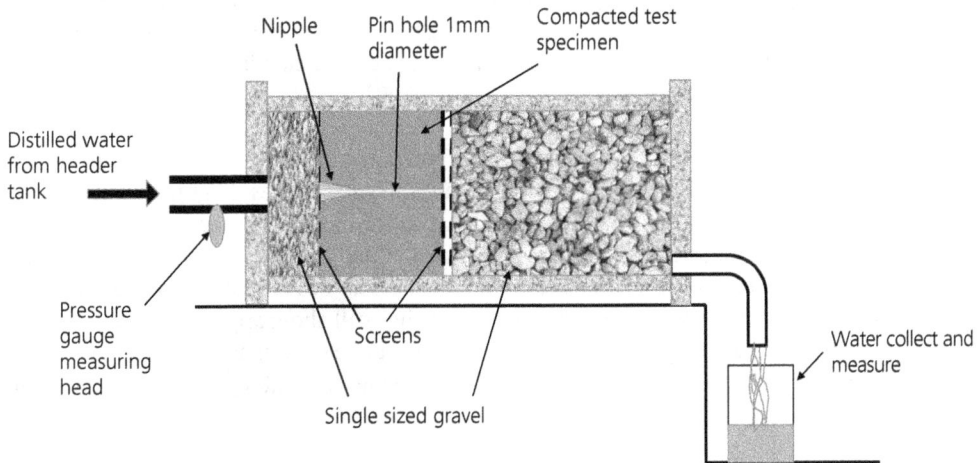

from ND1 and ND2 (essentially nondispersive) through ND3 and ND4 (slightly dispersive) to D2 (dispersive) and D1 (highly dispersive).

6.12.3 Crumb test

The crumb test is a simple test to identify dispersive soils such as loess. The test is an indication of entrained air and a silty composition. Soils that are highly dispersive will generally exhibit high settlement under relatively light loads and are prone to collapse when inundated by water. While this test provides a rapid assessment, other testing will be required to quantify settlement.

The test comprises observing cubical specimens of the soil that are submerged in a glass of either distilled water (ASTM D6572-21 (ASTM, 2000) or 0.001 M sodium hydroxide solution (BS 1377-5:1990, test 6.3 (BIS, 1990)). The soil reaction is observed for 5–10 min after which the material in colloidal suspension is noted and a class assigned. Visual observation of the reaction follows the pattern: the shape of the sample, changes in the structure, colour and turbidity of the water, forming a colloidal 'cloud' or clustering of colloidal particles. Depending on the reactions registered, there are four grades of dispersion: Grade 1 – no reaction; Grade 2 – slight cloudiness of fluid; Grade 3 – moderate cloudiness of fluid and Grade 4 – strong reaction and colloidal cloud over most of the bottom of the beaker.

This simple test can also be used as a rough guide to identify silt while logging a soil profile; in general terms, silty clay would be classified as Grade 1 while a clayey silt would sit between Grades 2 and 3, and silt would be Grade 4.

6.13. Frost susceptibility

This test method was first referred to in *Road Note 29* (1970), which was originally published in 1960, when the scope was widened to include light traffic roads as well as medium to heavily loaded roads. The frost susceptibility of the subgrade was further detailed in LR90 (Croney and Jacobs, 1968), a TRL report published in 1967. This was revised in SR829 (Roe and Webster, 1984), although the test is mentioned in BS 1377-5, test 7 (BSI, 1990), which refers to the full test method provided in BS 812-124 (BSI, 2009).

The test is conducted to assess the effect of ice formation within soils as the bound water expands on freezing. While the effect is generally small in coarse granular soils, finer soils such as fine sand and silt can be severely affected causing considerable damage to structures and paving. The usual approach to dealing with susceptible soils is to provide an adequate capping layer of non-susceptible soil above the susceptible material.

While grain size is a primary factor in the severity of freezing, other factors such as water content, mineralogy and the number of freeze-thaw cycles may influence the resultant heave. It has been identified that there are four factors of particular significance in affecting the amount of ice segregation during soil freezing, and these are: the pore size of the soil, the moisture supply, the rate of heat extraction and the confining pressure. Reliance on frost heave alone to predict the effect may not be sufficient since soils may exhibit little heave but still show appreciable weakening on thawing. To overcome these problems, methods have been developed that rely on more standard testing to identify frost-susceptible soils, principally focusing on grading, moisture content and plasticity testing.

The test that is usually considered to be the defining measurement of frost susceptibility is the frost heave test, as described in BS 1377-5, test 7 (BSI, 1990) and BS 812-124 (BSI, 2009). This consists of freezing a soil specimen in a cylindrical mould by cooling it at its top face while the base is kept above freezing, thus developing a temperature gradient through the specimen. The resulting amount of heave is measured directly, in a similar way to the method described for soaked CBR testing.

Peter Reading and Miles Martin
ISBN 978-1-83549-891-0
https://doi.org/10.1108/978-1-83549-890-320251012
Emerald Publishing Limited: All rights reserved

References

Alber M, Yaralı O, Dahl F *et al.* (2014) ISRM suggested method for determining the abrasivity of rock by the Cerchar abrasivity test. *Rock Mechanics and Rock Engineering* **47(1)**: 261–266.

ASTM (2007a) ASTM D4015-07: Modulus and damping of soils by resonant-column method. ASTM International, West Conshohocken PA, USA.

ASTM (2007b) ASTM D6528-07: Standard test method for consolidated undrained direct simple shear testing of cohesive soils. ASTM International, West Conshohocken PA, USA.

ASTM (2013a) ASTM 3999-91: Standard tests methods for the determination of the modulus and damping properties of soils using the cyclic triaxial apparatus. ASTM International, West Conshohocken PA, USA.

ASTM (2013b) ASTM D5311-92: Standard tests methods for load controlled cyclic triaxial strength of soil. ASTM International, West Conshohocken PA, USA.

ASTM (2014) ASTM D4428/D4428M-14: Standard test methods for cross-hole seismic testing. (Originally published 2000). ASTM International, West Conshohocken PA, USA.

ASTM (2017) ASTM D7400-17: Standard test methods for down-hole seismic testing. ASTM International, West Conshohocken PA, USA.

ASTM (2000) ASTM D6572-21: Standard test methods for determining dispersive characteristics of clayey soils by the crumb test. ASTM International, West Conshohocken PA, USA.

Barentsen P (1936) Short description of a field testing method with cone shaped sounding apparatus. *Proceedings of the 1st International Conference on Soil Mechanics and Foundation Engineering.* Cambridge, MA, USA, 1, B3.

Barnes GE (1987) The moisture condition value and compaction of stony clays. In *Compaction Technology.* Thomas Telford, London, UK.

Begemann HKSP (1974) The Delft continuous soil sampler. *Bulletin of the International Association of Engineering Geology* **10(1)**: 35–37.

Bishop (1948) A new sampling tool for use in cohesionless sands below ground water level. *Geotechnique* **1(2)**: 125–131.

Bishop AW and Henkel DJ (1962) *The Measurement of Soil Properties in the Triaxial Test*, 2nd edn. Edward Arnold, London, UK.

Black L and Lister N (1979) *The Strength of Clay Fill Subgrades: Its Prediction in Relation to Road Performance.* TRRL Report 889. Transport Research Laboratory, Crowthorne, UK.

BRE (Building Research Establishment) (1973) *BRE Digest 151. Soakaways.* Building Research Establishment, Garston, UK.

BRE (1991) *BRE Digest 365. Soakaway design.* Building Research Establishment, Garston, UK.

Broch E and Franklin JA (1972) The point-load strength test. *International Journal of Rock Mechanics and Mining Sciences & Geomechanics Abstracts* **9(6)**: 669–676.

Bromhead E (2008) Why bother with instrumentation? Powerpoint presentation at Geotechnica.

BSI (1990) BS 1377: Methods of test for soils for civil engineering purposes. BSI, London, UK.

BSI (2003) ISO 14686:2003: Hydrometric determinations – pumping test for water wells – Considerations and guidelines for design, performance and use. BSI, London, UK.

BSI (2004) BS EN 1997-1:2004: Eurocode7: Geotechnical design – Part 1: General rules. BSI, London, UK.

BSI (2005a) BS EN ISO 22476-2 + A1:2011: Geotechnical investigation and testing – field testing – Part 2: Dynamic probing. BSI, London, UK.

BSI (2005b) BS EN ISO 22476-3 + A1:2011: Geotechnical investigation and testing – field testing – Part 3: Standard penetration test. BSI, London, UK.

BSI (2007a) BS EN 1997-2: Eurocode 7 – Geotechnical design – Part 2: Ground investigation and testing. BSI, London, UK.

BSI (2007b) BS 6297:2007+A1:2008: Code of practice for the design and installation of drainage fields for use in wastewater treatment. BSI London, UK.

BSI (2009a) BS 6031: Code of practice for earthworks. 2nd rev. BSI, London, UK.

BSI (2009b) BS 812-124: Testing aggregates – Part 124: Method for determination of frost heave. BSI, London, UK.

BSI (2009c) BS EN ISO 22476-12: Geotechnical investigation and testing – field testing – Part 12: Mechanical cone penetration test. BSI, London, UK.

BSI (2012a) BS EN ISO 22476-4: Geotechnical investigation and testing – field testing – Part 4: Menard pressuremeter test. BSI, London, UK.

BSI (2012b) BS EN ISO 22476-7: Geotechnical investigation and testing – field testing – Part 7: Borehole jack test. BSI, London, UK

BSI (2012c) BS EN ISO 22282-1: Geotechnical investigation and testing – geohydraulic testing – Part 1: General rules. BSI, London, UK.

BSI (2012d) BS EN ISO 22282-2: Geotechnical investigation and testing – geohydraulic testing – Part 2: Water permeability tests in a borehole using open systems. BSI, London, UK.

BSI (2012e) BS EN ISO 22282-3: Geotechnical investigation and testing – geohydraulic testing – Part 3: Water pressure tests in rock. BSI, London, UK.

BSI (2012f) BS EN ISO 22282-4: Geotechnical investigation and testing – geohydraulic testing – Part 4: Pumping tests. BSI, London, UK.

BSI (2012g) BS EN ISO 22282-5: Geotechnical investigation and testing – geohydraulic testing – Part 5: Infiltrometer tests. BSI, London, UK.

BSI (2012h) BS EN ISO 22282-6: Geotechnical investigation and testing – geohydraulic testing – Part 6: Water permeability tests in a borehole using closed systems. BSI, London, UK.

BSI (2014a) BS EN ISO 17892-1: Geotechnical investigation and testing – laboratory testing of soil – Part 1: Determination of water content. BSI, London, UK.

BSI (2014b) BS EN ISO 17892-2: Geotechnical investigation and testing – laboratory testing of soil –Part 2: Determination of bulk density. BSI, London, UK.

BSI (2014c) BS EN ISO 17892-3: Geotechnical investigation and testing – laboratory testing of soil – Part 3: Determination of particle density. BSI, London, UK.

BSI (2014d) BS EN ISO 17892-4: Geotechnical investigation and testing – laboratory testing of soil – Part 4: Determination of particle size distribution. BSI, London, UK.

BSI (2014e) BS EN ISO 17892-5: Geotechnical investigation and testing – laboratory testing of soil – Part 5: Incremental loading oedometer test. BSI, London, UK.

BSI (2014f) BS EN ISO 17892-6: Geotechnical investigation and testing – laboratory testing of soil – Part 6: Fall cone test. BSI, London, UK.

BSI (2014g) BS EN ISO 17892-7: Geotechnical investigation and testing – laboratory testing of soil – Part 7: Unconfined compression test on fine grained soil. BSI, London, UK.

BSI (2014h) BS EN ISO 17892-8: Geotechnical investigation and testing – laboratory testing of soil – Part 8: Unconsolidated undrained triaxial test. BSI, London, UK.

BSI (2014i) BS EN ISO 17892-9: Geotechnical investigation and testing – laboratory testing of soil – Part 9: Consolidated triaxial compression tests on water saturated soils. BSI, London, UK.

BSI (2014j) BS EN ISO 17892-10: Geotechnical investigation and testing – laboratory testing of soil – Part 10: Direct shear tests. BSI, London, UK.

BSI (2014k) BS EN ISO 17892-11: Geotechnical investigation and testing – laboratory testing of soil – Part 11: Permeability tests. BSI, London, UK.

BSI (2014l) BS EN ISO 17892-12: Geotechnical investigation and testing – laboratory testing of soil – Part 12: Determination of the Atterberg limits. BSI, London, UK.

BSI (2015) BS 5930: Code of practice for ground investigations. BSI, London, UK.

BSI (2016) BS EN ISO 22476-15: Geotechnical investigation and testing – field testing – Part 15: Measuring while drilling. BSI, London, UK.

BSI (2017a) BS EN ISO 22476-10: Geotechnical investigation and testing – field testing – Part 10: Weight sounding test. BSI, London, UK.

BSI (2017b) BS EN ISO 22476-11: Geotechnical investigation and testing – field testing – Part 11: Flat dilatometer test. BSI, London, UK.

BSI (2018a) BS EN ISO 14688-1: Geotechnical investigation and testing – identification and classification of soil – Part 1: Identification and description. BSI, London, UK.

BSI (2018b) BS EN ISO 14688-2: Geotechnical investigation and testing – identification and classification of soil – Part 2: Principles for a classification. BSI, London, UK.

BSI (2018c) BS EN ISO 14689: Geotechnical investigation and testing – identification and classification of rock – Part 1: Identification and description. BSI, London, UK.

BSI (2018d) BS EN ISO 22476-6: Geotechnical investigation and testing – field testing – Part 6: Self boring pressuremeter test. BSI, London, UK.

BSI (2018e) BS EN ISO 22476-8: Geotechnical investigation and testing – field testing – Part 8: Full displacement pressuremeter test. BSI, London, UK.

BSI (2020a) BS EN ISO 22476-9 Geotechnical investigation and testing – field testing – Part 9: Field vane test. BSI, London, UK.

BSI (2020b) BS EN ISO 22476-14: Geotechnical investigation and testing – field testing – Part 14: Borehole dynamic probing. BSI, London, UK.

BSI (2021) BS EN ISO 22475-1: Geotechnical investigation and testing – sampling methods and groundwater measurements – Part 1: Technical principles for the sampling of soil, rock and groundwater. BSI, London, UK.

BSI (2022a) BS 1377-2: Methods for test for soils for civil engineering purposes. Classification tests and determination of geotechnical properties. BSI, London, UK.

BSI (2022b) PAS 128:2022: Underground utility detection, verification and location. BSI, London, UK.

BSI (2023a) BS EN ISO 22476-1: Geotechnical investigation and testing – field testing – Part 1: Electrical cone and piezocone penetration tests. BSI, London, UK.

BSI (2023b) BS EN ISO 22476-5: Geotechnical investigation and testing – field testing – Part 5: Flexible dilatometer test. BSI, London, UK.

Card G and Roche D (1988) *Penetration Testing in the UK*. Thomas Telford, London, UK.

Clayton C (1983) The influence of diagenesis on some index properties of chalk in England. *Geotechnique* **33(3)**: 225–241.

Clayton CRI (1995) *The Standard Penetration Test (SPT): Methods and Use*. Ciria Report 143 1995 (reprinted 2005). Construction Industry Research and Information Association (CIRIA), London, UK.

Clayton CRI and Siddique A (2001) Tube sampling disturbance – forgotten truths and new perspectives. *Geotechnical Engineering* **149(3)**: 195–200.

Clayton CRI and Smith DM (2013) *Site Investigation in Construction, Effective Site Investigation*. ICE Publishing, London, UK.

Clayton CRI, Simons NE and Matthews MC (1982) *Site Investigation.* Granada Publishing, London, UK (reprinted 1984).

Clayton C, Sididque A and Hopper CS (1998) Effect of sampler design on tube sampling disturbance –numerical analytical investigations. *Geotechnique* **48(6)**: 847–867.

Coulomb CA (1776) Essay on an application of the rules of maximis and minimis to some problems of relative statics, to the architecture. *Memoires Academie Royale des Sciences par savants divers (Academy Royale des Sciences, France)* 7: 343–387.

Cripps AC and McCann DM (2000) The use of the natural gamma log in engineering geological investigations. *Engineering Geology* **55(4)**: 313–324.

Croney D and Jacobs J (1968) *The Frost Susceptibility of Soils and Road Materials.* Ministry of Transport RRL, Report No. 90. Transport Research Laboratory, Crowthorne, UK.

Dennehy JP (1988) Interpretation of moisture condition value tests. *Ground Engineering* **(July)**.

Design Manual for Roads and Bridges (DMRB) National Highways. https//www.standardsforhighways.co.uk (accessed 25/02/2025).

DIN Media (2012) DIN 1813:2012-04: Soil – testing procedures and testing equipment – Plate load test (English translation). Deutsches Institut Fur Norung (German Institute for Standardisation), DIN Media, Berlin, Germany.

Dlubac K, Knight R, Y-Q Song *et al.* (2013) Use of NMR logging to obtain estimates of hydraulic conductivity in the high plains aquifer. *Water Resources Research* **49(4)**: 1871–1886.

Dumbleton W (1966) *Studies of the Keuper Marl (Mercia Mudstone) Geology and Geography RRL Report No. 39.* Road Research Laboratory, Crowthorne, UK.

Dunnicliff J (1999) Systematic approach to planning and monitoring programs using geotechnical instrumentation: an update. Field measurement in geomechanics. *Proceedings of the 5th International Symposium on Field Measurements in Geomechanics*, FMGM99, Singapore, pp. 12–30.

Emdal A, Gylland AS, Amundsen HA and Kåsin K (2016) Mini block sampler. *Canadian Geotechnical Journal* **53(8)**: 1235–1245.

Farnell L (1990) Operating instructions for the TRRL dynamic cone penetrometer (Model A2465).

Gasparre A (2005) *Advanced Laboratory Characterisation of London Clay.* PhD thesis, Imperial College London, London, UK.

Geodrilling (1979) Editorial. A scientific art or an artistic science? **1(January)**.

Hazen A (1892) Some physical properties of sands and gravels with special reference to their use in filtration. Massachusetts State Board of Health, 24th Annual report. Wright and Potter Printing, Boston, MA, USA pp. 534–539.

Head K (1992) *Manual of Soil Laboratory Testing Volume 1: Permeability Shear Strength and Compressibility Tests*, 2nd edn. Whittles Publishing, Dunbeath, UK.

Head K and Epps RJ (2011) *Manual of Soil Laboratory Testing Volume 2: Permeability Shear Strength and Compressibility Tests*, 3rd edn. Whittles Publishing, Dunbeath, UK.

Head K and Epps RJ (2014) *Manual of Soil Laboratory Testing Volume 3: Effective Stress Tests*, 3rd edn. Whittles Publishing, Dunbeath, UK.

HMSO (1991) NRSWA. New Roads and Street Works Act. HMSO, London, UK.

HSE (Health and Safety Executive) (2013) INDG 258: Confined spaces: A brief guide to working safely. HMSO, Norwich, UK.

HSE (2014) *HSG47 Avoiding Danger Form Underground Services*, 3rd edn. HMSO, Norwich, UK.

HSE (2015) *The Construction (Design and Management) Regulations (CDM).* HMSO, Norwich, UK.

Hvorslev (1951) Time lag and soil permeability in ground water observations. *Bulletin No. 36.* Waterways Experimental Station, Vicksberg, USA.

ICE (1998) Penetration testing in the UK. Proceedings of the Geotechnology Conference. Organised by the Institution of Civil Engineers, Birmingham, July 1988. Thomas Telford, London, UK.

ICE (2022) *UK Specification for Ground Investigation*, 3rd edn (AGS Working Group (eds)). Emerald Publishing, Leeds, UK.

Idel M, Muhs H and Von Soos P (1969) Proposal for 'quality classes' in soil sampling in relation to boring methods and sampling equipment. *Proceedings of Specialty Session 1 on Soil Sampling, 7th International Conference on Soil Mechanics and Foundation Engineering.* Sociedad Mexicanna de Mecanica desueos, Mexico City, Mexico, pp. 11–14.

Ingoldby H and Parsons A (1977) *The Classification of Chalk for Use as a Fill Material.* TRRL Laboratory Report 806. Transport Research Laboratory, Crowthorne, UK.

Jones CR and Rolt J (1986) Operating instructions for the TRRL dynamic cone penetrometer. Overseas unit information note (unpublished report available by personal application). Transport Research Laboratory, Crowthorne, UK.

ISRM (International Society for Rock Mechanics) (1972) *Suggested Methods of Determining the Point Load Strength Index.* ISRM Committee on Laboratory Tests. Pergamon Press, Oxford, UK.

ISRM (1985) Suggested method for determining point load strength. *International Journal of Rock Mechanics and Mining Sciences & Geomechanics Abstracts* **22**: 53–60.

ISRM (2007) *The Complete ISRM Suggested Methods for Rock Characterisation, Testing and Monitoring (1974–2006).* Commission on Testing Methods (Ulusay R and Hudson J (eds)). International Society for Rock Mechanics. Pergamon Press, Oxford, UK.

ISSMFE (International Society for Soil Mechanics and Foundation Engineering) (1989) Appendix A. International Reference Text Procedure for Cone Penetration Test (CPT). *Report of the ISSMFE Technical Committee on Penetration Testing of Soils – TC16 with Reference to Test Procedure.* Swedish Geotechnical Institute Linkoping, Information 7, pp. 6–16.

Jacobs C (1967) *Frost Susceptibility of Soils and Road Materials TRL LR 90.* Transport Research Laboratory, Crowthorne, UK.

Kaneko F, Kanemori T and Tonouchi K (1990) *Low Frequency Shear Wave Logging in Unconsolidated Formations for Geotechnical Applications Geophysical Applications for Geotechnical Investigations ASTM STP1101.* ASTM International, Philadelphia, PA, USA.

King C (1981) *The Stratigraphy of the London Clay and Associated Deposits.* Tertiary Research Special Paper No. 6.

Kinney (1979) *Laboratory Procedures for Determining the Dispersivity of Clayey Soils.* Report No. REC-ERC-79-19. Bureau of Reclamation, Denver Co, USA.

Krejci M, Lett M, Lloyd A, Hopper T, Neville and, Birt B (2018) *Groundwater Assessment in Coal Measure Sequence Using Borehole Magnetic Resonance. Australian Exploration Geoscience Conference*, Sydney, Australia 2018.

La Rochelle P, Roy and Tavenas F (1973) Field measurement of cohesion in Champlain clays. *Proceedings of the 8th International Conference on Soil Mechanics and Foundation Engineering*, Moscow, Volume 1.1. ISSMGE, London, UK, pp. 229–236.

La Rochelle PL, Sarrailh J, Tavenas F, Roy M and Leroueil S (1981) Causes of sampling disturbance and design of a new sampler for sensitive soils. *Canadian Geotechnical Journal* **18(1)**: 52–66.

Lamont-Black J and Mortimore RN (1996) Determination of the intact dry density of irregular chalk lumps: a new method. *Quarterly Journal of Engineering Geology* **29(3)**: 241–248.

Lefebvre G and Poulin C (1979) A new method of sampling insensitive clay. *Canadian Geotechnical Journal* **16(1)**: 226–233.

Lunne T, Berre T and Strandvik S (1997a) Sample disturbance effects in soft low plastic Norwegian clay. *Proceedings of Symposium on Recent Developments in Soil and Pavement Mechanics.* Rio de Janeiro, Brazil. AA Balkema, Rotterdam, the Netherlands.

Lunne T, Robinson PK and Powell JJM (1997b) *Cone Penetration Testing in Geotechnical Practice.* E & FN Spon, London, UK.

Mair W and Wood D (1987) *Pressuremeter Testing – Methods and Interpretation*. CIRIA Ground Engineering Report: In-situ Testing. Construction Industry Research and Information, Butterworths.

Marchetti S, Monaco P, Totan G and Calibrese M (2001) The flat pate dilatometer test in soils investigations. *Report to ISSMGE TC 16 Proc. Insitu 2001 International Conference on Insitu Measurement of Soil Properties Bali.*

Meigh AC (1987) *Cone Penetration Testing – Methods and Interpretation – Methods and Interpretation*. CIRIA Ground Engineering Report: In-situ Testing. Construction Industry Research and Information. Butterworths.

Ménard (1957) Mesures in situ des proprietes physiques des sols: Annales des Ponts et chausses 1.3: 357–376.

Mikkelsen PE (2002) Cement-bentonite grout backfill for borehole instruments. *Geotechnical Instrumentation News*. December.

Montague KN (1990) SPT and pile performance in upper chalk. *Proceedings of the International Chalk Symposium*, Brighton, Telford, London, UK, pp. 269–276.

Mortimer RN (2012) Making sense of chalk. *Quarterly Journal of Engineering Geology and Hydrogeology* **45**: 252–334.

National Highways (2017) *Manual of Contract Documents for Highway Works*, Volume 1 Specification for Highway Works.

National House Builders Council (NHBC) (2024) NHBC standards. Available online. NHBC, Milton Keynes, UK.

Norbury D (1986) The point load test. *Geological Society, London – Engineering Geology Special Publications* **2(1)**: 325–329.

Norbury D (2016) *Soil Ad Rock Description in Engineering Practice*. 2nd edn. Whittles Publishing, Dunbeath, UK.

Norbury D and Powell J (2022) Presentation at Eurocode and sample disturbance. Webinar hosted by the AGS, London 2022.

Nowak G and Gilbert P (2015) *Earthworks: A Guide*, 2nd edn. ICE Publishing, London, UK.

Palmer A (1972) Undrained plain strain expansion of a cylindrical cavity in clay: a simple interpretation of the pressuremeter test. *Geotechnique* **22(3)**: 451–457.

Parsons B (1979) *The Moisture Condition Test and its Potential Applications in Earthworks*. TRRL Supplementary Report 522. Transport and Road Research Laboratory, Crowthorne, UK.

Powell WD, Potter JF, Mayhew HC and Nunn ME (1984) *The Structural Design of Bituminous Roads*. TRRL Report LR 1132. TRRL, Crowthorne, UK.

Preene M (2019) Design and interpretation of packer permeability tests for geotechnical purposes. *Quarterly Journal of Engineering Geology and Hydrogeology* **52(2)**: 182–200.

Reading P and Martin M (2025) *ICE Handbook of Ground Investigation – Planning and Reporting, Geoenvironmental and Non-Intrusive Techniques*. Emerald, London, UK.

Road Note 29 (1970) *Guide to the Structural Design of Pavement for New Roads*, 3rd edn. HMSO, London, UK.

Robertson PK (1990) Soil classification using the cone penetration test. *Canadian Geotechnical Journal* **27(1)**: 151–158.

Roe PW and Webster D (1984) *Specification for the TRRL frost-heave test*. TRRL Supplementary Report 892. Transport and Road Research Laboratory, Crowthorne, UK.

Rowe PW (1971) Representative sampling in location, quality and size. *Proceedings of the Symposium on Sampling Soil and Rock*, Toronto. ASTM Special Technical Publication No. 483, pp. 77–106. ASTM, Philadelphia, PA, USA.

Rowe PW (1972) The relevance of soil fabric to site investigation practice: 12th Rankine lecture. *Geotechnique* **22(2)**: 195–300.

Schmertmann JH (1953) Estimating the true consolidation behaviour of clay. *Proceedings of the American Society of Civil Engineering 79, Separate Vol. No. 311.* ASCE, Philadelphia, PA, USA.

Schmertmann JH (1979) Statics of SPT. Journal of Geotechnical Division. *Proceedings of American Society of Civil Engineering (ASCE) GT5 May.* Volume 5. ASCE, Philadelphia, PA, USA, pp. 655–670.

Shukla SK (2014) *Core Principles of Soil Mechanics, ICE Textbooks.* ICE Publishing, London, UK.

Skempton AW (1944) Notes on the compressibility of clay. *Quarterly Journal of the Geological Society* **100**: 119–135.

Skempton AW (1986) Standard penetration test procedures and the effects in sand of overburden pressure, relative density, particle size, aging and over consolidation. *Geotechnique* **36(3)**: 425–447.

Soresen O and Okkels N (2013) Correlation between drained shear strength and plasticity index of undisturbed overconsolidated clays. *10th International Conference on Soil Mechanics and Geotechnical Engineering,* Paris. ISSMGE TC 101 session 11. ISSMGE, London, UK, pp. 423–428.

Standing J (2018) Identification and implications of the London Clay formation divisions from an engineering perspective. *Proceedings of the Geologists Association 131.* Elsevier Ltd, Amsterdam, the Netherlands, pp. 486–499.

Stringer M, Taylor ML and Cubrinovski M (2015) *Advanced Soil Sampling of Silty Sands in Christchurch N.Z.* Research Report 2015–16 Civil and Natural Resources Engineering. University of Canterbury, Christchurch, New Zealand.

Stroud MA (1974) The standard penetration test in insensitive clays and soft rocks. *Proceedings of European Symposium on Penetration Testing.* Statens Geoteckniska Institute, Stockholm, Sweden, pp. 367–375.

Terzaghi K (1926) Erdbaumechanik auf bodenphysikalischer grundlage (Soil Mechanics in Engineering Practice). (Published in English) John Wiley & Sons Inc., New York/London.

Terzaghi K (1936) Relations between Soil Mechanics and Foundation Engineering. Presidential address. *Proceedings of the 1st International Conference on Soil Mechanics and Foundation Engineering,* Volume 3. ICSMFE and Harvard University, Massachusetts, USA.

Terzaghi K (1947) Recent trends in subsoil exploration. *Proceedings of the 7th Texas Conference on Soil Mechanics and Foundation Engineering.* Special Publication No. 17. Bureau of Engineering Research and Department of Civil Engineering, University of Texas, Austin, TX, USA, pp. 1–15.

Terzaghi K (1955) Evaluation of coefficients of subgrade reaction. *Geotechnique* **5(4)**: 297–326.

Terzaghi K and Peck R (1948) *Soil Mechanics in Engineering Practice,* 1st edn. John Wiley, New York, NY, USA.

Terzaghi K and Peck R (1967) *Soil Mechanics in Engineering Practice,* 2nd edn. John Wiley, New York, NY, USA.

Thorburn S (1963) Tentative correction chart for the standard penetration test in non-cohesive soils. *Civil Engineering and Public Works Review* **(June)**: 752–753.

Tomlinson S (2021) *Manual of Rotary Drilling.* British Drilling Association (BDA), Brentwood, UK.

Zohrabi S and Scott PL (2003) *The Correlation Between the CBR and the Penetrability of Pavement Construction Materials.* TRL Report TRL587. Transport Research Laboratory, Crowthorne, UK.

ICE Handbook of Ground Investigation

emerald PUBLISHING ice

Peter Reading and Miles Martin
ISBN 978-1-83549-891-0
https://doi.org/10.1108/978-1-83549-890-320251008
Emerald Publishing Limited: All rights reserved

Index